13 IIIA	14 IVA	15 VA	16 VIA	17 VIIA	18 VIIIA
								helium 2 **He** 4.00
			boron 5 **B** 10.81	carbon 6 **C** 12.01	nitrogen 7 **N** 14.01	oxygen 8 **O** 16.00	fluorine 9 **F** 19.00	neon 10 **Ne** 20.18
10 VIII	11 IB	12 IIB	aluminum 13 **Al** 26.98	silicon 14 **Si** 28.09	phosphorus 15 **P** 30.97	sulfur 16 **S** 32.07	chlorine 17 **Cl** 35.45	argon 18 **Ar** 39.95
nickel 28 **Ni** 58.69	copper 29 **Cu** 63.55	zinc 30 **Zn** 65.39	gallium 31 **Ga** 69.72	germanium 32 **Ge** 72.61	arsenic 33 **As** 74.92	selenium 34 **Se** 78.96	bromine 35 **Br** 79.90	krypton 36 **Kr** 83.80
palladium 46 **Pd** 106.42	silver 47 **Ag** 107.87	cadmium 48 **Cd** 112.41	indium 49 **In** 114.82	tin 50 **Sn** 118.71	antimony 51 **Sb** 121.75	tellurium 52 **Te** 127.60	iodine 53 **I** 126.90	xenon 54 **Xe** 131.29
platinum 78 **Pt** 195.08	gold 79 **Au** 196.97	mercury 80 **Hg** 200.59	thallium 81 **Tl** 204.38	lead 82 **Pb** 207.2	bismuth 83 **Bi** 208.98	polonium 84 **Po** (209)	astatine 85 **At** (210)	radon 86 **Rn** (222)
ununnilium 110 **Uun** (269)	unununium 111 **Uuu** (272)	ununbium 112 **Uub** (277)						

gadolinium 64 **Gd** 157.25	terbium 65 **Tb** 158.93	dysprosium 66 **Dy** 162.50	holmium 67 **Ho** 164.93	erbium 68 **Er** 167.26	thulium 69 **Tm** 168.93	ytterbium 70 **Yb** 173.04	lutetium 71 **Lu** 174.97
curium 96 **Cm** (247)	berkelium 97 **Bk** (247)	californium 98 **Cf** (251)	einsteinium 99 **Es** (252)	fermium 100 **Fm** (257)	mendelevium 101 **Md** (258)	nobelium 102 **No** (259)	lawrencium 103 **Lr** (260)

PRINCIPLES OF
ENVIRONMENTAL
CHEMISTRY

Jones and Bartlett Titles in Physical Science

Astronomy Activity and Laboratory Manual
Hirshfeld

Climatology
Rohli

Environmental Oceanography: Topics and Analysis
Abel

Environmental Science, Eighth Edition
Chiras

Environmental Science: Systems and Solutions, Fourth Edition
McKinney

Essentials of Geochemistry, Second Edition
Walther

Igneous Petrology, Third Edition
McBirney

In Quest of the Universe, Fifth Edition
Koupelis

Invitation to Oceanography, Fifth Edition
Pinet

Invitation to Organic Chemistry
Johnson

Organic Chemistry, Third Edition
Fox

Organic Chemistry and Biochemistry Structure Visualization Workbook
Luceigh

Outlooks: Readings for Environmental Literacy, Second Edition
McKinney

Principles of Atmospheric Science
Frederick

Writing Science Through Critical Thinking
Moriarty

Principles of Environmental Chemistry

SECOND EDITION

James E. Girard
American University

JONES AND BARTLETT PUBLISHERS
Sudbury, Massachusetts
BOSTON   TORONTO   LONDON   SINGAPORE

World Headquarters

Jones and Bartlett Publishers
40 Tall Pine Drive
Sudbury, MA 01776
978-443-5000
info@jbpub.com
www.jbpub.com

Jones and Bartlett Publishers
Canada
6339 Ormindale Way
Mississauga, Ontario L5V 1J2
Canada

Jones and Bartlett Publishers
International
Barb House, Barb Mews
London W6 7PA
United Kingdom

Jones and Bartlett's books and products are available through most bookstores and online booksellers. To contact Jones and Bartlett Publishers directly, call 800-832-0034, fax 978-443-8000, or visit our website, www.jbpub.com.

Substantial discounts on bulk quantities of Jones and Bartlett's publications are available to corporations, professional associations, and other qualified organizations. For details and specific discount information, contact the special sales department at Jones and Bartlett via the above contact information or send an email to specialsales@jbpub.com.

Production Credits

Chief Executive Officer: Clayton Jones
Chief Operating Officer: Don W. Jones, Jr.
President, Higher Education and Professional Publishing:
  Robert W. Holland, Jr.
V.P., Sales and Marketing: William J. Kane
V.P., Design and Production: Anne Spencer
V.P., Manufacturing and Inventory Control:
  Therese Connell
Publisher, Higher Education: Cathleen Sether
Acquisitions Editor, Science: Molly Steinbach

Associate Editor, Science: Katherine Theroux
Editorial Assistant, Science: Caroline Perry
Associate Production Editor: Leah Corrigan
Senior Marketing Manager: Andrea DeFronzo
Composition: Circle Graphics
Cover Design: Scott Moden
Assistant Photo Researcher: Jessica Elias
Cover Image: Courtesy of NASA/USGS
Printing and Binding: Malloy, Inc.
Cover Printing: John Pow Company

Photo credits appear on pages 651-652, which constitute a continuation of the copyright page.

Library of Congress Cataloging-in-Publication Data
Girard, James.
    Principles of environmental chemistry / James E. Girard.—2nd ed.
        p. cm.
    Includes index.
    ISBN 978-0-7637-5939-1 (alk. paper)
1. Environmental chemistry. I. Title.
    QD33.2.G57 2010
    628.501'54—dc22
                                2009013468

6048
Printed in the United States of America
13 12 11 10 09   10 9 8 7 6 5 4 3 2

Dedicated to my wife, Connie Diamant,
the real environmentalist in our home.

BRIEF CONTENTS

CONTENTS

PREFACE

At present there is worldwide concern that many of our human activities are endangering—perhaps permanently—the quality of the environment, and that time is running out to address these problems. The public is becoming increasingly aware of the environmental damage caused by pesticides, toxic wastes, chlorofluorocarbons, nuclear radiation, oil spills, and the greenhouse effect, to name just a few issues. Environmental organizations like the Sierra Club, the National Wildlife Federation, and Friends of the Earth are gaining support—especially on college campuses—and are becoming a major influence in the political arena. Articles on environmental issues appear daily in the newspapers, and members of Congress are introducing legislation to combat threats to the environment.

I developed an environmental chemistry course, and subsequently wrote this book, to expose students to environmental issues from a perspective that appreciates the chemical reactions that drive natural environmental processes. Furthermore, I wanted to help students see the connection between natural environmental processes, human behavior, and the potential for the latter to cause environmental processes to go awry.

■ Objectives

The primary objective of this text is to enable students to understand environmental issues and the underlying chemistry. The text emphasizes that all parts of our environment are made up of chemicals, and that the natural processes continuously occurring in the environment all involve chemical reactions. Appropriate chemical analysis of the lithosphere, hydrosphere, and atmosphere helps illustrate to students what composes an unpolluted environment and sets a benchmark from which our stewardship of the Earth can be monitored. With a grasp of this information, students begin to comprehend the chemical basis of the changing world around them and the consequences of their actions.

■ Organization

This textbook describes the Earth's lithosphere, hydrosphere, atmosphere, and sources of energy. Like other environmental texts, this book focuses on important physical and chemical principles that define each of these parts of our Earth. However, the organization and approach of this text differ in several significant ways from other environmental chemistry books. First, this book emphasizes the role of the U.S. Environmental Protection Agency, EPA regulations for pollutants, and the limits the EPA sets for those pollutants. Next, it features the analytical methods and techniques that are used to measure pollutants. Throughout the text, the appropriate instrumental method that measures the concentration of specific pollutants is presented and described. In some cases the analyses presented may be as mundane as the yearly automobile tailpipe emissions tests that each automobile owner endures, or the concentration of pollutants in wastewater that are measured by environmental contractors, or as sophisticated as remote measurements of our atmosphere from satellites in space. In this way, students not only learn environmental chemistry, but also gain practical knowledge of instrumental and quantitative analysis, two subjects that are the foci of entire courses.

The second edition has been updated and revised. The first chapter provides a description of the Earth's lithosphere and its ecosystems. This early coverage of the dynamic nature of the Earth and its natural cycles not only establishes the importance of maintaining a sustainable natural world but also gives students a refresher in inorganic elements and their distribution.

The next five chapters (Chapters 2–6) concentrate on the chemistry of the atmosphere of the Earth. Chapter 2 discusses the major atmospheric layers and how energy from the sun is captured by Earth. A new Chapter 3 focuses on what may be the most important issue of our time—global warming. Chapter 4 presents the chemistry of the troposphere: CO, NO_x, SO_2, volatile organic chemicals, and photochemical smog. Chapter 5 describes the production and destruction of ozone in the stratosphere. Chapter 6 presents *in situ* and remote analytical methods for measuring the composition of the atmosphere.

Chapters 7–9 cover water—its properties, its importance to life on Earth, and the dangers of polluting and misusing it. Chapter 7 begins with a discussion of the distribution of water on Earth and the unique properties of water. It is followed by Chapter 8, which describes water pollution and water treatment. This section is completed by Chapter 9, which describes the analytical methods that are used to measure water pollution.

The next three chapters (10–12) focus on energy. Chapter 10 describes fossil fuels and their use as our major energy source, and the consequences of the depletion of these nonrenewable resources. Chapter 11 describes the chemistry of nuclear power and the nuclear fuel cycle. Chapter 12 explores other energy sources such as wind power and geothermal and solar energy. The use of hydrogen as a major fuel and fuel cell chemistry are also presented in Chapter 12.

The next four chapters (13–16) present specific environmental topics in more depth. Chapter 13 describes inorganic pollutants such as lead, mercury, and cadmium and the analytical methods that are used to measure these elements in the environment. Persistent organic pollutants (POPs) are introduced in Chapter 14 and students are taught how to use the EPA's PBT Profiler to determine if a chemical might be a persistent, bioaccumulative, or toxic organic pollutant. Different classes of insecticides, herbicides, and the analytical methods that are used to measure them in the environment are presented in Chapter 15. Alternative methods of insect control are also presented, including the use of juvenile hormones and sex pheromones.

The following chapter (16) introduces the student to toxicology and risk assessment. This chapter includes insightful discussions of how to measure the risks posed by chemicals. Chapter 17 describes asbestos: the different fiber types, asbestos disease, and analytical methods that are used to measure the amount of asbestos in the air. The last chapter (18) examines the laws governing the proper disposal of hazardous and radioactive chemicals, such as the Resource Conservation and Recovery Act (RCRA), and the Comprehensive Environmental Response, Compensation, and Liability Act (CERCLA). It also presents the EPA Superfund methods used for analysis of hazardous waste.

■ Chapter Elements

Examples and Exercises   Illustrative worked examples, each one accompanied by a challenging practice exercise, are included throughout the text, particularly in the chapters covering basic chemical principles.

Keywords and Concepts   Lists of keywords and concepts introduced in the chapter are included at chapter's end to help reinforce the most important information.

Questions and Problems   Each chapter includes a wide selection of problems and questions (40–50), with answers to all even-numbered ones given in an appendix. Quantitative, review, and discussion-type questions are included.

Additional Sources of Information   A bibliography provides sources for the material covered in the chapter and serves as a suggested list for further reading.

■ Course Use

Principles of Environmental Chemistry offers the flexibility to tailor a course to suit both instructors' preferences and the needs of particular audiences. The full text may be used for a comprehensive two-semester course in which the instructor has the time to explore the underlying chemical principles in detail. Appendix B contains a chapter on basic organic chemistry, which may be useful to cover early in the course to refresh the memory of your students.

The book may be used in several ways for a one-semester course. An option for a one-semester course is to use the first eight chapters, followed by selections from the remaining chapters on more advanced chemistry and environmental applications according to the teacher's preferences. Those who wish to teach a more traditional one-semester course, not emphasizing environmental analysis, should begin with Chapter 1 and proceed through the first 12 chapters in order, skipping Chapters 6 and 9, and then cover more in-depth environmental topics in the later chapters according to preference.

■ Instructors' Supplements

These supplements can be accessed online, via http://www.jbpub.com/science/chemistry.

Online Solutions Manual   Contains solutions to chapter-end exercises.

Online Image Bank   Provides a PowerPoint® library of all the art and tables in the text to which Jones and Bartlett owns the copyright or has digital print rights.

ACKNOWLEDGMENTS

I would like to express my gratitude and appreciation to a number of people who have contributed to this book:

To my students who have suffered through draft manuscripts of this text.

To the reviewers of the text for their helpful comments and suggestions:

Michael E. Ketterer, Ph.D.,
  Northern Arizona University

Chunlong Zhang, Ph.D., P.E.,
  University of Houston–Clear Lake

Marie de Angelis, Ph.D.,
  SUNY Maritime College

Michelle M. Ivey, Ph.D.,
  Florida Atlantic University

Robert Kerber, Ph.D., M.R.S.C.,
  Stony Brook University

Matthew Elrod, Ph.D.,
  Oberlin College

Thomas G. Chasteen, Ph.D.,
  Sam Houston State University

Brent L. Lewis, Ph.D.,
  Coastal Carolina University

Timothy L. Rose, Ph.D.,
  Brandeis University

To Nell Buell for all the discussion and hard work with background materials.

To Bill Hirzy of the US Environmental Protection Agency who reviewed and updated the data concerning the regulation of hazardous waste in Chapter 18.

To Ken Harvey of Horiba, Inc. who provided details of how automobile emission measurements are made.

To Kaanan Snirvasian of Dionex Corporation who facilitated acquisition of the ion chromatography figures.

To Wayne Neimayer at McCrone Laboratories for the SEM and EDX spectra in the asbestos chapter.

To the Jones and Bartlett team: Molly Steinbach, Caroline Perry, Leah Corrigan, Jessica Elias, Dean DeChambeau, and Cathy Sether.

Jim Girard

PRINCIPLES OF
ENVIRONMENTAL CHEMISTRY

CHAPTER

1

Planet Earth:
Rocks, Life, and History

TO UNDERSTAND HOW OUR **ENVIRONMENT** WORKS, WE MUST FIRST LOOK BACK BILLIONS OF YEARS AT THE TIME WHEN THE EARTH WAS BORN AND SEE HOW IT EVOLVED INTO THE LIFE-SUPPORTING PLANET THAT WE INHABIT TODAY. In this chapter, we consider the formation of the universe, including the origin of the galaxies, the stars, and our own planet, Earth. We look at how the oceans, the atmosphere, and the rocky surface on which we live were formed; examine the Earth's mineral resources; and discuss the ways in which society uses them. We see how life developed on Earth and how all living organisms interact with their physical surroundings and with each other, how all of these interactions are intertwined, and how a continuing flow of energy through all of its parts fuels the entire system.

■ The Formation of the Universe

If we gaze at the sky on a clear night, away from the lights of any city, we see myriads of stars. All of the stars that we see are a part of our galaxy, the **Milky Way**. This pinwheel-shaped body, which is made up of clouds of gas and cosmic dust and billions and billions of stars, includes our solar system: the sun and its nine orbiting planets. What we see is only a minute fraction of the entire **universe**. Beyond the Milky Way, extending into space for distances beyond our comprehension, are countless other galaxies. It was probably only when we humans first ventured into space in the 1960s that we began to appreciate the smallness and insignificance of our planet in relationship to the universe as a whole. The first photographs of the Earth taken from the moon showed us our planet suspended in the black vastness of space (Figure 1.1).

According to the most recent research, the universe began between 12- to 13.5-billion years ago. Although differences of opinion still exist, many scientists believe that all of the matter in the universe was once compressed into an infinitesimally small and infinitely dense mass that exploded with tremendous force. This explosion of unimaginable proportions—appropriately called the **big bang**—generated enormous amounts of light, heat, and energy, and released the cosmic matter from which the galaxies and stars were eventually formed. The universe began expanding in all directions and, according to most astronomers, has been expanding ever since.

Figure 1.1 The Earth seen from the surface of the moon.

Galaxies and Stars

As the universe expanded, it cooled very, very slowly, and cosmic matter gradually condensed to form the first galaxies. Atoms of hydrogen—the simplest and lightest of all of the elements—formed in the swirling clouds of condensing matter. Over billions of years, the galaxies gave birth to the early stars, which generated sufficient heat to cause hydrogen atoms to fuse (join) to form atoms of helium, the second lightest of the elements. The energy released during these **fusion reactions** initiated further fusion reactions, in which all 90 of the remaining naturally occurring elements found on Earth were formed. In the universe as a whole, 90% of all atoms are hydrogen, and 9% are helium, whereas the remaining 1% are atoms of all of the other elements. Scientists believe that subsequent explosions of the early stars scattered the elements and that our sun was born from the debris of one of these explosions. The sun, which to us appears very bright, is an average-sized star that is located toward the edge of the Milky Way.

The Planets in Our Solar System

Scientists still do not know with any certainty how the planets in our **solar system** developed (a solar system is a group of planets that revolve around a star), but it is generally believed that they began to form approximately 5-billion years ago from hot, mainly gaseous matter rotating about the sun. With time, the matter slowly cooled, and solid particles condensed from the gases. The particles gradually coalesced into clumps of matter. Larger clumps had stronger gravity and gradually drew in and retained additional particles, eventually forming the eight planets that revolve around the sun: Mercury, Venus, Earth, Mars, Jupiter, Saturn, Uranus, Neptune, along with the dwarf planet, Pluto (Figure 1.2).

The four planets closest to the sun—Mercury, Venus, Earth, and Mars—are called **terrestrial planets** and are small and dense. The more distant **giant planets**—Jupiter, Saturn, Uranus, and Neptune—are much larger and are of lower density than the terrestrial planets.

The Earth and the other terrestrial planets formed close to the sun and were so hot that lighter, easily evaporated materials could not condense and were swept away. Only substances with extremely high boiling points, such as metals and minerals, condensed on these planets. Mercury, the planet closest to the sun and therefore the hottest, is composed mainly of iron. On the Earth, which formed at a somewhat lower temperature, silicates and other metals besides iron were able to condense. (Silicates are minerals that are formed from the elements silicon, oxygen, and a variety of metals.) The larger planets, with their greater mass and thus a stronger gravitational pull, retained gases—mostly hydrogen and helium—in the atmospheres surrounding them. Some important features of the planets as they exist today are listed in Table 1.1.

The Sun

The sun is the ultimate source of energy for life on Earth. It makes up 99.9% of the mass of the solar system, and its diameter is approximately 110 times as great as that of the Earth. Scientists estimate that temperatures near the center of this immense rotating sphere of extremely hot gases reach almost 15,000,000°C (27,000,000°F). Fusion reactions occur at these incredibly high temperatures, continually releasing tremendous amounts of energy that are continually in the form of heat and light. These fusion reactions have

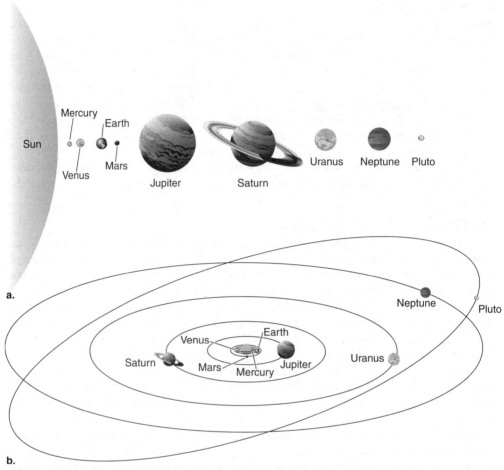

Figure 1.2 The solar system. (a) The relative sizes of the planets. (b) The planets in their orbits around the sun.

allowed the sun to shine brightly for billions of years and will allow it to continue for billions more.

■ Differentiation of the Earth into Layers

Exactly how the Earth evolved to its present state is not known, but Earth scientists believe that when the Earth was first formed approximately 4.7-billion years ago. It was homogeneous in composition: a dense, rocky sphere with no water on its surface and no atmosphere. Then, over time, the interior of the sphere gradually grew hotter, and the Earth became differentiated into layers, with each layer having a different chemical composition. This crucial period in the development of the Earth led to the formation of its magnetic field, atmosphere, oceans, and continents, and—ultimately—to life.

Table 1.1

Important Features of the Planets in Our Solar System

Planet	Diameter (km)	Diameter (mi)	Mass (Earth = 1)	Density (water = 1)	Gravity (Earth = 1)	Time for One Rotation on Axis (Earth days or hours)	Time for One Revolution around Sun (Earth years)	Distance from Sun (million km)	Distance from Sun (million mi)	Composition of Atmosphere
Terrestrial										
Mercury	4835	3004	0.055	5.69	0.38	59 days	0.24	57.7	36.8	None
Venus	12,194	7577	0.815	5.16	0.89	243 days	0.62	107	66.9	CO_2
Earth	12,756	7926	1.00	5.52	1.00	1.00 days	1.0	149	92.6	N_2, O_2
Mars	6760	4200	0.108	3.89	0.38	1.03 days	1.9	226	141	CO_2, N_2, Ar
Giant										
Jupiter	141,600	87,986	318	1.25	2.64	9.83 hours	12	775	482	H_2, He
Saturn	120,800	75,061	95.1	0.62	1.17	10.23 hours	29	1421	883	H_2, He
Uranus	47,100	29,266	14.5	1.60	1.03	23.00 hours	84	2861	1777	H_2, He, CH_4
Neptune	44,600	27,713	17.0	2.21	1.50	22.00 hours	165	4485	2787	H_2, He, CH_4

Heating of the Earth

Three factors are believed to have caused the Earth to heat. First, the cosmic particles that collided and clumped to form the Earth were drawn inward by the pull of gravity. As more particles collided with the developing planet, heat was released. Some of this heat was retained within the Earth; this heat gradually built up as increasing amounts of material accumulated.

As the Earth grew, material in the center was compressed by the weight of new material that struck the surface and was retained. Some of the energy that was expended in compression was converted to heat and caused a further rise in the temperature within the Earth.

The third and very significant factor in the warming of the Earth was the decay of radioactive elements within the interior that released energy in the form of heat. The atoms in radioactive elements are unstable and disintegrate spontaneously, emitting atomic particles and energy. In this process, which continues today, the radioactive elements are converted into atoms of other elements (discussed in Chapter 10). Only a very small percentage of naturally occurring elements has atoms that disintegrate in this way, and the heat generated with each disintegration is extremely small. Nonetheless, Earth scientists have calculated that the retention of this heat within the Earth over billions of years (together with the heat released as new material accumulated and was compressed) would have been sufficient to raise the temperature of the material at the center of the Earth to the point where it became molten.

It seems probable that this critical temperature was reached approximately 1-billion years after the Earth was born. Metallic iron, which melts at 1535°C (2795°F) and makes up over 30% of the mass of the Earth, began to melt. This heavy molten iron, together with some molten nickel, sank to the center of the Earth. As the molten iron sank, less dense material was displaced and rose toward the surface. As a result, the Earth ceased to be homogeneous and eventually became differentiated into three distinct layers: the **core**, the **mantle**, and the **crust** (Figure 1.3).

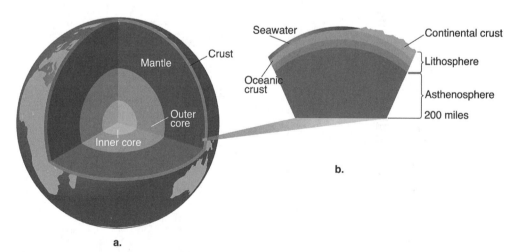

Figure 1.3 The structure of the Earth: (a) The Earth is differentiated into three distinct layers called the core, the mantle, and the crust. (b) The lithosphere, which comprises the continental and oceanic crust together with the solid upper part of the mantle, rests on the partially molten asthenosphere.

The Core

The Earth's core, which extends 3500 km (2200 miles) from the planet's center, is believed to be composed of iron and small amounts of nickel. The metals are thought to be in solid form in the inner core and molten in the surrounding outer core (Figure 1.3a). Because the core is inaccessible to us, there is no way to prove that it consists primarily of iron, but considerable indirect evidence supports this view. For example, analysis of light emitted by the sun and stars has revealed that iron is the most abundant metal in the universe, and most of the meteorites that have landed on the Earth from outer space are composed of iron. Furthermore, analysis of waves generated by Earthquakes has shown that the core is very dense, and iron is the densest metal found in any quantity on Earth.

The Mantle

The Earth's mantle, which lies between the core and the crust, is approximately 2900 km (1800 miles) thick (Figure 1.3b). The relatively thin upper part of the mantle is solid and rigid, but the layer below it—called the **asthenosphere**—although essentially solid, is able to flow extremely slowly, like a very thick, viscous liquid. In the deep mantle, below the asthenosphere, the rock is believed to be rigid.

The Crust

Above the mantle is the crust, which forms the thin outer skin of the Earth (Figure 1.3b). The crust is thicker beneath the continents than beneath the oceans. Its thickness ranges from 6 km (4 miles) under the oceans to 70 km (45 miles) under mountainous regions. Although the crust makes up a very small part of the Earth as a whole, we gather from it practically all of the resources that sustain our way of life.

Together the crust and the solid upper part of the mantle make up the relatively cool and rigid **lithosphere**, which floats on the hotter, partially molten asthenosphere. The boundary between the lithosphere and the asthenosphere is not caused by a difference in the chemical composition of their rocks but by a change in the physical properties of the rocks that occurs as temperature and pressure increase with depth.

Relative Abundance of the Elements in the Earth

By mass, the four most abundant elements in the Earth are iron, oxygen, silicon, and magnesium, which together account for approximately 93% of the Earth's mass (Figure 1.4a). Nickel, sulfur, calcium, and aluminum make up another 6.5%. The remaining 0.5% or so of the Earth's mass is made of the other 84 naturally occurring elements.

Primarily because most of the iron sank to the center of the Earth during the period of differentiation, the relative abundance of the elements in the crust differs greatly from that in the Earth as a whole (Figure 1.4b). Seventy-four percent of the crust consists of oxygen and silicon, whereas aluminum, iron, magnesium calcium, potassium, and sodium together account for 25%.

It might have been expected that as the Earth became differentiated into layers, the elements would have been distributed strictly according to mass, with the heavier elements falling to the Earth's center and the lighter ones rising to the surface. This distribution did not occur, however, because some elements combined with other elements to form compounds, and the melting points and densities of the compounds (rather than those of the elements from which they were formed) primarily determined how the elements were distributed in the Earth. For example, silicon, oxygen, and various metals combined to form

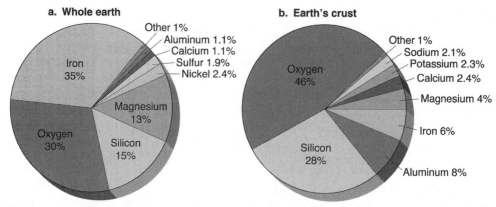

Figure 1.4 The relative abundance (by mass) of elements in the whole Earth and in the Earth's crust. Because of the differentiation that occurred early in the Earth's history, the percentage of iron in the crust (b) is lower than that in the whole Earth (a), and the percentages of aluminum, silicon, and oxygen (the elements that combine to form silicates) are higher.

silicates, which are relatively light compounds that melt at relatively low temperatures. When the Earth's interior was hot, these silicates rose to the surface. They are the most abundant minerals in the Earth's crust.

As a result of the chemical changes that occurred during the period of differentiation, the distribution of the elements on the Earth is very uneven. The relative abundance in the Earth's crust of the economically valuable elements is shown in Table 1.2. Of these, only four—aluminum, iron, magnesium, and potassium—are present in amounts greater than 1% of the total mass of the crust. It is fortunate for us that as the result of geologic processes that have been occurring for millions of years, the less abundant (but valuable) elements such as gold and silver are concentrated in specific regions of the world. If these elements had been distributed evenly throughout the Earth's crust, their concentrations would be too low to make their extraction technically or economically feasible.

■ Formation of the Oceans and the Atmosphere

It is generally accepted that there was no water on the Earth's surface for millions of years after the planet was formed. Then, as the interior of the Earth heated up, minerals below the Earth's surface became molten. The molten material rose to the surface, and oxygen (O) and hydrogen (H) atoms that were chemically bound to certain minerals escaped explosively into the atmosphere as clouds of water (H_2O) vapor. In these tremendous volcanic eruptions, which were widespread and numerous, carbon dioxide (CO_2) and other gases were also released from the planet's interior (Figure 1.5). The lighter gases escaped into space, but the heavier ones, including water vapor and carbon dioxide, were held by gravity and formed as a thick blanket of clouds surrounding the Earth. In time, as the Earth's surface cooled, the water vapor condensed, and the clouds released their moisture. For the first time, rain fell on the Earth. During the next several million years, volcanoes continued to erupt, and the oceans filled with water as more rain fell.

Table 1.2

The Relative Abundance of the Economically Valuable Elements in the Earth's Crust

Name	Chemical Symbol	Abundance in Crust (% by mass)
Aluminum	Al	8.00
Iron	Fe	5.80
Magnesium	Mg	2.77
Potassium	K	1.68
Titanium	Ti	0.86
Hydrogen	H	0.14
Phosphorus	P	0.101
Fluorine	F	0.0460
Sulfur	S	0.030
Chlorine	Cl	0.019
Chromium	Cr	0.0096
Zinc	Zn	0.0082
Nickel	Ni	0.0072
Copper	Cu	0.0058
Cobalt	Co	0.0028
Lead	Pb	0.00010
Arsenic	As	0.00020
Tin	Sn	0.00015
Uranium	U	0.00016
Tungsten	W	0.00010
Silver	Ag	0.000008
Mercury	Hg	0.000002
Platinum	Pt	0.0000005
Gold	Au	0.0000002

Source: Adapted from F. Press and R. Siever, *Earth,* 3rd ed. (New York: W. H. Freeman, 1982), p. 553.

The Earth's first atmosphere was quite different from the one that surrounds the Earth today. Volcanic eruptions continued to occur long after the Earth's surface had cooled to the point where water vapor began to condense to form the oceans. Evidence suggests that in addition to water vapor and carbon dioxide, the enormous volumes of gases emitted were mostly nitrogen, with smaller amounts of carbon monoxide, hydrogen, and hydrogen chloride—the same gases that erupting volcanoes emit today. Hydrogen gas, being very light, was lost into space, but the Earth's gravitational pull held other gases near the surface.

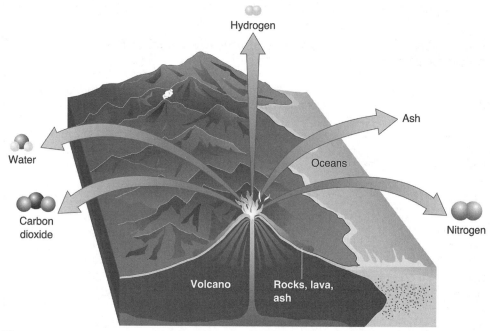

Figure 1.5 In volcanic eruptions, huge quantities of rocks, lava, ash, and gases (mainly water vapor, carbon dioxide, and nitrogen) are ejected. Early in the Earth's history, when volcanic eruptions were frequent and widespread, water vapor released to the atmosphere condensed and fell to Earth as rain and the oceans were formed; carbon dioxide and nitrogen became the main constituents of the early atmosphere.

After millions of years of volcanic activity, the atmosphere was rich in nitrogen and carbon dioxide but was completely devoid of oxygen. Today, the Earth's atmosphere is still rich in nitrogen (78%), but only 0.03% of the atmosphere is carbon dioxide, whereas oxygen accounts for 21%. There were two main ways in which the excess carbon dioxide was removed: First, when rain began to fall on the Earth, very large quantities of carbon dioxide dissolved in the oceans that were formed, and much of it combined with calcium in the water to form limestone (calcium carbonate). Second, approximately 3-billion years ago, the first primitive blue-green algae, or **cyanobacteria**, developed in shallow waters. Similar to the more advanced plants on Earth today, these organisms used energy from the sun to convert carbon dioxide and water into simple carbohydrates and oxygen by the process of **photosynthesis**. Oxygen escaped from the water and for the first time entered the atmosphere.

As cyanobacteria multiplied, increasingly larger amounts of carbon dioxide were removed from the atmosphere and replaced with oxygen. Eventually, when photosynthesis had been going on for millions of years, the Earth's atmosphere attained its present composition. Although photosynthesis was important in reducing the carbon dioxide content of the atmosphere, dissolution in the oceans followed by the formation of limestone was the major factor in its removal.

Since the beginning of the Industrial Revolution, humans have been pouring unprecedented amounts of carbon dioxide back into the atmosphere by burning carbon-containing fuels such as coal and petroleum. There is growing concern that the continuing increase in atmospheric carbon dioxide may lead to a rise in the Earth's temperature, which could have catastrophic consequences for our planet (discussed in Chapters 2 and 3).

■ Rocks and Minerals

The rocks that make up the hard surface of the Earth, the lithosphere, are composed of one or more substances called **minerals**. A mineral is a naturally occurring, usually crystalline, substance that has a definite composition or a restricted range of composition; it may be either an element or a compound.

Minerals form the **inorganic** part of the Earth's crust; materials derived from the decayed remains of plants and animals make up the **organic** part. The terms *inorganic* and *organic* were introduced in the 18th century to distinguish between compounds derived from non-living matter and compounds derived from plant and animal sources. At that time, chemists believed that the complex compounds that make up living matter—such as carbohydrates and proteins—could be produced only by living organisms. Once it was discovered that these compounds, which all contain carbon, could be synthesized in the laboratory, the definition of organic compounds was broadened to include all nonmineral compounds of carbon.

Some minerals have been known and used since ancient times. There is evidence that very early in human history flint and obsidian (a volcanic glass) were shaped to make weapons and primitive knives, and clay was formed into pottery vessels and bricks. Gold, silver, copper, and brightly colored minerals such as jade and amethyst were fashioned into jewelry and other objects, and pigments were made from red and black iron oxides.

Over 2500 distinct minerals have now been identified, but only a few of them are distributed widely over the Earth's surface. Many of the more valuable minerals are found in only a few limited regions of the world, where they became concentrated as a result of the upheaval and subsidence of crust materials and other rock-forming processes that have gone on for millions of years. The study of the composition of rocks has been an important factor in helping to explain how the Earth was formed.

■ Rocks as Natural Resources

Rocks and minerals are **natural resources**, which are anything taken from the physical environment to meet the needs of society. Such resources may be **renewable** or **nonrenewable**. Resources such as soil, natural vegetation, fresh water, and wildlife are all renewable; if not depleted too rapidly, they are replaced in natural recycling processes. Rocks and minerals—as well as oil, natural gas, and coal—are nonrenewable. They are present in the Earth in fixed amounts and are not replaced as they are used.

Rocks and minerals are quarried and used widely, often in modified form, in the construction and chemical industries and for making ceramics and many other products.

■ Ores and Metals

Minerals in which a particular metallic element occurs in a sufficiently high concentration to make mining and extracting it economically feasible are termed **ores**. Silicates, although very abundant, are seldom used as ores because extraction of their metallic elements involves high costs and technologic difficulties. Instead, most metals are extracted from sulfide, oxide, carbonate, chloride, and phosphate ores, which are found concentrated in a relatively small number of regions of the world. We now consider briefly the sources and uses of a few of the metals that are essential for our modern way of life.

Iron

Iron (Fe), the fourth most abundant element in the Earth's crust (see Figure 1.4), is the metal that industrialized nations use in the greatest quantity. The ores from which it is extracted usually contain a mixture of two iron oxides: hematite (Fe_2O_3) and magnetite (Fe_3O_4). Nearly all of the iron extracted from ores is used to manufacture **steel**, an alloy of iron with a small amount of carbon. The percentage of carbon determines the properties of steel. Low-carbon steel (less than 0.25% carbon) is relatively soft and suitable for making cans and wire. High-carbon steel (up to 1.5% carbon) is very hard and strong and is used for making tools and surgical instruments. Steels with a variety of properties and uses are made by alloying iron with small amounts of other metals.

Although world reserves of iron are still large—and the United States has an abundant supply within its own borders—the demand for iron ore and steel continues to rise, and known sources of ores must eventually run out.

Aluminum

Aluminum (Al) is the second most highly used metal in industrialized nations and is the third most abundant element in the Earth's crust (see Figure 1.4). Nearly all naturally occurring aluminum is a component of complex silicates, but currently, there is no economically viable way to extract it from these silicates. Instead, the source of practically all aluminum is **bauxite**, an ore rich in aluminum oxide that is found in quantity in only a few places in the world. Very large deposits of bauxite are found in Jamaica and Australia.

Aluminum is a light, strong metal that is used primarily for making beverage cans. In the building industry, it is used to make doors, windows, and siding; it is ideal for cooking utensils and many household appliances. Because it is a good conductor of electricity, aluminum is used extensively for high-voltage transmission lines. When alloyed with magnesium, it forms a light but strong material that is vital to the manufacture of airplane bodies.

Aluminum corrodes less easily than does iron; this feature is an important advantage for certain purposes such as the construction of homes, but it also means that aluminum cans are very slow to degrade. Tin cans (which are made of steel and coated with tin) eventually break down completely, but discarded aluminum cans remain in the environment for a very long time.

Aluminum oxides that include traces of certain metal impurities are valuable as gemstones. Rubies are crystalline aluminum oxide that is colored red with traces of chromium. Sapphires, which occur in various shades of yellow, green, and blue, owe their color to traces of nickel, magnesium, cobalt, iron, or titanium.

Copper

Another valuable and extensively used metal is copper (Cu), which today is obtained from low-grade copper sulfide ores. Because the copper content of these ores is 1% or less, the cost of obtaining the pure metal is high. Valuable by-products of copper production are gold and silver, which are frequently present in very small quantities in the original ores. Increasingly, pure copper is being obtained by recycling copper-containing materials.

Copper is an excellent conductor of electricity and is used extensively for electrical wiring. It is also used for plumbing fixtures and as a constituent of alloys. Brass is an alloy of copper and zinc, and bronze is an alloy of copper and tin. Copper and copper alloys are also used for coinage.

Strategic Metals

Because metal ores are very unevenly distributed in the Earth's crust, many countries must depend on imports for their supplies. Within the United States, for instance, there are no deposits of ores of many **strategic metals**, which are metals that are essential for industry and defense. Large reserves of chromium (Cr), manganese (Mn), and the platinum (Pt) group of metals (platinum [Pt], palladium [Pd], rhodium [Rh], iridium [Ir], osmium [Os], and ruthenium [Ru]), which are needed for the manufacture of specialty steels, heat-resistant alloys, industrial catalysts, and parts for automobiles and aircraft, are found in only South Africa and the former Soviet Union.

■ Mineral Reserves

The world's human population continues to grow at an ever-increasing rate, and with it grows the demand not only for food but also for material goods. To meet these material needs, metals and minerals (nonrenewable resources) are being consumed at a tremendous rate, one that is bound to rise as the developing nations become increasingly industrialized.

The industrialized nations currently consume a disproportionate amount of the Earth's mineral reserves. The North American continent, for example, has less than 10% of the world's population, but it consumes almost 75% of the world's production of aluminum. The same disproportionate usage rate holds true for many other metals. At this rate of consumption, supplies of many important metals will be severely depleted by the end of the first quarter of the 21st century.

It is possible that new mineral deposits will be discovered in the future. However, because geologists have already thoroughly explored most of the Earth's surface, it is unlikely that significant quantities of ores will be found. One source of minerals that has not yet been fully explored is the ocean floor. Areas where a new floor is being formed have been shown to be rich sources of manganese and polymetallic sulfides. As deep-sea mining technology develops, sites on the ocean floor may provide a much needed source of metals.

One of the most important ways to conserve our mineral supplies is to recover metals by recycling. More and more communities are collecting aluminum beverage cans (and also glass containers and paper) for recycling, which saves energy as well as conserves natural resources. Approximately half as much energy is required to make new aluminum cans from old cans as is needed to make them from bauxite.

■ The Origin of Life on Earth

Thus far in this chapter we have considered only the nonliving part of the Earth. We now turn our attention to the living creatures that inhabit the planet. It is generally agreed that life on Earth began between 3.5- and 4.0-billion years ago, but exactly how it began will probably never be fully understood. The early atmosphere is thought to have consisted mainly of CO_2 and nitrogen (N_2), with smaller amounts of ammonia (NH_3) and methane (CH_4), which would have dissolved to some extent in the early oceans. Some scientists believe that life began in tidal pools or lagoons where evaporation would have concentrated dissolved chemicals, making it possible for them to combine to form simple amino acids. **Amino acids,** which are basic building blocks of living tissues, might then have joined together to form simple proteins, and further reactions in the chemical "soup" of the ocean pools or lagoons could have produced other compounds that are essential for life. The essential compounds produced in this way might then have gradually clumped to form larger masses. Membranes formed around these masses, separating them from the surrounding environment. The organic matter gradually acquired the characteristics of living cells.

Other researchers believe that life is more likely to have begun near volcanic vents on the ocean floor where there was heat and protection from destructive ultraviolet (UV) radiation. Still others believe that the first living organisms did not arise on Earth but came from outer space in meteorites or interplanetary dust.

No matter how the first one-celled organisms were formed, the environment was devoid of oxygen. These early **anaerobic bacteria** (bacteria that require an oxygen-free environment) flourished until the development of oxygen-producing cyanobacteria. Oxygen was lethal to the anaerobic bacteria and, except in a few specialized locations, they gradually died.

■ The Uniqueness of the Earth

The Earth is unique, as it is the only planet in our solar system that developed an environment that is capable of supporting life as we know it. The position of the Earth relative to the sun made possible the formation of the atmosphere and the oceans, which together maintain the temperature on the Earth's surface within a very narrow range: a range that extends approximately from the freezing point of water (0°C, 32°F) to the boiling point of water (100°C, 212°F). If the Earth had formed a little closer to the sun, it would have been too hot to support life; if it were a little farther away, it would have been too cold.

The size of our planet is another important factor. If the Earth were much smaller, the pull of gravity would be too weak to hold the atmosphere around the Earth. Without an atmosphere, we would be exposed to life-destroying amounts of UV radiation from the sun. If the Earth were much larger, the atmosphere would be thicker and would contain more kinds of gases, possibly some that are poisonous.

■ The Environment

When speaking of the environment, we are referring to all of the factors, both living and nonliving, that in any way affect living organisms on Earth. The living, or **biotic,** factors include plants, animals, fungi, and bacteria. The nonliving, or **abiotic,** factors include physical and chemical components such as temperature, rainfall, nutrient supplies, and sunlight.

■ Ecosystems

For purposes of study and the sake of simplicity, it is useful to subdivide the environment, which comprises the entire Earth, into small functional units called **ecosystems.** An ecosystem consists of all of the different organisms living within a finite geographic region and their nonliving surroundings. It may be a forest, desert, grassland, marsh, or just a pond or a field (Figure 1.6). Interrelationships between the organisms and the surroundings are such that an ecosystem is usually self-contained and self-sustaining.

Producers and Consumers

Ecosystems are sustained by the energy that flows through them. The biotic part of any ecosystem can be divided into producers of energy and consumers of energy. Green plants and cyanobacteria (blue-green algae) are the **producers**; they are able to manufacture all of their own food. By means of photosynthesis, they absorb light energy from the sun and use it to convert H_2O and CO_2 from the air into the simple carbohydrate glucose ($C_6H_{12}O_6$). At the same time, O_2 is released to the atmosphere. By further reactions between glucose and chemicals obtained from water and soil, plants manufacture all of the complex materials that they need (Figure 1.7). The plant world includes trees, bushes, flowers, grasses, mosses, and algae.

Consumers are unable to harness energy from the sun to manufacture their own food and must consume plants or other creatures to obtain the nutrients and energy that they need. Consumers can be divided into four main groups according to their food source: herbivores, carnivores, omnivores, and decomposers. **Herbivores** feed directly on producers. Examples of herbivores are deer, cows, mice, and grasshoppers. **Carnivores** eat other animals and include spiders, frogs, hawks, and all cats (lions, tigers, and domestic cats). The animals that carnivores eat may be herbivores, carnivores, or omnivores. **Omnivores** are creatures, including rats, raccoons, bears, and most humans, that feed on both plants and animals.

Decomposers feed on **detritus**, the freshly dead or partly decomposed remains of plants and animals. Decomposers include bacteria, fungi, Earthworms, and many insects. Decomposers perform the very useful task of breaking down complex organic compounds in dead plants and animals into simpler chemicals and returning them to the soil for the producers to reuse. In this way, many nutrients are endlessly recycled through an ecosystem.

Figure 1.6 An ecosystem is a group of plants and animals interacting with one another and their surroundings. It may cover a small area, such as a pond.

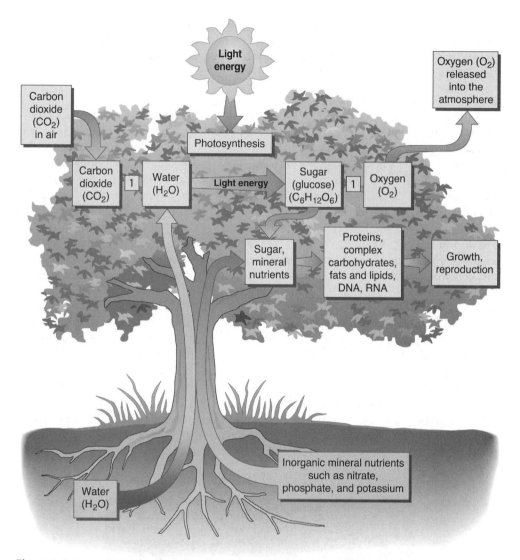

Figure 1.7 Green plants such as this tree are producers. In the process of photosynthesis, they use light energy from the sun to convert carbon dioxide and water to glucose and oxygen. The oxygen is released to the atmosphere; the glucose, together with mineral nutrients from the soil, is used to produce the complex organic compounds that make up plant tissues.

■ The Flow of Energy Through Ecosystems

All of the activities that go on in an ecosystem require **energy**. Without a constant flow of energy from the sun to producers and then to consumers, an ecosystem would not be able to maintain itself. Before you can understand these energy relationships, you need to have some appreciation of what is meant by *energy*. We all know that it takes enormous amounts of energy to maintain our industrialized society. Energy is needed to run automobiles, heat and cool buildings, supply light, and grow food. We receive energy from the food that we eat and use it when we perform any activity. Energy exists in many forms, including light, heat, electrical energy, nuclear energy, and chemical energy. The ultimate source of energy for our planet is the sun. But what exactly is energy?

What Is Energy?

Energy is usually defined as the ability to do work or bring about change. **Work** is done whenever any form of matter is moved over a distance. Everything that goes on in the universe involves work, in which one form of energy is transformed into one or more other forms of energy. Energy is the capacity to make something happen.

All forms of energy can be classified as either kinetic or potential. **Kinetic energy** is energy of motion. A moving car, wind, swiftly flowing water, and a falling rock all have kinetic energy and are capable of doing work. Wind, for example, can turn a windmill, and water rushing from a dam (Figure 1.8a) can turn a turbine to produce electricity. **Potential energy** is stored energy, which is converted to kinetic energy when it is released. Water held behind a dam (Figure 1.8b) or a rock poised at the edge of a cliff has potential energy because of its position. When the water is released or the rock falls off the cliff, the potential energy is converted to kinetic energy. Potential energy is also stored in chemical compounds, such as those present in food and gasoline. When food is digested, chemical bonds are broken, and energy that the body needs to function is released. Similarly, combustion of gasoline in a car engine releases energy to set the car in motion.

Energy Transformations

In the universe, energy transformations occur continuously. Stars convert nuclear energy into light and heat. Plants convert light energy from the sun into chemical energy in the bonds within sugar molecules, and animals convert the chemical energy in sugars into energy of motion. None of these transformations is 100% efficient; in every case, a part of the energy is converted to some useless form of energy, usually heat. In any transformation, however, no new energy is created, and no energy is destroyed. This is the **first law of thermodynamics**

Figure 1.8 (a) The kinetic energy of water rushing from a dam can be harnessed to do useful work. (b) The potential energy of the water in the lake behind the dam is converted to kinetic energy as it is released from the dam.

(also known as the *law of conservation of energy*): Energy can be neither created nor destroyed; it can only be transformed from one form to another.

The efficiency of any energy-transfer process is defined as the percentage of the total energy that is transformed into some useful form of energy. For example, the efficiency of an incandescent electric light bulb is very low. Only approximately 5% of the electrical input is converted into light energy; the remainder, as anyone who has touched a lighted bulb knows, is converted to heat (Figure 1.9). If the light bulb is in its normal surroundings in a home or office, there is no way for the heat energy to do useful work or be converted into some useful form of energy. It is essentially lost.

The loss of useful energy is summed in the **second law of thermodynamics**: In every energy transformation, some energy is always lost in the form of heat energy that thereafter is unavailable to do useful work. This statement means that all systems tend to run "downhill." In other words, whenever any work is done, high-quality energy is converted into lower quality energy. All of the life-sustaining processes that go on in the human body and in all other living organisms follow this pattern of energy flow.

Although energy is never destroyed, the fact that energy is lost as heat in all transformations means that, unlike many material resources, energy cannot be recycled. This fact has important implications for our society, which is dependent on so many energy-inefficient machines. The gasoline engine, for instance, is only 10% efficient.

Food Chains and Trophic Levels

Green plants (producers) are the only organisms that can take energy from the sun and use the process of photosynthesis to store some of that energy in chemical bonds in sugars, starches, and other large molecules. When an animal eats a plant, the sugars and other chemical substances in the plant are broken down in chemical reactions in the animal's body. Bonds that had connected the atoms in the plant's molecules are broken, and energy is released. The energy is used to power the many activities that enable the animal to grow and survive.

Figure 1.9 (a) When electrical energy passes through the filament of a light bulb, only about 5% of it is converted to light; the rest is lost as heat. (b) An automobile engine converts about 10% of the chemical energy in gasoline to mechanical energy that can be used to drive the vehicle; 90% is lost as heat.

Any ecosystem has innumerable feeding pathways, or **food chains**, through which energy flows. In one typical food chain, grasshoppers eat green leaves. Frogs eat grasshoppers, and fish eat frogs (Figure 1.10a). Each step in the chain is called a **trophic level.** In this example, there are four trophic levels: green leaves (plants) at the first trophic level, grasshoppers (herbivores) at the second trophic level, frogs (carnivores eating herbivores) at the third trophic level, and fish (carnivores eating carnivores) at the top of this food chain, at the fourth trophic level. A human eating a carrot would be at the second trophic level; a human eating beef that has been raised on corn would be at the third trophic level. Organisms at any one trophic level are dependent on the organisms at the level below them for their energy needs. Ultimately, all animals—including humans—are dependent on producers for their existence.

Energy and Biomass

At each trophic level in a food chain, organisms use the energy at that level to maintain their own life processes. Inevitably, as a result of the second law of thermodynamics, some energy is lost to the surroundings as heat. It is estimated that in going from one trophic level to the next, about approximately 90% of the energy that was present at the lower level is lost (Figure 1.10b). Thus, in any ecosystem, the energy at the second trophic layer (herbivores) is only approximately 10% of the energy at the first trophic level (producers). The energy at the third trophic level (carnivores) is a mere 1% of that at the first trophic level (producers). This progression has an important implication for humans. It means that it is much more efficient to eat grain than to eat beef that has been fed on grain. It also means that to support a given mass of herbivores requires a mass of producers that is 10 times as large. To support a given mass

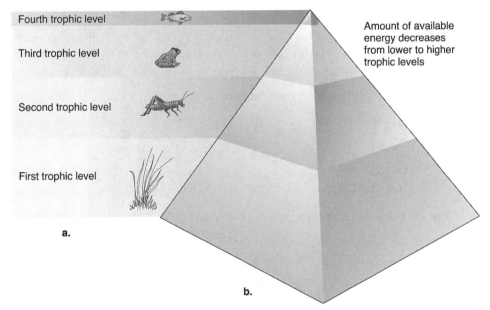

a.

b.

Fourth trophic level

Third trophic level

Second trophic level

First trophic level

Amount of available energy decreases from lower to higher trophic levels

Figure 1.10 (a) A typical food chain with four trophic levels. (b) As energy passes to a higher trophic level in a food chain, approximately 90% of the useful energy is lost. High trophic levels contain less energy and fewer organisms than lower levels.

of carnivores requires a mass of producers that is 100 times as large. Because of these mass requirements, food chains rarely go beyond four trophic levels.

■ Concentration Units

When describing an environmental process, it is important to know how much of each participating chemical is present. Because the chemicals of interest are often present in a mixture of solids, liquids, or gases, chemists need to know the exact **concentration** of each participating chemical before they can assess the process. In addition, government agencies, such as the Environmental Protection Agency (EPA) that regulates the release of pollutants, often establish regulations with rules that limit the concentration of pollutants in our air, water, and soil.

These mixtures can be considered solutions. A *solution* is a homogeneous mixture of two or more substances. The minor species in a solution is called the *solute,* and the major species the *solvent.* In water, that which is dissolved is the solute, and water is the solvent. There are various ways of expressing concentration; we consider molarity, parts per million (ppm) and parts per billion (ppb).

Molarity and Molar Solutions

For aqueous solutions, the unit of concentration that is most often used by chemists is **molarity**: the number of moles of solute per liter of solution.

$$\text{molarity} = M = \frac{\text{moles of solute}}{\text{liters of solution}}$$

A 1-M solution contains the molecular (or formula) mass of the solute dissolved in 1 L of solution. Molarity can also be expressed as millimoles of solute per milliliter of solution (mmol/mL).

Sea water, for instance, contains 1.06% of Na^+ (percentage wt/vol, weight of Na per 100 mL of solution). Thus, to convert 1.06% Na^+ to molarity of Na^+, do the following:

Determine how many grams of Na^+ would be present in 1 L of sea water.

$$1.06\% = \frac{1.06 \text{ g } Na^+}{100 \text{ mL}} \text{ so to find how much in 1 L multiply by 10}$$

$$1.06\% = \frac{10.6 \text{ g } Na^+}{1000 \text{ mL}}$$

Determine how many moles are present in 10.6 g of Na^+.

$$\text{Moles } Na^+ = \frac{10.6 \text{ g}}{23.0 \text{ g/mol}} = 0.46 \text{ moles}$$

Thus, seawater has a molarity (M) of Na^+ of 0.46 M.

For elements or chemicals that are present at low concentration, it is often more convenient to express the concentration in millimolar (millimoles [10^{-3}]/L), micromolar (micromoles [10^{-6}]/L), or nanomolar (nanomoles [10^{-9}]/L) units.

Parts per Million

The unit ppm is a convenient way to describe very dilute solutions. It is frequently used for stating concentrations of pollutants in water. Consider drinking water in which the concentration of dissolved lead is 1 ppm. This means there is one part of lead in every 1 million parts of water. Parts can be expressed in any unit of mass (e.g., ounces, tons, micrograms), but the same unit must be used for both solute and solvent. We use grams.

$$1 \text{ ppm} = \frac{1 \text{ g of solute}}{1 \text{ million g of water}}$$

Because it is more convenient to measure liquids by volume rather than by mass, we change the mass of water to a volume of water:

1 g of pure water has a volume of 1 mL (density = 1.00 g/mL)

Therefore,

$$1 \text{ ppm} = \frac{1 \text{ g of solute}}{1 \text{ million } (1{,}000{,}000) \text{ mL of water}}$$

We change milliliters to liters:

$$1 \text{ ppm} = \frac{1000 \text{ mL}}{1 \text{ L}} \times \frac{1 \text{ g}}{1{,}000{,}000 \text{ mL}}$$

$$= \frac{1 \text{ g}}{1000 \text{ L}}$$

We change grams to milligrams:

$$1 \text{ ppm} = \frac{1000 \text{ mg}}{1 \text{ g}} \times \frac{1 \text{ g}}{1000 \text{ L}}$$

Therefore,

$$1 \text{ ppm} = 1 \text{ mg per L}$$

Each liter of the drinking water contains 1 mg of lead.

To interconvert ppm concentration and molarity, determine how many moles of solute are present. Using the solution that contains 1 ppm lead:

$$1 \text{ ppm} = \frac{1 \text{ mg}}{1 \text{ L}}$$

To convert to moles, divide 1 mg by lead's gram atomic weight.

$$\frac{1 \text{ mg}}{207.2 \text{ g/mol}} = 0.00482 \text{ moles}$$

$$1 \text{ ppm lead} = 0.00482 \text{ mol/L} = 0.00482 \text{ M} = 4.82 \times 10^{-3} \text{ M}$$

Parts per Billion

For certain solutions, particularly water samples containing minute traces of contaminants, it is often more convenient to express concentration in ppb rather than ppm.

$$1\,\text{ppb} = \frac{1\,\text{ppm}}{1000}$$

$$= \frac{1\,\text{mg}}{1\,\text{L}} \times \frac{1}{1000}$$

We change milligrams to micrograms as follows:

$$1\,\text{ppb} = \frac{1000\,\mu\text{g}}{1\,\text{mg}} \times \frac{1\,\text{mg}}{1\,\text{L}} \times \frac{1}{1000}$$

Therefore, 1 ppb = 1 μg/L.

EXAMPLE **1.1**

The U.S. EPA set a limit for the concentration of lead in drinking water at 15 ppb. A laboratory finds the concentration of lead in a sample taken from a water fountain to be 18 μg/100 mL. Is this above or below the EPA limit? By how much?

Solution

Step 1. Given the concentration of lead in the sample

$$\text{concentration} = \frac{18\,\mu\text{g}}{100\,\text{mL}}$$

Step 2. To convert the concentration to μg/L (ppb), multiply the top and bottom by 10 to make the denominator 1000 mL (1 L)

$$\text{concentration} = \frac{18\,\mu\text{g}}{100\,\text{mL}} \times \frac{10}{10}$$

$$= \frac{180\,\mu\text{g}}{1000\,\text{mL}}$$

$$= \frac{180\,\mu\text{g}}{1\,\text{L}}$$

Remember, 1 ppb = 1 μg/L. Therefore,

$$\frac{180\,\mu\text{g}}{1\,\text{L}} = 180\,\text{ppb}$$

Step 3. This concentration is above the EPA limit of 15 ppb. How much above?

$$180 - 15 = 165\,\text{ppb}$$

The concentration (180 ppb) is above the EPA limit by 165 ppb.

The assumption that we made in these calculations is that the density of water is 1.00 g/mL. This is true in freshwater but not in seawater. The density of seawater is 1.025 g/mL. If measurements were being taken in the Chesapeake Bay, where Northern samples are freshwater and Southern samples are seawater, an error in the measured concentration would be introduced because of the difference in density between seawater and freshwater. It would be better to use a more unambiguous unit than mg/L. It is more preferable to describe the concentration as mass solute per mass solution. This means that the ppm concentration would be expressed as mg solute per kg solution and the ppb concentration as g solute per kg solution.

When expressing the concentration of solid samples, such as soil, sludge, rocks, and hazardous materials in ppm concentration, it is best to use mg/kg as the unit of measure. When expressing the concentration in ppb, it is better to use μg/kg.

■ Nutrient Cycles

To survive, a community of plants and animals in an ecosystem requires a constant supply of both energy and **nutrients**. The energy that sustains the system is not recycled. It flows endlessly from producers to consumers, entering as light from the sun and leaving as waste heat that cannot be reused. Nutrients, however, are continually recycled and reused. When living organisms die, their tissues are broken down, and vital chemicals are returned to the soil, water, and atmosphere.

Analysis of tissues from living organisms shows that more than 95% of the mass of the tissues is made from just 6 of the Earth's 92 naturally occurring elements: carbon (C), hydrogen (H), oxygen (O), nitrogen (N), sulfur (S), and phosphorus (P). These six elements are the building blocks for the manufacture of carbohydrates, proteins, and fats. These compounds, along with water, make up almost the entire mass of all living organisms. Plants are composed primarily of carbohydrates, whereas animals are composed primarily of proteins.

Small amounts of other elements are also required in order for plants and animals to survive and thrive. Iron (Fe), magnesium (Mg), and calcium (Ca) together make up most of the remaining 5% of the mass of living organisms. In many animals, iron is bound to hemoglobin, the protein in the blood that supplies oxygen to all parts of the body. In green plants, magnesium is bound to chlorophyll, the protein that absorbs light from the sun and that is a vital part of the process of photosynthesis. Animals with skeletons need calcium as well as phosphorus to make bones and cartilage.

Trace amounts of approximately 16 other elements are also required. Copper (Cu) and zinc (Zn), for example, are essential components of certain enzymes, specialized proteins that facilitate many vital chemical reactions within the animals' bodies.

Plants obtain the essential elements from the soil and the atmosphere. Animals obtain them from their food: plants and other animals. Consider the cycles by which supplies of oxygen, carbon, and nitrogen are constantly renewed.

The Carbon Cycle

The **carbon cycle** is illustrated in Figure 1.11. The major sources of carbon for our planet are the carbon dioxide gas in the atmosphere and the carbon dioxide dissolved in the oceans. Enormous quantities of carbon are also present in rocks, tied up in carbonates such

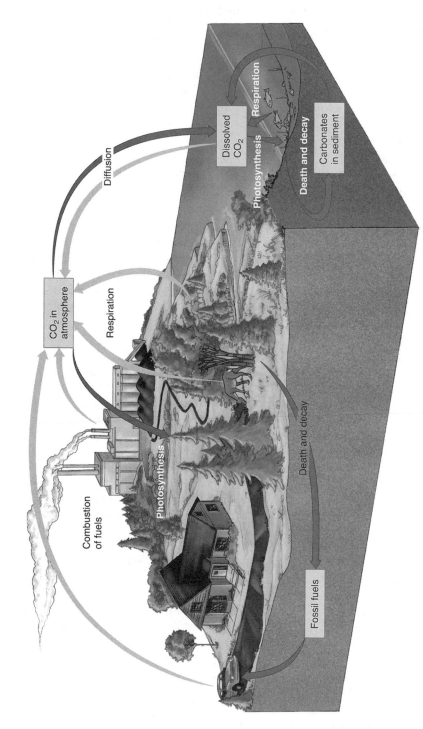

Figure 1.11 The carbon cycle: Atmospheric carbon dioxide is consumed by green plants in photosynthesis. Respiration in animals and plants and combustion of fossil fuels return carbon dioxide to the atmosphere. Carbon also cycles through water; dissolved carbon dioxide reacts with minerals and water to form carbonates, which are deposited in sediments.

as limestone, but this source recycles so slowly that it is not available to plants and animals for their daily needs.

Atmospheric carbon dioxide, although it makes up only 0.036% (360 ppm) of the atmosphere by volume, is the starting material on which all living organisms depend. In the process of photosynthesis, carbon dioxide is taken into the leaves of green plants, where it combines with water to form sugar (glucose) and oxygen, as shown in this equation:

$$\text{Solar energy} + \underset{\text{carbon dioxide}}{6\,CO_2} + \underset{\text{water}}{6\,H_2O} \xrightarrow{\text{chlorophyll}} \underset{\text{glucose}}{C_6H_{12}O_6} + \underset{\text{oxygen}}{6\,O_2}$$

Plants on land obtain the needed water from the soil; aquatic plants obtain it from their surroundings. The water is absorbed through the plant's roots and is then transported to the leaves. Part of the sugar that is formed is stored in the leaves, and part is converted into the complex carbohydrates and other large molecules that make up plant tissues. The oxygen is released to the atmosphere.

Photosynthesis is a very complex process that is still not fully understood. It involves many chemical reactions in which the green pigment chlorophyll plays an important role. We consider only the overall result of photosynthesis, as summarized by the previous equation.

When an animal such as a deer or rabbit eats a green plant, the carbohydrates in the plant are digested and broken into simple sugars, including glucose. Glucose is absorbed into the bloodstream and carried to the cells of the animal's body, where in the process of **respiration,** it reacts with oxygen in the blood. Carbon dioxide and water are formed, and energy is released.

$$C_6H_{12}O_6 + O_2 \rightarrow 6\,CO_2 + 6\,H_2O + \text{energy}$$

Part of the energy released in respiration is used to power the many activities that go on in living cells, and part is lost as heat.

Notice that the overall reaction for respiration is the same as the overall reaction for photosynthesis written backward. However, although cellular respiration is essentially the reverse of photosynthesis, the complex intermediate steps that are involved in the two processes are very different.

Plants engage in both photosynthesis and respiration. During the day, photosynthesis is the dominant process. At night, when there is no sunlight, respiration is dominant. Decomposers also play a part in the carbon cycle. They feed on the dead remains of plants and animals and via respiration release carbon dioxide and water to the atmosphere.

Approximately 300-million years ago, huge quantities of dead and decaying plant and animal remains became buried deeply under sediments before they could be completely broken down. Over time, the remains were compressed, and chemical reactions gradually transformed them into the fossil fuels: coal, oil, and natural gas. When these fuels are burned to release the chemical energy stored within their molecules' bonds, oxygen from the atmosphere is used to convert their carbon into carbon dioxide.

$$C + O_2 \rightarrow CO_2 + \text{energy}$$

The combustion of the fossil fuels that powers our industrialized society therefore forms an integral part of the carbon cycle. Humans also intervene in the carbon cycle when they cut down more trees than they replace, thereby decreasing the amount of carbon dioxide that otherwise would be taken from the atmosphere and converted to nutrients. The climatic implications of these two activities, both of which tend to increase carbon dioxide concentration in the atmosphere, are discussed in Chapter 3.

Another important part of the carbon cycle is the continual exchange of carbon dioxide between the atmosphere and the oceans, a process that is important in maintaining the carbon dioxide concentration of the atmosphere at a constant level. The amount of carbon dioxide that dissolves in the oceans depends mainly on the temperature of the ocean water and the relative concentrations of carbon dioxide in the atmosphere and the water. When the temperature falls or when the carbon dioxide concentration in the water becomes relatively low, more atmospheric carbon dioxide dissolves. Only the surface layer of the ocean is at equilibrium with the atmosphere. A very small portion of the dissolved carbon dioxide reacts with chemicals in the water, such as calcium, to form carbonates. The carbonates, primarily limestone ($CaCO_3$), are insoluble (do not dissolve) in water and settle on the ocean floor.

Rock formation and weathering are other aspects of the carbon cycle. Sedimentary rocks such as limestone and dolomite ($CaMg[CO_3]_2$) were formed millions of years ago from the skeletal remains of coral and other marine creatures that were rich in calcium carbonates. When chemically weathered by rain with a slight natural acidity, limestone rocks very gradually dissolve, releasing carbon dioxide into the atmosphere. This recycling of carbon through rock is a very slow process.

The **solubility product** is the equilibrium constant for the reaction in which a solid salt dissolves to give its constituent ions in solution. The solution is saturated; that is, some of the undissolved solute is in contact with the solution. For example, when limestone (calcium carbonate) is placed in water, the dissociation process is described with this equation:

$$CaCO_3 \rightleftharpoons Ca^{2+} + CO_3^{2-}$$

The equilibrium constant, K for this reaction, can be written as follows:

$$K = \frac{[Ca^{2+}][CO_3^{2-}]}{[CaCO_3(s)]} \quad \text{remember that [ ] means "molar concentration"}$$
$$M \text{ in mol/L or mmol/mL}$$

The solid $CaCO_3$ is not dissolved. It is considered to be in a standard state and is omitted from the equilibrium constant. The solubility product equilibrium constant, K_{sp} for this reaction, is defined as follows:

$$K_{sp} = [Ca^{2+}][CO_3^{2-}]$$

This equation describes the equilibrium concentration of the calcium and carbonate ions in solution. The K_{sp} for various solid compounds are listed in Appendix A.

If calcium carbonate ($CaCO_3$) is placed in water, the equilibrium concentration of calcium ion in solution is calculated as follows:

The K_{sp} of $CaCO_3$ is 4.9×10^{-9}

This equilibrium shows that every $CaCO_3$ that dissolves produces one Ca^{2+} and one CO_3^{2-}.

$$[Ca^{2+}] = [CO_3^{2-}]$$

Therefore,

$$K_{sp} = [Ca^{2+}]^2 = 4.9 \times 10^{-9}$$

Take the square root of both sides.

$$[Ca^{2+}] = 6.8 \times 10^{-5} \text{ M}$$

This is a molar concentration, M, expressed as mol/L or mmol/mL.

The mass of dissolved calcium ion is (40.08 g/mol) (6.8×10^{-5} mol/L) = 2.8×10^{-3} g/L or 2.8 mg/L. This corresponds to 2.8 ppm (mg/L). The Ksp indicates that the equilibrium concentration of calcium ion should be 2.8 ppm. This means that if a limestone rock is in contact with pure water (carbon dioxide free), it will dissolve until the concentration of calcium ion reaches 2.8 ppm. Once this concentration is attained, dissolution will stop and the limestone will no longer dissolve, even though it is in contact with the solution. This calculation ignores the fact that dissolved carbon dioxide can react with calcium carbonate.

Calcium ion is the most common metal ion in rivers and lakes. When measured in lakes, the concentration is much higher than the 2.8-ppm concentration that the Ksp indicates. In natural waters, Ca^{2+} is produced by the dissolution of calcium carbonate by the action of dissolved CO_2 to produce 2 moles of bicarbonate (HCO_3^-) for each mole of Ca^{2+}.

$$CaCO_3(s) + CO_2(aq) + H_2O \quad Ca^{2+} + 2\,HCO_3^-$$

If the pH is near neutral, most of the product is bicarbonate, not CO_3^{2-} or H_2CO_3. Measurements of Ca^{2+} and HCO_3^- in many rivers confirm that there is one calcium ion for every two bicarbonate ion. This process raises the $[Ca^{2+}]$ to 20 ppm (for an average atmospheric carbon dioxide concentration). As can be seen in Figure 1.12, the concentration of calcium and bicarbonate in river water has been measured across the world. Rivers such as the Congo, the Mississippi, and the Danube all show a relationship of $2[Ca^{2+}] = [HCO_3^-]$ and appear to be saturated with calcium carbonate. Rivers such as the Nile and the Amazon for which $2[Ca^{2+}]$ is less than $[HCO_3^-]$ are not saturated with calcium carbonate.

In principle, it may be possible to use the Ksp of a solid to estimate the solubility of a solid. In practice, the solubility of the solid is affected by the presence of other solutes. As was shown previously here, the presence of an acid can greatly increase the solid solubility. The presence of complexing agents or common ions also dramatically affects the solubility of solids.

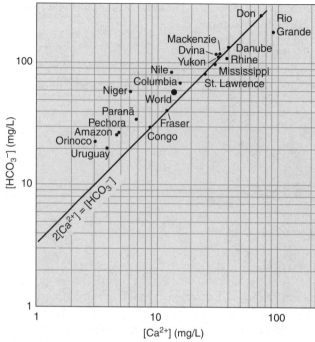

Figure 1.12 Concentrations of bicarbonate and calcium in many rivers conform to the mass balance for the reaction: $[HCO_3^-] = 2[Ca^{2+}]$. Data from W. Stumm and J. J. Morgan, *Aquatic Chemistry*, 3rd ed. (New York: Wiley Interscience, 1996), p. 189, and H. D. Holland, *The Chemistry of the Atmosphere and Oceans.* (New York: Wiley Interscience, 1978).

The Nitrogen Cycle

The **nitrogen cycle** is illustrated in Figure 1.13. Nitrogen is an essential component of proteins and of the genetic material that makes up DNA (deoxyribonucleic acid) and is a constant supply that is vital for all living organisms. Although 7% of the Earth's atmosphere is composed of nitrogen gas (N_2), plants and animals cannot use this nitrogen directly. Atmospheric nitrogen must be converted to other nitrogen compounds before it can be absorbed through the roots of plants. This change is achieved by **nitrogen fixation**, a process that is carried out by specialized bacteria that have the ability to transform atmospheric nitrogen into ammonia (NH_3). Some nitrogen-fixing bacteria live in soil, whereas others live in nodules on the roots of leguminous plants such as peas, beans, clover, and alfalfa. The ammonia produced in root nodules is converted into a variety of nitrogen compounds that are then transported through the plant as needed.

Another means by which atmospheric nitrogen is converted to a usable form is lightning. The electric discharges in lightning cause nitrogen and oxygen in the atmosphere to combine and form oxides of nitrogen, which in turn react with water in the atmosphere to form nitric acid (HNO_3). The nitric acid, which reaches the Earth's surface dissolved in rainwater, reacts with materials in soil and water to form nitrates (NO_3^-) that are directly absorbed through plant roots. Compared with biological fixation, lightning accounts for only a small fraction of the usable nitrogen in soil.

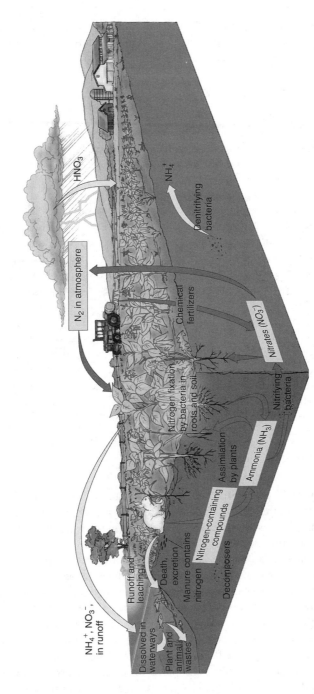

Figure 1.13 The nitrogen cycle: In nitrogen fixation, specialized bacteria convert atmospheric nitrogen to ammonia and nitrates, which plants absorb through their roots. Some nitrogen gas is fixed by lightening. Animals obtain the nitrogen they need to make tissues from plants. The wastes animals produce and their dead bodies return nitrogen to the soil in forms plants can use. In denitrification, bacteria convert nitrates in soil back to nitrogen gas.

Although some trees and grasses can absorb ammonia produced by nitrogen-fixing bacteria directly from the soil, most plants can use only nitrogen that is in the form of nitrates. The transformation of ammonia into nitrates is carried out by specialized soil bacteria in the process known as **nitrification**. A soil pH of 6.5 to 8 is optimum, and the reaction rate decreases when the pH is below 6. The two-step nitrification reaction is as follows:

$$2\,NH_4^+ + 3\,O_2 + 2\,H_2O \rightarrow 2\,NO_2^- + 4\,H_3O^+$$

$$2\,NO_2^- + O_2 + H_2O \rightarrow 2\,NO_3^-$$

When these two reactions are added together, they yield this overall reaction:

$$NH_4^+ + 2\,O_2 + H_2O \rightarrow NO_3^- + 2\,H_3O^+$$

Plants convert ammonia or nitrates that they get from soil or root nodules into proteins and other essential nitrogen-containing compounds. Animals get their essential nitrogen supplies by eating plants. When plants and animals die and decompose, the nitrogen-containing compounds in their tissues are broken down by decomposers; ammonia is eventually formed and returned to the soil. Nitrogen is also returned to the soil in animal wastes. Both urine and feces have a high content of nitrogen-containing compounds. In this way, nitrogen is continually cycled through food chains.

Not all of the ammonia and nitrates that are formed in soil by the processes just described become available for plants. Both ammonia and nitrates are very soluble in water. As rainwater percolates downward through the ground, these compounds are leached out of topsoil. They are often carried away in runoff into nearby streams, rivers, and lakes, where they are recycled through aquatic food chains. Another process that removes nitrates from the soil is **denitrification**, in which bacteria carry out a series of reactions that convert nitrates back to nitrogen gas. One of the denitrification reactions can be written as follows:

$$4\,NO_3^- + 5\,CH_2O + 4\,H_3O^+ \rightarrow 2\,N_2 + 5\,CO_2 + 11\,H_2O$$

In a few locations around the world, nitrates have accumulated in large mineral deposits. In Chile, for example, as mountain streams originating in the Andes Mountains flowed across dry, hot desert toward the sea over thousand of years, much of the water evaporated, leaving behind huge deposits of sodium nitrate.

In the natural environment, a balance is maintained between the amount of nitrogen removed from the atmosphere and the amount returned. However, because most soils contain an insufficient amount of nitrogen for maximum plant growth, farmers frequently apply synthetic inorganic fertilizers containing ammonia and nitrates. As a result of runoff from fertilized farmland, the extra nitrogen-containing compounds reaching rivers and lakes may upset the natural balance, sometimes with damaging consequences for the environment.

An alternative to the use of synthetic fertilizers for replenishing nitrogen in the soil is to plant a nitrogen-fixing crop, such as clover, and plow it back into the soil. Another method is to spread manure and allow the natural soil bacteria to degrade it and release the nitrogen-containing compounds that plants can absorb.

The Oxygen Cycle

Oxygen is all around us. As O_2, it makes up 21% of the atmosphere and is a component of all of the important organic compounds in living organisms. Oxygen is a very reactive element that combines readily with many other elements. It is a component of CO_2, nitrate (NO_2^-), and phosphate (PO_4^{3-}), and thus is an integral part of the recycling of carbon, nitrogen, and phosphorus. Oxygen is also a constituent of most rocks and minerals, including silicates, limestone ($CaCO_3$), and iron ores (Fe_2O_3, Fe_3O_4). The **oxygen cycle** is very complex and is interconnected with many other cycles. Here we briefly consider only some of the more important pathways in the cycle.

Photosynthesis and respiration are the basis of both the carbon cycle and the oxygen cycle. Oxygen is released during photosynthesis and is consumed during respiration. The oxygen and carbon cycles are also interconnected when coal, wood, or any other organic materials are burned. During burning, oxygen is consumed, and carbon dioxide is released.

Another part of the oxygen cycle is the constant exchange of oxygen between the atmosphere and bodies of water, especially oceans. Oxygen dissolved at the surface of water is carried to deeper levels by currents. Dissolved oxygen is essential for fish and other aquatic life.

The Phosphorus Cycle

The **phosphorus cycle** is illustrated in Figure 1.14. Phosphorus is a component of many important biological compounds, including DNA and enzymes that play an essential role in

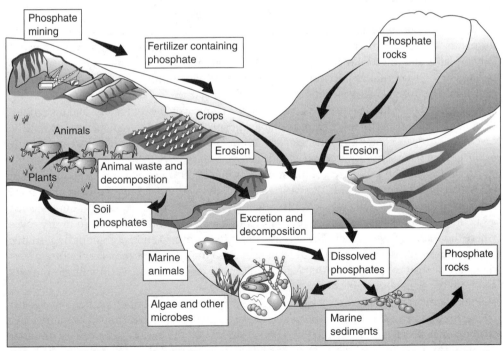

Figure 1.14 The phosphorus cycle. The main reservoir of phosphorus is phosphate rock. Phosphate released to the soil during weathering is absorbed through plant roots. Animals obtain phosphorus form plants and return it to soil in wastes and dead tissue. Phosphates leached from the soil enter waterways and are carried to the oceans.

the transfer of energy in living cells. It is a major constituent of cell membranes and is present in high concentration in bones, teeth, and shells.

Unlike the carbon and nitrogen cycles, the phosphorus cycle does not include an atmospheric phase, and the ultimate source of phosphorus for plants and animals is rock, nearly all of which contains small amounts of phosphorus, mostly in the form of phosphates (PO_4^{3-}). Two examples are apatite, $Ca_5(PO_4)_3(F)$, and vivianite, $Fe_3^{2+}(PO_4)_2 \cdot 8H_2O$. The phosphorus cycle is primarily a sedimentary cycle. As rocks weather, phosphates are very slowly dissolved and released to the soil. The crustal average abundance of apatite is 1.2 mg/g.

Plants absorb dissolved phosphate directly from the soil through their roots and transport it to their leaves, where it is incorporated into large biological molecules. Animals get their phosphorus by eating plants. Not all of the phosphate formed from rock by weathering becomes available to plants. Some of it becomes tightly bound to elements such as aluminum, iron, and calcium in compounds that are not very soluble in water.

In water, phosphorus solubility is largely controlled by pH. The phosphate ion has three forms: $H_2PO_4^-$, HPO_4^{2-}, and PO_4^{3-}.

$$H_3PO_4 \xrightarrow{\text{p}K_a\ 2.15} H_2PO_4^- + H^+ \xrightarrow{\text{p}K_a\ 7.2} HPO_4^{2-} + H^+ \xrightarrow{\text{p}K_a\ 12.3} PO_4^{3-} + H^+$$

At a pH of less than 4, insoluble iron and aluminum phosphates are formed. At a pH of greater than 8, almost insoluble calcium phosphates are formed. Iron phosphate, $Fe_3^{2+}(PO_4)_2 \cdot 8H_2O$, has a solubility product constant (K_{sp}) of 1×10^{-36}, and calcium phosphate, $Ca_3(PO_4)_2$, has a K_{sp} of 1×10^{-24}. At a pH of between 4 and 8, phosphorous has its maximum solubility.

Renewal of phosphorus through rock cycling is an extremely slow process; the main route for plants is through the recycling of the phosphorus in dead and decomposing organisms. When plants and animals die, microorganisms break down the organic phosphorus compounds in the dead tissues into inorganic phosphates, which immediately become available to plants. The average phosphorous content of plants is 0.1%.

Soil is often poor in the types of phosphates that plants can absorb readily, and on agricultural land, synthetic fertilizers rich in available phosphate are used to replenish it. As was the case with nitrogen-containing fertilizers, runoff of large quantities of phosphate-containing fertilizers can have serious effects on aquatic ecosystems.

An important natural source of phosphates is **guano**. Droppings from thousands of sea birds that breed on islands off the west coast of South America have built up into huge deposits of phosphate-rich material called guano that because of low rainfall are not washed away. Large guano deposits also have been found in Arizona and New Mexico in dry caves where thousands of bats gather. These guano deposits are mined and are an important source of phosphorus for the manufacture of fertilizers. Also important for the manufacture of fertilizer are deposits of phosphate that formed in certain regions of the world millions of years ago from the skeletal remains of sea creatures.

Phosphorus is continually recycled through plants and animals, but inevitably, some is lost by leaching and erosion of soil into streams and rivers. Human activities such as mining and farming often accelerate this loss. When phosphates reach the oceans, they react with other chemicals in sea water, and most are converted to insoluble phosphates that sink to the ocean floor and, for all practical purposes, are permanently lost. It can take millions of years for phosphate lost to the oceans to be replenished by weathering of rock. Humans add to the problem of phosphorus loss by disposing of large amounts of their wastes—which contain significant quantities of phosphorus—into waterways instead of recycling them and returning them to the land.

Reserves of phosphorus in high-grade phosphate ores are estimated to be sufficient for several hundred years. We do not have an immediate problem, but we would be wise to conserve the reserves of phosphorus that we still have and reduce phosphorus loss as far as possible.

Nature's Cycles in Balance

Life on Earth depends on a continual supply of energy from the sun and a continual recycling of materials. As you have learned here, the numerous and complex processes that sustain life are interlinked and interdependent. Because all aspects of an ecosystem are so completely interwoven, human intervention in any part of the system can easily lead to widespread disturbances of the natural balance in the environment.

■ Additional Sources of Information

Chiras DD. *Environmental Science: Creating a Sustainable Future*, 8th ed. Sudbury, MA: Jones and Bartlett Publishers, 2010.

Grotzinger J, Jordan TH, Press F, Seiver R. *Understanding Earth*, 5th ed. New York: W. H. Freeman & Co., 2006.

Kuhn KF, Koupelis T. *In Quest of the Universe*, 5th ed. Sudbury, MA: Jones and Bartlett Publishers, 2007.

Pinet PR. *Invitation to Oceanography*, 5th ed. Sudbury, MA: Jones and Bartlett Publishers, 2009.

Walther JV. *Essentials of Geochemistry*, 2nd ed. Sudbury, MA: Jones and Bartlett Publishers, 2010.

Zuckerman B. *The Origin and Evolution of the Universe*. Sudbury, MA: Jones and Bartlett Publishers, 1996.

■ Keywords

abiotic
amino acids
anerobic bacteria
asthenosphere
bauxite
big bang
biotic
carbon cycle
carnivores
concentration
consumers
core
crust
cyanobacteria
decomposers
denitrification

detritus
ecosystems
energy
environment
first law of thermodynamics
food chains
fusion reactions
giant planets
guano
herbivores
inorganic
kinetic energy
lithosphere
mantle
Milky Way
minerals

molarity
natural resource
nitrification
nitrogen cycle
nitrogen fixation
nonrenewable resource
nutrients
omnivores
ores
organic
oxygen cycle
phosphorus cycle
photosynthesis
potential energy

ppb (parts per billion)
ppm (parts per million)
producers
renewable resource
respiration
second law of thermodynamics
solar system
solubility product
steel
strategic metals
terrestrial planets
trophic level
universe
work (energy)

■ Questions and Problems

1. What percentage of the mass of the solar system is accounted for by the mass of the sun?
   a. Less than 1%
   b. Approximately 10%
   c. Nearly 30%
   d. More than 99%
2. Why does the large number of hydrogen atoms in the universe suggest that other elements were built from hydrogen rather than by larger elements breaking into smaller ones?
3. Make a diagram showing a cross-section of the Earth and the three main layers into which the Earth can be divided. What element predominates at the center of the Earth?
4. By weight, which are the four most abundant elements in the whole Earth? Explain why elements are not uniformly distributed throughout the Earth.
5. By weight, which are the four most abundant elements in the Earth's crust?
6. How did water form on the Earth's surface?
7. Why does Earth's sister planet, Venus, not support life?
8. The early Earth is thought to have been devoid of oxygen. How did oxygen first appear in the Earth's atmosphere?
9. What are anaerobic bacteria?
10. If aluminum cans and steel cans are thrown into the city dump, which will degrade first? Briefly explain your answer.
11. Name the mineral from which most of the world's aluminum is extracted. From what countries does the United States obtain its supply of aluminum?
12. Give three uses for the metal copper.
13. Making an aluminum can from recycled aluminum rather than starting with aluminum ore produces an energy savings of
    a. 10%
    b. 50%
    c. 75%
14. Give three uses for the metal aluminum.

15. What valuable metals are by-products of copper production?
16. What is the purpose of adding carbon to steel?
17. What is meant by a strategic metal?
18. Significant quantities of which strategic metal are found in
    a. South America
    b. Africa
    c. Europe
    d. Australia
19. Discuss each of the following claims:
    a. New discoveries of mineral deposits will provide all we need in the future.
    b. Our children will have sufficient mineral resources.
    c. The ocean will supply all our mineral needs in the future.
    d. Recycling scarce mineral resources will guarantee a continued ample supply of those minerals.
20. What is meant by biotic and abiotic? Give examples of each.
21. Living organisms can be divided into producers and consumers. Describe how producers make their food and how consumers get the food that they need.
22. Consumers can be divided into groups according to their food source. Name the main groups and give two examples of each.
23. Organisms that consume the remains of dead plants and animals are necessary for a balanced ecosystem. Explain their purpose.
24. Define what is meant by energy. Name four forms of energy.
25. State the first law of thermodynamics. Give an example of the conversion of one form of energy to another.
26. State the second law of thermodynamics. Give an example that illustrates this law.
27. Of what type is the unusable energy that is released during almost all energy transformations?
28. If 10,000 units of energy are available to organisms at the first trophic level of a food chain, how many units of energy will be available to organisms that occupy the third trophic level in the food chain?
29. Why do most food chains contain only three or four trophic levels?
30. Why are people with limited food resources often herbivores?
31. Each time energy flows from a tropic level to the one above it, approximately what percentage of useful energy is lost?
32. What four elements are most abundant in the bodies of plants and animals? Name three other elements that are essential for life.
33. What is the molar concentration (molarity) of a 1.0-L solution that contains 156 g of nitric acid, HNO_3, or another 1.0-L solution that contains 7.8 g?
34. Seawater contains the following dissolved ions. Convert the concentration of each from ppm to molarity.
    a. 19,000 ppm Cl^-
    b. 2600 ppm SO_4^{2-}
    c. 1300 ppm Mg^{2+}
35. Seawater contains 2.7 g of NaCl per 100 mL.
    a. What is the concentration of NaCl expressed as a percentage (wt/vol)?
    b. What is the molarity of NaCl in seawater?

36. The molarity of $MgCl_2$ in seawater is 0.54.
   a. How many grams of $MgCl_2$ are there in every 100 mL of seawater?
   b. What is the concentration of $MgCl_2$ expressed as a percentage (wt/vol)?
   c. How many mmol/mL of $MgCl_2$ does seawater contain?
37. Express 0.000056 M as
   a. A millimolar concentration
   b. A micromolar concentration
   c. A nanomolar concentration
38. What is the molarity of a 20.0-L solution that contains 40 g of sodium hydroxide (NaOH)?
39. Waste discharge water from a paper mill is sampled for dioxin, a toxic substance that is formed in the bleaching of paper pulp. Analysis of the waste water from the stream shows a dioxin concentration of 0.010 μg/1.0 mL. Express this concentration in ppb. The EPA standard for dioxin is 1 ppb. Is this waste water in violation of the EPA standard?
40. Which is the higher concentration in each pair?
   a. 100 ppb or 0.05 ppm?
   b. 500 ppb or 250 ppm?
   c. 10 ppb or 1 ppm?
41. If a 100 g water sample contains 1.5 mg of arsenic, what is the concentration of arsenic in the sample in ppm?
42. A laboratory has measured the amount of mercury in a sewage sludge sample to be 0.0005 g per kilogram. What is the concentration of mercury in ppm? In ppb?
43. An ore sample contains 0.0000068 g of gold. What is the concentration of gold in ppm and ppb?
44. Calculate the following:
   a. The molarity of a sample of freshwater that contains 10-ppm NH_4^+
   b. The ppm concentration of calcium in a freshwater sample that contains 5.4×10^{-6} M $CaCO_3$
45. Calculate the following:
   a. The molarity of a sample of freshwater that contains 10 ppm NO_3^-
   b. The ppm concentration of Cu in a freshwater sample that contains 5.4×10^{-4} M $CuCl_2$
   c. The ppm concentration of Cl in a freshwater sample that contains 5.4×10^{-4} M $CuCl_2$
46. What is the main source of carbon for living things? Describe the process by which plants use this source of carbon to make their food.
47. Make a diagram that illustrates one of the following nutrient cycles:
   a. carbon
   b. nitrogen
48. What part do decomposers play in the carbon cycle?
49. What part does each of the following play in the carbon cycle?
   a. The oceans
   b. Limestone
   c. Crude oil
50. What is the name of the class of compounds formed, in addition to oxygen, as products in the photosynthesis reaction?
51. Name the process by which the nitrogen of the atmosphere is converted into a form that can be used by plants.

    a. Where are the bacteria that bring about this process found?

    b. What chemical compound is produced by these bacteria?

52. Explain the process by which lightning can convert atmospheric nitrogen to a form that plants can use.

53. Why do nitrates not occur as large deposits in many places in the crust of the Earth?

54. List two ways streams and rivers receive excessive amounts of nitrogen-containing compounds because of the activities of people.

55. What is meant by *nitrification*? How is the nitrogen in animal waste recycled by nature?

56. Describe two ways by which a farmer might replenish the nitrogen in her fields without using inorganic fertilizers.

57. What two parts of the oxygen cycle are intertwined with the carbon cycle?

58. List these three salts in order of decreasing solubility.

CuBr	$K_{sp} = 5.2 \times 10^{-9}$
CuCl	$K_{sp} = 1.2 \times 10^{-6}$
CuI	$K_{sp} = 1.1 \times 10^{-12}$

59. Which of the following lead salts will produce the highest concentration of dissolved lead in water? Assume 1 mol of each is placed in 1 L of water.

Lead sulfate	$PbSO_4$	$K_{sp} = 1.6 \times 10^{-8}$
Lead chromate	$PbCrO_4$	$K_{sp} = 1.8 \times 10^{-14}$
Lead sulfide	PbS	$K_{sp} = 7 \times 10^{-28}$

60. Silver ion Ag^+ is an effective disinfectant for the water in swimming pools when it is kept at a concentration of 10 to 100 ppb (ng/L). A silver ion concentration higher than 300 ppb in drinking water, however, is considered unhealthy. One manufacturer of pool disinfectants provides a slightly soluble silver salt in pellet form that releases an equilibrium concentration of Ag^+ to the pool water, which is effective as a disinfectant but is not a danger to the swimmer's health. Which of the following silver salts would be the best candidate for this task?

AgCl	$K_{sp} = 1.8 \times 10^{-10}$
AgBr	$K_{sp} = 5.0 \times 10^{-13}$
AgI	$K_{sp} = 8.3 \times 10^{-17}$

61. What is the primary phosphate species at the following pH?

pH = 4, pH = 10, pH = 13

62. The concentration of arsenic in the soil around an abandoned pest control business is found to be 0.02 mg per 100 g of soil. Express the concentration in ppm and ppb.

CHAPTER

2

The Earth's Atmosphere

THE ATMOSPHERE IS A THIN BLANKET OF GAS THAT ENVELOPS THE EARTH. It provides the carbon dioxide that plants need for photosynthesis and the oxygen that animals need for respiration. It is also the ultimate source of nitrogen for plant growth. Freshwater reaches the Earth from the atmosphere as dew, rain, and snow. The atmosphere shields us from the sun's cancer-causing ultraviolet (UV) radiation and also moderates the Earth's climate. Without it, the Earth would experience the extremes of hot and cold that are found on planets that have little or no atmosphere.

The atmosphere is obviously vital for human existence, but we have nevertheless been polluting it for years. By the end of the 19th century, huge quantities of coal were being burned to fuel the Industrial Revolution, and smokestacks belching great brown clouds into the atmosphere became a sign of prosperity. By the middle of the 20th century, the automobile had become another significant source of air pollution. Today, in the United States, pollution from these sources has been greatly reduced as a result of legislation, but we face other problems. Our continued dependence on fossil fuels for energy is introducing increasingly large quantities of carbon dioxide into the atmosphere. In this chapter and in Chapter 3, we examine how this practice may be causing the atmosphere to become warmer, a trend that could have disastrous consequences for the world's climate.

Here we first examine the major layers of the atmosphere and their composition, followed by the important balance of energy reaching the Earth from the sun. Finally, the absorption of infrared radiation by atmospheric gases, the increased concentration of these gases in the atmosphere, and their effects on **global warming** are discussed.

■ The Major Layers in the Atmosphere

The gases that make up the atmosphere are held close to the Earth by the pull of gravity. With increasing distance from the Earth's surface, the temperature, density, and composition of the atmosphere gradually change. On the basis of air temperature, the atmosphere can be divided vertically into four major layers: **troposphere**, **stratosphere**, **mesosphere**, and **thermosphere** (Figure 2.1).

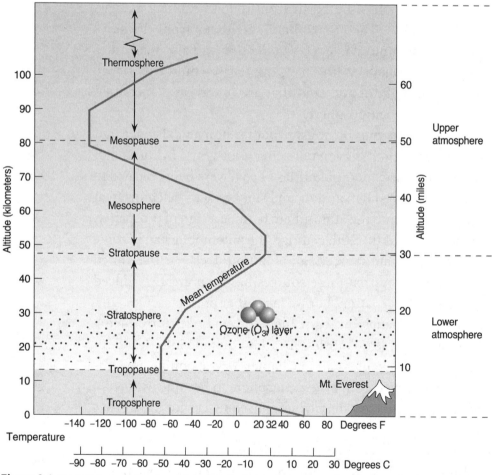

Figure 2.1 The Earth's atmosphere is subdivided vertically into four major regions based on the air-temperature profile. The ozone layer that protects us from the sun's UV radiation is in the stratosphere.

Temperature Changes in the Atmosphere

As shown in Figure 2.1, the Earth's atmosphere is stratified as a result of temperature and density relationships that result from the interaction of physical and photochemical (induced by sunlight) processes. The troposphere extends above the Earth to a distance of 10 to 16 kilometers (6 to 10 miles). The temperature of the troposphere decreases steadily as the distance from the Earth's warm surface increases until it reaches approximately −57°C (−70°F).

The lower part of the troposphere (0 to 3000 meters), which interacts directly with the surface of the Earth, is the part of the troposphere known as the **boundary layer.** Pollutants emitted near the ground accumulate in the boundary layer. The temperature of the air in the boundary layer responds to changes in ground temperature in less than one hour. Most weather occurs in the boundary layer of the troposphere.

The temperature of the **free troposphere** (the upper part of the troposphere), on the other hand, responds to changes in ground temperatures over a longer period. The temperature of the free troposphere decreases with rising altitude (approximately 6.5°K/km). The temperature of this region decreases with increasing altitude because of a number of reasons. First, the top of the free troposphere continuously radiates energy upward, cooling the upper troposphere. Second, the troposphere itself does not efficiently absorb solar radiation. Third, this region receives warmed air that rises from the surface. Because atmospheric pressure also decreases with increased altitude, the warm air enters a region of lower pressure. The warm air expands and as a result cools, resulting in a decrease of temperature with increasing height.

The region called the **tropopause** is at the top of the troposphere. The low temperature (−57°C) of this region serves as barrier that freezes water vapor as ice crystals that fall back to the surface of the Earth. If water vapor was able to rise above this layer, it could reach higher altitudes where it could be photodissociated by intense UV radiation. If this happened, the H_2 and O_2 generated would be lost into space, and as a result, the amount of water on Earth would be constantly decreasing.

Above the troposphere is the stratosphere, which extends to approximately 50 km (30 miles) and includes the ozone layer. The temperature remains constant in the lower part of the stratosphere but begins to rise with increasing altitude, reaching a maximum of approximately −1°C (30°F) at the **stratopause**, which is the boundary between the stratosphere and the mesosphere. The formation of ozone in the stratosphere, which absorbs UV radiation from incoming solar radiation and converts the radiant energy into heat, causes this rise in temperature. The chemistry of ozone formation is in more detail in Chapter 4. Together, the troposphere and stratosphere are called the **lower atmosphere.**

The **upper atmosphere** extends beyond the stratosphere and is divided into the **mesosphere** and the thermosphere. Continuing outward through the mesosphere, the temperature again falls. At an altitude of between approximately 80 and 90 km (50 and 56 miles), the lowest temperature in the atmosphere, approximately −90°C (−130°F), is reached. Above the mesosphere, the temperature rises once more and reaches a maximum of approximately 1200°C (2192°F) in the thermosphere. This rise in temperature is caused by the few gaseous molecules in the thermosphere absorbing the most energetic radiation emanating from the sun. The chemical reactions that occur in this region are discussed in Chapter 4.

Pressure and Density Changes in the Atmosphere

The gases in the atmosphere exert a pressure on the surface of the Earth. Although we are not aware of it and have adapted to it, humans and everything else on the Earth's surface are

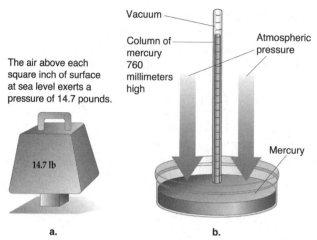

Figure 2.2 (a) At sea level, the air above each inch of the Earth's surface exerts a pressure of 14.7 pounds. (b) At sea level and 0°C, the average pressure of the atmosphere supports a column of mercury 740 to 770 mm in height.

constantly subjected to this pressure (Figure 2.2). We are adversely affected by relatively small variations in atmospheric pressure. People flying in a jet aircraft through the stratosphere, where the air is thin, could not survive if cabin pressure were not adjusted to match the air pressure normally found on the Earth's surface.

With increasing distance from the Earth, the pull of gravity becomes less, and air density (mass per unit volume) decreases. The troposphere and stratosphere together account for 99.9% of the mass of the atmosphere; almost half of this mass is concentrated within 6 km (3.6 miles) of the Earth's surface. As the air becomes thinner with increasing distance from the Earth's surface, atmospheric pressure decreases rapidly. At an altitude of 6 km (3.6 miles), atmospheric pressure is reduced to approximately 50% of the value at sea level. Although the actual atmospheric pressure at any location depends on weather conditions and altitude, the average value used is 1 atmosphere (atm), 760 torr or 101,325 Pa. The standard unit for pressure is the **pascal (Pa)**. A related unit sometimes used to report pressures is the bar (one bar equals 10^5 Pa).

■ Composition of the Atmosphere

The major components in the atmosphere are nitrogen (N_2) and oxygen (O_2), which make up approximately 78% and 21% of the volume of the atmosphere, respectively (Table 2.1). Minor components are the noble gas argon (0.93%) and carbon dioxide (0.038%). Smaller amounts of the other noble gases (neon, helium, and krypton) and methane are present. The percentage of carbon dioxide (0.038%) in the atmosphere is extremely small, but carbon dioxide is the essential raw material for photosynthesis; thus, this very small amount is vital for life itself on Earth. As we shall see later in Chapter 3, carbon dioxide also plays an important role in maintaining the Earth's heat balance.

Water vapor is not included in Table 2.1 because its concentration in the air is variable. Depending on temperature, precipitation, rate of evaporation, and other factors at a particular

Table 2.1

Composition of Pure Dry Air at Ground Level

Gas	Percent by Volume	Parts per Million
Nitrogen (N_2)	78.08	780,840
Oxygen (O_2)	20.94	209,440
Argon (Ar)	0.93	9,340
Carbon dioxide (CO_2)	0.04	370
All other gases	0.01	10

location, the percentage of water vapor in the atmosphere may be as low as 0.1% or as high as 5%. It generally lies between 1% and 3%, making water the third most abundant constituent of the air. The amount of water in the atmosphere depends on the temperature, as the vapor pressure of water increases with temperature. Table 2.2 shows the partial pressure exerted by water as a function of temperature.

The partial pressures, pH_2O, are equilibrium values and represent the maximum pressure that water vapor can exert at that temperature. If the pressure of all gases in the atmosphere is 1.0 atmosphere pressure, you can see that the percentage of water vapor varies from less than 1% at 0°C to over 5% at 35°C. Notice that the vapor pressure of water at its freezing point is not zero. This explains why even snow can evaporate. Usually, the atmosphere is not saturated with water vapor. The **relative humidity** is a term that expresses just how saturated with water vapor the atmosphere is. Using the data in Table 2.2, the partial pressure of water vapor can be calculated if the relative humidity is also known.

Table 2.2

Saturated Partial Pressure of Water Vapor in Air

Temperature °C	pH_2O, atm
−10	0.00257
−5	0.00396
0	0.00603
10	0.01683
20	0.02307
25	0.03126
30	0.04187
35	0.05418

EXAMPLE 2.1

RELATIVE HUMIDITY
Calculate the partial pressure of water at 25°C if the relative humidity is 80%.

Step 1: Obtain from Table 2.2 the saturated partial pressure of water vapor in air at 25°C.

Step 2: Multiply the saturated partial pressure of water vapor by the relative humidity expressed as a fraction.

$$pH_2O = (0.03126\,atm)(0.80) = 0.0250\,atm$$

Because complete mixing of the troposphere would take several years, water is very unevenly distributed in the atmosphere. The distribution of water is consistent with variations in the weather from place to place.

If water vapor is excluded, the concentration of the major components of the air is remarkably constant. In the absence of pollution, no matter where you may be on the surface of the Earth, the air that you breathe is the same. This homogeneity results from the mixing that is brought about by the continuous circulation of the air in the troposphere.

In addition to water vapor and gases, the atmosphere contains many airborne particles, which are the center around which ice crystals and water droplets form. Under appropriate conditions, the droplets coalesce to produce clouds and ultimately rain. Airborne particles range in size from those that are visible, such as dust, to others that can be seen only with a high-powered microscope. Minute particles with diameters of less than approximately 10 μm are termed **aerosols**; larger particles are called **particulates**. Both types of particles can be either liquids or solids.

Relative to their size, small particles have very large surface areas that act as sites for chemical interactions. Depending on the nature of the particle and of the impacting molecule (or other species), chemical reactions may occur at the surface of a particle or within it. If impacting molecules become attached to the particle's surface, the process is termed **adsorption**. If, in the case of liquid particles, molecules are drawn inside and dissolved, the process is termed **absorption**.

Particulates are studied in more detail when we consider their role as pollutants later in this chapter.

■ Units Used to Describe Atmospheric Chemistry

Parts per Million, Parts per Billion, and Parts per Trillion

The most commonly used units of concentration for atmospheric gases are the "parts per million" (ppm), the "parts per billion" (ppb) and "parts per trillion" (ppt) classifications. Unlike solids and liquids, where the "parts per _____" are expressed as mass, in gases, they are expressed as the number of molecules of a pollutant in air. The **ideal gas law** (PV = nRT) states that the volume of gas is proportional to the total number of molecules present. The terminology used to emphasize that the relationship is based on number of molecules, or volume, is to place the letter "v" as part of the unit. Thus, an ozone concentration of 10 ppmv indicates that there are 10 ozone molecules for every one-million air molecules; this measurement of molecules was made by volume. This means there are 2 L of pollutant in one-million liters of air, assuming that the temperature and atmospheric pressure of each are the same. Because the partial pressure of gas is proportional to the number of moles, 2 ppmv would have 2×10^{-6} atmospheres partial pressure of pollutant if the total atmospheric pressure was 1 atmosphere.

Molecules per Cubic Centimeter

The concentrations of many atmospheric pollutants are expressed in pollutant molecules per cubic centimeter (cm^3) of air. One liter equals $1000\ cm^3$. To convert a concentration of 5-ppm carbon monoxide to molecules of CO per cubic centimeter (cm^3) at one atmosphere pressure and 25°C, use the ideal gas law:

$$PV = nRT$$

Because we know there are 5 CO molecules for every one-million air molecules, we need to calculate the volume, in cm^3, that one-million molecules of air would occupy. First calculate the number of moles:

$$n = \frac{1.0 \times 10^6\ \text{molecules}}{6.023 \times 10^{23}} = 1.66 \times 10^{-18}\ \text{mol}$$

Rearrange the ideal gas law and substitute:

$$V = nRT/P = \frac{\left(1.66 \times 10^{-18}\right)\left(0.082\ \text{L-atm/mol K}\right)\left(298\ \text{K}\right)}{1.0\ \text{atm}}$$

$$V = 4.06 \times 10^{-17}\ \text{L}$$

Convert liters into cubic centimeters:

$$\left(4.06 \times 10^{-17}\ \text{L}\right)\frac{\left(1000\ cm^3\right)}{1.0\ \text{L}} = 4.06 \times 10^{-14}\ cm^3$$

Because we know that five molecules of CO occupy the volume of one million air molecules, we can now complete the calculation:

$$\frac{5\ \text{molecules CO}}{4.06 \times 10^{-14}\ cm^3} = 1.23 \times 10^{14}\ \text{molecules}/cm^3$$

Micrograms per Cubic Meter

Another unit that is often used to express the concentration of a pollutant in air is micrograms per cubic meter ($\mu g/m^3$). To convert the micrograms per cubic meter unit to ppm or ppb, all that is needed is the mass of the pollutant. As an example, assume a measurement has shown the atmosphere to contain 230 µg/m³ of nitrogen dioxide (NO_2) on a day that was 27°C and 1-atm atmospheric pressure. Express this concentration in ppb. The units for ppb are as follows:

$$\text{ppb } NO_2 \; = \; \frac{\text{molecules of } NO_2}{\text{1 billion molecules of air}}$$

First we have to compute how many NO_2 molecules there are in 230 µg:

$$230 \times 10^{-6} \,\text{g } NO_2 \; \times \; \frac{1 \text{ mol } NO_2}{46.0 \text{ g } NO_2} \; \times \; \frac{6.02 \times 10^{23} \text{ molecules of } NO_2}{1 \text{ mol } NO_2}$$

$$= 3.01 \times 10^{18} \text{ molecules of } NO_2$$

Next, using the ideal gas law, we can compute how many molecules of air are in 1 m³ of air. Use the conversion $1.0 \text{ L} = 10^{-3} \text{ m}^3$.

$$n \; = \; PV/RT \; = \; \frac{\left(1.0 \text{ atm}\right)\left(1.0 \text{ m}^3\right)\left(1.0\,L\big/10^{-3}\,m^3\right)}{\left(0.082 \text{ L-atm}\big/\text{mol K}\right)\left(300 \text{ K}\right)} \; = \; 40.7 \text{ mol}$$

Now find how many air molecules are in 40.7 mol.

$$\left(40.7 \text{ mol}\right)\left(6.023 \times 10^{23}\right) = 2.45 \times 10^{25} \text{ molecules of air}$$

So,

$$\frac{3.01 \times 10^{18} \text{ molecules of } NO_2}{2.45 \times 10^{25} \text{ molecules of air}} \; = \; \left(1.23 \times 10^{-7}\right) \frac{\left(10^9\right)}{\left(10^9\right)} \; = \; \frac{123 \text{ molecules of } NO_2}{10^9 \text{ molecules of air}}$$

$$= 123 \text{ ppb}$$

■ Energy Balance

Electromagnetic Radiation

The energy source that sustains all life on Earth is the sun. It provides the light energy that plants need for photosynthesis and the heat that warms the Earth. It controls our climate, drives our weather systems, and regulates the life cycles of all plant and animal species. The radiant energy—electromagnetic radiation (EMR)—that the sun continually transmits through space reaches the Earth in many forms. The most familiar forms are light and radiant heat; other forms include cosmic rays, X-rays, UV radiation, microwaves, and radio waves (Figure 2.3). The sun has a surface temperature of approximately 5800 K and acts as a black body emitter. As can be seen in Figure 2.4, as the temperature of a black body emitter rises, its emission of EMR

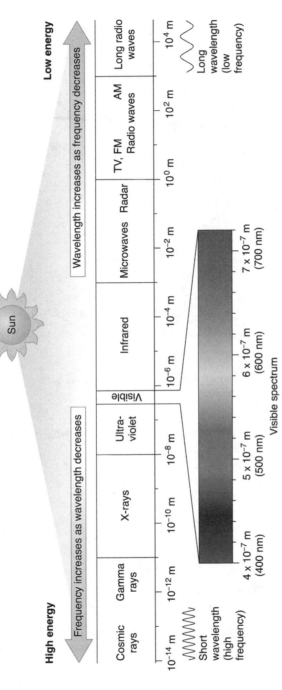

Figure 2.3 The electromagnetic spectrum. As the wavelength of EMR increases from cosmic rays to radio waves, the energy and frequency of the waves decrease. The visible wavelengths make up a very small part of the electromagnetic spectrum (1 nm is equal to 10^{-9} m).

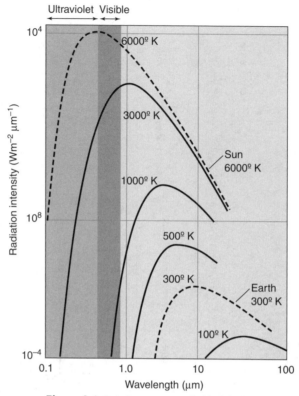

Figure 2.4 Emission spectrum of a black body.

increases, and the wavelength of the EMR shifts to increasingly shorter wavelengths (λ). Planck has described the energy (E) of the EMR.

$$E \ = \ \frac{hc}{\lambda}$$

Where

E = energy in Joules (J)
λ = wavelength in meters
h = 6.626 × 10⁻³⁴ Js, Planck's constant
c = 2.998 × 10⁸ m/sec, the speed of light

As the EMR shifts to increasingly shorter wavelengths, the energy of the EMR increases. To approximate the wavelength of maximum emission, **Wein's displacement law** is used:

$$\lambda_{max} \ = \ \frac{2897}{T}$$

Where

λ_{max} is in micrometers (µm)
T is temperature in K

The emission from the 5800 K sun has a maximum at approximately 500 nm, which is in the visible spectrum.

EXAMPLE 2.2

Calculate the wavelength of maximum radiative emission from the Earth.

Solution
1. The average surface temperature of the Earth is 288 K.
2. Use Wein's displacement law,

$$\lambda_{max} = \frac{2897}{288} = 10.1\ \mu m$$

Only approximately 69% of the total solar radiation or solar flux reaching the Earth is actually absorbed at the Earth's surface. Approximately 31% of the **solar flux** is reflected back into space. The solar flux that is reflected is referred to as Earth's **albedo**. Many factors, including clouds, dust, smoke, and volcanic ash, cause the albedo. The albedo of different surface types can be seen in Figure 2.5. The reflectivity of the Earth's surface also varies depending on location. Of the 69% of the solar flux that is absorbed, 23% is absorbed by water droplets in clouds and other gaseous molecules such as ozone in the atmosphere. The remaining 46% is absorbed at the Earth's surface and is used as an energy source for biomass growth and for thermal warming of the Earth's surface. The average solar flux reaching the Earth (at the top of the stratosphere) is 1368 W/m².

All of the energy that the Earth absorbs from the sun is eventually lost as EMR. The average surface temperature of the Earth is 288 K (15°C). The Earth loses EMR as a black body reradiating energy. The previously mentioned equation allows us to calculate the maximum wavelength of this reradiating EMR, which is 10 µm, which corresponds to the emission line for a 288 K black body emitter (see Figure 2.4). Figure 2.4 shows that the Earth reradiates EMR as infrared.

The Earth's Heat Balance

Solar energy is transmitted through space as EMR. The amount of solar energy that reaches the outer limits of the Earth's upper atmosphere is enormous; if all of it penetrated to the Earth's surface and were retained, the very high temperature would have prevented the development of any life on Earth. Fortunately, as solar radiation travels through the atmosphere, interactions with gases and particulates prevent approximately half of it from penetrating to the Earth's surface. The 69% of solar energy that eventually reaches the Earth includes the entire visible region of the spectrum together with smaller portions of the adjacent UV and infrared regions (IR). This incoming radiation is largely absorbed at the surface and is then

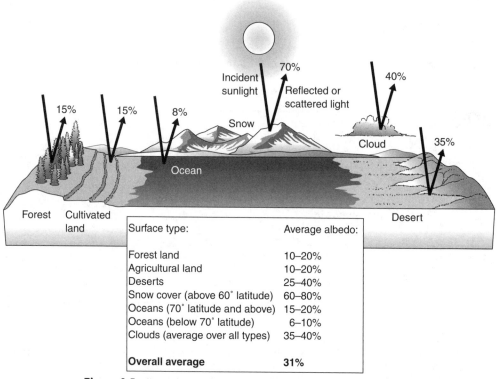

Figure 2.5 Albedo varies depending on cloud cover and the type of surface.

Surface type:	Average albedo:
Forest land	10–20%
Agricultural land	10–20%
Deserts	25–40%
Snow cover (above 60° latitude)	60–80%
Oceans (70° latitude and above)	15–20%
Oceans (below 70° latitude)	6–10%
Clouds (average over all types)	35–40%
Overall average	**31%**

reradiated back to space. If this radiation away from the surface did not occur, the Earth would become increasingly warm as solar energy continued to flow in. Outgoing radiation from the Earth is in the longer wavelength IR region.

As was stated before, the solar flux (Fs) is 1368 W/m². To estimate the amount of energy reaching the Earth, consider an area of πr^2, where r is the Earth's radius. The total amount of solar energy absorbed by the Earth (E_{in}) is equal to the solar flux times the area of the Earth minus the portion that is reflected back into space. The reflected portion is the albedo (A). Albedo varies greatly depending on the surface. Snow and ice reflect more incoming radiation than does a parking lot paved with black asphalt. The incoming radiation that is absorbed must be balanced with the radiation that is emitted to calculate an equilibrium temperature for Earth.

$$E_{in} = Fs(1 - A)(\pi r^2)$$

Where

E_{in} = total amount of solar energy absorbed by the Earth
A = albedo

The Earth gives off radiation as a sphere. The energy reradiating from the entire area of the Earth's surface (E_{out}) can be estimated by the **Stefan Boltzmann law.**

$$E_{out} = 4\pi r^2 S_b T^4$$

Where

T = the Kelvin temperature of the Earth

$S_b = 5.67 \times 10^{-8} \, Wm^{-2} \, K^{-4}$, the Stephan Boltzmann constant

E_{out} = energy reradiating from the entire area of the Earth's surface

Over time, the steady-state total energy that the Earth absorbed from the sun equals the energy that it reradiates.

$$E_{in} = E_{out}$$

$$Fs(1 - A)(\pi r^2) = 4\pi r^2 S_b T^4$$

rearranging this equation as follows

$$T = \left(\frac{(1-A) Fs}{4 \, S_b} \right)^{1/4}$$

knowing that

Fs = 1368 Wm^{-2}

A = 0.31

$$T = \left(\frac{(1 - 0.31)(1368 \, Wm^{-2})}{4 \, (5.67 \times 10^{-8} \, Wm^{-2} \, K^{-4})} \right)^{1/4}$$

$$T = 254 \, K \text{ or } -19°\, C$$

This calculation allows us to predict an average global surface temperature of −19°C. The experimentally measured average temperature of the Earth is 15°C (288 K), and this temperature is 34°C higher than calculated. This calculation assumes that all radiation leaves the Earth and is lost into space. If some of the IR radiation is absorbed by gases in the atmosphere and is not lost, then the Earth's temperature will be expected to be higher.

If the same equation is used to calculate the temperature of other nearby planets and the results compared with the actual temperature, an interesting correlation is revealed. Table 2.3 shows the difference between the calculated and actual temperatures on the surfaces of several other nearby planets and Earth's moon. The calculated and actual temperatures for Mars and the Earth's moon closely agree. Both of these objects have very thin atmospheres that could not absorb infrared radiation emitting from the surface of the planet.

The surface temperature of Venus is more than 475 K warmer than expected from the calculated temperature. Venus is closer to the sun than the other planets and receives more incident radiation. Early in its evolution, liquid water and ice were melted and evaporated. The water vapor was exposed to intense UV radiation from the sun, which photolyzed the water to atomic hydrogen and the hydroxyl radical.

$$H_2O + h\nu \rightarrow H^{\bullet} + OH^{\bullet}$$

$$2\,H^{\bullet} \rightarrow H_2$$

Table 2.3

Calculated and Actual Temperatures of the Surfaces of Planets and the Moon

Planet	Distance from Sun, 10^9 m	Calculated Temperature K	Actual Temperature K	ΔT
Venus	108	252	730	+478
Earth	150	255	288	+34
Earth's moon	150	270	274	+4
Mars	228	217	218	+1

Source: Adapted from M. Z. Jacobson. *Atmospheric Pollution* (Cambridge, UK: Cambridge University Press, 2002), p. 314.

Over time, the atomic hydrogen formed hydrogen gas (H_2), which was able to escape Venus's gravity and was lost into space. Because of the loss of hydrogen, water could not reform. At the same time, volcanic out gassing was releasing CO_2. Because there was no water left, there was no means to dissolve the CO_2 and convert it to carbonate rock, a process that took place on Earth during the same time period. As a result, CO_2 built up in the atmosphere, producing a runaway greenhouse effect. Today, Venus has a surface atmospheric pressure that is 90 times greater than that of Earth, and this atmosphere is mostly CO_2. It is this high concentration of CO_2 that absorbs reradiated IR radiation from the surface of Venus and renders the planet more than 475°K hotter than expected.

The Earth is a distance from the sun that makes it an ideal place to support life. The comparison of Earth to Venus and Mars produces the **Goldilock's hypothesis.** That is, Venus is too hot because it is too close to the sun, and its atmosphere has too high of a concentration of greenhouse gases. Mars is too cold because it is too far away from the sun and has little atmosphere. The Earth is the ideal distance from the sun and has a low concentration of greenhouse gases that warm the planet enough to allow water to exist as a gas, liquid, and solid and to support life.

Although the amount of energy reaching the top of the Earth's atmosphere facing the Sun is 1368 Watts per square meter per second, the average amount of energy reaching the entire planet is about a quarter of this or 342 Wm^{-2}. As can be seen in Figure 2.6, about 30% of the incoming solar radiation (107 Wm^{-2}) is reflected back to space. Light-reflecting areas of the Earth's surface, such as snow, ice or deserts, reflect about one-third of the energy (30 Wm^{-2}), while aerosols suspended in the atmosphere reflect the rest (77 Wm^{-2}). The amount of light reflected by aerosols can be dramatically altered by natural and anthropogenic events that suspend light-scattering material in the troposphere. Volcanic eruptions, for instance, eject dust very high into the atmosphere and are a natural way the Earth's albedo can be increased. The dust ejected by the eruption of Mount Pinatubo in 1991 reflected sunlight for several years, and the increased albedo caused a slight drop in globally averaged temperature at the ground.

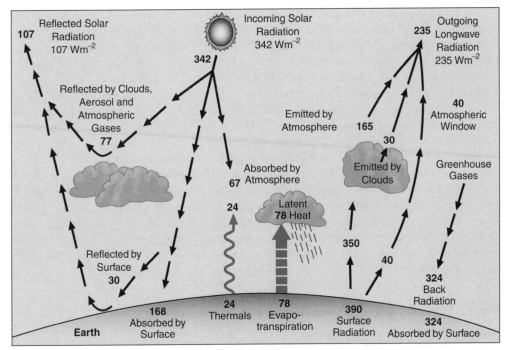

Figure 2.6 An estimate of the Earth's energy balance. *Source:* Adapted from J.T. Kiehl and K.E. Trenberth, *Bull. Amer. Meteor. Soc.* 78 (1997): 197–208.

The same mechanism also can operate on a smaller scale. Where large quantities of smoke are released, local areas can be cooler during the daytime than they would otherwise because of the enhanced scattering of sunlight back into space. These are **anthropogenic** events because they are caused by human activity.

The energy not reflected is absorbed by the Earth's surface and atmosphere (235 Wm^{-2}). The Earth keeps an energy balance by reradiating the same amount of energy back into space in the form of infrared radiation. The calculation we just completed showed that to emit 235 Wm^{-2}, a surface would have to have a mean temperature of −19°C. This is much colder than the Earth's global mean surface temperature, which is 14°C. A closer inspection of the Earth's atmosphere indicates that the atmosphere's temperature at 5 km above the surface is −19°C.

The Earth's surface is warmer than expected because of the presence of greenhouse gases in the atmosphere that absorb the infrared radiation that would have been lost into space. These gases are called **greenhouse gases** because they warm the atmosphere much the same way as a greenhouse warms the air inside the structure so that plants can grow. A greenhouse, which is made from glass panels, is designed to allow incoming radiation from the sun to fall on plants inside. Because infrared radiation cannot pass through glass, inside the greenhouse it is trapped while outside it is reflected back to space. Thus, the greenhouse remains warm even in the cold of winter. Chapter 3 focuses more on the greenhouse effect.

The Earth is a sphere and more solar energy strikes a square meter of the surface in the tropics than at higher latitudes because of the inclination of the Earth to the Sun. Energy is distributed from the tropical regions to higher latitudes by atmospheric and oceanic circulation as well as by storms. The energy required to evaporate water from the sea or other bodies of water is called *latent heat* (see Figure 2.6) and latent heat is released when water vapor condenses into clouds. Circulation of the atmosphere is set into motion primarily by the release of latent heat. Atmospheric circulation then induces circulation of water in the ocean by the action of winds on the surface. Due to the Earth's rotation, movement of the atmosphere tends to be more in an east–west direction.

There are climatic feedback mechanisms that can amplify (and are called *positive feedback*) or diminish (*negative feedback*) the mechanisms that affect global energy. One such feedback loop is the *ice-albedo feedback loop.* This loop begins with increasing concentrations of greenhouse gases that cause more infrared energy to be absorbed by the atmosphere, which in turn heats the atmosphere and causes more ice to melt. When the ice melts, darker surfaces below the ice are exposed and they absorb more solar radiation than does ice. This causes more warming and a reinforcing cycle is created. The understanding of climatic feedback loops has been a major focus of scientists for the past 20 years.

■ Particles in the Atmosphere

Suspended Particulate Matter

Natural sources of airborne particles, which may be solid or liquid, include smoke and ash from forest fires and volcanic eruptions, dust, sea salt spray, pollen grains, bacteria, and fungal spores. Aerosols and particulates that are liquid are generally called *mist,* which includes fog and raindrops. A significant quantity of harmful particulate matter is also emitted as a result of human activities (Figure 2.7).

Eventually, all particulate matter is deposited on the Earth's surface. Relatively large particles (diameters of greater than 10 μm) settle under the influence of gravity within one to two days. Medium-sized particles (diameters of 1 to 10 μm) remain suspended for several days. Fine particles (diameters of less than 1 μm) may remain in the troposphere for several weeks and in the stratosphere for up to five years. Aerosols, acting as nuclei for the formation of droplets in clouds, reach the ground when the droplets condense and fall as rain or snow. Fine particles can be transported considerable distances by winds before settling to the ground or being washed out by falling rain or snow.

Aerosol Particles

The effect of particulate matter on the heat flux of the atmosphere depends mostly on particle size and not as much on the total concentration of particles. Large, dark particles absorb light and add to the warming of the atmosphere. Small particles, regardless of the color, scatter incident sunlight and increase the albedo of the atmosphere.

Natural sources of light-scattering aerosols are estimated to produce 50% to 75% of all atmospheric aerosols. There are two major natural sources of aerosols. The first is ammonium sulfates generated during microbial degradation of decaying biomass and organic matter in soil and water. The other is reactive organic molecules that are released from natural sources. A good example of this is the release of a group of organic molecules called *terpenes,*

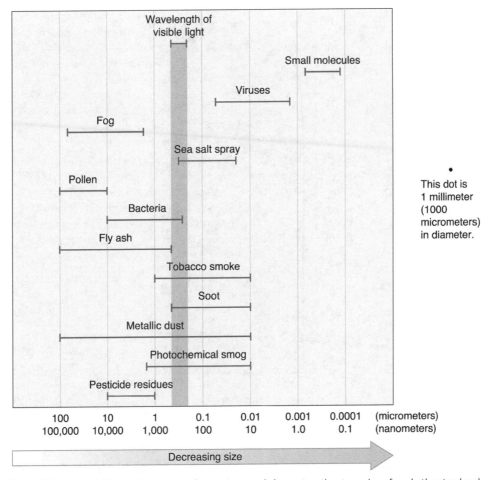

Figure 2.7 Suspended particulate matter of many types and sizes enters the atmosphere from both natural and anthropogenic sources.

which are from coniferous trees that include pine, spruce, and fir trees. One of these terpenes, α-pinene, produces the smell that we associate with pine forests.

Light scattering by aerosols with dimensions that are significantly smaller than the wavelength of the radiation is called **Rayleigh scattering**. Its intensity is proportional to the inverse fourth power of the wavelength ($S = 1/\lambda^4$). Blue light, which has a shorter wavelength, is scattered more than red. This is why the sky, which we see in scattered light, appears blue, and the sunset, which we see in transmitted light, appears red. The blue haze of the Great Smoky Mountains of Tennessee is due to light scattering from small aerosols formed by the oxidation of terpenes emitted from the conifer forests.

As can be seen in Figure 2.8, the size distribution of particles in the atmosphere is wide, but there are a larger number of smaller particulates with a radius of 0.01 to 0.1 μm. Larger particles of sea spray, pollen, and dust settle out of the atmosphere more quickly than do the smaller particles.

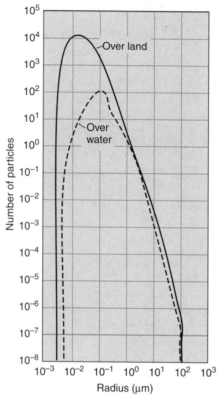

Figure 2.8 Distribution (by size) of particles in the lower atmosphere.

One way to measure the total amount of suspended particulate matter is the **total suspended particulate (TSP)** test (described in detail in Chapter 6). Air is drawn through a preweighed filter at a rate of $1 \, m^3 \, h^{-1}$. The total time that air is pumped is recorded so that the total volume of air passing through the filter will be known. At the end of the test, the filter is weighed again so that the total weight of particulate matter is known. TSPs up to 200 μg/m³ have been measured in large cities; individual measurements as high as 500 μg/m³ have been measured in very dusty locations such as the desert. Rural farming areas average 10 to 50 μg/m³. The effects of airborne particulates on health and the environment depend on both the size and the nature of the particles. Very fine particles (diameters of less than approximately 1 μm) are the most hazardous to human health. They are not filtered by hairs and mucus in the nose but are drawn deep into the lungs, where they can remain indefinitely, causing tissue damage and contributing to the development of the lung disease emphysema. They take with them any toxic chemicals that are attached to their surfaces or are dissolved within them, such as sulfuric acid.

The **Clean Air Act** Amendments of 1970 for the first time set U.S. standards for that fraction of particles that are easily inhaled and deposited in the lung. Inhalable particulates were defined as those particles with a diameter of smaller than 10 μm and are referred to as PM_{10}.

In 1997, in a revision of the Clean Air Act, an additional standard was added for particles of diameter 2.5 μm or less, the $PM_{2.5}$. The current federal standard for $PM_{2.5}$ is 65 μg/m³ for a 24-hour average concentration.

Epidemiologic studies in the 1970s found a link between short-term increases in the PM_{10} of 10 μg/m³ and an increase in hospitalization and healthcare visits for respiratory and cardiovascular disease and enhanced outbreaks of asthma and coughing. Although studies have found an association between exposure to outdoor air pollution and health, we spend most of our time indoors. The concentration of particles indoors is often greater than outdoors.

Anthropogenic Sources of Particulate Matter

Suspended particulate matter is the most visible form of air pollution. The major source of such particles is the combustion of coal by electric power-generating plants (Figure 2.9). One product of the incomplete combustion of coal is **soot**, a finely divided, impure form of carbon. Soot is an impure form of elemental carbon in which the structure contains a series of benzene rings that are fused. Graphite, another form of elemental carbon, has fused benzene rings, forming a flat, layered structure, whereas soot particles are close to spherical. Soot is also present in black smoke that is emitted by diesel-powered trucks. Diesel engines that operate at lower temperatures than gasoline engines produce 3 g of soot per kg of diesel fuel, whereas gasoline engines produce only 0.1 g of soot per kg of gasoline. Because soot is black, suspended soot particles absorb incoming radiation from the sun. Other suspended particulates are near white in color and reflect incoming radiation from the sun.

Other products include metallic and nonmetallic oxides that are formed from minerals present in coal. These materials, called **fly ash**, are swept up smokestacks in drafts from roaring furnaces and are emitted into the atmosphere. The methods used to trap and measure fly ash are described in Chapter 5. Fly ash particles are small enough to be respirable. Their composition depends on the quality of the coal being burned and the temperature of the combustion. These particles are composed primarily of the following metallic and nonmetallic oxides: SiO_2, Al_2O_3, Fe_2O_3, and CaO. They also are known to sometimes contain toxic metals such as arsenic, lead, and cadmium.

Other anthropogenic sources of particulate emissions are solid-waste incineration, mining and ore processing, construction sites, and agricultural activities. Toxic metal particles, cement dust, and pesticide and fertilizer residues that are released in these activities are all potentially hazardous.

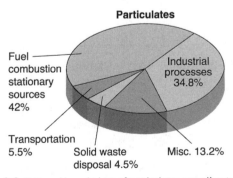

Figure 2.9 Nationwide emissions of particulates according to source.

Residence Times of Particles

The most important physical process that determines the residence time of an aerosol is the **settling rate** of the particles. Settling occurs because the force of gravity pulls the particles and deposits them on the surface of the Earth. For particles in which the diameter exceeds 1 μm, the settling velocity v is given by **Stokes law.**

$$v = \frac{gd^2(p_1 - p_2)}{18\eta}$$

Where

g = 9.80 m/s² the acceleration caused by gravity
d = diameter of particle in meters
p_1 = density of the particle expressed in g/m³
p_2 = density of air in g/m³

Because $p_1 >>> p_2$, then $(p_1 - p_2)$ is approximately equal to p_1, and $\eta = 1.9 \times 10^{-2}$ g/ms, which is the viscosity of air at atmospheric pressure and 25°C.

The settling rate for a particle of fly ash that is 2 μm in diameter and has a density (p_1) of 1.0 g/mL is calculated as follows:

Step 1: Convert the density of the particle into g/m³:

$$p_1 = \frac{1.0\,g}{mL} = \left(\frac{1.0\,g}{mL} \times \frac{1 \times 10^6\,mL}{1.0\,m^3}\right) = 1.0 \times 10^6\,\frac{g}{m^3}$$

Step 2: Plug values into equation:

$$v = \frac{gd^2p_1}{18\eta} = \frac{\left(9.80\,m/s^2\right)\left(2 \times 10^{-6}\,m\right)^2\left(1.0 \times 10^6\,g/m^3\right)}{18\left(1.9 \times 10^{-2}\,g/ms\right)}$$

Where

v = 114.6 × 10⁻⁶ m/s
v = 0.011 cm/s

This particle has a settling velocity of 0.011 cm/s.

Step 3: The settling rate is the distance that it will drop in a specific unit of time. The settling rate per day will be as follows:

$$\left(8.6 \times 10^4\,s/day\right)\left(114.6 \times 10^{-6}\,m/s\right) = 9.8\,m/day$$

If the fly ash particle described here was released into the atmosphere from a 100-meter tall smokestack, it would take 10 days for it to settle on the ground. Stoke's law predicts that the residence time of a particle in air increases as the radius decreases. The residence times of particles in tropospheric air are often estimated by assuming that they fall 10 m/day.

Control of Particulate Emissions

One method that industry uses to control particulate emissions is **electrostatic precipitation** (Figure 2.10a). In this process, gases and particulate matter are passed through a high-voltage chamber before leaving the chimney stacks. A negatively charged central electrode imparts a negative charge to the particles, which are then attracted to the positively charged walls of the chamber. As their charges are neutralized, the particles clump together and fall to the bottom, where they can be collected.

A second method for controlling particulate emissions is **bag filtration**, in which lined-up fabric bags function essentially like the bag in a household vacuum cleaner (Figure 2.10b). Gases can pass through the finely woven bags, but particulate matter cannot; particles with diameters of 0.01 to 10 µm are effectively filtered. The bags are sensitive to temperature and humidity and can become clogged by fine particles. The bags are shaken at intervals to dislodge particles, which are collected in a hopper located underneath the bags. Although both methods remove more than 98% of particulates, they fail to remove the finest ones, which are the most dangerous.

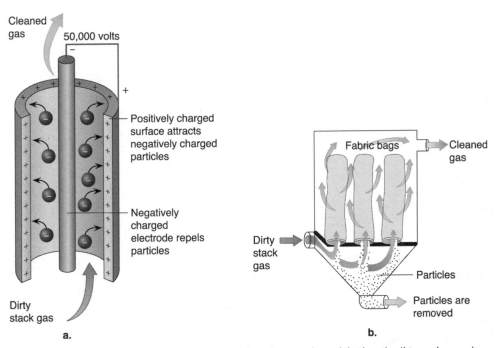

Figure 2.10 Two methods for controlling emissions: (a) In electrostatic precipitation, the dirty gas is passed though a chamber, where particles acquire a negative charge from a negatively charged central electrode. The charged particles are attracted to the positively charged wall of the chamber, where they are deposited. (b) In bag filtration, the dirty gas is forced through fabric bags, which trap particles. The bags are shaken at intervals, causing the particles to fall into a hopper.

Another less efficient method is **cyclone separation**, in which the particle-laden gases are subjected to a centrifugal force. The spiraling particles hit the walls, settle, and fall to the bottom, where they are collected. Stoke's law determines how many of the airborne particulates will be removed. The centrifugal force of the cyclone greatly increases the settling rate. Particles that are larger than 1 μm in diameter are more efficiently removed than are smaller particles. All three methods produce solid wastes, which must be disposed of and which mostly end in landfills.

Although the 1970 Clean Air Act mandated standards for particulate emissions, many areas failed to meet the requirements because of difficulties in enforcing the law. Under the 1990 amendments to the Act, areas not in compliance must take steps to meet the standards.

■ Additional Sources of Information

Brasseur GP, Orlando JJ, Tyndall GS. *Atmospheric Chemistry and Global Change*. New York: Oxford University Press, 1999.

Environmental Protection Agency. Homepage. Accessed November 17, 2008 from http://www.epa.gov.

Frederick JE. *Principles of Atmospheric Science*. Sudbury, MA: Jones and Bartlett, 2008.

Jacobson MZ. *Atmospheric Pollution*. Cambridge, UK: Cambridge University Press, 2002.

Rohli RV, Vega AJ. *Climatology*. Sudbury, MA: Jones and Bartlett, 2008.

Wayne RP. *Chemistry of Atmospheres*, 3rd ed. New York, Oxford University Press, 2000.

■ Keywords

absorption
adsorption
aerosols
albedo
anthropogenic
atmosphere
bag filtration
boundary layer
Clean Air Act
cyclone separation
electromagnetic radiation (EMR)
electrostatic precipitation
fly ash
free troposphere
global warming
Goldilock's hypothesis
greenhouse gases
ideal gas law
lower atmosphere
mesosphere

particulates
pascal (Pa)
PM_{10}
$PM_{2.5}$
Rayleigh scattering
relative humidity
settling rate
solar flux
soot
Stefan Boltzmann law
Stokes law
stratopause
stratosphere
thermosphere
total suspended particulate (TSP)
tropopause
troposphere
upper atmosphere
Wein's displacement law

■ Questions and Problems

1. What is the concentration of each of the following gases in clean, dry air? Express the concentration in percent and parts per million.
   a. O_2
   b. CO_2
   c. N_2
   d. Ar
2. Draw a diagram showing the four main layers into which the atmosphere can be divided. Label your diagram to show the following:
   a. The distance the two layers closest to the Earth extend out from the Earth's surface.
   b. The changes in temperature with increasing distance from the surface of the Earth.
   c. The location of the ozone layer.
3. Explain why the stratosphere, which is more than 20 miles thick, has a smaller total mass than the troposphere, which is less than 10 miles thick.
4. Calculate the partial pressure of water in the Earth's atmosphere if the temperature is 20°C and the relative humidity is 50%.
5. The percentage of water vapor in the atmosphere is variable. It generally lies in the range of which of the following:
   a. 0.05% to 1%
   b. 1% to 3%
   c. 5% to 10%
   d. 10% to 20%
6. Define the term *relative humidity*.
7. Calculate the relative humidity for the following atmospheric conditions:
   a. A temperature of 25°C and a partial pressure of 0.0275 atm for water
   b. A temperature of 20°C and a partial pressure of 0.0205 atm for water
8. Draw a diagram to show how atmospheric pressure varies with altitude.
9. If the surface of Venus is 730°K, what is the wavelength of the radiation it emits as a black body? Does it emit infrared radiation?
10. If the temperature of the sun increased from 5800°K to 8500°K, how would the wavelength of its emission change? What effect would this have on plants and mammals?
11. Define the term *albedo*. Arrange the following surfaces in order of lowest albedo to highest albedo: desert, ice, grassland, forest, water, asphalt.
12. Describe the difference between an aerosol and a particulate.
13. Define the following:
    a. mist
    b. fog
    c. soot
    d. fly ash
    e. smog
14. Give four examples of particulate matter that arises from the following:
    a. Natural sources
    b. Anthropogenic sources
15. Describe Raleigh scattering by particulates.

16. Why is the sky blue?

17. Describe the TSP test.

18. Define the following:
  a. $PM_{2.5}$
  b. PM_{10}

19. Explain how the surface areas of particulates and aerosols are a factor in causing them to be health hazards.

20. In what way does the size of a particle influence its affect on health?

21. Calculate the settling rate for the following:
  a. A particle of fly ash that is 2.5 µm in diameter with a density of 1.2
  b. A particle of soot that is 1.0 µm in diameter with a density of 0.7
  c. Which of the two particles would settle out of the air first?

22. Estimate the residence time of a fly ash particle that is released from a 300 meter smoke stack. Assume the particle has a 2 µm diameter and a density of 1.0 g/ml.

23. Describe three different methods for removing particulate matter from industrial emissions. What is done with the solid particles after they are removed from the air?

24. How does the unit ppbv differ from ppb?

25. The airborne concentration of NO_2 over Chicago on a 25°C and 1.0 atm day is measured to be 46 ppmv. How many µg/m^3 of NO_2 are there in the atmosphere?

26. A tropospheric (25°C, 1.00 atm) measurement of ozone indicates there is 50 ppbv. Express this concentration:
  a. In ppmv
  b. In µg/m^3
  c. In molecules/cm^3

27. The atmospheric concentration of methane over Iowa in the summer is measured to be 9×10^{13} molecules per cm^3. Express this concentration of methane in ppmv.

28. Calculate the temperature of the Earth if anthropogenic airborne particulates increased the average albedo of the Earth to 0.38. Assume there is no greenhouse effect. Compare this temperature to the temperature calculated in Table 2.3 and describe what effect the increased tropospheric particulate concentration has on global temperature.

29. Why is the variation between calculated temperature and actual temperature for the Earth greater than that for the Earth's moon? Both bodies are the same distance from the sun.

30. Describe the runaway greenhouse effect that took place on Venus.

31. Draw the structure of a hydroxyl radical.

32. The following species are all present in the atmosphere. Which are free radicals?

O_3, NO, N_2O, ClO, N_2O, OH

33. Describe the Goldilocks hypothesis.

34. Describe what is meant by a "positive" atmospheric feedback mechanism. Give an example.

35. Describe what is meant by a "negative" atmospheric feedback mechanism. Is a negative feedback mechanism, by definition, detrimental to the atmosphere?

36. Assume that polar ice caps have an albedo of 80% and that the sea near the poles has an albedo of 20%. Explain how a small rise in global ambient temperature could cause a rapid melting of much of the polar ice caps.

37. Assume that the snow pack in northern Canada has an albedo of 75%. Explain how the deposition of soot from coal-burning power plants would affect global ambient temperature.

38. Reflected solar radiation from the Earth has been measured by satellite to be 107 Wm^{-2}. There are two major sources that contribute to this total. Describe each and the amount that it contributes.

39. Explain how volcanic eruptions can cause year-long cooling in the mean global surface temperature. Why does the effect of an eruption last so long?

40. Because the Earth is a sphere, more solar energy strikes a given surface area in the tropics or northern latitudes? Explain.

CHAPTER

3

Global Warming and Climate Change

ALTHOUGH SCIENTISTS HAVE STUDIED EARTH'S ATMOSPHERE SINCE THE START OF THE 20TH CENTURY, DURING THE PAST 20 YEARS THEIR UNDERSTANDING OF THE PROCESSES THAT TAKE PLACE IN THE ATMOSPHERE HAS GROWN MORE SOLID AND THE CLIMATE MODELS MORE SOPHISTICATED. Evidence that human behavior is affecting the climate of the Earth has become more certain and the possibility of even greater global warming is now expected in the future. The certainty that global warming is a result of human activity and that even more human-induced climate change is on the way is presented in the latest report of the Intergovernmental Panel on Climate Change (IPCC), which was released in February 2008 (see http://www.ipcc.ch).

The panel released a "Summary of Policymakers," bluntly stating that "scientists are more confident than ever that humans have interfered with the climate and that further human-induced climate change is on the way." Clearly, we have caused global warming, but the report does offer a note of optimism. Although we have harmed the climate of the Earth with our activities, the magnitude of global warming in the future depends on how seriously we choose to limit greenhouse gas emissions.

Four topics are presented in this chapter. What changes have been observed in the Earth's climate? What components of our atmosphere are considered to be greenhouse gases? How do they cause climate change, and what does the future hold?

■ Global Temperature from the Ice Ages to Present Time

As we saw in the last chapter, global climate is determined by energy. There are three different ways the amount of energy in the Earth's atmosphere can be changed: (1) changes in solar radiation reaching the Earth (changes in the Sun's output or the Earth's orbit), (2) changes in the fraction of solar radiation that is reflected (*albedo*), and (3) changes in the infrared radiation that is radiated back into space.

Measurements of the output of the Sun over recent decades show that solar output remains fairly constant (varies only about 0.1%). Ice ages have occurred in regular cycles for the past 3 million years and it appears as if they are linked to regular variations in the Earth's orbit around the Sun, which are known as *Milankovitch cycles*, rather than any variation in solar output. The Earth has gone through four glacial periods during the past 450,000 years and they are caused by gravitational attraction between the Earth and other planets in the solar system. It appears that if the inclination of the Sun is not optimal, then the amount of sun striking the northern continents in the summer is not sufficient to melt the snow that fell that year. If the snow from the past winter doesn't melt, then an ice sheet starts to grow and glaciation begins. Computer climate models confirm that Ice Ages can be started this way and these models also predict that the next such Ice Age will begin in about 30,000 years.

Once the Earth has absorbed solar radiation in order to maintain an energy balance it must radiate, on the average, the same amount of energy back into space. The Earth does this by emitting infrared (IR) radiation. A considerable amount of the outgoing IR radiation does not escape into space but is reabsorbed by gases in the atmosphere called **greenhouse gases** and then reradiated back to Earth. As a result of this absorption and reradiating, the atmosphere is warmed (see Figure 2.6). The warming effect caused by the absorption and reradiating of IR radiation by the greenhouse gases is commonly called the **greenhouse effect.** This name arose because the gases act somewhat like a pane of glass in a greenhouse: Visible light passes through the glass of a greenhouse and is absorbed by the objects inside. The objects are warmed and emit heat energy in the form of IR radiation, which cannot pass through glass. It is trapped and the temperature rises inside the greenhouse.

Atmospheric carbon dioxide concentration also plays an important role in ice ages. Ice from polar regions can provide a history of atmospheric temperature and atmospheric CO_2 concentration at the time each ice layer was deposited. Antarctic ice core data, which can be seen in Figure 3.1, shows that atmospheric CO_2 concentration is low (~190 ppm) in the cold glacial times and high (~280 ppm) in the warm interglacial times. It appears that small increases in surface temperature cause dissolved CO_2 to move from the ocean water into the atmosphere. Increased CO_2 in the atmosphere further increases warming by the greenhouse effect. When the surface cools, the atmospheric CO_2 redissolves in ocean water, which causes further cooling as the greenhouse effect is reduced. In this way, surface temperature and atmospheric CO_2 concentration have tracked each other for almost a half a million years.

Instrumental observations from 1850 to the present show that the global mean temperature of the Earth has risen during this period. As can be seen in Figure 3.2, there wasn't much change in temperature from 1850 to 1915, but since then global warming has occurred in two

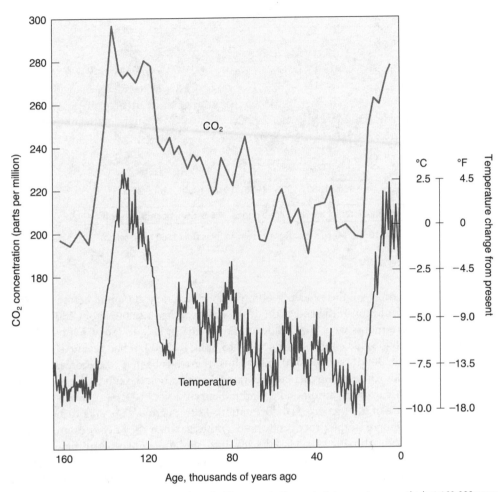

Figure 3.1 Variations in atmospheric carbon dioxide concentration and air temperature over the last 160,000 years were revealed by analysis of Antarctic ice cores. Air temperature has risen and fallen in step with increases and decreases in the carbon dioxide concentration.

phases. During the first phase (1910 to 1940) the global mean temperature rose 0.35°C and during the second phase (1970 to present) the global mean temperature rose 0.55°C. Eleven of the 12 warmest years have occurred during the second phase, and 1998 and 2005 are the two hottest years on record. Measurements above the surface have shown that the troposphere has warmed at a slightly greater rate that the surface, while the stratosphere has cooled. Global warming has been confirmed by measurements that show the ocean temperature and sea level to be rising, and diminished snow cover in the northern hemisphere.

The Increase in Atmospheric Carbon Dioxide

As we have seen, analysis of air trapped in samples of ancient ice shows that the average concentration of carbon dioxide in the atmosphere has been increasing for thousands of years.

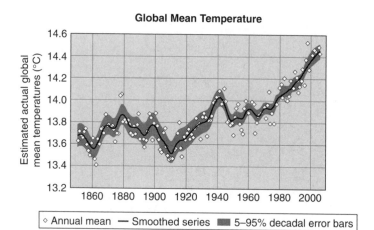

Figure 3.2 Annual global mean temperatures from 1850 to present.

Twenty thousand years ago, the concentration was approximately 200 ppm; before the start of the Industrial Revolution at the end of the 19th century, it had risen to about 280 ppm. In 1958, when measurements were first made at Mauna Loa in Hawaii (see Chapter 6), the atmospheric carbon dioxide concentration was 315 ppm; by 2000, it had reached 370 ppm (Figure 3.3) and in 2008 it reached 380 ppm. Thus, there has been a 30% increase in just 100 years. The major cause of the increase in atmospheric carbon dioxide is the burning of fossil fuels by electric utilities, automobiles, and industry.

A secondary cause of the increase is deforestation. As trees grow, they take carbon dioxide from the atmosphere in the process of photosynthesis; when they decay or are burned in forest fires, carbon dioxide is released back into the atmosphere. Any new growth in a cleared area, whether crops or grasses, takes much less carbon dioxide from the atmosphere than does a mature forest. By 2000, 2% of all rain forests were being cut and burned each year. Destruction is particularly severe in tropical rain forests, especially in Brazil.

Human activities produce four principal greenhouse gases: carbon dioxide (CO_2), methane (CH_4), nitrous oxide (N_2O), and halocarbons. As can be seen in Table 3.1, these gases accumulate in the atmosphere and their concentration increases over time. The concentration of these gases in the atmosphere has increased significantly during the industrial era (Figure 3.4) and their releases are directly related to human activities. Although their combined total concentration is considerably less than that of carbon dioxide, the other greenhouse gases absorb and re-emit more IR radiation, molecule for molecule, than does carbon dioxide. Methane, for example, is approximately 25 times more effective than carbon dioxide at trapping heat. If the levels of the other greenhouse gases continue to rise at the present rate, it is estimated that their warming effect will soon equal that of carbon dioxide.

The steady rise in methane during the later part of the 20th century is attributed primarily to a worldwide increase in the number of cattle and the number of rice paddies, both of which are **sources** of methane. Other anthropogenic sources of methane include landfills and

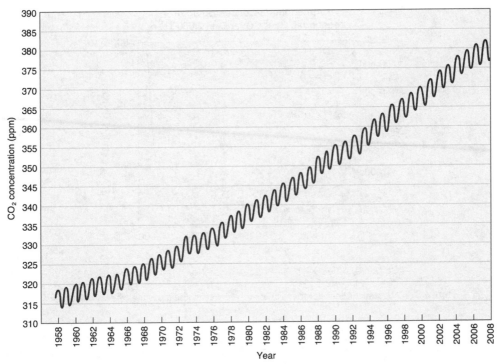

Figure 3.3 Since the first measurements were made at Mauna Loa, Hawaii, in 1958, the carbon dioxide concentration in the atmosphere has risen dramatically. The yearly seasonal variations are caused by the removal of carbon dioxide from the atmosphere by growing plants in the summer and the return of carbon dioxide to the atmosphere in the winter when plants decay.

Table 3.1

Properties of Anthropologic Greenhouse Gases

	CO_2	CH_4	N_2O	Freon-11	Freon-23
Atmospheric concentration	ppmv	ppbv	ppbv	pptv	pptv
Preindustrial (1750–1800)	~280	~700	~270	0	0
Current	370	1745	314	268	14
Current rate of change/year*	1.5	7.0	0.8	−1.4	0.55
(% increase/year)	0.41	0.40	0.25	−0.52	3.92
Atmospheric lifetime (years)	5 to 200†	12	114	45	260

* Rate is calculated over the period 1990–1999.
† No single lifetime can be defined for CO_2 because of the different rates of uptake by different removal processes.

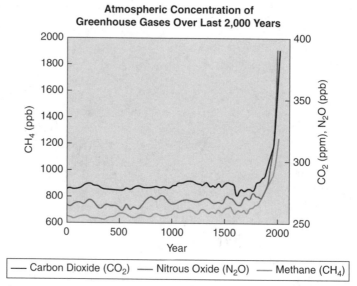

Figure 3.4 Atmospheric concentrations of long-lived greenhouse gases over the last 2000 years.

coal mines. Nitrous oxide is formed naturally in the soil during microbiological processes associated with nitrogen fixation. Its increase in the atmosphere is believed to be caused primarily by the increased use of nitrogen-containing fertilizers.

■ Infrared Absorption and Molecular Vibrations

Infrared radiation is not energetic enough to break covalent bonds or to cause electronic transitions, but it can change the vibrational or rotational motion of a molecule. In order to absorb IR radiation, a molecule must undergo a net change in **dipole moment** as a result of its vibrational or rotational motion. Monoatomic inert gases, such as argon, which is the third most abundant atmospheric gas (with a concentration of 0.9%) do not have covalent bonds and, hence, cannot absorb IR radiation. No net change in dipole moment occurs during the vibration or rotation of homonuclear species, such as N_2 and O_2. Because these molecules are symmetrical, no matter how much the covalent bonds are stretched there is no change in dipole moment. Because of this, these molecules, which are the principal constituents of the atmosphere, also cannot absorb infrared radiation. Other diatomic molecules, which contain two different atoms such as CO, do absorb IR radiation. The frequency at which they absorb it can be estimated assuming that the atoms in the molecule are balls and the bond that connects them is a spring. The frequency of oscillation depends on the weight of the two atoms and the strength of the bond connecting them. This calculation assumes the molecule is a harmonic oscillator (a worked-out example can be found in Chapter 6, Example 6.1). All of the gases that contribute to the greenhouse effect have three or more atoms (polyatomic molecules) and, although it is possible to extend the harmonic oscillator equation to estimate the wavelength of absorption for polyatomic molecules, a different type of analysis will be used to determine the absorption of atmospheric polyatomic molecules.

Carbon dioxide (CO_2), for example, is a linear triatomic molecule with a carbon atom in the middle and two oxygen atoms on its left and right each bonded by a double bond (O=C=O). Upon first examination you might think there would be only one vibrational mode for CO_2, a stretching (and compressing) of the carbon oxygen double bond. But because each double bond stretches independently, CO_2 has more than one IR absorption. As can be seen in Figure 3.5, CO_2 has two stretching vibrations and two bending vibrations. The arrows show the direction of motion of each atom. A **symmetric stretch** occurs when the two oxygen atoms both move outward and inward together while the carbon remains fixed. An **asymmetric stretch** occurs when one double bond is compressed while the other is stretched. Besides stretching vibrations, CO_2 can also vibrate by bending. One bending vibration, which can be seen in Figure 3.5, occurs when bending occurs in-plane of the linear molecule, the other occurs when the atoms bend out-of-plane. IR radiation causes most polyatomic molecules, including greenhouse gases H_2O, CH_4, and N_2O, to vibrate and absorb IR radiation.

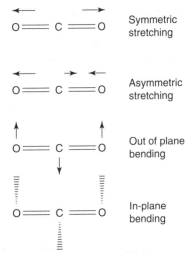

Figure 3.5 Vibrations of CO_2 molecule. Symmetric and asymmetric stretching and in-plane and out-of-plane bending.

The two most important greenhouse gases are the polyatomic molecules CO_2 and H_2O. Approximately 89% of the 34°K temperature increase resulting from natural greenhouse warming can be attributed to water. Carbon dioxide accounts for approximately 7.5% of the greenhouse effect. As can be seen in Figure 3.6, carbon dioxide represents the largest fraction of all greenhouse gas emissions in the United States and the world with fossil fuel emissions representing the vast majority of CO_2.

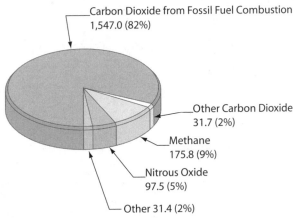

Figure 3.6 U.S. anthropogenic greenhouse gas emissions by gas in 2001 (million metric tons of carbon equivalent).

■ Residence Time of Atmospheric Gases

To assess the effect of any atmospheric gas on global warming, it is important to know how long the gas will remain in the atmosphere before it decomposes by one pathway or another. Each atmospheric gas has a **residence time** (T_{res}), which describes the average amount of time one of its molecules spends in the atmosphere. If the steady-state atmospheric concentration (C) is known and the rate of input (flux, F) of that gas is known, the residence time can be calculated.

$$T_{res} = C/F$$

The average residence time of the CO_2, N_2O, and CFCs is more than a century, and thus, these gases, when emitted, will have an effect for a long, long time. Methane, on the other hand, has a residence time of only approximately a decade.

EXAMPLE 3.1

What is the residence time of CH_4 if the current atmospheric concentration is 1.7 ppmv and the total release of methane from all sources is 550 million tons per year?

Solution
Either mass or concentrations of the gas can be used. Because we know the atmospheric concentration in ppm, we can express the rate in ppm by dividing the yearly input amount by the total mass of the atmosphere (5.27×10^{18} kg). The average mass of atmospheric gases is estimated to be 29 g/mol (an average of N_2 and O_2 molecular weights).

Step 1: Convert 550 million tons of methane released per year to kilograms (flux):

$$\frac{550,000,000 \text{ tons/year} \left(2000 \text{ pounds/ton}\right)}{2.2 \text{ pounds/kg}} = 5.0 \times 10^{11} \text{ kg/year}$$

Step 2: Compute the number of moles of gas in the atmosphere:

$$5.27 \times 10^{18} \text{ kg} = 5.27 \times 10^{21} \text{ g}$$

$$\frac{5.27 \times 10^{21} \text{ g}}{29 \text{ g/mol}} = 1.82 \times 10^{20} \text{ mol}$$

Step 3: Compute the mass of methane in the atmosphere:

$$\left(1.82 \times 10^{20} \text{ mol}\right) \left(1.75 \times 10^{-6}\right) = 3.19 \times 10^{14} \text{ mol methane}$$

$$\left(3.19 \times 10^{14} \text{ mol}\right) \left(16 \text{ g/mol}\right) = 5.1 \times 10^{15} \text{ g} = 5.1 \times 10^{12} \text{ kg}$$

Step 4: Plug into the equation:

$$T_{res} = C/F = \frac{5.1 \times 10^{12} \text{ kg}}{5.0 \times 10^{11} \text{ kg/year}} = 10 \text{ years}$$

Atmospheric Water Vapor

For Earth, water vapor is actually the most important greenhouse gas because it absorbs strongly in the infrared (IR) region. As can be seen in Figure 3.7 gaseous water strongly absorbs in two distinct IR regions: 2.5 to 3.4 μm and 10 to 13 μm. The infrared spectrometer that makes these measurements is described in Chapter 6. Water vapor is not listed in Table 3.1 as a constituent of air because its concentration in the atmosphere varies greatly depending on location and temperature. Water vapor on the average has approximately a 0.4% concentration in Earth's atmosphere. It is involved in two important feedback processes that oppose one another: A **positive feedback** occurs when global warming increases evaporation from oceans that leads to higher concentrations of water vapor in the air. The increased amount of water vapor causes more infrared to be absorbed, and that increases warming of the Earth's surface. A **negative feedback** occurs as the troposphere becomes cloudier, and that causes more reflection of incident solar flux. This causes cooling of the Earth's surface. Because there are no **anthropogenic** (made by humans) activities that will directly cause and increase in the concentration of water

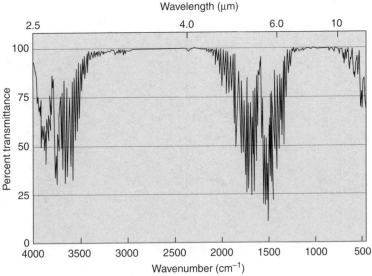

Figure 3.7 Infrared absorption spectrum of water [cm^{-1} = 10,000/λ(cm)].

vapor in Earth's atmosphere, it is not usually listed among the greenhouse gases. Other greenhouse gases have their own feedback loops.

Atmospheric Carbon Dioxide

There is no doubt that carbon dioxide contributes to greenhouse warming. As can be seen in Figure 3.8, it absorbs in three IR regions. It has a very strong absorbance in the 14- to 19-μm range, a strong absorbance in the 4- to 4.5-μm range, and a weak absorbance in the 3-μm range. If the Earth's IR emission spectrum is superimposed with the IR absorption spectra of carbon dioxide and water vapor, the result can be seen in Figure 3.9. The combined IR absorption of carbon dioxide and water coincide with most of the IR emission from the Earth. There is an unobstructed region of the spectrum between 7.5 and 13 μm, through which IR radiation from the Earth's surface can still escape. This section of the IR is called the **atmospheric window**.

Emissions from the burning of fossil fuels are the primary anthropologic source of CO_2. The carbon stored in fossil fuels has been sequestered for millions of years beneath the surface of the Earth until the CO_2 was released through the combustion process. Natural sources of atmospheric CO_2 include the decomposition of organisms, plant and animal respiration, and some oceanic processes

Several important CO_2 **sinks**, which are mechanisms that remove CO_2 from the atmosphere, operate on planet Earth. Sedimentary rock formation (*lithification*) to form limestone ($CaCO_3$) represents by far the largest sink, but because the process of lithification requires an enormously long time to remove the CO_2 from the atmosphere, we will not consider this process to be important in mitigating the anthropological greenhouse effect.

On a human time scale, the oceans represent the largest sink. The oceans are a tremendous reservoir in the global carbon cycle. The first way oceans remove CO_2 from the atmosphere

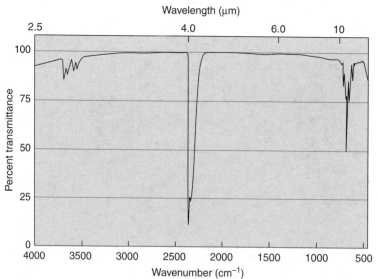

Figure 3.8 Infrared absorption spectra of carbon dioxide [cm^{-1} = 10,000/λ(cm)].

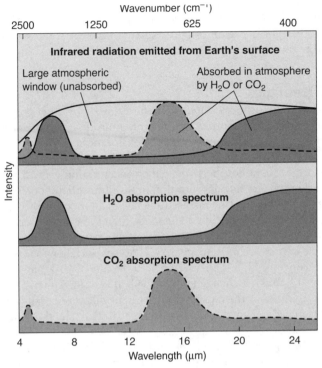

Figure 3.9 Absorption of radiation emitted from the Earth's surface by carbon dioxide and water vapor.

is by dissolving the gas (solubility pump). The cold polar ocean water at the surface dissolves atmospheric CO_2. The CO_2-rich surface water sinks into the deep ocean where it sequesters the CO_2 for several hundred years. This sink is expected to become less important as polar regions become warmer. The second way the oceans remove atmospheric CO_2 is biologically (*biological pump*). Marine phytoplankton take in atmospheric CO_2 and subsequently release carbon in a form that sinks to the ocean floor where it is sequestered indefinitely through the lithification of sedimentary rocks. This process is estimated to reduce the atmospheric CO_2 concentration by one-third of what it would be otherwise. Some scientists have suggested that fertilizing the ocean surface with iron sulfate will allow marine phytoplankton to reproduce more rapidly and remove even more CO_2. Although laboratory-scale iron fertilization experiments have verified that Fe is a limiting nutrient, its effect is short-lived, and the amount of new production exported from the surface layer is small. In addition, this approach would require continuous seeding of vast areas with outrageous amounts of Fe, which may cause unknown harm to existing marine ecosystems.

Other atmospheric sinks are minor compared to the oceans. Soils sequester atmospheric CO_2 by containment of partially decomposed organic matter known as humus. While the total amount of CO_2 sequestered in soil is unknown, improved soil conservation practices are anticipated to help mitigate the buildup of atmospheric CO_2 at least slightly. The third sink is the nonsoil component of the biosphere, vegetation, which absorbs CO_2 during photosynthesis and stores carbon in its biomass as it continues to grow. Only 30% of the Earth's surface is land and only 30% of that is covered by forests, thus the uptake by vegetation is a far smaller sink than the ocean and slightly smaller than the soils. Within the biosphere, the most critical region for sequestration is the tropical rain forest and these are rapidly disappearing.

Atmospheric Methane

The IR radiation in the atmospheric window can be absorbed by other polyatomic molecules. Because these molecules are composed of different atoms, they vibrate at different frequencies and as a result will absorb IR in different regions of the spectrum.

Methane (CH_4) is a polyatomic molecule with a tropospheric concentration of 1.75 ppmv (1745 ppbv). Before 1750, atmospheric methane concentration was 0.75 ppmv (750 ppbv) and during the industrial era, its concentration in the atmosphere increased steadily. In the mid-20th century its concentration increased at a rate of approximately 0.5% per year. In the early 1990s its rate of increase declined and since then the rate of increase has been zero. The rise in atmospheric CH_4 has been associated with human activity and fossil fuel use. The reason why the rate of increase of atmospheric CH_4 today has fallen to zero is not well understood. Because its concentration is affected by both the rate of CH_4 release (sources) and the rate of CH_4 removal (sinks), a change in either will affect atmospheric CH_4 concentration. Neither the rate of release nor the rate of removal can be measured easily and this reflects the overall difficulty facing atmospheric scientists trying to understand the complicated processes acting in our atmosphere.

As can be seen in Figure 3.10, methane absorbs IR in the 3- to 4-μm and 7- to 8.5-μm regions. Methane's absorption at 3 to 4 μm is not as important because the radiation from the Earth (black body radiation; see Chapter 2) hardly has any emission in this region. The absorption between 7 and 8.5 μm is within the atmospheric window, and through this absorption, methane can contribute to the greenhouse effect.

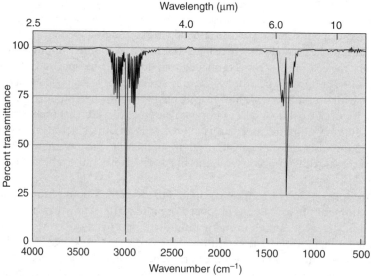

Figure 3.10 Infrared absorption spectrum of methane [$cm^{-1} = 10,000/\lambda(cm)$].

The sources of atmospheric methane have been understood only for the last three decades (Figure 3.11). Methane is released to the atmosphere through anaerobic decomposition, which is the decomposition of organic material in the absence of oxygen.

During the process of anaerobic decomposition, cellulose is converted into CH_4 and CO_2. Anaerobic decomposition occurs readily in a waterlogged environment. On a global basis, about 23% of the atmospheric CH_4 is released naturally from wetlands, including bogs, tundra, and swamps, where methane-producing bacteria thrive in the rich organic matter (with a low oxygen content). The next leading source at about 20% is from the flooding of rice fields, where certain types of bacteria also thrive. A third major source of CH_4 is from the bacteria present in the anaerobic digestive process of animals, particularly ruminants (cattle, camels, sheep) and certain species of termites. The inefficient digestion of ruminants releases 16% of the atmospheric CH_4, whereas termite mounds release another 4%. Methane emission from ruminants and rice fields are considered to be anthropogenic since ruminants are raised for and rice is cultivated for human needs.

There are several anthropogenic sources of methane in the troposphere. CH_4 is released when natural gas pipelines rupture or leak. It is estimated that 1.5% of all methane carried in pipelines today is lost to the atmosphere. Although this may seem like a very high leakage rate, it is lower than before. Twenty years ago, natural gas pipelines in the former Soviet Union leaked methane at a much higher rate than today. At oil refineries, methane

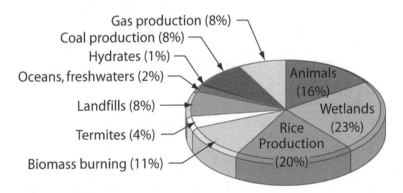

Figure 3.11 Sources of net global emission of atmospheric methane. *Source:* Data from Reeburg, W.S. *Global methane budget studies.* 1996. IGAC*tivities* Newsletter No. 6, September 1996. The International Global Atmospheric Chemistry (IGAC) Project. NOAA/OAR/PMEL.

that has been dissolved in the crude oil for thousands of years is released during the refining process and it can be released into the atmosphere. Also, during the mining of coal, CH_4 that was trapped for centuries in the solid coal is released into the air. Another major anthropogenic source (8%) of atmospheric CH_4 comes from anaerobic decomposition of garbage in landfills.

The most important sink for CH_4 is its oxidation by hydroxyl radicals (Chapter 4) in the troposphere.

$$CH_4 + OH^. \longrightarrow CH_3^. + H_2O$$

Secondary sinks include soils that take up CH_4 and the transfer of CH_4 from the troposphere to the stratosphere.

There is great concern that the rate of methane release could increase in the future as a result of a temperature increase from the greenhouse effect. As the greenhouse effect raises the temperature of the Earth, the rate of anaerobic decay in swamps would increase and, thus, more methane would be released. In turn, the additional methane in the atmosphere would absorb IR radiation and itself cause a further rise in temperature. This is an example of a positive feedback, which describes a series of events that accelerates the rate of change.

There is also a large quantity of methane frozen in the permafrost of the subarctic regions. It was produced by the decay of plant material during warmer periods that became trapped by glaciation. An increase in global temperature could melt the permafrost and release the trapped methane, which in turn would further accelerate the rate of global warming as it absorbs IR radiation. Even larger quantities of methane are trapped at the bottom of the oceans as methane hydrate, which has the formula $CH_4 \cdot 6\,H_2O$. Under high pressure and low temperature, the CH_4 is surrounded by a cage of water molecules. $CH_4 \cdot 6\,H_2O$ is an example of a **clathrate**, where a small molecule occupies a vacant space in the center of a three-dimensional cage of water (ice) molecules. Greenhouse warming of the oceans, if extended to their bottoms, could cause the release of methane from methane hydrate. It is possible that these positive-feedback mechanisms could combine to release large amounts of methane that would begin a runaway greenhouse effect that would threaten all life on Earth.

Atmospheric Nitrous Oxide

Figure 3.12 shows that nitrous oxide absorbs IR radiation in the 3- to 5-μm and 7.5- to 9-μm regions. The 7.5- to 9-μm absorption falls in the atmospheric window. This concentration of nitrous oxide in the atmosphere is 314 ppbv and is increasing by approximately 0.3% per year.

The major source of nitrous oxide is release from soil, lakes, and oceans by microbial denitrification of nitrate. It is difficult to determine how much of its release is natural and which portion is caused by anthropogenic events. The application of synthetic nitrogen fertilizers increases the concentration of nitrate in the soil, and this facilitates the microbial decomposition reaction. There are no important troposphere sinks for N_2O and as a result it has a long tropospheric residence time, an estimated 120 years.

Atmospheric Chlorofluorocarbons

Chlorofluorocarbons (CFCs) are also greenhouse gases. (Chapter 5 discusses their role in stratospheric ozone depletion.) Figure 3.13 shows that CFCs such as Freon 12 absorb strongly

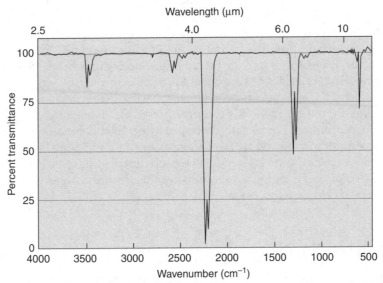

Figure 3.12 Infrared adsorption spectrum of nitrous oxide [cm^{-1} = 10,000/λ(cm)]. © Bio-Rad Laboratories, Inc., Informatics Division, Sadtler Software & Databases, 1980–2008. All rights reserved.

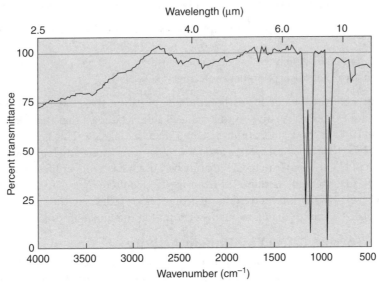

Figure 3.13 Infrared adsorption spectrum of CFC-12 [cm^{-1} = 10,000/λ(cm)]. © Bio-Rad Laboratories, Inc., Informatics Division, Sadtler Software & Databases, 1980–2008. All rights reserved.

in the atmospheric window at 9.5 and 11.5 μm. The concentration of Freon 12 in the atmosphere is over 1 ppbv, and up until the 1987 **Montreal Protocol**, its concentration was increasing at a rate of 5% per year. Because of the Montreal Protocol, some of the more developed countries (including the United States) have banned the manufacture and use of Freon 12, and its rate of increase has slowed. The Montreal Protocol allows lesser developed countries, such as Mexico, to continue the manufacture and use of CFCs. Because of this, Freon 12 is now one of the most smuggled materials into the United States, lagging only behind illicit drugs. The Montreal Protocol has called for a phase out of the manufacture of CFCs by 2010.

Fully fluorinated CFCs do not easily decompose, having lifetimes of thousands of years in the atmosphere. Even if the rate of CFC release decreases, these molecules will be in our atmosphere for generations. The recently developed hydrochlorofluorocarbons (HCFCs) also absorb in the same IR window and are greenhouse gases. The C—H bond in the HCFC is easier to break because it has a lower enthalpy (413 kj/mol) than does the C—F bond (485 kj/mol), so the HCFC decomposes more quickly. Fortunately, their resonance times in the troposphere are considerably shorter than that of the CFCs.

■ Radiative Forcing

All of the greenhouse gases discussed are present at some concentration in the Earth's atmosphere, and they all absorb IR. Each absorbs IR at different wavelengths but with different efficiencies, however. To assess the effect each greenhouse gas exerts on the atmosphere, the relative **radiative forcing** is a unit used to compare the relative contribution to the infrared absorption per molecule added to the atmosphere. Three factors determine radiative forcing:

1. The concentration of the gas in the atmosphere today (and its residence lifetime in the atmosphere).
2. The wavelengths at which the gas molecules absorb.
3. The intensity of the absorption per molecule.

The relative radiative value for carbon dioxide is arbitrarily set at 1, and the values for other gases are expressed relative to this value. Table 3.1 shows that the relative radiative forcing from CFCs is thousands of times greater than that of CO_2. This is because the intensity of the CFCs' IR absorption is large and CFCs have a very long atmospheric lifetime. The contribution of N_2O is greater than CH_4 mainly because it has a longer lifetime. In Chapter 4, we will learn that methane is removed from the troposphere by a reaction with hydroxyl radicals, whereas N_2O is not removed in the troposphere; it drifts to the stratosphere, where it is decomposed when irradiated by intense UV rays at the upper altitudes.

Radiative Forcing Caused by Human Activity
The effect of human activities on radiative forcing since the start of the Industrial Revolution in 1750 can be seen in Figure 3.14. Because they all absorb IR radiation, the forcing for all greenhouse gases are positive. Decreases in stratospheric ozone concentration have caused cooling, while increases in troposphere ozone concentration have caused warming. Aerosol particles affect radiative forcing in two ways: 1) aerosols can reflect incoming solar radiation, which causes negative forcing; 2) they can absorb IR radiation

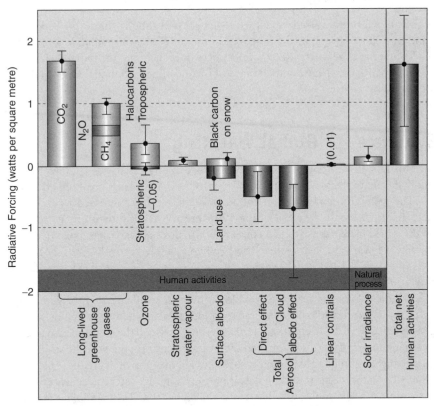

Figure 3.14 Radiative forcing of the climate between 1750 and 2005.

that would be lost from the surface of Earth, which causes positive forcing. Some aerosols such as soot (which is mostly carbon) strongly absorbs, whereas others such as sulfate aerosols are highly reflective. If radiative forcing for all aerosols is summed, the effect of aerosols is negative. Human activity, which has produced more soot, also has changed the albedo of the surface of the planet through changes in farming and forests. Human activity probably has changed the reflectivity of ice and snow as well. Experts estimate that it is likely that more solar radiation is now being reflected from the surface of the Earth (negative forcing). Aircraft exhaust (**contrails**) can reflect incoming solar radiation while absorbing outgoing infrared radiation. Aircraft contrails increase the Earth's clouds and have a small positive radiative forcing.

Radiative Forcing Caused by Nature

Changes in solar radiation and volcanic eruptions are the two major events that affect radiative forcing. Solar output has increased gradually during the industrial era causing a small positive radiative forcing. Volcanic eruptions have historically caused a negative forcing for two to three years after an eruption. The last major volcanic eruption was

Mt. Pinatubo in 1991, and at the present time there is no volcanic aerosol causing radiative forcing.

The radiative forcing caused by nature at the present time is about the same as it was in 1750, and this affect is very small compared to the differences in radiative forcing that have resulted from human activities. In our atmosphere, it is clear that the radiative forcing caused by human activities contributes much more to future climate change than does the radiative forcing caused by nature.

■ Evidence for Global Warming

Evidence for global warming is considerable. Glaciers all over the world have been retreating steadily for many years, and unusually large icebergs have been breaking off Antarctic ice shelves. The average temperature at the Earth's surface has increased by 0.3°C to 0.6°C (0.5°C to 1.1°F) since the last half of the 19th century. During the same period, sea levels around the world have risen 10 to 25 cm (4 to 10 inches), as water has expanded and ice has melted. The 14 warmest years since 1860 (when reliable recordkeeping began) have all occurred since 1980; 1998 and 2005 were the hottest years on record (Figure 3.2). The rate of temperature increase in the 20th century was greater than at any time since 1200 AD.

Recent studies of ancient ice have provided convincing evidence that the warming trend is likely to continue if carbon dioxide emissions continue to rise. Scientists took mile-long ice cores, dating back 160,000 years, from Antarctic glaciers and analyzed the air trapped in pockets in the ice for carbon dioxide. They also estimated the air temperature at the time the air became locked in the ice, using a technique based on measurements of the ratio of deuterium ($^2_1$H) to ordinary hydrogen ($^1_1$H) in the ice surrounding the air pockets. Lighter water molecules evaporate more readily than heavy water molecules. Therefore, the higher the $^2_1$H/$^1_1$H ratio in the frozen water, the higher the temperature. From their findings, the scientists were able to estimate temperatures at the time the different layers of ice were formed. As Figure 3.1 shows, a remarkable correlation exists between carbon dioxide concentration and air temperature. The air temperature rises and falls in step with the increases and decreases in the carbon dioxide concentration.

Computer-generated climate models also predict a warming trend. These models are based on numerous complex mathematic equations representing the many variables that affect climate. The information is fed into supercomputers, which then project temperature changes that are likely to occur if atmospheric carbon dioxide increases. Because many factors, such as the influence of clouds and the role of deep ocean currents, are not well understood and are not included in the computer models, some scientists have been skeptical about global warming. However, in 2007, the Intergovernmental Panel on Climate Change (IPCC), a group of 2500 scientists from 100 countries, concluded that "the balance of evidence suggests that there is a discernible human influence on global climate."

Two recent advancements have improved our ability to predict global climate changes. The first is the coordination of 18 different groups of climate scientists from around the world, each of which has created their own computer model for the factors that control the Earth's climate. Some climatic processes, such as the flow of the ocean and propagation sunlight and heat, are well understood and well represented by traditional models. Others, such as the formation and movement of clouds and the flow of ocean eddies, are still being studied and their effects need to be estimated. Models approximate the effect of each of these using simplified equations called **parameterizations.** The use of so many different models (18) by the IPCC is unprecedented in climate studies and allowed the IPCC to quantify the effects of uncertainties in climate processes and use these data to improve their confidence in predicting global climate changes. The second advance is the fact that climate scientists now better understand the complex processes that occur on Earth. The behavior of aerosols, movement of sea ice, and the exchange of water and energy between land and atmosphere are now much better understood than they were 20 years ago.

The IPCC studies show two major patterns that attribute global warming to human activity. The first pattern of change is the observation that there is greater warming of the atmosphere over land than over the ocean and that the surface of the ocean is warmer than deeper ocean water. This pattern is consistent with the warming caused by greenhouse gases in the atmosphere above. A second pattern of change is the observation that although the troposphere has warmed, the stratosphere just above it has cooled. If an increase in radiation from the sun was the cause of global warming, then both the stratosphere and troposphere would be expected to warm. The observed warming of the troposphere is consistent with an increase of the concentration of greenhouse gases there and the observed cooling of the stratosphere is consistent with the destruction of ozone there. The collected climatic data, when statistically analyzed, provides the base for the increased confidence that human activities are the cause of global warming.

■ Effects of Global Warming

The effects of global warming could be devastating. The IPCC predicts that if nothing is done to lower emissions of greenhouse gases, by 2100 the average global surface air temperature will increase from 1°C to 3.5°C (1.8°C to 6.3°F), and world sea levels will rise 15 to 90 cm (6 to 36 inches). These may seem like small increases, but we should remember that at the time of the last Ice Age, the Earth's temperature was only 5°C lower than it is today. A sea level rise of only 15 cm (6 inches) would cause flooding in many coastal areas.

With rising temperatures, air circulation, ocean currents, and rainfall patterns would change, causing generally violent weather. As a result, some regions of the world, including much of the United States, would experience droughts while other regions would become much wetter. Ecosystems all over the world would be disrupted, and some species might face extinction. Climate-related diseases such as malaria might attack areas where they are currently unknown.

The IPCC also points out that in addition to global temperature increases there will be other effects on our ecosystem. Those effects are:

1. Water
   a. increased water availability in moist tropics and high latitudes
   b. decreasing water availability and increasing drought in mid-latitudes and semi-arid latitudes
   c. glaciers, polar ice caps, and polar sea ice melting
   d. rising sea levels threaten to flood Pacific Islands
2. Ecosystems
   a. up to 30% of species at increasing risk of extinction
   b. increased coral bleaching
   c. more species will shift their ranges to cooler locals
   d. increased wildfire risk

3. Food
   a. negative impact on subsistence farmers
   b. cereal productivity will decrease in low latitudes
   c. negative impact on subsistence fishers
4. Coasts
   a. increased damage from floods and storms
   b. about 30% of all coastal wetland will be lost
   c. increased risk of coastal flooding
5. Health
   a. increasing burden from malnutrition, diarrhea, and infectious diseases
   b. increased morbidity from heat waves, floods, and droughts
   c. shift in location of disease vectors (mosquito-borne diseases)

■ Slowing Global Warming

Because of its potential to change world climate, global warming is viewed by scientists in many nations as the major environmental problem of the 21st century. The first treaty to address this problem was signed at the United Nations Conference on the Environment (the Earth Summit) in Rio de Janeiro, Brazil, in June 1992. The treaty set goals to reduce emissions of carbon dioxide and other greenhouse gases by industrialized nations to 1990 levels by 2000. However, because the proposed reductions were not binding, they were not met.

At a 1998 conference on climate change held in Buenos Aires, Argentina, 106 nations (including the United States) agreed to begin implementing the agreement reached the year before at Kyoto, Japan. The market-based Kyoto Protocol committed the 38 industrialized nations, including the republics of the former Soviet Union, to reducing global greenhouse gas emissions to at least 5% below 1990 levels by 2012. Different nations have different targets and timetables. The major flaw in the Kyoto Protocol is that it did not require cuts by the developing nations, most of which, including India and China, are opposed to making any reductions in the near future. An encouraging sign, however, was Argentina and Kazakhstan's voluntary agreement to reduce their greenhouse gas emissions. Although developing countries currently produce one-tenth as much carbon dioxide per person as industrialized countries, in the last 10 years, as their economies have grown, their emissions have increased 75%. It is predicted that even if the industrialized nations reach their targets, the total concentration of greenhouse gases in the atmosphere will continue to rise.

Taking steps to reduce emissions will be both difficult and costly. In the United States, there is opposition from the oil and coal industries and their supporters in Congress, many of whom continue to deny the evidence for global warming. As a result, the Kyoto Protocol has not been ratified by the U.S. Senate. Recently, a group of major corporations launched the nation's first trading program for air pollution credits earned by firms that exceed emission-reduction goals. Launched in 2003, the Chicago Climate Exchange marks an expansion of the market-based steps to reduce greenhouse gas emissions. Patterned after commodity exchanges and similar in design to those in Britain and Denmark, companies that exceed reduction goals could sell excess reduction to other companies that have not met average greenhouse gas emission standards. The Chicago Climate Exchange has proposed establishing ways of measuring greenhouse gas reductions. Some methods are direct, such as changes in the industrial process that reduce carbon dioxide emissions. Others are indirect and involve controversial "offsets," such as increased planting of forests or farms that absorb carbon dioxide. Critics of this approach say that by buying credits, big carbon dioxide emitters can avoid the reductions mandated by the Kyoto Protocol.

We can respond to climate change in one of two ways: We can adapt and learn how to live in a much warmer world or we can act to mitigate the rise in temperature by reducing the release of greenhouse gases into the atmosphere. Because global temperatures are already on the rise, continued unabated release of greenhouse gases would cause catastrophic increases in global temperature. A combination of **adaptation** and **mitigation** will be necessary to control the rise in global temperature.

The IPCC considered six scenarios of economic expansion, population growth, and fossil-fuel use for its 2007 report. These studies concluded that the concentration of greenhouse gases will rise to levels from 445 ppm to 1130 ppm and that this would cause a corresponding increase in temperature from 2.0°C to 6.1°C over preindustrial levels. To keep the temperature increase to a minimum, the IPCC recommends that the world must stabilize atmospheric greenhouse gases at 445 ppm by 2015, or the world will face severe flooding in some local regions and severe drought in others, extinction of some species, and an overall economic disaster.

The IPCC report examined the most promising policies for holding the concentration of greenhouse gases at 445 ppm. The IPCC recommends the following:

1. Energy supply
   a. reduce fossil fuel subsidies
   b. taxes (carbon charges) on fossil fuels
   c. tax incentives for renewable energy technology
2. Transport
   a. mandatory fuel economy standards for road transportation
   b. taxes on truck purchase, registration, motor fuel, and road use
   c. invest in public transport and non-motorized means of transport
3. Buildings
   a. improve appliance efficiency
   b. increase energy efficiency in building codes
4. Industry
   a. improvement in performance standards
   b. subsidies and tax credits for greenhouse gas reduction
   c. trading of carbon credits
5. Agriculture and forestry
   a. financial incentives for improved land management
   b. efficient use of fertilizers and irrigation
   c. financial incentives to increase forest area and reduce deforestation
6. Waste management
   a. financial incentives for improved waste and wastewater management
   b. use of waste as a renewable source of energy
   c. improve waste management regulations

■ Additional Sources of Information

Brasseur GP, Orlando JJ, Tyndall GS. *Atmospheric Chemistry and Global Change.* New York: Oxford University Press, 1999.

Environmental Protection Agency (EPA). Homepage. Accessed November 18, 2008 from http://www.epa.gov.

Frederick JE. *Principles of Atmospheric Science.* Sudbury, MA: Jones and Bartlett, 2008.

Intergovernmental Panel on Climate Change (IPCC). IPCC Reports. Accessed November 18, 2008 from http://www.ipcc.ch/ipccreports/index.htm.

Jacobson MZ. *Atmospheric Pollution.* Cambridge, UK: Cambridge University Press, 2002.

Rohli RV, Vega AJ. *Climatology.* Sudbury, MA: Jones and Bartlett, 2008.

Wayne RP. *Chemistry of Atmospheres,* 3rd ed. New York: Oxford University Press, 2000.

■ Keywords

adaptation

anthropogenic

asymmetric stretch

atmospheric window

clathrate

contrail

dipole moment

greenhouse gases

greenhouse effect

mitigation

Montreal Protocol

negative feedback

parameterizations

positive feedback

radiative forcing

residence times (T_{res})

sinks

sources

symmetric stretch

■ Questions and Problems

1. List the three ways the amount of energy in the Earth's atmosphere can be changed.
2. What is a Milankovitch cycle and how is it involved in the development of an Ice Age?
3. Describe what is meant by the greenhouse effect.
4. The concentration of CO_2 in the atmosphere was
   a. 20,000 years ago
   b. measured in Mauna Loa in 1958
   c. measured in Manua Loa in 2008
5. List five greenhouse gases.
6. Which gas is considered the most important greenhouse gas?
7. What historical evidence exist that links atmospheric carbon dioxide concentration with global temperature?
8. List the two most important anthropogenic reasons that tropospheric carbon dioxide concentrations are increasing.
9. What is meant by the term *sequestered carbon*? Give two examples.
10. Why do not molecular oxygen and nitrogen, the primary constituents of the troposphere, absorb infrared radiation?

11. Why is water vapor not considered a greenhouse gas even though it absorbs infrared radiation?
12. Describe what is meant by the *atmospheric window*.
13. Describe the molecular characteristics of CH_4 and CO_2 that makes them able to absorb infrared radiation.
14. What is the residence time of pollutant XY if it has an atmospheric concentration of 6.0 ppm and 300-million tons of it are being released into the atmosphere yearly? XY has a molecular weight of 50 g/mol.
15. If the residence time of pollutant ZZ is 100 years and its input rate is 5×10^8 kg/year, what is the total amount of ZZ in the atmosphere? ZZ has a molecular weight of 62 g/mol.
16. The amount of water vapor in the troposphere varies depending on the temperature and location. Describe what is meant by the terms *positive feedback* and *negative feedback*.
17. List two natural sources of tropospheric CO_2 and one anthropogenic source.
18. List three sinks for tropospheric CO_2. Which is the most important?
19. Coal with a high sulfur content burns to produce the expected combustion products from carbon, but it also produces sulfate from burning of the sulfur. Sulfate emitted from smoke stacks creates tropospheric sulfate clouds. What two effects would emissions from this coal have on global warming?
20. List two natural sources of tropospheric methane and one anthropogenic source.
21. List three sinks for tropospheric methane. Which is the most important?
22. What is methane hydrate?
    a. Where can it be found?
    b. What is a clathrate?
    c. What series of events could lead to the release of methane from methane hydrate?
23. Describe events that could lead to the following:
    a. Increased releases of methane from swamps
    b. Increased releases of methane from methane hydrate
    c. Runaway greenhouse effect
24. What is the major source of tropospheric nitrous oxide?
25. What is the major sink for tropospheric nitrous oxide? What is its atmospheric lifetime?
26. What is the major anthropogenic source of atmospheric CFCs?
27. How do the molecular structures of HCFCs differ from CFCs?
28. For infrared radiation, give the wavelength range in μm and cm^{-1}.
29. As can be seen in Figure 3.10, methane absorbs at two regions in the infrared: 3 to 4 μm and 7 to 8.5 μm. Which of these absorptions is considered to cause the greenhouse effect? Why?
30. Which of the following molecules would you expect to absorb infrared radiation and be greenhouse gases if they were released into the atmosphere by human activity: H_2S, H_2, or Rn?
31. Using information in Table 3.1, compare the greenhouse effect caused today by past releases of methane and Freon 12. Which of the two has the greater effect? Why?
32. What three factors contribute to radiative forcing?
33. During the industrial period, have the following caused a positive or negative radiative forcing?
    a. carbon dioxide
    b. atmospheric aerosols
    c. land use

34. Give two examples of human activities that cause radiative forcing other than the release of greenhouse gases.
35. What natural events cause radiative forcing? Are any of these events important today?
36. Aircraft produce persistent linear trail of condensation which are called contrails. Go to http://www.ipcc.ch and find out if contrails cause positive or negative forcing.
37. The manufacture and sale of which of the greenhouse gases has been banned by the Montreal Protocol?
38. Draw the molecular formula of nitrous oxide. Why is this molecule able to absorb infrared energy? What is the major source of nitrous oxide in the troposphere?
39. From ice cores taken from Antarctic glaciers, scientists are able to estimate the global temperature at the time that the ice was frozen. Describe how this is done.
40. There have been estimates of global temperatures in the year 2100 if greenhouse gas emissions are not lowered. What temperature increase is expected, and what will be the suspected consequences of that increase?
41. Describe the goals of the Kyoto Protocol with respect to greenhouse gases.
42. How does the Chicago Climate Exchange work?
43. What two recent advancements have improved climate scientists' ability to predict global climate change?
44. Global climate change will cause significant changes in several ecosystems. Go to http://www.ipcc.ch and find changes expected in three different ecosystems.
45. Global climate change is predicted to negatively affect human health. Go to http://www.ipcc.ch and find four different ways that human health is expected to be negatively impacted.
46. What was the atmospheric concentration of CO_2 in 2008? The IPCC report recommends that atmospheric CO_2 concentration must be kept to what ceiling in the future?
47. There are several actions suggested to increase (or speed) the Earth's sinks for greenhouse gases. One such suggestion is to seed the warmer parts of the Pacific Ocean with zinc and iron compounds. Explain what effect on atmospheric greenhouse gas concentration this might have.
48. Ozone is found in both the stratosphere and troposphere. What effect on global warming would each of these changes have?
    a. increasing stratospheric ozone concentration
    b. increasing tropospheric ozone concentration
49. Rice fields are an important part of human nutrition but also a major source of greenhouse gas. First determine what greenhouse gas is released from rice paddies and then go to http://www.ipcc.ch and investigate how changing the soil on which rice is grown might reduce the greenhouse effect.
50. Essay question: Identify four of your behaviors that release greenhouse gases. Explain what gas is released and how this action releases the gas into the troposphere; include products you consume on a regular basis. Suggest behavior modification that could reduce the greenhouse effect.
51. Essay question: What specific steps should the U.S. government take to reduce greenhouse warming? Does a global institution, such as the United Nations, or a group of industrial nations, such as the G8, need to coordinate action, or should our government act independently to lead the world?

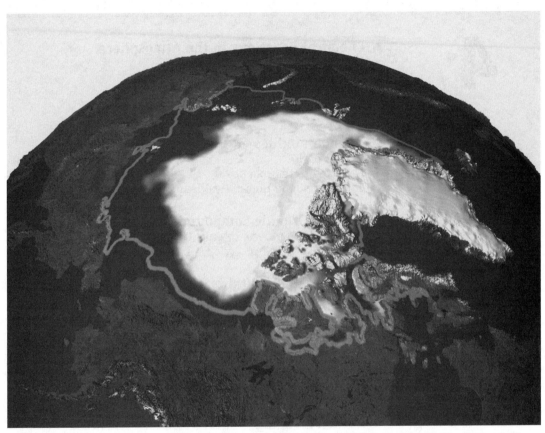

Perennial sea ice cover. Courtesy of Josefino Comiso and NAA/Goddard Space Flight Center Scientific Visualization Studio.

CHAPTER

4

Chemistry of the Troposphere

- Indoor Air Pollution

- Additional Sources of Information

- Keywords

- Questions and Problems

AN **AIR POLLUTANT** IS DEFINED AS A SUBSTANCE THAT IS PRESENT IN THE ATMOSPHERE AT A CONCENTRATION THAT IS SUFFICIENT TO CAUSE HARM TO HUMANS, OTHER ANIMALS, VEGETATION, OR MATERIALS. Each day humans inhale approximately 20,000 L (5300 gallons) of air. If harmful gases or fine toxic particles are present in the air, they are also drawn into the lungs, where they may cause serious respiratory diseases and other health problems.

Approximately 90% of all air pollution in the United States is caused by five **primary air pollutants**: carbon monoxide (CO), sulfur dioxide (SO_2), nitrogen oxides (NO_x), **volatile organic compounds** (**VOCs**; mostly hydrocarbons [HCs]), and suspended particles. Their major sources and the relative contribution of each to air pollution nationwide are shown in Figure 4.1. Emissions of all five are now regulated in the United States.

The transportation industry is responsible for nearly 50% of all air pollution from anthropogenic sources. In addition to CO, automobiles emit NO_x and HCs. The burning of fossil fuels by stationary sources (power plants and industrial plants) accounts for approximately one-third of air pollutants, mainly

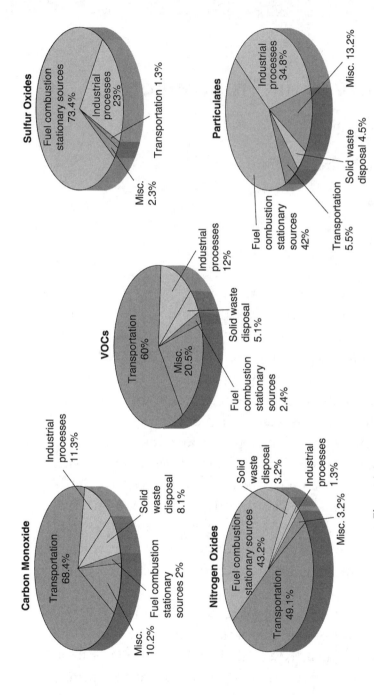

Figure 4.1 Nationwide emissions of primary air pollutants according to source.

in the form of sulfur oxides. Other industrial activities, along with a variety of processes, including incineration of solid wastes, contribute smaller amounts.

If air pollutants were distributed evenly over the entire country, their harmful effects would be greatly reduced. Because the pollutants tend to be concentrated in urban areas, where industry is more common and automobile traffic is congested, large segments of the population are exposed to their harmful effects, particularly during daily rush hours.

In addition to the five primary air pollutants, the atmosphere is contaminated with **secondary air pollutants**, which are harmful substances produced by chemical reactions between primary pollutants and other constituents of the atmosphere. Secondary pollutants include sulfuric acid, nitric acid, sulfates and nitrates (which contribute to acid deposition), and **ozone** and other photochemical oxidants (which contribute to photochemical smog).

This chapter first identifies the sources and fate of the major pollutants of the troposphere: CO, NO_x, VOCs, and SO_2. Next, the chemistry of photochemical smog and importance of the hydroxyl radical are presented. Finally, indoor air pollution and the impact of government regulation on air pollution are presented.

■ Chemical Reactions in the Atmosphere

In the atmosphere, species called **free radicals** (often referred to simply as radicals) are often formed under the influence of solar radiation. Radicals are uncharged fragments of molecules that, unlike ordinary chemical species, have an unpaired electron. As a result, radicals are highly reactive and are very short-lived. They are responsible for many of the complex, often poorly understood reactions that occur in the normal and polluted atmosphere.

Central to the chemistry of the troposphere is the **hydroxyl radical**. The hydroxyl radical ($\cdot$ OH) is uncharged and thus quite different from the negatively charged hydroxide ion (OH^-). It is written with a dot ($\cdot$) beside it to indicate an unpaired electron.

$:\overset{..}{O}:H^-$ $:\overset{..}{O}:H$ ← unpaired electron
hydroxide ion   hydroxyl radical

Hydroxyl radicals are continually formed and consumed in the troposphere and are produced as the result of a series of complex reactions primarily involving ozone, water, and nitrogen dioxide (NO_2), which are described later in the chapter. They play a role in the removal of

CO and HCs from the atmosphere and in the formation of nitric acid, sulfuric acid, and photochemical smog from atmospheric gases.

■ Carbon Monoxide (CO)

Sources of CO

The main anthropogenic source of CO is the combustion of gasoline in automobile engines (Figure 4.1). Gasoline is a complex mixture of HCs. If it is ignited in an adequate supply of oxygen, the products are carbon dioxide and water, as shown here for octane (C_8H_{18}), a representative gasoline HC.

$$2 \, C_8H_{18} \; + \; 25 \, O_2 \; \rightarrow \; 18 \, H_2O \; + \; 16 \, CO_2$$

In the confined space of the internal combustion engine, however, atmospheric oxygen is in limited supply, and combustion is incomplete. CO is formed and released to the atmosphere in automobile exhaust.

$$2 \, C_8H_{18} \; + \; 17 \, O_2 \; \rightarrow \; 18 \, H_2O \; + \; 16 \, CO$$

Since the introduction of the catalytic converter (described later in this chapter), CO emissions have been greatly reduced. In addition to the automobile, other anthropogenic sources of CO are combustion processes that are used by the electric power industry, various industrial processes, and solid-waste disposal.

It is perhaps surprising to discover that natural sources release approximately 10 times more CO into the atmosphere than all of the anthropogenic sources combined. The main natural source is methane gas, which is released during the anaerobic decay of plant materials in swamps, rice paddies, and other wetlands, where vegetation is submerged in oxygen-depleted water. Methane is also produced in the stomachs of ruminants (cattle and sheep) and the intestines of termites. Cattle, as they digest food, produce methane in their intestines. The gas then enters the bloodstream, and when the blood reaches the lungs, the methane is released and exhaled in normal breathing. Oxygen in the atmosphere oxidizes the methane to CO.

$$2 \, CH_4 \; + \; 3 \, O_2 \; \rightarrow \; 2 \, CO \; + \; 4 \, H_2O$$

Unlike anthropogenic sources, natural emissions of CO are dispersed over the entire surface of the Earth. Two mechanisms are believed to be at work to maintain the average global level constant at approximately 0.1 parts per million (ppm): (1) the conversion of CO to carbon dioxide in reactions involving hydroxyl radicals and (2) the removal of CO from the atmosphere by microorganisms in soil. In cities, where soil has been largely replaced with asphalt and concrete and where emissions are very concentrated, nature's natural defense mechanism is overwhelmed, and atmospheric CO levels increase.

Effects of CO on Human Health

Although CO is the most abundant air pollutant, it is not very toxic at the levels usually found in the atmosphere; however, if allowed to build up in a confined space, it can cause serious health problems.

CO interferes with the oxygen-carrying capacity of blood. Normally, **hemoglobin (Hb)** in red blood cells combines with oxygen in the lungs to form oxyhemoglobin (HbO_2). The HbO_2 is carried in the bloodstream to the various parts of the body, where the oxygen is released to the tissues.

CO binds much more strongly to Hb than oxygen. If CO is present in the lungs, it displaces oxygen from Hb and thus reduces the amount of oxygen that can be delivered to the tissues.

$$HbO_2 \quad + \quad CO \quad \rightarrow \quad HbCO \quad + \quad O_2$$
oxyhemoglobin   carboxyhemoglobin

Treatment for CO poisoning is inhalation of pure oxygen, which reverses the direction of this reaction. The symptoms of CO poisoning are those of oxygen deprivation: headache, dizziness, impaired judgment, drowsiness, slowed reflexes, respiratory failure, and eventually loss of consciousness and death. Prolonged exposure to CO levels as low as 10 ppm can be harmful. The danger from CO is heightened by the fact that the gas is colorless, tasteless, and odorless; people succumb to its effects before they are aware of its presence. On busy city streets, the CO concentration may reach 50 ppm and may be much higher in underground garages and traffic jams.

■ Nitrogen Oxides (NO_x)

Nitrogen dioxide (NO_2) is the major NO_x pollutant in the atmosphere and is formed from nitric oxide (NO). Collectively, these related **nitrogen oxides** are designated as NO_x.

Sources of NO_x

Practically all anthropogenic NO_x enter the atmosphere from the combustion of fossil fuels by automobiles, aircraft, and power plants (see Figure 4.1). At normal atmospheric temperatures, nitrogen and oxygen—the two main components of air—do not react with each other; however, at the very high temperatures that exist in the internal combustion engine and in industrial furnaces, normally unreactive atmospheric nitrogen reacts with oxygen. In a series of complex reactions, the two gases combine to form NO:

$$N_2 + O_2 \rightarrow 2\,NO$$

When released to the atmosphere, NO combines rapidly with atmospheric oxygen to form NO_2. NO is another molecule that like the hydroxyl radical has an unpaired electron and can be written as $NO\cdot$.

$$2\,NO + O_2 \rightarrow 2\,NO_2$$

As is the case with CO, far more NO_x are released to the atmosphere by natural processes than by human activities. During electrical storms, atmospheric nitrogen and oxygen react to form NO, which then rapidly combines with more atmospheric oxygen to form NO_2, as shown in the previous equations. Bacterial decomposition of nitrogen-containing organic matter in soil is another natural source of NO_x. Because emissions from natural processes are widely dispersed, they do not have an adverse effect on the environment.

Fate of Atmospheric NO_x

NO_2, regardless of its source, is ultimately removed from the atmosphere as nitric acid and nitrates in dust and rainfall (discussed in Chapter 7). In a series of complex reactions involving hydroxyl radicals, NO_2 combines with water vapor to form nitric acid. The simplified overall reaction can be written as follows:

$$4\ NO_2\ +\ 2\ H_2O\ +\ O_2\ \rightarrow\ 4\ HNO_3$$

Much of the nitric acid in the atmosphere is formed within aqueous aerosols. If weather conditions are right, the aerosols coalesce into larger droplets in clouds, and the result is **acid rain**. Some of the nitric acid formed reacts with ammonia and metallic particles in the atmosphere to form nitrates. Ammonium nitrate is formed as follows:

$$HNO_3\ +\ NH_3\ \rightarrow\ NH_4NO_3$$

Nitrates dissolve in rain and snow or settle as particles. The combined fallout contributes to acid deposition.

Effects of NO_x on Human Health and the Environment

NO_2 is a red-brown toxic gas that has a very unpleasant acrid odor. It can cause irritation of the eyes, inflammation of lung tissue, and emphysema. Even in badly polluted areas, however, its concentration in the atmosphere is rarely high enough to produce these symptoms. NO_x is a serious health problem because of its role in the formation of the secondary pollutants associated with photochemical smog (discussed later in this chapter).

Emissions from stationary fuel combustion sources are difficult to control. Lowering the combustion temperature of the furnace decreases formation of NO, but it decreases efficiency at the same time. Most research has concentrated on reducing automobile emissions by means of the catalytic converter.

■ Volatile Organic Compounds (VOCs)

A great variety of VOCs, including many HCs, enter the atmosphere from both natural and anthropogenic sources (see Figure 4.1). Most are not pollutants themselves but create problems when they react with other substances in the atmosphere to form the secondary air pollutants associated with photochemical smog.

The petroleum industry is the main anthropogenic source of HCs in the atmosphere. Gasoline is a complex mixture of many volatile HCs, and in urban areas, gasoline vapors can escape into the atmosphere in several ways: when gas is pumped at gas stations, during filling of storage tanks, and as unburned gasoline in exhaust from automobiles and small combustion engines, such as weed whackers and lawn mowers.

In the natural world, the pleasant aroma of pine, eucalyptus, and sandalwood trees is caused by the evaporation of VOCs called **terpenes** from their leaves. Natural sources account for approximately 85% of total emissions of volatile HCs. The remaining 15% comes from anthropogenic sources, which are of concern because unlike natural sources they are not evenly distributed but are concentrated in urban areas.

Automobile Four-Cycle Internal Combustion Engine

In more developed countries, the gasoline-powered **four-cycle internal combustion engine** is predominant. To understand why pollutants such as NO_x, CO, and HCs are produced by the automobile's four-cycle gasoline engine requires an insight into how this engine works. As shown in Figure 4.2, there are four steps in one complete cycle (two revolutions of the crank shaft) of the gasoline engine. If the combustion of gasoline were carried out in an open vessel exposed to the atmosphere, there would be much less air pollutants produced than are produced in the automobile engine. In the confined space of the internal combustion engine, however, atmospheric oxygen is in limited supply, and combustion is incomplete. CO is formed and released to the atmosphere in automobile exhaust.

The combination of the high temperature and pressure of the internal combustion engine increases the pollutants that are emitted from the engine. In the combustion cylinder, normally unreactive atmospheric nitrogen reacts with oxygen. In a series of complex reactions, the two gases combine to form NO. Not all of the gasoline vapor is burned. Walls of the combustion cylinder are cooler than the rest of the cylinder and cause the explosive flame to be extinguished in the area adjacent to the cylinder walls. The unburned HCs are emitted with the burned gas as pollutant VOCs.

The air/fuel ratio has a dramatic effect on the emission of pollutants from the four-cycle engine. During an engine tune-up, adjustment of the carburetor or the fuel injection system

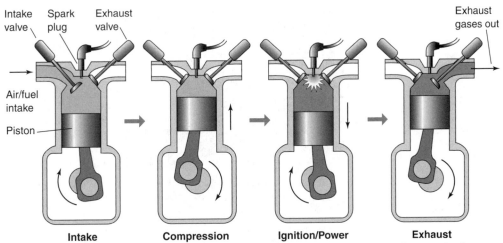

Figure 4.2 The steps in one complete cycle of a four-cycle internal combustion engine. Gasoline vapor has been mixed with the intake air.

can greatly reduce emission of pollutants. Figure 4.3 shows that the ideal, stoichiometric air/fuel ratio is 14.5. Adjustment to a more fuel-rich (lower air/fuel) ratio makes the car easier to start and decreases the emissions of NO_x and HCs. Unfortunately, it also decreases fuel efficiency, and the car will get fewer miles per gallon. The emission of CO also is increased with a fuel-rich ratio because less air is available. Adjustment to a more air-rich (higher air/fuel) ratio will make the car harder to start. Although it increases fuel efficiency, the emission of NO_x is increased. The ideal ratio is a compromise between fuel efficiency and the emission of the lowest collective amount of pollutants.

Gasoline Powered Two-Cycle Engines

The **two-cycle engine** is used to power motor scooters, mopeds, snowmobiles, marine outboard motors, and lawncare equipment such as weed whackers and leaf blowers. Because it is a less complicated device than the four-cycle engine, it can be made to be much lighter than a four-cycle engine. It is also much less expensive; however, and unlike automobile engines, which have been continuously modified for decades to increase efficiency and reduce emissions, the two-cycle gasoline engine has not been improved significantly since it was introduced in the 1940s.

The two-cycle gasoline engine, which can be seen in Figure 4.4, takes in fuel, releases exhaust in the same stroke, and emits from 25% to 30% of the fuel consumed as unburned HCs. Notice that the two-cycle engine has no valves and that power is produced in one of two steps rather than one in four steps. This means that the two-cycle engine runs faster (higher rpm) and hotter than the four-cycle engine.

Gasoline that is used in two-cycle engines is mixed with motor oil in a 40:1 ratio. The fuel is also used to lubricate the internal parts of the engine. Motor oil has a higher molecular weight than gasoline; gasoline is a mixture of C_6 to C_{10} HCs, whereas motor oil is a mixture of C_{18} to C_{25} HCs. Another facet of the two-cycle engine is that combustion takes place at a lower temperature than the four-cycle engine. Because of its high molecular weight, the

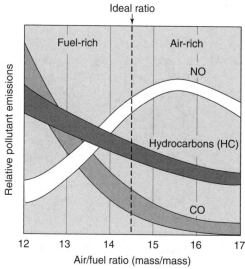

Figure 4.3 Emission of air pollutants from four-cycle internal combustion engine as the air/fuel ratio is varied.

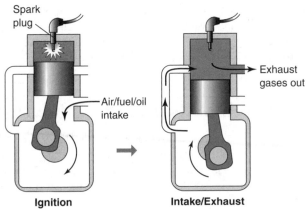

Spark plug

Air/fuel/oil intake

Exhaust gases out

Ignition   **Intake/Exhaust**

Figure 4.4 The steps in one complete cycle of a two-cycle internal combustion engine. Lubricating-oil is added to gasoline in a 1:40 ratio.

motor oil is not vaporized and burned as efficiently as the gasoline. The two-step cycle requires that the exhaust gases from combustion are leaving the cylinder through the exhaust port while the fresh air/fuel mixture for the next power stroke is simultaneously entering the cylinder from the side arm. All of these factors act together to produce exhaust gases that have a very high unburned HC concentration.

A comparison of the emissions from marine outboard two- and four-cycle engines of the same horsepower was made. The results of those tests (Table 4.1) show that the exhaust of a two-cycle engine contained more than 12 times the amount of HCs than a four-cycle engine of the same power.

The U.S. Environmental Protection Agency (EPA) estimates that gasoline-burning engines in recreational vehicles such as snowmobiles, all-terrain vehicles, boats, and personal watercraft are responsible for approximately 13% of the HCs that mobile sources emit. In 2008, EPA issued new regulations intended to reduce emissions from small engines used in

Table 4.1

Emissions from Marine Outboard Engines

Type of Marine Outboard Engine	CO(g)	NO$_x$(g)	HC(g)
Two-cycle engine	165	0.3	89
Four-cycle engine	127	0.7	7

Source: Juttner FD et al. Emissions of two- and four-stroke outboard engines-1: quantification of gases and VOC. *Water Research* 1995;29:1976–1982.

newly manufactured lawnmowers, power garden equipment, and outboard marine engines by 35%. The new standards will take effect in 2011 for smaller engines or 2012 for larger engines. EPA estimates that by 2030, the annual reduction to be 600,000 tons of HCs, 130 tons of NO_x and 5,500 tons of particulates ($PM_{2.5}$) as a result of the new regulation.

■ Automobile Pollutants and the Catalytic Converter

Motor vehicles are a major source of CO, NO_x, and volatile HCs. Since 1975, when all new cars in the United States were required by law to be equipped with a catalytic converter, emissions of those pollutants have been reduced significantly. Table 4.2 shows that today's cars emit 95% less pollutants than pre-1970 vehicles, despite the fact that the number of miles traveled has almost doubled in the last 20 years.

NO_x emissions are difficult to reduce. As was stated before, the amount of NO_x produced by a four-cycle engine is near its maximum at the ideal, stoichiometric air/fuel ratio. One approach to lowering both NO and HC is by a two-stage combustion process. The first step is to operate the system rich in fuel, and the second step is rich in air. This system burns the fuel completely, but not at a high enough temperature to produce as much NO_x. The "stratified-charge" engine uses this design to make modest reductions in NO_x emissions.

Another way to reduce emissions is by use of the three-way **catalytic converter** (in use since 1981), so named because it simultaneously reduces the amount of HC, NO_x, and CO in the exhaust stream. Hot exhaust gases from the engine pass through the converter before they enter the muffler. The converter is a very fine honeycomb structure made of ceramic coated with the precious metals platinum (Pt), palladium (Pd), and rhodium (Rh), which act as catalysts (Figure 4.5).

Table 4.2

Emission Standards for Light-Duty Vehicles (Passenger Cars)

Year	Federal Standards				California Standards			
	HCs	CO	NO_x	Evaporative HCs	HCs	CO	NO_x	Evaporative HCs
<1970	10.6	84	4.1	>45	10.6	84	4.1	>45
1970	4.1	34			4.1	34		
1975	1.5	15	3.1	2	0.9	9	2.0	2
1980	0.41	7.0	2.0	2	0.39	9	1.0	2
1985	0.41	3.4	1.0	2	0.39	7	0.4	2
1990	0.41	3.4	1.0	2	0.39	7	0.4	2
1993	0.41	3.4	1.0	2	0.25	3.4	0.4	2
2000	0.41	3.4	0.4	2	0.25	3.4	0.4	2

All values reported in grams per mile except for evaporative HCs, which are expressed as grams per test.

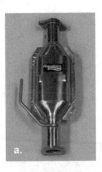

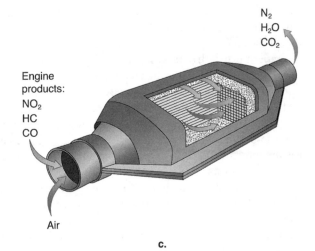

Figure 4.5 (a) The catalytic converter, which has been standard equipment in automobiles sold in the United States since 1975, reduces engine emissions. (b) A three-way catalytic converter consists of a ceramic honeycomb coated with the precious metals platinum, palladium, and rhodium. (c) The metals catalyze the conversion of nitrogen oxides and carbon monoxide to nitrogen and carbon dioxide and the conversions of hydrocarbons to carbon dioxide and water.

The catalytic converter has two chambers in succession. As the gases enter, Rh catalyzes the reduction of NO_x to nitrogen gas by hydrogen, which is generated at the surface of the Rh catalyst by the reaction of water on unburned HC molecules.

$$HCs + H_2O \rightarrow H_2 + CO$$

$$2 NO + 2 H_2 \rightarrow N_2 + 2 H_2O$$

Then air is injected into the exhaust stream to provide oxygen that, in the presence of the Pt and Pd catalysts, oxidizes CO to carbon dioxide and HCs to water and carbon dioxide.

$$2 CO + O_2 \rightarrow 2 CO_2$$
$$HC + 2 O_2 \rightarrow CO_2 + 2 H_2O$$

The overall reaction for the reduction of NO and the oxidation of CO can be written as follows:

$$2 NO + 2 CO \xrightarrow{\text{Rh, Pt, Pd catalysts}} N_2 + 2 CO_2$$

The oxidation of a typical gasoline HC, octane, occurs as follows:

$$2 C_8H_{18} + 25 O_2 \xrightarrow{\text{Pt, Pd catalysts}} 16 CO_2 + 18 H_2O$$

Automobiles using catalytic converters must have their air/fuel ratio set at 14.8:1. If the amount of air is increased (leaner), then CO and HCs are converted efficiently, but NO_x is not. If the amount of air is reduced (richer), the reverse is true.

Under normal driving conditions, the exhaust gases heat the catalytic converter. It operates at optimum efficiency in a temperature range of 350°C to 600°C. At 25°C, the efficiency of the catalytic converter is near zero. This means, of course, that the catalytic converter is ineffective when starting a cold engine and that its efficiency is low during engine warm-up.

Pt and Rh are very expensive, but because they are not used up, the small amounts needed last a long time. Replacement of a catalytic converter can cost more than $1000. Lead-free gasoline must be used in cars fitted with catalytic converters because lead coats and inactivates the catalysts.

Federal standards for automotive emissions became law in 1970. Table 4.2 lists the emission standards and the year that they were required. Notice that the amount of emissions allowed in the year 2000 for HC and CO are only one-tenth of what the original 1970 standard allowed. The catalytic converter has been credited with bringing significant reductions in automotive emissions. Today's catalytic converters remove 96% of CO and HCs and 76% of NO_x from auto exhausts. Even better results will have to be achieved in the future to meet the requirements of the 1990 Clean Air Act Amendments.

■ Sulfur Dioxide (SO_2)

Sources of SO_2

The release of SO_2 to the atmosphere is the primary cause of acid rain in the United States. Fossil fuel combustion at electric power-generating plants accounts for about 70% of the emissions; industrial sources contribute approximately 23% (Figure 4.1).

Coal, oil, and all other fossil fuels naturally contain some sulfur because the plant materials from which they were formed included sulfur-containing compounds. Coal frequently contains additional sulfur in the form of the mineral pyrite (FeS). Coal mined in the United States is typically between 1% and 4% (by weight) sulfur. The percentage is lower in coal from the western states than in coal mined in the central and eastern states. When sulfur-containing coal is burned, the sulfur is oxidized to SO_2:

$$S + O_2 \rightarrow SO_2$$

Natural sources account for approximately half of all SO_2 emissions. Table 4.3 shows that hydrogen sulfide produced as an end product of the anaerobic decomposition of sulfur-containing organic matter by microorganisms is the main source. After entering the atmosphere, hydrogen sulfide is oxidized to SO_2:

$$2 H_2S + 3 O_2 \rightarrow 2 SO_2 + 2 H_2O$$

Volcanic eruptions are another more localized natural source of SO_2. It has been estimated that the eruption of Mt. Pinatubo in the Philippines in June 1991 (Figure 4.6) injected as much as 25 million tons of SO_2 into the stratosphere, where it was converted into sulfuric acid aerosols.

Table 4.3

End Products of Decomposition of Organic Compounds under Aerobic and Anaerobic Conditions

Element in Organic Compound	End Product(s) of Decomposition	
	Aerobic conditions	Anaerobic conditions
Carbon (C)	CO_2	CH_4
Nitrogen (N)	NO_3^-	NH_3 and amines
Sulfur (S)	SO_4^{2-}	H_2S
Phosphorus (P)	PO_4^{3-}	PH_3 and other phosphorus compounds

Fate of Atmospheric SO_2: Acid Rain

SO_2 in the atmosphere reacts with oxygen to form sulfur trioxide (SO_3), which then reacts readily with water vapor or water droplets to form sulfuric acid. The mechanism involves hydroxyl radicals, and the sequence of reactions can be written as follows:

$$SO_2 + OH^• \rightarrow HSO_3^•$$

$$HSO_3^• + O_2 \rightarrow SO_3 + HOO^•$$

$$SO_3 + H_2O \rightarrow H_2SO_4\,(g)$$

$$H_2SO_4 + H_2O \rightarrow H_2SO_4\,(aq)$$

Sulfuric acid in the atmosphere becomes concentrated near the base of clouds, where pH levels as low as 3 (approximately the same pH as orange juice) have been recorded. Thus,

Figure 4.6 The eruption of Mt. Pinatubo in the Philippines in June 1991 ejected huge quantities of sulfur dioxide and ash into the atmosphere.

cloud-enshrouded, high-altitude trees and vegetation may be exposed to unusually high acidity. Because rain is made of moisture from all cloud levels, it is less acidic than moisture at the lower cloud levels.

Some of the atmospheric SO_2 dissolves if there is a significant amount of water in the air. In this case, most of the oxidation of SO_2 to H_2SO_4 occurs in the liquid phase rather than the gas phase. When SO_2 dissolves in water, some of it forms sulfurous acid (H_2SO_3):

$$SO_2(g) + H_2O(aq) \rightleftharpoons H_2SO_3(aq)$$

The concentration of H_2SO_3 is determined by the equilibrium constant for this reaction. Whenever gases are dissolved in water, the equilibrium constant is expressed as a Henry's law constant K_H, which for this case is equal to 1.0 M/atm at 25°C and is defined as follows:

$$K_H = \left[H_2SO_3(aq)\right]/P$$

Where P is the atmospheric partial pressure of SO_2 (which is usually approximately 0.1 ppm), which at atmospheric pressure is a partial pressure of 1.0×10^{-7} atm. Thus,

$$1.0 \times 10^{-7} \text{ M} = \left[H_2SO_3(aq)\right] = PK_H$$

Although technically a weak acid, H_2SO_3 has a large enough Ka (1.7×10^{-2}) that in the atmospheric aerosols it subsequent ionizes to HSO_3^-, bisulfite ion:

$$H_2SO_3(aq) \rightleftharpoons HSO_3^-(aq) + H^+(aq)$$

Because of the equilibrium between gaseous SO_2 and dissolved $H_2SO_3(aq)$, the $[H_2SO_3(aq)]$ corresponds to only the H_2SO_3 that does not ionize to bisulfite, and it remains at a constant 1.0×10^{-7} M.

If the equilibrium xy is the only source of acid, then it follows that

$$\left[HSO_3^-\right] = [H^+]$$

and

$$Ka = 1.7 \times 10^{-2} = \frac{\left[HSO_3^-\right][H^+]}{\left[H_2SO_3\right]} = \frac{\left[HSO_3^-\right]^2}{1.0 \times 10^{-7}}$$

$$\left[HSO_3^-\right] = 4.1 \times 10^{-5} \text{ M}$$

Thus, the ratio of HSO_3^- to H_2SO_3 is 410:1 ($4.1 \times 10^{-5}/1.0 \times 10^{-7}$). Because

$$\left[HSO_3^-\right] = [H^+] = 4.1 \times 10^{-5} \text{ M}$$

The pH of the aerosol droplets is therefore 4.4.

On the other hand, if strong acids are present in the aerosol, they will control the pH. If strong acids are also present in the droplet, the bisulfite concentration can be easily calculated:

$$\left[HSO_3^-\right] = \frac{\left(1.7 \times 10^{-2}\right)\left(1.0 \times 10^{-7}\right)}{\left[H^+\right]}$$

because $[H^+]$ is controlled by the release of H^+ from strong acids and the $[HSO_3^-]$ is inversely proportional to $[H^+]$.

The dissolved SO_2 is oxidized by trace amount of hydrogen peroxide (H_2O_2) and ozone that are also present in the aerosol droplets to sulfate ion (SO_4^{2-}). In the next section, ozone and H_2O_2 are the products of photodissociation reactions in photochemical smog.

The bisulfite is oxidized by either H_2O_2 or ozone to produce bisulfate (HSO_4^-) ion. The reaction with H_2O_2 can be written as follows:

$$HSO_3^- + H_2O_2 \rightarrow H_2O + HSO_4^-$$

This reaction is acid catalyzed. The reaction with ozone, on the other hand, is not acid sensitive:

$$HSO_3^- + O_3 \rightarrow O_2 + HSO_4^-$$

Oxidation of bisulfite in acidic aerosols (pH of less than 5) proceeds primarily by the acid-catalyzed oxidation by H_2O_2. In aerosols with higher pH, the oxidation is accomplished by O_3.

Ash particles are usually emitted together with SO_2 from electric power-generating plants. Sulfur oxides become adsorbed onto particle surfaces and may be carried many miles from their source before settling or being washed out by precipitation. Like nitric acid and nitrates formed from NO_x, sulfuric acid, sulfates, and particulates, all contribute to acid deposition.

Effects of SO_2 on Human Health and the Environment

SO_2 is a colorless, toxic gas with a sharp, acrid odor. Exposure to it causes irritation of the eyes and respiratory passages and aggravates symptoms of respiratory disease. Children and the elderly are especially susceptible to its effects.

SO_2 is also harmful to plants. Crops such as barley, alfalfa, cotton, and wheat are particularly likely to be adversely affected. In Sudbury, Ontario, tall smokestacks (1200 feet high) are used to protect local agriculture from the release of SO_2. Unfortunately, this only produces a problem elsewhere; SO_2 emissions from North America have been detected as far away as Greenland.

Methods for Controlling Emissions of SO_2

As a result of the Clean Air Act of 1970 and amendments made to it in 1990, coal-fired electric power plants were required to make significant reductions in their emissions of SO_2. Reductions can be achieved in two ways: (1) Sulfur can be removed from coal before combustion, or (2) SO_2 can be removed from the smokestack after combustion—but before it reaches the atmosphere. The second, cheaper approach is generally chosen.

The most commonly used method is **flue-gas desulfurization (FGD)** in which sulfur-containing compounds are washed out (or scrubbed) by passing the chimney (flue) gases through a slurry of water mixed with finely ground limestone ($CaCO_3$) or dolomite [$Ca \cdot Mg(CO_3)_2$] or both. On heating, the basic calcium carbonate reacts with acidic SO_2 and oxygen to form calcium sulfate ($CaSO_4$):

$$2\ SO_2\ +\ 2\ CaCO_2\ +\ O_2\ \rightarrow\ 2\ CaSO_4\ +\ 2\ CO_2$$

Scrubbers, which remove up to 90% of the SO_2 in the flue gas, can be quite easily and inexpensively retrofitted onto existing power plants.

A promising newer method is **fluidized bed combustion (FBC)**, a process in which a mixture of pulverized coal and powdered limestone is burned, with air being introduced to keep the mixture in a semifluid state. The limestone is converted to $CaSO_4$ according to the previous equation. In this process, however, because the coal is so finely divided, the reaction occurs at a lower temperature than in FGD, and as a result, the quantity of NO_x emitted is much lower. The disadvantage of FBC is that it cannot be added to existing power plants, but it is the preferred technology for installing in a new power plant. Both FGD and FBC have the problem of disposing of large quantities of $CaSO_4$.

Legislation to Control Emissions of SO_2

Before the passage of the Clean Air Act Amendments of 1990, the electric utility industry was unwilling to take steps to reduce its emissions of SO_2 because of cost considerations. The law required that total emissions of SO_2 be reduced to 10 million tons per year for the next ten years. Power companies could meet this requirement by installing efficient scrubbers or by switching to low-sulfur western coal, a move that would mean a loss of jobs and great economic hardship for the high-sulfur coal miners of the Ohio Valley and West Virginia. The law gave each power plant an emissions allowance that permits it to release a certain amount of SO_2 per year. To make it easier for the industry to reach the required overall reduction, the EPA introduced a free-market system of emissions trading, which allowed a plant emitting less SO_2 than its allowance to sell the difference to a plant that is emitting more than its allowance. This trading of pollution reduction credits allowed the lower overall emission levels required by law to be achieved, but the disadvantage is that the oldest and dirtiest plants can continue to release unacceptably high levels of pollution by buying credits from newer, cleaner plants.

Between 1983 and 2002, emissions of SO_2 from burning fossil fuels fell by almost 50% in Europe and by 33% in the United States. During this same period, however, emissions in the developing countries more than doubled and are expected to rise further as populations in these countries increase and they become more industrialized. China released 25 million tons of SO_2 in 2005 and is now the largest producer of atmospheric SO_2. China's SO_2 emissions have increased 27% in the period 2000 to 2006.

■ Industrial Smog

Particulate matter and SO_2 can be a deadly combination. Released into the atmosphere together when coal is burned, they can form **industrial smog** (sometimes called *London smog*), a mixture of fly ash, soot, SO_2, and some VOCs.

In the 19th and 20th centuries, industrial smog was common in the industrial centers of Europe and the United States. It formed in winter, typically in cities where the weather was cold and wet. Visibility was often reduced to a few yards, and people in factory towns lived under a pall of black smoke (Figure 4.7).

■ Photochemical Smog

The origin of **photochemical smog** is quite different from that of industrial smog. Typically, photochemical smog develops as a yellow-brown haze in hot sunny weather in cities, such as Los Angeles, where automobile traffic is congested (Figure 4.8). The reactions that led to its formation are initiated by sunlight and involve the HCs and NO_x emitted in automobile exhaust. NO_2 is responsible for the brownish color of the haze.

Figure 4.7 Industrial smog is created by the burning of coal and is made up primarily of fly ash, soot, and sulfur dioxide.

Production of Hydroxyl Radicals

It is well understood that solar energy is required for smog production. By the time sunlight reaches the surface of the Earth, all of the high-energy UV light has been absorbed in the stratosphere. As can be seen in Figure 4.9, NO_2 is the only automobile emission that is capable of absorbing visible light that reaches the Earth's surface.

NO_2 + sunlight (less than 320 nm) $\rightarrow$ NO + O

photodissociation reaction

Figure 4.8 Photochemical smog forms in cities such as Los Angeles when nitrogen oxides and hydrocarbons in the air interact in the presence of sunlight to form a yellow-brown haze.

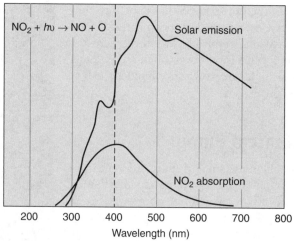

$NO_2 + h\upsilon \rightarrow NO + O$ Solar emission

NO$_2$ absorption

Wavelength (nm)

Figure 4.9 At λ < 400 nm, NO$_2$ absorbs sunlight to produce oxygen atoms.

The O atoms react immediately with atmospheric oxygen, O_2, producing ozone, O_3, in the lower stratosphere.

$$O + O_2 \rightarrow O_3$$

Because each molecule of ozone formed requires a NO$_2$ to photodissociate, this pathway cannot produce ozone concentrations that are higher than that of the NO$_2$ itself. Furthermore, the NO produced from the photodissociation reaction can react with the ozone to reduce its concentration even more as it reacts to produce more NO$_2$.

$$O_3 + NO \rightarrow NO_2 + O_2$$

The ozone produced absorbs light in the blue region of the visible spectrum (less than 310 nm) and photodissociates:

$$O_3 \rightarrow O_2 + O$$

The oxygen atom produced (which has six electrons) reacts with water vapor in the atmosphere and abstracts a hydrogen atom (with its electron), producing a hydroxyl radical (seven electrons).

$$O + H_2O \rightarrow 2 \,{}^{\cdot}OH \left(\text{hydroxyl radical}\right)$$

By this pathway, one NO$_2$ molecule produces two hydroxyl radicals.

The concentration of hydroxyl radicals does not continue to increase out of control because there are termination reactions that remove it from the troposphere. It can react with other radical species in the troposphere:

$$\cdot OH + \cdot NO_2 \rightarrow HNO_3$$

$$\cdot OH + HOO\cdot \rightarrow H_2O + O_2$$

$$2\cdot OOH \rightarrow H_2O_2 + O_2$$

The products of these reactions are very soluble in water and are removed from the troposphere during precipitation. Because hydroxyl radical production is a photochemical reaction, nightfall also causes the reaction sequence to stop.

Unburned HCs in automobile exhaust (represented by RCH_3 in the following equation) react with hydroxyl radicals to form a number of secondary pollutants, including the HC radical $RO_2\cdot$. This radical then reacts with NO to form aldehydes and the hydroperoxide radical ($HO_2\cdot$). The step wise mechanism is as follows:

$$RCH_3 + \cdot OH \rightarrow RCH_2\cdot + H_2O$$

$$RCH_2\cdot + O_2 \rightarrow \underset{\text{peroxyalkyl radical}}{RCH_2OO\cdot}$$

$$RCH_2OO\cdot + NO \rightarrow \underset{\text{alkoxy radical}}{RCH_2O\cdot}$$

$$RCH_2O\cdot + O_2 \rightarrow \underset{\text{aldehyde}}{RCHO} + \underset{\text{hydroperoxyl}}{\cdot OOH}$$

$$HOO\cdot + NO \rightarrow NO_2 + \cdot OH$$

Each step in this reaction produces a radical. The overall reaction is obtained by summing these individual steps.

$$RCH_3 + 2\,O_2 + 2\,NO \rightarrow RCHO + 2\,NO_2 + H_2O$$

One NO_2 produces two hydroxyl radicals by the following equation:

$$NO_2 + H_2O \rightarrow NO + 2\cdot OH$$

Multiply this equation by 2 and add it to the previous equation and the following is obtained:

$$RCH_3 + 2\,O_2 + H_2O \rightarrow RCHO + 4\cdot OH$$

This reaction produces four hydroxyl radicals for every HC reacted. This is a catalytic reaction. A very small number of radicals can produce a large amount of product through the production of four radicals per cycle.

Reactions of Hydroxyl Radicals with HCs

Abstraction of Hydrogen
Hydroxyl radicals will react with certain unburned HCs from the automobile's exhaust depending on the number and type of C—H bonds in the HC. Not all C—H bonds are equally

reactive. To compare the reactivity of HC molecules, the dissociation energy of the C—H bond must be known. The dissociation reaction can be written as follows:

$$R — H \rightarrow R^{\cdot} + H^{\cdot}$$

The stability of the R radical depends on its structure. An unpaired electron on a carbon atom is stabilized if that carbon atom is attached to other carbon substituents. For this reason, a tertiary C—H bond is more easily dissociated than is a secondary C—H bond, and the secondary is more easily dissociated than a primary C—H bond. Table 4.4 lists the dissociation energy of certain C—H bonds.

Aromatic HCs, such as benzene, have a high C—H dissociation energy because of hybridization. The hybridization on the aromatic carbons is sp^2, not the sp^3, of aliphatic carbons. The aromatic sp^2 C—H bonds are shorter and stronger because they have more character.

Because the dissociation energy is the highest for benzene and HCs with mostly methyl C—H (ethanol, methanol, ethane) bonds, the reaction rate with hydroxyl radical is the lowest for these HCs. As the number of secondary C—H bonds increase (more methylenes), the rate of the reaction increases, and the rate of reaction follows the trend, n-butane < pentane < n-hexane < n-heptane < n-octane. Higher reaction rates are observed for methyl substituted benzenes, such as toluene and xylene. The methylene radicals formed by the abstraction of hydrogen by the hydroxyl radical are stabilized by the aromatic ring. The free electron on the methyl carbon, which is adjacent to the benzene ring, is delocalized over the entire benzene pi orbital system.

Addition to Double Bonds

The reaction of hydroxyl radicals with alkenes proceeds at even a faster rate than the hydrogen abstraction reactions for HCs. This reaction is not a hydrogen atom extraction but rather an addition of the hydroxyl radical to the double bond. The pi electrons in the double bond are not as tightly bound and offer a site for interaction with radicals. The mechanism shown here is for the addition reaction of hydroxyl radical with propene (propylene).

Table 4.4

Dissociation Energy of C—H Bonds

Compound	Bond	Energy (kJ/mol)
Methane	$H_3C—H$	427
Ethane	$H_3CH_2C—H$	406
Propane (methylene)	$(H_3C)_2HC—H$	393
Methanol	$HOH_2C—H$	393
Benzene	$H_5C_5C—H$	427
Toluene	$H_5C_6H_2C—H$	326

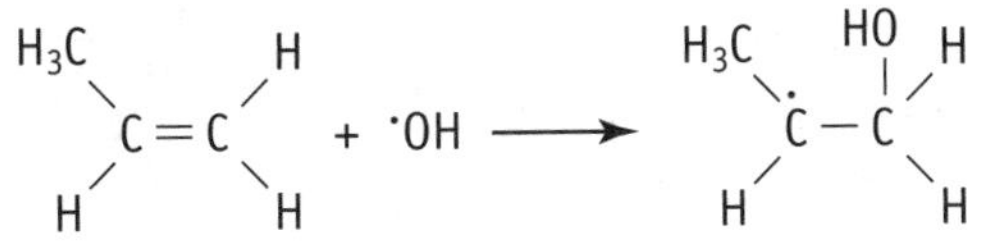

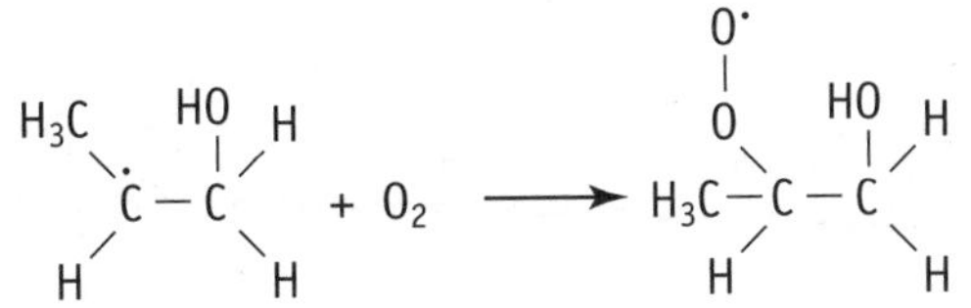

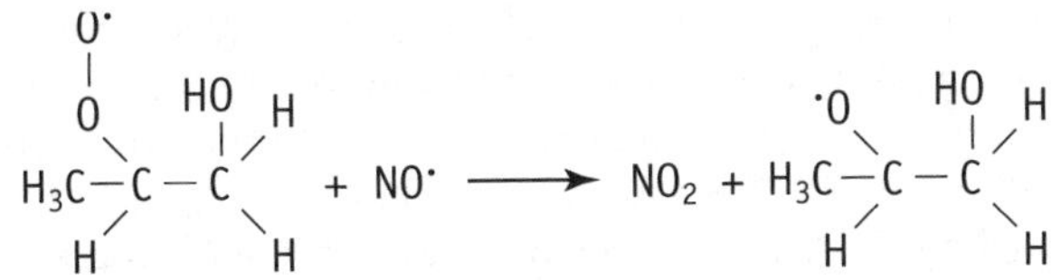

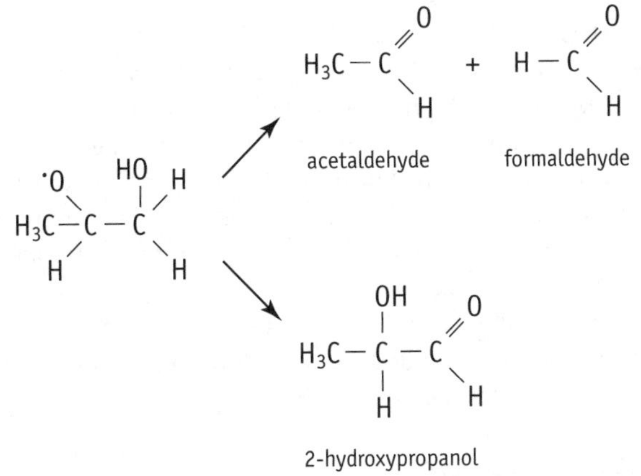

H_3C H
$C = C$ $+$ $\cdot OH$ $\longrightarrow$ H_3C HO H
H H $\overset{\cdot}{C} - C$
H H

H_3C HO H $O\cdot$
$\overset{\cdot}{C} - C$ $+$ O_2 $\longrightarrow$ O HO H
H H $H_3C - C - C$
H H

$O\cdot$
O HO H $\cdot O$ HO H
$H_3C - C - C$ $+$ $NO\cdot$ $\longrightarrow$ NO_2 $+$ $H_3C - C - C$
H H H H

$H_3C - C \overset{O}{\underset{H}{\diagdown}}$ $+$ $H - C \overset{O}{\underset{H}{\diagdown}}$

acetaldehyde   formaldehyde

$\cdot O$ HO H
$H_3C - C - C$
H H

OH O
$H_3C - C - C$
H H

2-hydroxypropanol

The NO_2 produced in this reaction can go on to make more ozone, and the products of this addition reaction, acetaldehyde, formaldehyde, and 2-hydroxypropanol, all go on to form other pollutants in secondary smog-forming reactions.

Secondary Smog-Forming Reactions

In the previous reaction sequences, aldehydes were one of the products produced by the attack of hydroxyl radicals on HCs. Once formed, these primary reaction products can undergo further reaction in the troposphere. The following reaction sequence takes place for all aldehydes formed. Acetaldehyde is used as an example.

$$CH_3CHO + {}^{\bullet}OH \rightarrow CH_3\overset{\bullet}{C}O + H_2O$$

$$CH_3\overset{\bullet}{C}O + O_2 \rightarrow CH_3COOO^{\bullet}$$

$$CH_3COOO^{\bullet} + {}^{\bullet}NO2 \longleftrightarrow CH_3COONO_2$$
PAN

The drawn-out structure of **peroxyacetyl nitrate (PAN)** is as follows:

$$CH_3-\overset{\overset{\textstyle O}{\|}}{C}-O-O-NO_2$$

PAN is the component of smog that causes major eye irritation. PANs are relatively stable molecules and have long lifetimes in cooler air. Because of this, they may travel long distances. In warmer climates, the PANs break down to release NO_2 and can begin the cycle described previously producing additional ozone and hydroxyl radicals. In this way, PANs can be considered a reservoir for NO_x species. Ozone, aldehydes, and PANs all contribute to the harmful effects of photochemical smog, but ozone—the pollutant produced in greatest quantity—causes the most serious problems.

Ozone: A Pollutant in the Troposphere

Ozone in the stratosphere protects us from damaging ultraviolet radiation from the sun, but ozone in the troposphere is a dangerous pollutant. Ozone is a powerful oxidizing agent. It is a colorless, pungent, very reactive gas that irritates the eyes and nasal passages. People with asthma or heart disease are particularly susceptible to its harmful effects. Exposure to ozone levels as low as 0.3 ppm for one to two hours can cause fatigue and respiratory difficulties. In Los Angeles, school children are kept indoors if ozone levels reach 0.35 ppm. Fortunately, because of pollution controls, today these levels are rarely reached.

Ozone is also very toxic to plants. In California, crop damage caused by ozone and other photochemical pollutants costs the state millions of dollars a year. Ozone also damages fabrics and the rubber in tires and windshield wiper blades.

■ Temperature Inversions and Smog

Certain meteorologic and geographic conditions favor the formation of both industrial and photochemical smog. Normally, air temperature in the troposphere decreases with increasing altitude (see Figure 2.1). Warm air at the Earth's surface expands, becomes less dense, and rises. As it does so, cooler air from above flows in to replace it. In turn, the cooler air is warmed and rises. Through this process, the air is continually renewed, and pollutants are dispersed by vertical currents and prevailing winds (Figure 4.10a).

A reversal of the usual temperature pattern, called a **temperature inversion**, sometimes occurs; after an initial decrease, the air temperature, instead of continuing to decrease with increasing altitude, begins to increase. A lid of warm air forms over cooler air near the Earth's surface (Figure 4.10b). The cooler, denser layer cannot rise through the warm lid of air above it and becomes trapped, sometimes for days. There is no vertical circulation, and pollutants accumulate.

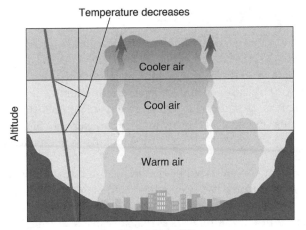

a. Normal conditions

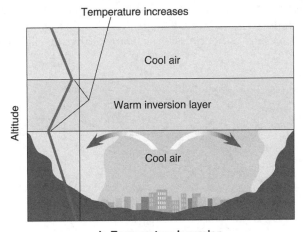

b. Temperature inversion

Figure 4.10 (a) Under normal atmospheric conditions, air warmed at Earth's surface rises and mixes with cooler air above. Any pollutants are dispersed upward. (b) When a temperature inversion occurs, a layer of warm air settles over cooler air, preventing it from rising. Any pollutants become trapped under the warm lid until atmospheric conditions change.

A particularly serious incident occurred in 1948 in the Donora Valley, an industrial area in Pennsylvania. As a result of a temperature inversion, industrial smog settled over the valley for five days. Many people died, and almost half of the population suffered from respiratory ailments. In 1952, a similar incident in London resulted in 4000 deaths. In that case, smoke from coal burned for heat in homes and workplaces was the main cause of the air pollution.

If a temperature inversion occurs in an area partly surrounded by mountains (e.g., Los Angeles, Denver, or Salt Lake City), photochemical smog buildup is particularly serious. Because of the encircling mountains, the pollutants cannot be dispersed horizontally. They remain in a blanket over the city until the weather changes and the wind disperses the polluted air.

Regulating Air Pollution

The Clean Air Act of 1970 mandated air quality standards for five air pollutants: suspended particles, SO_2, CO, NO_x, and ozone. A few years later, lead was added to the list. Until a phaseout began in 1975, lead was added to gasoline to prevent knocking, and large quantities of lead were released to the atmosphere in engine exhausts, causing serious problems in urban areas. Despite great improvements in air quality, including a dramatic drop in lead emissions (Figure 4.11), many urban areas were not meeting the desired standards by the late 1980s. Thus, in 1990, tougher standards were established in the Clean Air Act Amendments.

A significant problem not addressed by the original Act was the emission by industry of thousands of tons of unregulated hazardous chemicals, many of which are suspected of being carcinogens. The 1990 amendments require that industries emitting any of 189 specifically named toxic chemicals install control devices that are capable of reducing these emissions by at least 90%.

As noted earlier, the 1990 amendments include other important provisions: Coal-burning power plants must reduce annual SO_2 emissions to 10 million tons by 2000, and new cars will have to meet increasingly strict emission standards. Currently, light trucks (sport utility

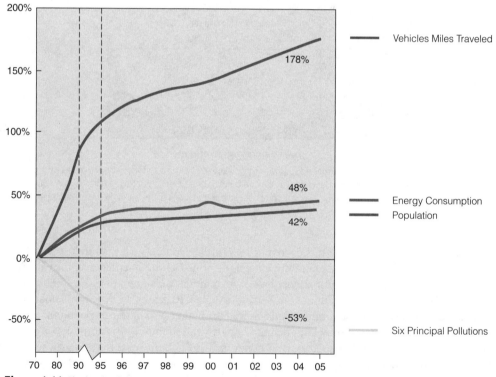

Figure 4.11 The levels of six principal pollutants have dropped in the United States despite increasing population, energy consumption, and increasing vehicle miles traveled. (Source: EPA.)

vehicles, minivans, and pickups) emit three times as much pollution as the average passenger car because they are not required to meet the same emissions standards. In May of 2007, President George W. Bush issued an executive order that would cut greenhouse gas emissions (GHGs) released by motor vehicles. The executive order was issued in response to a Supreme Court order to the EPA requiring the EPA to take action under the Clear Air Act to regulate GHGs from motor vehicles. The Bush order proposed a 20-in-10 regulation, which is a mandate to reduce gasoline consumption by 20% over the next ten years. The Bush proposal would set a mandatory fuel standard that requires 35 billion gallons of renewable (ethanol) and alternate fuels to be available by 2017, which would replace 15% of the projected annual gasoline use in the United States. In addition, he proposed to increase fuel efficiency by "reforming and modernizing" the Corporate Average Fuel Economy (CAFE) standards for cars and by extending the current Light Truck Rule. These rule changes were intended to bring another 5% reduction in gasoline consumption. Late in 2008, the White House announced the Bush administration would not implement the executive order by the end of the Bush presidency. This announcement caused the *Washington Post* to suggest that the acronym EPA should stand for the "Emitters Protection Agency."

Since 1992, cities with unacceptably high emissions of CO have been required to sell gasoline containing 2.7% oxygen during the winter months to reduce these emissions. The oxygen content can be increased by adding ethanol or methanol, but MTBE (methyl *t*-butyl ether) is preferred because of its higher octane rating. As a result of its increased oxygen content, this reformulated gasoline emits less CO when it is burned than does regular gasoline. However, MTBE—a known animal carcinogen—has been detected in groundwater in several states after gasoline spills and leaks from underground storage tanks. Both California and Maine began phasing out the use of MTBE, and in 1999, the EPA recommended a substantial reduction in its use because of the threat to human health.

The Clean Air Act and its amendments required the EPA to establish National Ambient Air Quality Standards for the major pollutants (Table 4.5). For each pollutant, these standards specified a concentration, averaged over a specified time period, which must not be exceeded. At the end of 1996, the EPA proposed even more stringent standards for ozone and particulates: The allowable level for ozone was lowered from 0.12 to 0.08 ppm, and the size (diameter) of particulates covered by the standards was reduced from 10 to 2.5 μm. Because of the costs involved and questions about the potential health benefits, the tougher standards are opposed by industry, state governors, and cities, and have not been adopted.

In 1998, the EPA announced further new rules, this time aimed at reducing the flow of smog-causing emissions from 22 industrial eastern states across state borders. Power plants in the targeted states would have to limit emissions of oxides of NO_x, which are precursors of both ozone and photochemical smog.

Air pollution is a worldwide problem that is far more serious in many other countries than it is in the United States. For example, in Mexico City, São Paulo (Brazil), and many industrial cities in China, Eastern Europe, and the former Soviet Union, people live under a pall of toxic smog. The sun is obscured. Severe respiratory ailments are common, and infant mortality rates are high; often there are few complaints because the smog and pollution mean jobs.

Cleaning the atmosphere will be costly, and economic priorities will need to be readjusted, particularly for the developing countries of the world. However, the alternative of continuing to pour pollutants into the air will have far more serious consequences.

Table 4.5

National Ambient Air Quality Standards

Pollutant	Time Period*	Limit Set in 1992	2009 Limit†
Carbon monoxide	8 hours	9 ppm	no change
	1 hour	35 ppm	no change
Lead	3 months	1.5 µg/m³	no change
Nitrogen oxides	1 year	0.05	no change
Ozone	1 hour	0.12 ppm	no change
	8 hours	none	0.08 ppm
Sulfur dioxide	1 year	0.03 ppm	no change
	24 hours	0.14 ppm	no change
Inhalable Particulates	1 year	50 µg/m³	no change
10 µm diameter or less (PM10)	24 hours	150 µg/m³	no change
Inhalable Particulates	1 year	none	15 µg/m³
2.5 µm diameter or less (PM10)	24 hours	none	35 µg/m³

* Period over which the concentrations are measured and averaged.
† New standards set in 2006.
Source: www.epa.gov.

■ Indoor Air Pollution

You might expect to be safer from air pollutants indoors, but in today's well-sealed homes and offices, this is often not the case. In buildings where there is little or no circulation of fresh air, pollutants may accumulate to dangerous levels. Figure 4.12 shows sources of the major indoor pollutants.

Smoking is a particularly dangerous cause of indoor air pollution. In addition to nicotine, **environmental tobacco smoke** contains high levels of all the primary pollutants (CO, NO_2) and particulates associated with combustion. Tobacco smoke contains many VOCs such as aldehydes (formaldehyde), ketones (acetone), HCs, and organic acids. Because the tobacco is tightly paced inside a paper wrapper, air cannot reach the burning tobacco easily, and the combustion process takes place at a lower temperature unless the smoker draws the air through the cigarette. Consequently, sidestream smoke (emitted when the cigarette is burning but not being drawn on) actually contains more products of incomplete combustion than does the smoke that is inhaled. For this reason, many localities have issued smoking bans in restaurants and public buildings to limit the exposure of their customers to second hand smoke. Of increasing concern is the risk of smoking to nonsmokers.

Because of the low temperature of combustion, cigarette smoke contains tar, which contains particulates consisting of large HC molecules and nicotine. Smoking also contributes

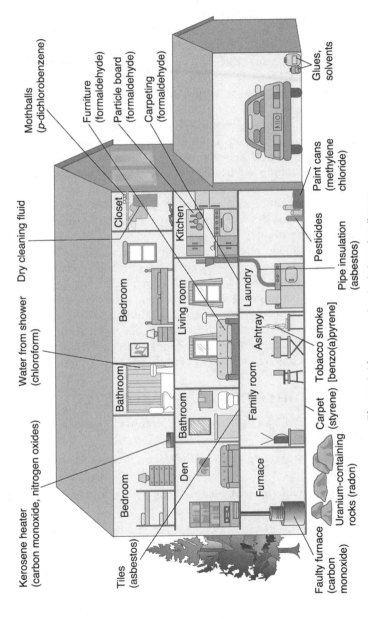

Kerosene heater
(carbon monoxide, nitrogen oxides)

Tiles
(asbestos)

Water from shower
(chloroform)

Dry cleaning fluid

Mothballs
(p-dichlorobenzene)

Furniture
(formaldehyde)

Particle board
(formaldehyde)

Carpeting
(formaldehyde)

Glues,
solvents

Paint cans
(methylene
chloride)

Pesticides

Pipe insulation
(asbestos)

Tobacco smoke
[benzo(a)pyrene]

Carpet
(styrene)

Uranium-containing
rocks (radon)

Faulty furnace
(carbon
monoxide)

Closet

Kitchen

Bedroom

Living room

Laundry

Bathroom

Ashtray

Family room

Bathroom

Den

Bedroom

Furnace

Figure 4.12 Sources of major indoor air pollutants.

to the amount of respirable particulate matter inside a room. Studies have established a relationship between the level of indoor $PM_{2.5}$ (particulate matter 2.5 μm or less in diameter from all sources) and respiratory illness. Young children of smokers, for example, are more likely to suffer from asthma and bronchitis than are children of nonsmokers. Smoking causes emphysema, lung cancer, and coronary heart disease.

Gas stoves, kerosene heaters, wood stoves, and faulty furnaces are potential sources of NO_x and CO. Polynuclear aromatic hydrocarbons (PAHs) are also released by the combustion of organic material (biomass). Because there are many individual PAH molecules, measuring the concentration of one PAH, benzo(a)pyrene, is often used to indicate the collective concentration of all PAHs in the air. Homes with wood stoves or open fireplaces often have elevated levels benzo(a)pyrene.

Paint, paint strippers and thinners, gasoline, and pesticides, which many people store in their basements, release harmful vapors and dust particles. Formaldehyde, a toxic and irritating gas, is released from the polymers that are used to manufacture certain types of insulation foam and furniture stuffing and from newly installed carpeting and paneling. Clothes brought home from the dry cleaners may also cause a problem, as traces of harmful volatile solvents used in the cleaning process are retained by the garments and later released into the atmosphere. Even taking a hot shower or bath may be harmful because chloroform can be released from chlorine-treated water.

Another particularly insidious indoor air pollutant is radon (discussed in Chapter 11). It is released through the foundation walls into basements from uranium-containing minerals in the ground. Asbestos (discussed in detail in Chapter 17) is an inhalation hazard that is still present in some older homes, where it was used as an insulating material for furnaces and pipes.

Apart from preventing pollutants from entering a building in the first place, one way to control indoor air pollution is to install air-to-air heat exchangers that circulate fresh air without adversely upsetting the temperature of the indoor air. Air conditioners, smoke removers, and vacuum cleaners all help to reduce indoor air pollutants.

■ Additional Sources of Information

Brasseur GP, Orlando JJ, Tyndall GS. *Atmospheric Chemistry and Global Change*. New York: Oxford University Press, 1999.

Environmental Protection Agency (EPA). Homepage. Accessed November 2008 from http://www.epa.gov.

Hemond HF, Fechner-Levy EJ. *Chemical Fate and Transport in the Environment*. London, UK: Academic Press, 1994.

Jacobson MZ. *Atmospheric Pollution*. Cambridge, UK: Cambridge University Press, 2002.

Wayne RP. *Chemistry of Atmospheres*, 3rd ed. New York: Oxford University Press, 2000.

■ Keywords

acid rain
air pollutant
carbon monoxide
catalytic converter

environmental tobacco smoke
flue-gas desulfurization (FGD)
fluidized bed combustion (FBC)
four-cycle internal combustion engine

free radicals
hemoglobin (Hb)
hydroxyl radical
industrial smog
nitrogen oxides
operation of the two-cycle engine
operation of the four-cycle internal
  combustion engine

ozone
peroxyacetyl nitrate (PAN)
photochemical smog
primary and secondary air pollutants
sulfur dioxide
temperature inversion
terpenes
volatile organic compounds (VOCs)

■ Questions and Problems

1. Define primary air pollutant. List the five primary air pollutants.
2. List the five substances that are responsible for more than 90% of air pollution in the United States. What is the major source for each of these pollutants?
3. What are the major air pollutants produced by the following industries?
   a. Trucking
   b. Electric power generation
4. Define a secondary air pollutant. List five secondary air pollutants.
5. What is a free radical? How does hydroxyl radical differ from hydroxide ion?
6. Write chemical equations to show how CO and CO_2 are formed during the burning of fossil fuels. Which gas (CO or CO_2) is produced in each of the following situations?
   a. Gasoline is burned in an automobile engine.
   b. Coal is burned in an open fireplace.
7. Assume that gasoline is octane, C_8H_{18}. Calculate the stoichiometric mass of air that is required to burn 1.0 kg of octane. Assume that air is 20% oxygen. Compute the optimum fuel/air ratio.
8. Do anthropogenic or natural sources release more CO into the troposphere?
   a. List the natural sources of CO.
   b. Describe natural processes that remove CO from the troposphere.
   c. List two reasons that CO concentrations are higher in cities than rural areas.
9. Why do almost all of the anthropogenic NO_x emissions come from combustion in internal combustion engines? Write reactions that show how NO_2 is formed from O_2 and N_2 in the engine.
10. Draw the Lewis electron dot structure for N_2O. Should it be written $N_2O^{\cdot}$?
11. Draw the Lewis electron dot structure for NO. Why is it written $NO^{\cdot}$?
12. Using chemical reactions, describe the fate of tropospheric NO_2.
13. Using schematics, describe the following:
    a. a four-cycle internal combustion engine
    b. a two-cycle internal combustion engine
14. Compare a two-cycle versus a four-cycle engine:
    a. revolutions per minute
    b. temperature of operation
    c. HC emissions
    d. Cost
15. What is the ideal air/fuel ratio for a four-cycle engine?
    a. What happens to HC emissions if the ratio is made more lean?

b. What happens to NO_x emissions if the ratio is made more rich?

c. What happens to fuel economy if the ratio is made more lean?

16. You are having trouble starting you car. Your auto mechanic suggests that he or she should adjust your engine so that it has a more fuel rich mixture, which will make it start easier. What effect will this have on your engine's emission of HC, NO_x, and CO?

17. Compare the emission of HC, NO_x, and CO coming from two- and four-cycle engines.

18. Why is the catalytic converter called a three-way catalytic converter?

19. Draw a diagram of a catalytic converter and describe how it reduces pollutants in automobile exhausts. Catalytic converters reduce automobile emissions by 95%. Why are there calls for devices with greater efficiency?

20. Write the following catalytic chemical reactions that occur in the catalytic converter:

a. The removal of CO from the exhaust stream

b. The removal of unburned octane from the exhaust stream

c. The removal of NO from the exhaust stream

21. List the catalysts that are used in the catalytic converter. Why are they coated on a ceramic honeycomb structure?

22. Why will the use of leaded gasoline in an automobile destroy the effectiveness of its catalytic converter?

23. By how much have catalytic converters reduced the amount of CO, NO_x, and HC in automobile exhaust since 1975?

24. What are the main anthropogenic sources of SO_2? Why does the burning of coal release SO_2? Write equations to show how SO_2 reacts in the atmosphere.

25. Calculate the pH of rainwater that is in equilibrium with air that has an SO_2 concentration of 2 ppm.

26. You have measured the pH of rainwater on your campus to be 4.2. Assuming that the acidity in the rainwater is cause by SO_2, calculate the concentration of SO_2 that must be in the air in the vicinity of your campus.

27. What is the pH of an aerosol that is saturated with air that has a $[CO_2]$ concentration of 250 ppm? The $K_H = 3.4 \times 10^{-2}$ mol/L atm at 25°C.

28. Calculate the pH of rainwater that is in equilibrium with air that has an SO_2 concentration of 5.

29. What adverse effects does SO_2 have on human health?

30. Describe two methods for removing SO_2 from smokestack emissions.

31. In your opinion, is the installation of very tall smokestacks a good way to deal with hazardous industrial emissions? Explain.

32. Describe how the flue gas desulfurization process removes SO_2 from chimney gases.

33. What is FBC?

34. List two uses for the $CaSO_4$ produced from the FGD process.

35. Describe how the SO_2 emissions changed during the 1980s

a. in the United States

b. in Europe

c. in developing countries

36. What two air pollutants are responsible for photochemical smog?

37. Using chemical reactions, show how tropospheric ozone is formed from the reaction of NO_2 and sunlight. Next show how the hydroxyl radical is formed.

38. How many hydroxyl radicals are produced for every HC reacted in the troposphere?
39. Write two reactions that show how the hydroxyl radical is removed from the troposphere.
40. If the concentration of OH· in air is 5×10^6 molecules/cm³, calculate its concentration in ppbv and pptv (parts per trillion).
41. Write the mechanism for the production of hydroxyl radical in unpolluted air.
42. Assume that there is unburned pentane (C_5H_{12}) in automobile exhaust. Write the reaction that would occur between pentane and hydroxyl radical. Is the reaction that you wrote an addition or an abstraction reaction?
43. Which of the following HCs would react faster with hydroxyl radical? Why?
    a. Butane
    b. Octane
44. Which of the following aromatics would react faster with hydroxyl radical?
    a. Benzene
    b. Xylene
45. Assume that there is unburned 2-pentene in automobile exhaust. Write the stepwise mechanism of the reaction that would occur between 2-pentene and hydroxyl radical. What reaction products are formed?
46. Does the reaction of 2-octene and a hydroxyl radical proceed at a faster rate than the reaction between octane and hydroxyl radical? If 2-octene does react faster, by how much is the rate increased and why?
47. Write the reaction that occurs between propanol and hydroxyl radical. What products are formed?
48. Draw the structure of PAN.
49. Describe the following about PAN:
    a. Lifetime in the troposphere
    b. Breakdown products in warmer air
    c. Why it is a reservoir for NO_x?
50. What chemical is responsible for giving smog an orange brown color?
51. Use a diagram to show how a temperature inversion occurs. What geographic features promote the formation of an inversion?
52. In what year was the Clean Air Act first enacted? It mandated air pollution standards for what five air pollutants?
53. The 1990 amendments to the Clean Air Act tightened emission standards for what air pollutants?
54. What are the National Ambient Air Quality Standards?
55. Since 1992, what new EPA regulations have been mandated for cities with high levels of tropospheric CO.
56. List three chemicals that can be added to gasoline to increase its oxygen content.
57. Draw the structure of MTBE. Why is it no longer being used in gasoline in California?
58. List four classes of chemicals that are components of cigarette smoke.
59. Describe the risk to nonsmokers who inhale secondhand cigarette smoke.
60. How is the PAH concentration in indoor air measured?
61. Formaldehyde is known to irritate the eyes and cause bronchial problems. List two sources of formaldehyde in indoor air.

CHAPTER

5

Chemistry of the Stratosphere

THE OZONE LAYER THAT LIES AT THE TOP OF THE EARTH'S STRATO-
SPHERE BEGAN TO FORM AFTER BLUE-GREEN ALGAE BEGAN PRODUC-
ING OXYGEN THROUGH PHOTOSYNTHESIS APPROXIMATELY 2.3 BILLION
YEARS AGO. It was not until green plants evolved, approximately
400 million years ago, that the ozone layer fully developed.
Today, the ozone layer is considered the Earth's natural sun-
screen because it filters harmful ultraviolet (UV) radiation
before it can reach the surface of the Earth. A substantial reduction in the amount of ozone in the ozone layer could threaten
all life on Earth.

In this chapter, we first examine the relationship between
energy and the wavelength of sunlight and the common names
for regions of the UV spectrum. We then study the reactions
that produce a steady-state concentration of ozone in the
stratosphere and those that cause its catalytic destruction.
Next, the importance of **polar stratospheric clouds (PSCs)** in
the destruction of ozone over Antarctica is explained. Finally,
the Montreal Protocol and efforts to reduce the emission of
ozone-depleting chemicals are discussed.

■ Dobson Unit

The **Dobson unit (DU)** is used to describe the amount of ozone in the
stratosphere. It is named after Gordon M. B. Dobson, the Oxford University
researcher who built the first instrument to measure the total ozone abundance from the ground. One DU is 2.7×10^{16} ozone molecules per square
centimeter. One DU refers to a layer of ozone that would be 0.001 cm thick
under conditions of standard temperature (0°C) and pressure (the average
pressure at the surface of the Earth). For example, 300 DUs of ozone
brought to the surface of the Earth at 0°C would occupy a layer only 0.3 cm

thick in a 1-cm square column. In the past, the average amount of ozone covering the Earth was more than 270 DU. In 2000, for instance, the average abundance of ozone from 90°N to 90°S was 293.4 DUs. The amount of ozone over the Antarctic, however, has been much lower than other parts of the Earth and the **ozone hole** above the Antarctic is defined as the area in which the ozone is less than 220 DU.

The global average ozone concentration varies from month to month and region to region. Figure 5.1 shows the variation of the ozone concentration as a function of latitude and time of year. Many natural factors affect the fluctuation of the ozone concentration. Anthropogenic events greatly affect the concentration of the ozone in specific locations, such as at the polar ice caps. These issues are discussed in more detail in this chapter.

■ The Production of Ozone in the Stratosphere

In the absence of pollution, ozone (O_3) is not present to any appreciable extent in the troposphere, but it occurs naturally in the stratosphere. As can be seen in Figure 5.2, its concentration is greatest at an altitude of 20 to 30 km (12 to 19 miles) from the Earth's surface. The **ozone layer** is formed when ordinary molecules of oxygen gas (O_2) in the stratosphere absorb UV radiation from the sun with wavelengths of less than 240 nm, which causes them to dissociate into single oxygen atoms (O).

$$O_2 + h\nu\,(\lambda < 240\,nm) \rightarrow O + O\ slow \qquad \Delta H = 498\,kJ/mol$$

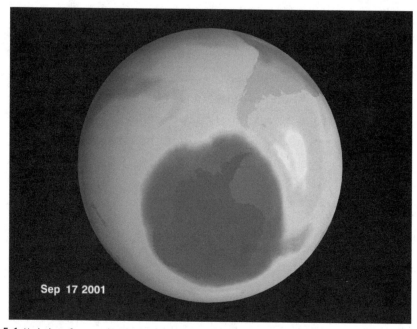

Sep 17 2001

Figure 5.1 Variation of ozone abundance by latitude and month during 2000. Data provided by the Total Ozone Mapping Spectrometer (TOMS) which is located on a NASA satellite.

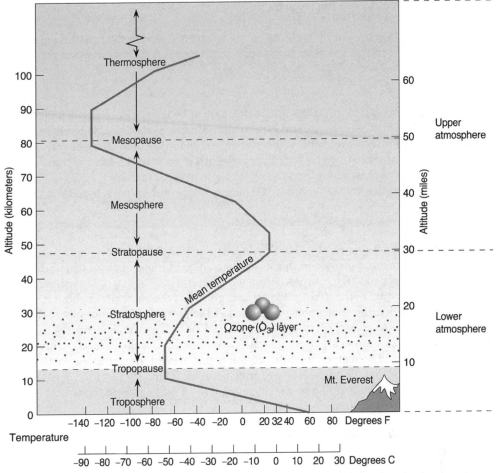

Figure 5.2 The Earth's atmosphere is subdivided vertically into four major regions based on the air-temperature profile. The ozone layer that protects us from the sun's ultraviolet radiation is in the stratosphere.

The dissociation of molecular oxygen into monatomic atoms requires an enthalpy change of 495 kJ/mol; this is the amount of energy that must be supplied (at stratospheric temperature and pressure) to make the reaction go. To determine the wavelength of sunlight that will have sufficient energy to dissociate oxygen use Planck's equation. The energy (E) of the light is related to its frequency v and the wavelength λ by this formula:

$$E = hv \quad \text{and} \quad c = \lambda v \quad \text{so} \quad v = c/\lambda$$

Planck's constant (h) is 6.63×10^{-34} Js, and the speed of light (c) is 3.0×10^{8} m/s.

$$E = \frac{hc}{\lambda} \quad \text{rearranging} \quad \lambda = \frac{hc}{E} \quad \text{for 1 mole of } O_2, \text{ multiply by Avogadro's number, N}$$

and then converting the λ to nm (nanometers)

$$\lambda = \frac{hcN}{E} = \frac{\left(6.63 \times 10^{-34}\ Js\right)\left(2.99 \times 10^8\ m/s\right)\left(10^9\ nm/m\right)\left(6.02 \times 10^{23}/mol\right)}{498,000\ J/mol}$$

$$\lambda = 240\ nm$$

This calculation shows that sunlight with a wavelength of 240 nm or less has sufficient energy to dissociate O_2. This region of the sun's spectrum is known as the Far-UV. Remember that sunlight with a shorter wavelength is more energetic than that with a longer wavelength. Atmospheric scientists have given specific wavelength regions in the sun's spectrum common names. A list of those names and the wavelengths is listed in Table 5.1.

Single oxygen atoms are very reactive and immediately combine with O_2 to form O_3:

$$O + O_2 + M \rightarrow O_3 + M + heat \qquad fast$$

In this reaction, M designates a third molecule that is present in the atmosphere, usually an N_2, the most abundant molecule in the atmosphere. The M molecule is necessary to carry away the heat generated in the collision between O and O_2. The heat that this reaction releases causes the temperature of the stratosphere to be higher than the air below or above it. As can be seen in Figure 5.2, the stratosphere is defined as the portion of the atmosphere that lies between two lower temperature regions. Because cold air is denser than warmer, it does not rise to mix with the warmer air above it, and vertical mixing is a slow process. This causes the air to be stratified by temperature; this region is thus named the stratosphere.

At the same time that the ozone is being formed, it is absorbing UV radiation very strongly with wavelengths of 220 to 330 nm, which causes it to be broken down to form an oxygen molecule and an oxygen atom:

$$O_3 + hv\left(\lambda\ 240 - 320\ nm\right) \rightarrow O_2 + O$$

Also, when O_3 encounters an O atom, the two can combine to form two O_2 molecules.

$$O_3 + O \rightarrow 2\ O_2$$

Table 5.1			
Common Names for Regions of the UV Spectrum			
λ (nm)	Name	Species Absorbing	Location
10–240	Far-UV	O_2, N_2	Thermosphere, mesosphere
250–290	UV-C	O_3	Stratosphere
290–320	UV-B	O_3	Stratosphere, troposphere
320–380	UV-A	NO_2	Polluted troposphere
400–750	Visible	Many	Earth's surface

A dynamic equilibrium is established, maintaining a fairly constant concentration of ozone in the stratosphere. The concentration varies with season and latitude but averages 10 parts per million. Although it is low, this concentration of ozone is sufficient to screen 95% to 99% of the sun's dangerous UV radiation.

Determining the Steady-State Concentration of Ozone

The steady-state concentration of ozone is calculated from the rates of the reactions discussed previously here. In 1930, English physicist Sidney Chapman suggested that ozone in the stratosphere must be produced from UV irradiation of molecular oxygen. The following reactions comprise the **Chapman cycle**, a series of reactions that simulate stratospheric ozone chemistry. The reaction rate for each individual reaction is found by multiplying the reactant concentration by the rate constants, K_x. The rates for the four reactions discussed previously can be written as follows:

$$O_2 + hv\left(\lambda < 240\,nm\right) \rightarrow O + O$$

The rate of reaction $1 = K_1\,[O_2]$

$$O + O_2 + M \rightarrow O_3 + M$$

The rate of reaction $2 = K_2\,[O][O_2][M]$

$$O_3 + hv\left(\lambda\ 240\ to\ 320\ nm\right) \rightarrow O_2 + O$$

The rate of reaction $3 = K_3[O_3]$

$$O_3 + O \rightarrow 2O_2$$

The rate of reaction $4 = K_4[O][O_3]$

Ozone is produced in only the second reaction and is destroyed in reactions 3 and 4. The ozone steady state can be expressed as follows:

Rate of reaction 2 = rate of reaction 3 + rate of reaction 4

$$K_2\left[O\right]\left[O_2\right]\left[M\right] = K_3\left[O_3\right] + K_4\left[O\right]\left[O_3\right] \tag{5.1}$$

To determine the steady state, we need to assess which of the concentrations we know and which we need to compute. Laboratory experiments have determined the concentration of oxygen $[O_2]$, air $[M]$, and the rate constants for these equations. What is not known is the concentration of oxygen radical $[O]$ and ozone $[O_3]$. We assume that the concentration of oxygen radicals $[O]$ is also at steady state. The rate of O formation is twice the rate of the reaction of reaction 1 plus the rate of reaction 3; the rate of reduction is the rate of reaction 2 plus the rate of reaction 4.

$$2\,(\text{rate of } 1) + \text{rate } 3 = \text{rate of } 2 + \text{rate of } 4$$

substituting

$$2\,K_1\!\left[O_2\right] + K_3\!\left[O_3\right] = K_2\!\left[O\right]\!\left[O_2\right]\!\left[M\right] + K_4\!\left[O\right]\!\left[O_3\right] \tag{5.2}$$

subtraction of Equation 5.1 from Equation 5.2 gives

$$2\,K_2\!\left[O\right]\!\left[O_2\right]\!\left[M\right] = 2\,K_1\!\left[O_2\right] + 2\,K_3\!\left[O_3\right] \tag{5.3}$$

This equation can be simplified further. The highest energy region of solar radiation is absorbed by oxygen atoms at the upper regions of the stratosphere. The influx of solar radiation with λ of less than 240 is greatly reduced as it proceeds from the upper to the lower portion of the stratosphere. This means that there are too few photons with sufficient energy to induce reaction 1. A lower energy solar radiation ($\lambda = 240$–330 nm) is not absorbed and is available to carry our reaction 3. This allows us to assume that $K_1[O_2] <<<< K_3[O_3]$. Thus, we can simplify Equation 5.2 to

$$2\,K_2\!\left[O\right]\!\left[O_2\right]\!\left[M\right] = 2\,K_3\!\left[O_3\right] \tag{5.4}$$

rearranging

$$\left[O\right] = \frac{K_3\!\left[O_3\right]}{K_2\!\left[M\right]\!\left[O_2\right]} \tag{5.5}$$

Addition of Equations 5.1 and 5.2 gives

$$2\,K_1\!\left[O_2\right] = 2\,K_4\!\left[O\right]\!\left[O_3\right] \tag{5.6}$$

substitute Equations 5.5 into 5.6

$$2\,K_1\!\left[O_2\right] = \frac{2\,K_4 K_3\!\left[O_3\right]^2}{K_2\!\left[M\right]\!\left[O_2\right]}$$

$$\frac{\left[O_3\right]}{\left[O_2\right]} = \left(\frac{K_1 K_2\!\left[M\right]}{K_3 K_4}\right)^{1/2} \tag{5.7}$$

The numerical value obtained for this ratio depends on the altitude. As the altitude increases, the concentration of air [M] decreases, and the solar flux and temperature increase. This causes K_1 and K_3 to increase and K_2 and K_4 to decrease. At an altitude of 40 km, the concentration of air is approximately 10^{16} molecules/cm^3 and $K_1 = 10^{-11}$, $K_2 = 10^{-32}$, $K_3 = 10^{-3}$, and $K_4 = 10^{-15}$. These values when substituted into Equation 5.7 give an $[O_3]/[O_2] = 10^{-4}$.

Even the ozone layer has many more oxygen than ozone molecules.

This theoretical ratio is approximately a factor of 10 higher than what is experimentally measured (approximately 10 ppm). This leads to the conclusion that there must be some other reactions that are causing ozone destruction that we have not already considered.

■ Catalytic Destruction of Ozone

Ozone is a very reactive molecule and is termed a **meta-stable molecule**, meaning that isolated molecules will decompose slowly, but when an ozone molecule comes into contact with a molecule of another gas, it will react quickly. The equilibrium concentration of ozone in the stratosphere is low, but there are very few other molecules present in the stratosphere for it to react with.

In the 1960s, chemists realized that there were additional pathways for ozone destruction in the stratosphere that were not considered before. They realized there were a number of species (which are described in the following sections) that reacted efficiently by abstracting an oxygen atom from the ozone molecule. Generally, the letter X designates the reactive species. The abstraction reaction is as follows:

$$X + O_3 \rightarrow XO + O_2$$

If this reaction takes place in a region of the stratosphere that has atomic oxygen present, the XO molecule will react with it:

$$XO + O \rightarrow X + O_2$$

These reactions are additional pathways for ozone destruction that had not been considered before. The steady-state concentration of ozone will be lowered by these oxygen-abstraction reactions.

The sum of these two reactions is

$$O_3 + O \rightarrow 2O_2 \quad \text{Decomposition reaction}$$

Notice that this overall reaction does not contain the X species. Because X is not consumed in the reaction, it acts as a catalyst for the destruction of ozone. X lowers the energy of activation (Ea) for the reaction and increases the efficiency of the decomposition reaction.

The reaction requires atomic oxygen to regenerate X through reaction with the XO intermediate:

$$XO + O \rightarrow X + O_2$$

In the lower stratosphere (15 to 25 km), where most ozone is found, little UVC radiation (100–290 nm wavelength) from the sun penetrates. The concentration of molecular oxygen is high in this region. Because of this, atomic oxygen is rapidly converted into ozone. This keeps the concentration of atomic oxygen very low. This regeneration reaction will be very slow in the lower stratosphere because of the scarcity of atomic oxygen. Most ozone destruction by catalytic species (X) occurs in the upper and middle stratosphere (25 to 50 km). Four species have been identified as important in the global destruction of ozone: the catalytic species (X) generically describes either the hydroxyl radical, nitric oxide (NO), chlorine, or bromine atoms. Each is described in the next sections.

Hydroxyl Radical Cycle

Hydroxyl radical (˙OH) can be produced by two different photochemical processes: The first is a hydrogen abstraction reaction with either water or methane.

$$O + H_2O \rightarrow 2\,˙OH$$

$$O + CH_4 \rightarrow ˙OH + ˙CH_3$$

The second is the photolysis of water.

$$H_2O + hv \rightarrow ˙H + ˙OH$$

The hydroxyl radical is responsible for nearly one-half of the total ozone destruction in the lower stratosphere. The ozone destruction reaction is shown here. The hydroxyl radical is able to accept oxygen from ozone:

$$˙OH + O_3 \rightarrow ˙OOH + O_2$$

$$˙OOH + O \rightarrow ˙OH + O_2$$

Net reaction: $O + O_3 \rightarrow 2\,O_2$

The **hydroxyl radical** participates in the **catalytic ozone destruction cycle**, which is important in the lower stratosphere. The chain reaction shown here can have as many as 40 cycles (one hydroxyl radical will catalyze the destruction of 40 ozone molecules).

Nitric Oxide Cycle

Automobile and truck engines release large quantities of nitric oxide (NO) into the troposphere. Almost all of the NO is oxidized to NO_2 and then converted to nitric acid (HNO_3). Rainfall washes the nitric acid from the troposphere before it can reach the stratosphere. On the other hand, nitrous oxide (N_2O) is much less reactive than NO and eventually reaches the stratosphere. N_2O is also a tailpipe pollutant, but is also released by soil during denitrification by anaerobic bacteria.

Above 30 km, the N_2O can absorb high-energy photons to produce molecular nitrogen and an excited oxygen atom (O*). The excited oxygen atom has an electron that is in an excited state.

$$N_2O + hv \rightarrow N_2 + O*$$

Below 30 km, in the stratosphere, the excited-state oxygen reacts with the N_2O to produce NO.

$$N_2O + O* \rightarrow 2\,NO˙$$

NO can act as an X catalytic species.

$$NO^{\cdot} + O_3 \rightarrow {}^{\cdot}NO_2 + O_2$$

$$NO_2^{\cdot} + O \rightarrow {}^{\cdot}NO + O_2$$

Net reaction: $O + O_3 \rightarrow 2O_2$

This sequence of reactions is called the **NO$_x$ catalytic ozone destruction cycle**. The chain reaction shown previously here can have as many as 10^5 cycles (1 NO$_x$ radical will catalyze the destruction of 10,000 ozone molecules). The term NO_x is used to refer collectively to all nitrogen oxides (NO, N$_2$, NO$_2$) that may be present.

In the 1960s, there was considerable interest in building a supersonic transport airplane that would be capable of flying in the stratosphere at altitudes of 17 to 20 km. Because of the thin atmosphere in the stratosphere, supersonic planes would be capable of flying much faster than conventional airplanes. Supersonic aircraft would burn 20,000 kg/hr of jet fuel and consequently emit 25,000 kg of water vapor, thereby potentially forming ${}^{\cdot}$OH radicals and 160 kg of NO$_x$ during each hour of flight. Fears that these aircraft would further deplete the ozone layer led the U.S. government to scrap the development of the supersonic transport.

The Chlorine Cycle

Few natural sources of chlorine and bromine atoms exist in the stratosphere. Although the expansive oceans of Earth all contain significant quantities of chloride and bromide ions, sea spray is cleared from the troposphere by natural mechanisms. There are also many natural organochlorine and organobromine compounds released; hydroxyl radicals clear most from the troposphere before they can reach the stratosphere. Only methyl bromide (CH$_3$Br) and methyl chloride are stable and long-lived enough to reach the stratosphere and contribute to halogen-induced ozone destruction.

The major sources of stratospheric chlorine and bromine have been the anthropogenic **chlorofluorocarbons (CFCs)** and bromine containing halons. CFCs have been used as refrigerants, propellants for aerosol sprays, and solvents for cleaning electric circuits. Halons are used in commercial fire extinguisher systems; the heavy bromine-containing molecules provide a blanket of gas that covers a fire and keeps oxygen from reaching the flame. CFCs and halons are nontoxic and nonflammable, which make them superior to other gases that were used for the same application.

Thomas Midgley, who was trying to find a replacement for the existing refrigerant gas, ammonia, invented CFCs in 1928. Ammonia, a flammable and toxic gas, would occasionally leak from 1930s vintage refrigerators, causing an unpleasant experience for the owners of the refrigerator. CFCs were shown to be superior coolants for refrigerators and air conditioners (including those in automobiles) and as blowing agents in the production of polymer foam insulation. They are unsurpassed as solvents for cleaning electronic microcircuits and were used for many years as propellants for aerosol spray cans. In all of these uses, except refrigeration, CFCs are released directly into the atmosphere. Midgley is the same chemist that invented tetraethyl lead, the octane-increasing additive for gasoline. Leaded gasoline is partly responsible for global lead pollution. Some suggest that Midgley's inventions led to the two greatest environmental disasters of the 20th century.

Commercially, the most important CFCs are the halogenated methanes Freon 11 and Freon 12 (Freon is a DuPont trade name). They contain only carbon, chlorine, and fluorine. A new group of fluorinated molecules developed in 1995 also contain hydrogen (besides carbon, chlorine, and fluorine); these molecules are designated **hydrochlorofluorocarbons (HCFCs)**. To determine the chemical structure of a CFC (or Freon), add 90 to the CFC number. The resulting three digits correspond to the number of carbon, hydrogen, and fluorine atoms present in the molecule, respectively.

$$
\begin{array}{cc}
\quad\ \ \text{Cl} & \quad\ \ \text{Cl} \\
\quad\ \ | & \quad\ \ | \\
\text{Cl}-\text{C}-\text{F} & \text{Cl}-\text{C}-\text{F} \\
\quad\ \ | & \quad\ \ | \\
\quad\ \ \text{Cl} & \quad\ \ \text{F}
\end{array}
$$

CFC-11   CFC-12

EXAMPLE 5.1

1. Deduce the formula of CFC-113.
   a. Add 90 to 113 = 203
   b. The first digit represents the number of carbons. There are two carbons.
   c. The second digit represents the number of hydrogens. There are no hydrogens.
   d. The third digit represents the number of fluorines. There are three fluorines.
   e. For saturated carbon chains, the total number of substituents is 2n + 2, where n is the number of carbons. Thus, a two-carbon chain will have six attached substituents. Because there are no hydrogens and only three fluorines, three chlorines must be attached.
   f. The formula must be $C_2F_3Cl_3$.

2. What is the Freon number of the CFC with the chemical formula CF_2ClCF_3?
   a. Count the number of carbons. There are two carbons.
   b. Count the number of hydrogens. There are no hydrogens.
   c. Count the number of fluorines. There are five fluorines.
   d. Assemble the digits 205.
   e. Subtract 90 from 205. The remainder is 115.
   f. Name the Freon. It is CFC-115.

3. What is the formula of HCFC-141? The answer is $C_2H_3FCl_2$.

Because CFCs are so unreactive, they do not break down when released and can persist in the troposphere for more than 100 years. Over time, air currents carry them into the stratosphere.

As early as 1974, two chemists at the University of California at Irvine, F. Sherwood Rowland and Mario Molina, predicted that when exposed to UV radiation in the stratosphere, CFCs would break down to form chlorine radicals ($Cl^{\bullet}$):

Step 1. $CF_2Cl_2 + h\nu \rightarrow CF_2Cl + Cl^{\bullet}$

Each chlorine radical would then destroy a molecule of ozone with the formation of a chlorine monoxide radical ($ClO^{\bullet}$) and a molecule of oxygen (Step 2). The chlorine monoxide radical in turn would react with an oxygen atom to yield another molecule of oxygen and another chlorine radical, which could then start the chain reaction over again (Step 3). The net result would be the destruction of a molecule of ozone.

UV

Step 2. $Cl^{\bullet} + O_3 \rightarrow ClO^{\bullet} + O_2$

Step 3. $ClO^{\bullet} + O \rightarrow Cl^{\bullet} + O_2$

Net result: $O_3 + O \rightarrow 2\ O_2$

It has been estimated that each chlorine radical involved in a chain reaction has the potential to destroy 100,000 molecules of ozone before winds carry it back to the troposphere.

There are more industrial chlorine-containing compounds in use than those that contain bromine. Bromine-containing compounds, on the other hand, are more efficient at destroying ozone. The primary source of bromine in the stratosphere is CH_3Br, which farmers use to fumigate crops.

Null Cycles

Other processes occur in the stratosphere that compete with the catalytic cycles. These reactions further complicate our ability to predict the extent of ozone depletion under certain stratospheric conditions. **Null cycles** (also known as **holding cycles**) are processes that prevent certain species from taking part in the catalytic cycles. One example of a null cycle begins with the reaction of NO_2 with O_3 that results in the production of NO_3:

$$NO_2 + O_3 \rightarrow NO_3 + O_2$$
$$NO_3 + h\nu \rightarrow NO_2 + O$$

Net reaction: $O_3 + h\nu \rightarrow O_2 + O$

Some of the NO_3 can react to produce N_2O_5:

$$NO_3 + NO_2 + M \rightleftharpoons N_2O_5 + M$$

As has been seen before, the M in the equation is a third molecule that participates in the reaction but is not incorporated into the product. M can be any molecule that happens to be present. In the atmosphere, it is most likely to be a molecule of nitrogen. The N_2O_5 is a relatively unreactive molecule and is not a catalyst for ozone destruction. It can be considered an unreactive reservoir of NO_x. At any given time it accounts for up to 10% of the total stratospheric NO_x. The formation of N_2O_5 can be considered a holding cycle for NO_x because it temporarily limits its ability to participate in the catalytic cycles.

Nitric and hydrochloric acid are formed in the stratosphere and are important reservoirs for ozone-depleting NO_x and chlorine species.

$$\cdot NO_2 \ + \ \cdot OH \ + \ M \ \rightarrow \ HNO_3 \ + \ M$$

$$\cdot Cl \ + \ CH_4 \ \rightarrow \ HCl \ + \ \cdot CH_3$$

Almost 70% of the stratospheric chlorine is present as hydrochloric acid, and 50% of NO_x is stored as nitric acid. The reservoirs that the null cycles produced are important in the chemistry of Antarctic ozone depletion.

■ Depletion of the Protective Ozone Layer in the Stratosphere

Since 1985, satellite images have revealed a "hole" in the ozone layer each spring over the South Pole (remember that spring in the Southern Hemisphere is autumn in the Northern Hemisphere) (Figure 5.3). The largest **ozone hole** ever recorded (29.5-million square kilometers) was recorded in September of 2006. The 2000 Antarctic ozone hole, which was recorded in September of 2000, is the second largest ozone hole area on record (29.4 million square kilometers, an area that is greater than all of North America). The 2007 ozone hole covered a much smaller total area (25 million square kilometers) and was seventh largest in the thirty years satellites have been monitoring ozone. There is convincing evidence that CFCs, a class of synthetic organic compounds, are responsible for this destruction of the ozone layer.

An ozone hole occurs as a result of the harsh winter weather conditions in the lower stratosphere. During the long, dark winter (June to September) in the Antarctic in the intense cold ($-80°C$), a stream of air is drawn toward the South Pole because of the rotation of the Earth. A giant vortex with speeds exceeding 300 km/hr is created, and the area within the vortex acts like a chemical reaction chamber. The outside air cannot penetrate the vortex, and at the South Pole, the vortex continues through the winter into spring (October). Condensation of gases into liquid droplets (clouds) or solid crystals does not usually occur in the stratosphere over Antarctica because the concentration of water vapor in the air is so low. Inside the vortex, however, stratospheric clouds form as a result of the exceptionally low temperatures and the absence of sunlight.

The crystals produced inside the vortex form PSCs. In comparison to tropospheric clouds, PSCs have few particles per unit volume of air and are optically thin. Two major types of PSCs exist. When polar temperatures drop to below 195°K ($-78°C$), nitric acid and water vapor grow on small sulfuric acid–water aerosol particles. Initially, it was thought that nitric acid and water

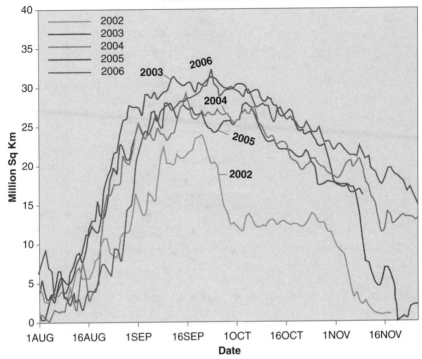

Figure 5.3 The ozone hole over the South pole was the largest ever recorded in September 2006. Courtesy of NASA.

molecules deposited in the solid phase in a ratio of 1:3. These ice crystals were found to have the composition $HNO_3 \cdot 3\ H_2O$ and are called nitric acid trihydrate crystals. More recently, it was found that these particles, which average 1-μm in diameter, also contain a variety of acids, including nitric acid dehydrate, sulfuric acid, and nitric acid. These nitrate-containing clouds are called **type I PSCs**. When the temperatures drop below the frost point of water, 187°K (−86°C), a second type of cloud forms. These clouds contain pure ice crystals, which are 20-μm in diameter and are **type II PSCs**. Usually, approximately 90% of PSCs are type I, and 10% are type II.

Once the PSC particles form in the vortex, chemical reactions begin to take place on their surfaces. Figure 5.4 shows how the vortex acts as a reaction chamber for these reactions, which are **heterogeneous** and occur only after the reacting gas has adsorbed on the surface of a particle. **Adsorption** is a process that proceeds through a collision between the gas and the surface in which the gas bonds to the surface.

Gas-phase HCl attaches itself to the surface of the ice crystals of both type I and type II PSCs, as is shown in Figure 5.5. When pollutants collide with the adsorbed HCl, the heterogeneous reaction takes place, producing Cl_2, chlorine gas.

$$ClONO_2(g) + HCl(s) \rightarrow Cl_2(g) + HNO_3(s)$$
$$HOCl(g) + HCl(s) \rightarrow Cl_2(g) + H_2O(s)$$

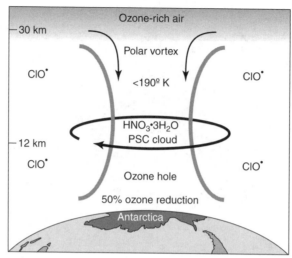

Figure 5.4 The polar vortex containing polar stratospheric clouds (PSC) that destroy ozone and create the ozone hole.

In the polar winter, chlorine-containing molecules accumulate; in the spring, when sunlight returns to the area, relatively inactive forms of chlorine in the stratosphere, such as $ClONO_2$ and HCl, are converted to photochemically active forms, such as Cl_2. The reactions mentioned here are known as **chlorine activation**. The Cl_2 produced can go on to attack ozone.

$$Cl_2 + h\upsilon \rightarrow 2\,^{\bullet}Cl$$

$$2\,^{\bullet}Cl + 2\,O_3 \rightarrow 2\,ClO^{\bullet} + 2\,O_2$$

$$ClO^{\bullet} + ClO^{\bullet} \rightarrow ClOOCl$$

$$ClOOCl + h\upsilon \rightarrow ClOO^{\bullet} + \,^{\bullet}Cl$$

$$ClOO^{\bullet} \rightarrow \,^{\bullet}Cl + O_2$$

Net reaction: $2\,O_3 + h\upsilon \rightarrow 3\,O_2$

In the **chlorine cycle** that we studied earlier, $ClO^{\bullet}$ attacked an oxygen atom to produce a molecule of O_2. In the lower stratosphere, where the polar ozone levels are the lowest, not enough oxygen atoms exist to sustain the reaction; the $ClO^{\bullet}$ concentration builds up (parts per billion level) until ClOOCl dimers are formed. As can be seen in these reactions, the ClOOCl dimers are then photodissociated, completing the cycle.

Each November, the Antarctic warms enough that the polar vortex breaks down and the PSC solids melt. Ozone from adjacent regions to the Antarctic diffuses over the area

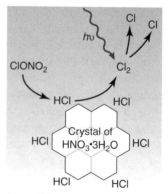

Figure 5.5 Type I PCS particle with HCl adsorbed on the surface catalyzes the reaction that produces an active form of chlorine ($^{\bullet}$Cl) from an inactive form ($ClONO_2$).

where the vortex was and re-establishes the ozone layer. The ozone hole is an annual, regional event that is controlled by the temperature of the polar stratosphere and the presence of chlorine and bromine.

Convincing field evidence for the involvement of CFCs in ozone depletion came in 1987. Between August and September of that year, a NASA research plane, equipped with sophisticated analytical instruments, flew 25 missions in the region of the ozone hole in the stratosphere over Antarctica. Data collected showed conclusively that as the ozone concentration decreased, the concentration of the chlorine monoxide radical (ClO˙) rose (Figure 5.6). The chlorine monoxide radical has been dubbed the "smoking gun" of ozone depletion. This chain reaction is now thought to account for approximately 80% of the ozone loss in the stratosphere.

A number of related bromine-containing compounds called **halons** also have been implicated in ozone destruction. As mentioned earlier, halons, which include CF_2ClBr and CF_3Br, are very effective as fire-extinguishing agents. Like CFCs, they break down in the stratosphere but form bromine radicals (Br˙), which initiate the same type of chain reaction as shown for chlorine radicals (see Steps 1–3 in the chlorine cycle).

The importance of the work of Rowland and Molina in elucidating the role of CFCs in ozone destruction was recognized in 1995. Along with Paul Crutzen, a Dutch scientist who had demonstrated the influence of nitrogen oxides in maintaining normal stratospheric ozone concentration, they were awarded that year's Nobel Prize in chemistry.

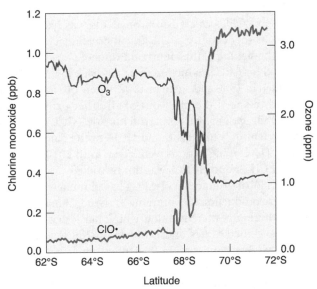

Figure 5.6 In September 1987, instruments aboard a NASA research plane simultaneously measured ozone and chlorine monoxide concentrations as the plane flew southward from Chile toward Antarctica. As the plane entered the ozone hole, chlorine monoxide concentration increased rapidly to about 500 times its normal atmospheric level, while ozone concentration fell dramatically. From D. H. Meadows, D. L. Meadows, and J. Randers. Beyond the Limits (White River Junction, VT: Chelsea Green Publishing, 1992) p. 152. Used by permission of Chelsea Green Co.

Effects of Ozone Depletion on Human Health and the Environment

Depletion of the ozone layer leaves the Earth vulnerable to the damaging effects of UV radiation. **UVB radiation** (wavelength of 280 to 320 nm), because it transmits more energy, is more damaging to living organisms than longer wavelength UVA radiation (320 to 400 nm). In humans, exposure to UVB radiation causes cataracts, tanning and sunburn, and suppression of the immune system. By damaging DNA, UVB radiation can cause several types of skin cancer, including deadly melanoma. It also harms crops and kills phytoplankton, the microscopic photosynthetic organisms that are at the lowest level of ocean food chains. It has been calculated that for every 1% decrease in stratospheric ozone there is a 2% increase in the amount of UVB radiation reaching the Earth's surface.

Ozone Loss Over the Arctic and the Middle to High Latitudes

A springtime depletion of ozone over the Arctic was first observed in 1995; however, because North Pole is not as cold as the South Pole and because the northern polar vortex is weaker, ozone loss over the Arctic has not been as great as over the Antarctic. But because three-quarters of the world's population lives in the Northern Hemisphere, continued depletion of ozone over the Arctic could have even more serious consequences than depletion over the Antarctic. In the past, some ozone depletion has occurred over the middle to high latitudes in both hemispheres, but in 2006, atmospheric scientists reported that ozone levels have stabilized in the middle to high latitudes due to the Montreal Protocol.

The Montreal Protocol

As early as 1978, the use of CFCs as aerosol propellants was banned in North America, although it was not in most other countries. The first international effort to protect the ozone layer came in 1987 with the signing of the **Montreal Protocol** on Substances that Deplete the Ozone Layer, which called for CFC production to be cut back to 5% of 1986 levels by 1998. This Protocol was amended in 1990 and again in 1992, when 140 nations agreed to end CFC production by 1995 and to speed the phaseout of all other ozone-depleting chemicals. Targeted chemicals include halons: carbon tetrachloride (CCl_4), an important solvent; CH_3Br, a widely used agricultural fumigant; and HCFCs, the compounds presently being used as CFC substitutes. Developing nations were given until 2010 to halt their production of CFCs, and their use of CH_3Br and HCFCs was not restricted.

Figure 5.7 shows that global production of CFCs has plummeted, but the drop has been less than hoped for because production is continuing to rise in China, India, Russia, and other developing nations. Furthermore, there is a flourishing black market in CFCs, as they are being smuggled into the United States and other industrialized countries, mainly from Mexico and the former East Bloc countries. In the United States, only the drug trade is more profitable than the traffic in CFCs.

Because CFCs are so long-lived in the atmosphere, even if all nations abide by their commitments, the ozone hole is expected to mend very slowly. The overall lifetimes of CFC-11 and CFC-12 (between release at the surface and destruction in the stratosphere) are approximately 55 and 116 years, respectively.

The Montreal Protocol, despite its limitations, is an encouraging example of international cooperation in the interest of solving a major global environmental problem. It can be cred-

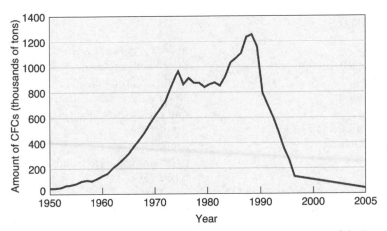

Figure 5.7 The worldwide production of CFCs has fallen dramatically since the signing of the Montreal Protocol in 1987. Used by permission of World Watch Institute, Washington, DC. www.worldwatch.org.

ited with preventing numerous cases of skin cancer and halting widespread damage to the environment.

Alternatives to CFCs

The first CFC substitutes to be introduced were HCFCs such as CF_3CHCl_2 and CHF_2Cl, compounds that have fewer chlorine atoms than do CFCs. HCFCs break down more readily in the troposphere than CFCs and thus are less likely to reach the stratosphere; however, because they can cause some ozone destruction, they are scheduled to be phased out by 2030.

Much better substitutes for CFCs are the **hydrofluorocarbons (HFCs)**, which contain no chlorine. One of these, CF_3CH_2F, has been used successfully as a refrigerant and since 1994 has replaced Freon in nearly all car air conditioners. In the electronics industry, soapy water followed by rinsing and air drying is now used instead of CFCs to clean microcircuits.

Unfortunately, a serious problem is associated with the long-term use of hydrofluorocarbons. Like CFCs and HCFCs, they contribute to climate change. Research is ongoing to find chemicals that are both efficient refrigerant and environmentally friendly.

■ Additional Sources of Information

Brasseur GP, Orlando JJ, Tyndall GS. *Atmospheric Chemistry and Global Change.* New York: Oxford University Press, 1999.

Hemond HF, Fechner-Levy EJ. *Chemical Fate and Transport in the Environment,* 2nd ed. London, UK: Academic Press, 2000.

Jacobson MZ. *Atmospheric Pollution.* Cambridge, UK: Cambridge University Press, 2002.

Wayne RP. *Chemistry of Atmospheres,* 3rd ed. New York: Oxford University Press, 2000.

■ Keywords

adsorption
catalytic ozone destruction cycle
Chapman cycle
chlorine activation
chlorine cycle
chlorofluorocarbons (CFCs)
Dobson unit (DU)
halons
heterogeneous
hydrochlorofluorocarbons (HCFCs)
hydrofluorocarbons (HFCs)

hydroxyl radical cycle
meta-stable molecule
Montreal Protocol
NO_x catalytic ozone destruction cycle
null cycles (holding cycles)
ozone hole
ozone layer
polar stratospheric clouds (PSCs)
type I PSCs
type II PSCs
UVB radiation

■ Questions and Problems

1. For what atmospheric measurement is a DU used? Explain how it is used.
2. The stratospheric ozone concentration has been measured to average 250 DUs over a 1000 square mile area. How many ozone molecules are in that area?
3. Report the following in DUs:
   a. The average abundance of ozone from 90°N to 90°S in the year 2000
   b. Ozone abundance over the Antarctic
4. The ozone layer is formed in the (stratosphere *or* stratopause), which is part of the Earth's (upper or lower) atmosphere. The altitude of the ozone layer is (20 to 30 km *or* 30 to 40 km) from the Earth's surface?
5. Match the following and arrange them in the order of increasing energy:

UVA	400 to 750 nm
UVB	290 to 320 nm
UVC	320 to 380 nm
Visible	250 to 290 nm

6. The dissociation reaction $NO_2 \rightarrow NO + O$ has a $\Delta H = 306$/mol. Calculate the λ of light that would have the minimum energy to initiate this reaction. Identify its region of the UV spectrum.
7. The dissociation reaction $O_3 \rightarrow O_2 + O$ has a $\Delta H = 105$/mol. Calculate the λ of light that would have the minimum energy to initiate this reaction. What region of the electromagnetic spectrum is this?
8. The dissociation of molecular N_2 to monoatomic nitrogen atoms requires an enthalpy change of 945 kJ/mol. Determine the wavelength of sunlight that will have sufficient energy to dissociate nitrogen. Why is O_2 easier to dissociate than N_2?
9. Why does the dissociation of O_2 require sunlight with $\lambda < 240$ nm? Why does O_3 dissociate in sunlight with λ of 240 to 320 nm? Write both reactions.
10. Why is there more UVB in solar radiation at the top of the stratosphere than at the surface of the Earth?

11. Write the four reactions that comprise the Chapman cycle.
12. What is meant by a "steady-state concentration of ozone in the stratosphere?"
13. Estimate the steady-state $[O_3]/[O_2]$ ratio at 30 km where the ozone photolysis reaction is slower than at higher altitudes and the concentration of air is higher.
14. Estimate the steady-state $[O_3]/[O_2]$ ratio at 20 km where the ozone photolysis reaction is much slower than at higher altitudes and the concentration of air is much higher.
15. The Chapman cycle assumes that ozone is produced by only one reaction. Write that reaction and its rate equation.
16. The Chapman cycle assumes that ozone is destroyed by only two reactions. Write these reactions and their rate equations.
17. How does a free radical differ from an ordinary molecule?
18. Why is the catalytic destruction of ozone considered to be such an important problem?
19. There are two ways that the hydroxyl radical can be formed in the stratosphere from atomic oxygen. Write these reactions.
20. Hydroxyl radical can be formed from the photolysis of water. Write the reaction.
21. The hydroxyl radical is responsible for what fraction of the total ozone destruction in the lower stratosphere?
22. The hydroxyl radical is a catalyst for ozone destruction. Show why it is considered a catalyst.
23. NO is released from automobile tailpipes. How does it get into the stratosphere to destroy ozone?
24. Is N_2O more or less reactive than NO?
25. Write a reaction that shows how NO is formed in the stratosphere.
26. Write reactions that show how NO catalytically destroys ozone.
27. How many ozone molecules will one molecule of NO destroy in the stratosphere?
28. How would supersonic airplanes affect the ozone layer?
29. What does CFC stand for? Write the chemical formula of Freon 12. List three uses for CFCs.
30. Write the formulas of the following:
    a. CFC-114
    b. CFC-115
    c. HCFC-22
    d. HCHC-142
31. Why do CFCs not react with ozone in the troposphere?
32. Write the CFC numbers for the following:
    a. $CHCl_3$
    b. CF_3CF_2Cl
    c. CHF_2CCl_3
33. If HCHC-22 were to replace CFC-11, would its lifetime from emission to chemical destruction be longer or shorter than CFC-11? Why?
34. The overall lifetime of CFC-11 in the atmosphere from emission to chemical destruction is 55 years, and CFC-12's lifetime is 116 years. Would you expect CFC-114 to have a longer or shorter lifetime than these two CFCs? Why?

35. Write reactions that show how chlorine catalytically destroys ozone. When CH_3Br is released into the stratosphere, it destroys ozone. Write a mechanism for this reaction.
36. What is a null cycle? Give an example.
37. Where do PSCs form?
38. List the following for type I PSC:
    a. Formed at what temperature?
    b. Chemical composition?
    c. Size?
39. List the following for type II PSC:
    a. Formed at what temperature?
    b. Chemical composition?
    c. Size?
40. Why are the reactions that occur on the surface of the PSC particles described as *heterogeneous reactions*?
41. What is the *chlorine activation* that takes place in PSCs? Give an example.
42. Write the five-step mechanism by which Cl_2 destroys ozone over Antarctica in the spring.
43. Could some gas be injected into the polar vortex that would react with the ozone-destroying Cl˙ to slow or stop its action but not trigger other undesirable reactions?
44. Why is ClOOCl formed in PSCs and not in the upper stratosphere?
45. What happens in the stratosphere over Antarctica in the winter?
46. Describe the evidence that implicates CFCs in the destruction of the ozone layer.
47. Describe the reactions that chemists now believe account for the destruction of the ozone layer by CFCs. Include the chemical reactions.
48. What is a *halon*? Cite an example with a chemical structure and its intended use.
49. For the following chain reaction:

$$Cl˙ + O_3 \rightarrow ClO˙ + O_2$$
$$ClO˙ + O \rightarrow Cl˙ + O_2$$

    a. What is the catalyst?
    b. Why is it a chain reaction?
    c. Write the overall net reaction.
50. For the following reaction:

$$NO_2 + O_3 \rightarrow NO_3 + O_2$$
$$NO_3 + hv \rightarrow NO_2 + O$$

    a. What is the catalyst?
    b. Why is it a chain reaction?
    c. Write the overall net reaction.
51. Where is the ozone *hole* located? Why is its formation cause for alarm? Why is the depletion of ozone not as great at the North Pole?

52. What international agreement regulates substances that deplete the ozone layer?

53. List two different classes of compounds that are being offered as alternatives to CFCs.

54. Hydrofluorocarbons are being used as CFC replacements. As compared with CFCs, how would the following compare and why?

    a. Lifetimes from emission to chemical destruction

    b. Their ozone depletion potential

    c. Flammability

CHAPTER

6

Analysis of Air and Air Pollutants

- **Additional Sources of Information**

- **Keywords**

- **Questions and Problems**

ANALYTIC MEASUREMENTS FORM THE FOUNDATION ON WHICH OUR UNDERSTANDING OF THE ATMOSPHERE IS BUILT. New measurement capability has often resulted in a significant expansion or revision of our understanding of atmospheric chemistry. Experimental measurements provide the information for testing theoretical models. In atmospheric chemistry, one of the most fundamental measurements is the most difficult: What is the composition of the atmosphere? Nitrogen gas represents more than 78% of the atmosphere, whereas reactive species, such as ozone, are present in parts per million (ppm) concentrations. The range in concentration of atmospheric constituents is extremely wide, as large as 10^{14}.

Measurements to determine the concentration of atmospheric constituents may be made in two different ways: **in situ**, which means the sample of the atmospheric gases must be placed inside the device (spectrometer) that is making the measurement, and **remotely** (remote-sensing technique) by passing a beam of energy that originates on a satellite, an aircraft, the space shuttle, or the ground through a portion of the atmosphere that is to be studied. Although these measurement techniques involve different instrumentation, the underlying principles are the same.

Automobile and smokestack emissions are the major sources of anthropogenic pollutants into the atmosphere, both of which are regulated by the Environmental Protection Agency (EPA). Measurements of the gases emitted from automobiles and smokestacks are difficult to make because the exhaust gases are hot and are traveling at a high velocity. The most common methods used to measure automobile exhaust and some of the EPA methods used for monitoring smokestack emissions are discussed.

■ In Situ Absorption Measurements

Spectroscopic measurements rely on the fact that different chemical compounds absorb electromagnetic radiation at different wavelengths. When a molecule absorbs a photon of electromagnetic radiation, the energy of the molecule increases. Figure 6.1 shows the interaction that occurs when a molecule absorbs electromagnetic radiation of different frequencies. For example, when a molecule absorbs microwave radiation, which is electromagnetic radiation of a relatively low energy, it only stimulates the rotational motion of the molecule. Infrared radiation, which has higher energy than microwaves, stimulates the vibrations of molecules that absorb it. Ultraviolet (UV) radiation, which has even more energy than infrared, causes

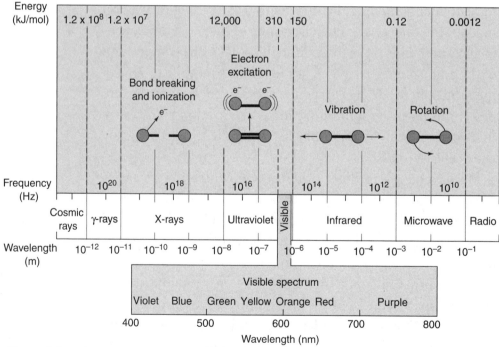

Figure 6.1 An electromagnetic spectrum showing the type of interaction that occurs when electromagnetic radiation in each energy region is absorbed by a molecule.

electrons in the molecules absorbing it to be promoted into higher energy orbitals; the molecule is said to be in the excited state. The lowest energy state of a molecule is called the **ground state.** Very high-energy electromagnetic radiation, such as an X-ray, has enough energy to break chemical bonds and ionize molecules.

When the sample absorbs a beam of electromagnetic radiation, the **irradiance** of the beam is decreased. Irradiance (P), which is sometimes referred to as *intensity or radiant power,* is the energy per second per unit area of the light beam. A spectrophotometric measurement is illustrated in Figure 6.2. Electromagnetic radiation is passed through a monochromator (a prism, grating or filter) to select one wavelength of electromagnetic radiation. The light of a single wavelength is said to be monochromatic, which means "one color." The monochromatic light, with irradiance P_0, passes into a sample of length b. The irradiance of the beam emerging from the other side of the sample is P. Some of the light may be absorbed by the sample, and thus, $P \leq P_0$.

Transmittance (T) is defined as the fraction of the original light that has passed through the sample.

$$T = \frac{P}{P_0}$$

T has the range of 0 to 1. The percentage of transmittance is 100T and has a range from 0% to 100%. **Absorbance** (A) is defined as follows:

$$A = \log\left(\frac{P_0}{P}\right) = -\log T$$

When no light is absorbed, $P = P_0$ and $A = 0$. Absorbance is important because it is directly proportional to concentration (c) of the absorbing molecules in the sample.

The relationship between the concentration of the absorbing molecule and absorbance has been described in the Beer-Lambert law or simply **Beer's law:**

$$A = \epsilon bc$$

Absorbance is dimensionless; thus, all of the constants and variables on the right side of the equation must have units that cancel. The concentration of the sample can be expressed in many ways: moles per liter (M/L), parts per million (ppm), parts per billion (ppb), or milligrams per meter cubed (mg/m^3). The pathlength (b) can be very small (cm) or very large (km). The quantity ϵ (epsilon), called the **molar absorptivity** (or extinction coefficient), is the characteristic of a molecule that indicates how much light it will absorb at a particular wavelength. The ϵ can be expressed by using many different units. Depending how the concentration and pathlength are expressed, ϵ must have units that make the products of (ϵbc) dimensionless.

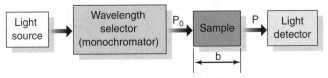

Figure 6.2 Diagram of single-beam spectrophotometric instrument. P_0, irradiance of beam entering sample; P, irradiance of beam emerging from sample; b, length of path through sample.

In Situ Ozone Measurements

Ultraviolet absorption has been used to measure the concentration of ozone in stratospheric air. Figure 6.3 shows that ozone has a strong absorption in the UV region. Other atmospheric gases such as O_2, N_2, and H_2O do not absorb UV radiation and thus do not have to be removed from the air sample because they do not interfere with the ozone measurement. The concentration of ozone in the stratosphere is less than 10 ppm, so the spectrophotometer that is used to make the measurement must be capable of measuring a small absorbance. This is most easily done by making the sample cell very long. Because Beer's law tells us that absorbance is proportional to the sample pathlength and because we can get a lot of sample easily, a long sample pathlength will greatly improve our ability to measure ppm levels of ozone.

A UV spectrometer has been designed to measure the amount of ozone present in stratospheric air. It is placed in a high-altitude aircraft, which is flown to the exact position where the measurement is to be made. Air that contains the ozone flows into the instrument through a tube (an in situ measurement). A schematic diagram of the instrument is shown in Figure 6.4. The UV spectrometer uses a low-pressure mercury lamp to produce the electromagnetic radiation. The output of the mercury lamp has a maximum at 254 nm, which is very near to the absorbance maxima for ozone (Figure 6.3). The tube carrying the air is split into two streams. In one stream there is an ozone scrubber that removes only ozone. The air passing through this side travels into the lower sample cell and acts as a reference. The other stream (which still contains ozone) travels into the upper sample cell. Each stream flows through one of the 50-cm sample cells. Light from the mercury lamp is sent down each cell, and a silicon photodiode at the end of each cell detects the intensity at the end of the path. An ozone concentration of 1 ppm in the stratosphere produces an absorption of 0.2. By comparing the absorbance of sample with ozone to the absorption sample without ozone, the difference in the absorbance of the two cells gives the ozone absorption independent of lamp intensity.

In Situ Carbon Dioxide Measurements

Since 1958, a measurement of atmospheric carbon dioxide (CO_2) concentration has been made at the Mauna Loa Observatory on the island of Hawaii. This observatory is located at a site that is far from major population centers (Figure 6.5). At an elevation of 3350 m on the

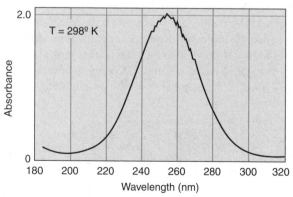

Figure 6.3 UV absorption spectrum of ozone. Adapted from L.T. Molina and M.J. Molina, *J. Geophys. Res.* 91 (1986): 14501–14508.

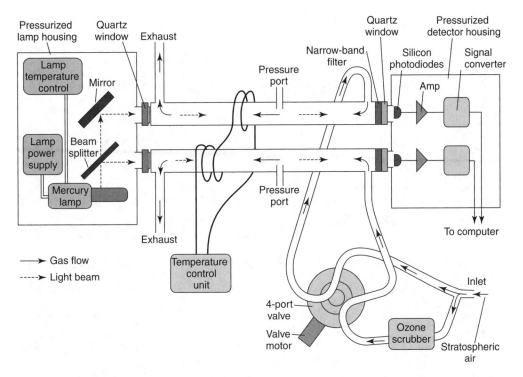

Figure 6.4 UV spectrometer that has been used to measure ozone concentration. Adapted from M.H. Proffitt and R.J. McLaughlin, *Rev. Sci. Instrum.* 54 (1983): 1719–1728.

flank of Mauna Loa volcano, it is an ideal site for making atmospheric CO_2 measurements. No nearby vegetation exists, and the prevailing nighttime, down-slope winds give a representative sampling of midtropospheric air from the central North Pacific Ocean. The record of CO_2 concentration is taken as a reliable index of long-term CO_2 growth.

The concentration of CO_2 in the atmosphere has been presented in a graphic form (Figure 6.6) that is widely known as the *Keeling curve*. The Mauna Loa atmospheric CO_2 concentration measurements, taken since 1958 and extending through the present, constitute the longest continuous record of atmospheric CO_2 concentrations available in the world. These measurements also are considered to be reliable indicators of the regional trend in the concentration of atmospheric CO_2 in the middle layers of the troposphere. The methods and equipment used to obtain these measurements have remained essentially unchanged during the 40-year monitoring program.

The CO_2 concentrations taken at Mauna Loa Observatory are obtained using a **nondispersive, dual-detector, infrared detector (NDIR)**. The components and operational principles of the NDIR detector are discussed later in this chapter. Air samples at Mauna Loa are collected continuously from air intakes at the top of four 7-m towers and one 27-m tower. Four air samples are collected each hour for the purpose of determining the CO_2 concentration. This analyzer registers the concentration of CO_2 in a stream of air flowing at approximately

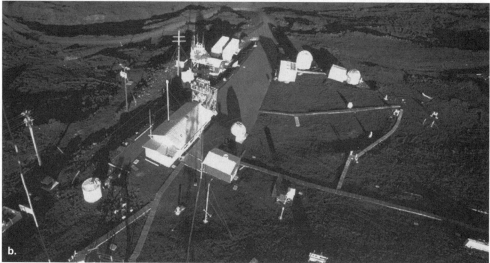

Figure 6.5 Picture of the Mauna Loa Observatory (a) and the Mauna Loa Observatory from the air (b).

0.5 L/min. Every 20 minutes the flow is replaced by a stream of calibrating gas or *working reference gas*. These calibration gases and other reference gases are compared periodically to determine the instrument sensitivity and to check for possible contamination in the air-handling system.

Because virtually all molecules (except inert gases and homonuclear diatomic molecules such as N_2 and H_2) absorb infrared radiation at specific wavelengths, infrared spectroscopy should be a useful technique for measuring atmospheric constituents. Infrared (IR) spectroscopy is relatively insensitive, and IR spectrometers require very long sample cells (10 to 100 meters) to detect ppb concentrations of atmospheric gases. An IR spectrometer that is more than 100 meters long is impractical, and because of this, IR is not used to make in situ measurements.

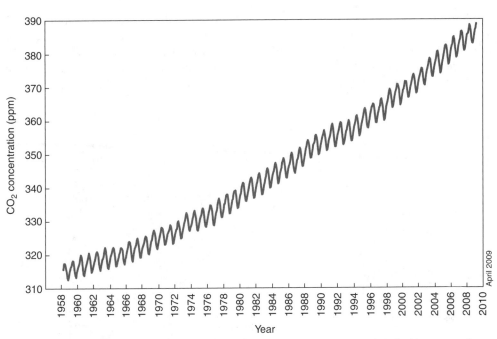

Figure 6.6 Because the measurements were made at Mauna Loa, Hawaii, in 1958, carbon dioxide concentrations in the atmosphere have risen dramatically. The yearly seasonal variations are caused by the removal of carbon dioxide from the atmosphere by growing plants in the summer and the return of carbon dioxide to the atmosphere in the winter when plants decay. Courtesy of NOAA.

■ Infrared Spectrometry

Molecules selectively absorb specific IR frequencies that correspond to the frequencies of the vibrational oscillations of the atoms, which are connected by covalent bonds. When a molecule absorbs IR radiation, the amplitudes of these vibrations increase. The absorption corresponding to these oscillations appears in certain definite wavelength regions of the spectrum.

The frequency (v) at which a characteristic IR absorption occurs depends on the mass of the atoms involved in the vibration and the strength of the bond connecting the atoms. Planck's equation relates the energy (E) of the absorbed radiation to its frequency.

$$E = hv \quad h = 6.626 \times 10^{-34} \text{ Js, Planck's constant}$$

Frequency is inversely proportional to the wavelength (λ) of the absorbed radiation.

$$E = hc\left(1/\lambda\right)$$

The quantity ($1/\lambda$) is call the wavenumber, $\bar{v}$, and is usually expressed in units of cm^{-1}.

$$E = hc\bar{v}$$

An IR band is normally reported as a position of maximum absorption in wavenumbers $(\bar{\nu})$ in units of reciprocal centimeters (cm^{-1}). Figure 6.7 shows that the most intense IR band for CO_2 is between 2200 to 2400 cm^{-1}, and Figure 6.8 shows that for N_2O the most intense IR bands lie in the region of 2100 to 2400 cm^{-1}. The absorbance of the pollutant CFC-12, which can be seen in Figure 3.13, has an intense absorbance in the 1200 to 800 cm^{-1} region.

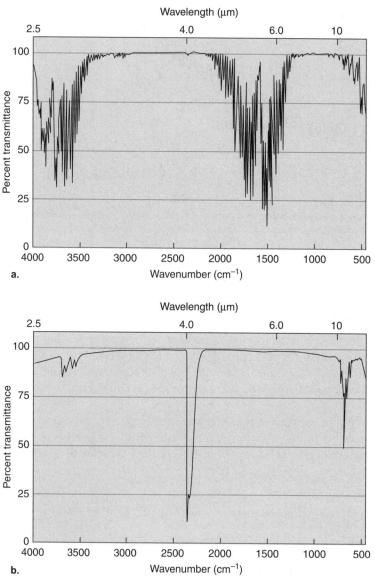

Figure 6.7 Infrared spectrum of the atmospheric gases water (a) and CO_2 (b). ©Bio-Rad Laboratories, Inc., Informatics Division, Sadtler Software & Databases, 1980–2008. All rights reserved.

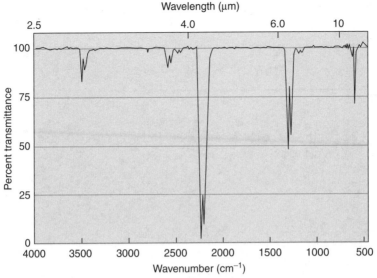

Figure 6.8 Infrared spectrum of pollutant gas N_2O. © Bio-Rad Laboratories, Inc., Informatics Division, Sadtler Software & Databases, 1980–2008. All rights reserved.

▪ Infrared Vibrational Frequencies

The absorption that is observed in the infrared region can be described by using a very simple model that calls for the atoms in the molecule to be considered balls and the bonds that connect them to be springs. As can be seen in Figure 6.9, for a diatomic molecule, the two atoms have masses (M_1 and M_2), and the spring has a force constant (k).

Two atoms that are connected by a spring would naturally settle into a specific motion when disturbed. The masses vibrate with a frequency that depends on the masses and the strength of the spring. Such vibrations, which occur at characteristic frequencies, are what we measure in vibrational spectra.

For diatomic molecules, the frequency of the vibration can be described as a harmonic oscillator and $\bar{\nu}$ by the following equation:

$$\bar{\nu} = \frac{1}{2\pi c}\sqrt{\frac{k}{\mu}} = 5.3 \times 10^{-12}\sqrt{\frac{k}{\mu}}$$

Where

$\bar{\nu}$ is the wavenumber of an absorption peak in cm^{-1}.
k is the force constant of the bond in Newtons per meter (N/m).

5×10^2 N/m is the average for single bonds.
1×10^3 N/m is the average for double bonds.
1.5×10^3 N/m is the average for triple bonds.

C is the velocity of light in cm/sec.

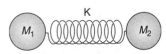

K is the force constant
of the spring

Figure 6.9 Molecular vibrations of two atoms connected by a bond.

μ is the reduced mass in kilograms, which can be computed by

$$\mu = \frac{M_1 M_2}{M_1 + M_2}$$

EXAMPLE 6.1

Calculate where in the IR the characteristic absorption of the C≡O bond in carbon monoxide (CO) would occur.

1. Compute the reduced mass:

   for carbon

   $$M_1 = \frac{12 \times 10^{-3}\,kg/mol}{6.02 \times 10^{23}\,atoms/mol} = 2.0 \times 10^{-26}\,kg$$

   for oxygen

   $$M_2 = \frac{16 \times 10^{-3}\,kg/mol}{6.02 \times 10^{23}\,atoms/mol} = 2.7 \times 10^{-26}\,kg$$

   $$\mu = \frac{\left(2.0 \times 10^{-26}\,kg\right)\left(2.7 \times 10^{-26}\,kg\right)}{\left(2.0 + 2.7\right) \times 10^{-26}\,kg} = 1.1 \times 10^{-26}\,kg$$

2. Substitute into the equation the following:

   $$\bar{v} = \frac{1}{2\pi c}\sqrt{\frac{k}{\mu}} = 5.3 \times 10^{-12}\sqrt{\frac{k}{\mu}}$$

   $$= 5.3 \times 10^{-12}\sqrt{\frac{1.5 \times 10^3\,N/m}{1.1 \times 10^{-26}\,kg}}$$

   $$= 1.96 \times 10^3\,cm^{-1} = 1960\,cm^{-1}$$

This is a good approximation for the absorbance. CO (C≡O) experimentally shows an absorbance between 1900 to 2250.

■ Remote Measurements of Atmospheric Composition

A variety of spectroscopic techniques have been used to measure the Earth's atmosphere from space. Spectroscopic instruments look along the edge of the Earth's stratosphere (**limb paths**), downward (**nadir**) paths, or a combination of viewing geometry for occultation, emission, and scattering experiments. In an **occultation experiment**, the sun or a bright star is observed as the line of sight from the object to the viewing instrument passes through the atmosphere. In an **emission experiment**, the radiation emitted by the atmosphere is observed. In a **scattering experiment**, solar radiation, which is scattered as it passes through the atmosphere, is observed. Space-based instruments have probed the atmosphere using UV, visible, infrared, and microwave radiation.

The geographic coverage of a satellite instrument depends on the orbit as well as the orientation of the satellite instrument with respect to the orbital plane. By adjusting the plane of the orbit, the period of precession can be made to be one year. In this way, the precession keeps up with the movement of the Earth in its orbit around the sun, and the orbit maintains a constant relationship to the sun. In this sun-synchronous mode, the observations can be made, for example, at local noon and local midnight around the orbit. Many of the atmospheric experiments use this mode so that geographic variation can be distinguished for diurnal (daily) variation. Figure 6.10 illustrates how satellites and Earth-based instruments are deployed to make measurements of atmospheric gases.

Atmospheric Trace Molecular Spectroscopy

Atmospheric trace molecular spectroscopy (ATMOS) was first flown on the Spacelab 3 Shuttle mission in 1985. On that first flight, 40 different gases were detected, and a database was constructed of trace gases in the stratosphere. The instrument also was able to detect and measure wind patterns in the mesosphere and troposphere. ATMOS has since flown several more shuttle missions. In 1992, ATMOS provided information about the Mount Pinatubo volcanic eruption and its effect on the atmosphere. On later flights, ATMOS detected dramatic increases in CFC gases that occurred in the stratosphere since its first flight in 1985.

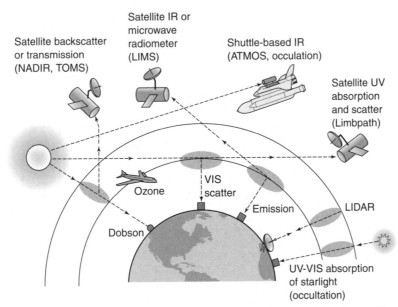

Figure 6.10 Some spectrometers are based on the Earth, others are positioned on satellites.

Figure 6.11 shows a schematic of the ATMOS instrument, which is a high-resolution Fourier Transform IR spectrometer. The ATMOS has a sun tracker that keeps the instrument's field of view on the sun, and a telescope that collects the IR radiation to be processed by the spectrometer. The spectrometer simultaneously measures the concentrations of gases in the stratosphere at altitudes of between 10 and 150 km (6 to 96 miles). Because gas molecules absorb specific wavelengths of incoming solar radiation by determining the wavelengths that have been absorbed, the composition of the stratosphere can be determined. Because ATMOS has the ability to detect gases in concentrations lower than 1 ppb, its data have been used to test theoretic modes that describe the chemistry of the stratosphere.

The principal disadvantage of ATMOS for the study of stratospheric chemistry is the fact that it can make measurements at only sunrise and sunset. This means that at most 24 sun-

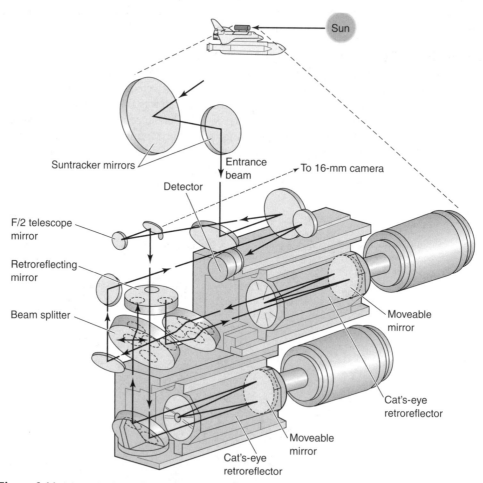

Figure 6.11 Schematic of the ATMOS IR spectrometer flown on the Space Shuttle. This spectrometer measures high-resolution atmospheric transmission spectra by observing sunlight that has passed through the atmosphere. Adapted from R.S. Zanders, et al., *Geophys. Res. Lett.* 23 (1996): 2353.

rises and sunsets are observed each day and that the measurement is local, not global. Many gases have diurnal variation because their production is dependent on photolysis by sunlight.

These disadvantages can be overcome by measuring atmospheric emission rather than absorption. Emission measurements may be made at any point on Earth and at any time of day.

Limb Infrared Monitor of the Stratosphere

The Nimbus 7 satellite, which was first placed in 1978, used a **limb infrared monitor of the stratosphere (LIMS)** to measure the IR emission of the Earth. This polar-orbiting satellite is able to complete a picture of the entire stratosphere in one day. The IR emission depends on atmospheric temperature as well as on the composition of the atmosphere. The temperature is determined by measurement of the emission from a gas of known concentration, such as CO_2. The radiance of the thermal emission of the Earth's atmosphere is much less than that of the sun; therefore, it is necessary to use broader spectral bands for the measurement rather than a high-resolution spectrum.

A schematic of the LIMS instrument is illustrated in Figure 6.12. It used a telescope to collect infrared radiation from a limb view (at the edge) of the stratosphere. The IR radiation from five spectral bands was measured. This enabled LIMS to measure global distributions of trace gases with concentrations in the ppb range for the first time. The five gases measured were CO_2, O_3, H_2O, HNO_3, and NO_2.

Total Ozone Mapping Spectrometer

Unlike the previous two instruments, which operate in the infrared, the **total ozone mapping spectrometer (TOMS)** measures the total atmospheric ozone column by measuring the UV region in a range in which few other atmospheric molecules absorb strongly. The atmosphere

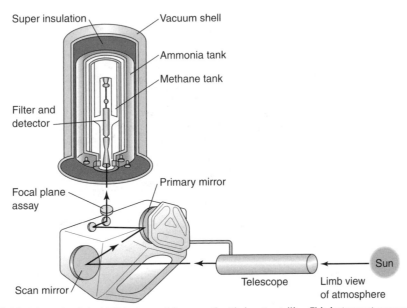

Figure 6.12 Schematic of the LIMS instrument flown on the Nimbus 7 satellite. This instrument measures the emission from the atmosphere in six infrared bands. Adapted from J.C. Gille and J.M. Russell III, *J. Geophys. Res.* 89 (1984): 5125–5140.

scatters the UV radiation from the sun strongly; the combination of scattering and absorption allows the instrument to determine the concentration of ozone from the spectrum of the radiation reflected back into space.

The TOMS instrument (Figure 6.13) is a UV instrument that measures the albedo of the Earth at six wavelengths between 213 and 380 nm; a comparison of the radiation scattered back from the atmosphere with the incident solar radiation permits measurement of the ozone column.

The original TOMS gave the most complete global picture of the distribution of ozone and its change in concentration in the years from 1979 until its failure in 1992. Subsequent TOMS instruments are maintaining the long-term ozone record. Another important atmospheric gas (sulfur dioxide [SO_2]) also has strong absorption in the UV region. TOMS has successfully tracked stratospheric clouds containing SO_2 from several volcanic eruptions.

Light Detection and Ranging

Light detection and ranging (**LIDAR**) is similar to radar. A radar transmitter sends radio waves, and the radar detector measures the time that it takes for them to bounce (scatter) off an object and return to the detector. LIDAR operates in the visible and infrared regions. Different types of physical processes in the atmosphere are related to different types of light scattering. The LIDAR system is usually Earth based, although a LIDAR system has been taken into space and has been operated on the space shuttle.

A simplified block diagram of a LIDAR system is shown in Figure 6.14. The LIDAR transmitter is a laser, and the receiver is an optical telescope that is focused on an extremely sensitive photomultiplier detector. This detector records the photons that reach it and converts

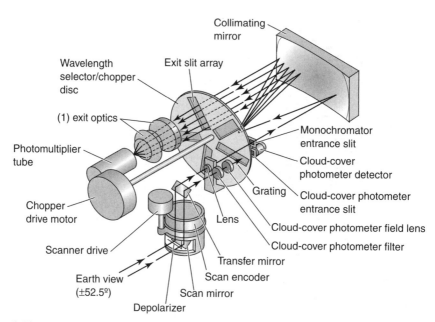

Figure 6.13 The TOMS ultraviolet photometer used to measure global distribution of ozone from the Nimbus 7 satellite. Adapted from D.F. Heath, et al., *Opt. Eng.* 14 (1975): 323–331.

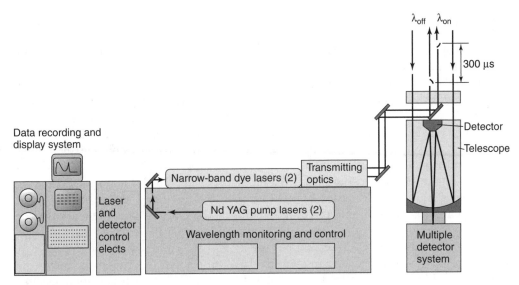

Figure 6.14 The LIDAR system. Adapted from E.V. Browell, *Proc. IEEE* 77 (1989): 419.

the response to an electronic signal that is many times stronger. Some LIDAR systems use multiple lasers that operate at several different wavelengths simultaneously. The laser beam is shot into the atmosphere, and the detector measures the light that is scattered back. In one type of experiment called *differential absorption LIDAR*, a technique that is used to measure atmospheric ozone, the laser is sequentially shot into the atmosphere at two different wavelengths. The first laser pulse is at a wavelength at which ozone absorbs weakly. In the second wavelength, ozone absorbs strongly. The difference in the absorption of light at the two wavelengths can be used to determine the amount of ozone.

One of the most common applications of LIDAR is to detect aerosol particles in the atmosphere. Particles have a much larger scattering cross-section than do molecules. Thus, if there are a substantial number of airborne particles, then the fraction of light scattered back to the detector is much larger than the return from gas molecules alone. In regions with high concentrations of particles, LIDAR determines the *backscatter ratio*, which is the ratio of observed backscattered signal to what would have been obtained from molecular scattering alone. Conversely, at altitudes in which the aerosol concentration is negligible, the variation of backscatter can be used for measurements of temperatures in the upper stratosphere and mesosphere, a height that weather balloons cannot reach. Measurement of the Doppler shift between the transmitted and received signals permits the measurement of wind velocity along the path of the laser beam.

■ Monitoring Automotive Emissions

Mobile sources of air pollution have a very large impact on our nation's air quality. For example, nationwide, highway motor vehicles contribute one-quarter of all volatile organic compound emissions. However, in some cities, such as Washington, DC, the contribution is even higher, approximately one-half of all volatile organic compounds. Motor vehicles also contribute almost one-third of the nitrogen oxide (NO_x) emissions nationwide and up to 62% of the CO emissions. It is clear that successful strategies for controlling mobile sources are essential to achieving clean air.

One of the major challenges that we are facing is that more cars are on the road today and they are being driven more miles every year. Vehicle miles traveled (VMT) increased from 1.4- to 3.7-trillion miles (2.6 times) between 1990 and 2001, the last year statistics were provided by the U.S. Department of Transportation. Also, according to a recent government study, over 25% of all trips are less than one mile, suggesting that many of us are becoming even more dependent on cars and trucks.

According to EPA modeling, since 1970 the statistics for vehicle emissions, except for NO_x, have significantly decreased despite that VMT increased by 149% during the same time period. NO_x emissions increased by 16% during the years 1970 to 1996, mostly due to emissions from trucks. The EPA model, shown in Figure 6.15 predicted that although VMT would continue to rise in the future, emissions of NO_x and VOCs would continue to fall as a result of stricter engine emission standards. In reality, NOx emissions from vehicles increased 35.8% in the period 1990–2000, due to the increasing number of light-duty

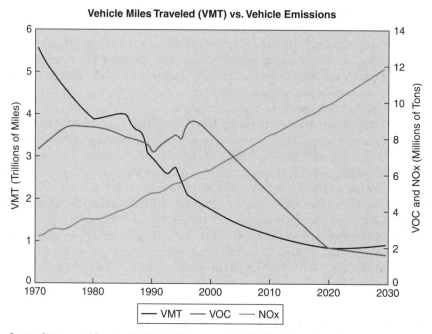

Source: Statement of Senator Bob Smith, Environment & Public Works Committee Hearing on Transportation & Air Quality, July 30, 2002

Figure 6.15 Growth in VMT and NO_x emissions.

trucks on the road and the retirement of low emitting pre-1983 vehicles with newer, higher emitting vehicles.

The second major challenge that we are facing is the problem of in-use deterioration. Although all new vehicles offered for sale must be designed and produced to meet emissions standards, actual in-use performance is not as good. Broken or malfunctioning parts or simply a lack of proper maintenance or repair too often results in vehicle emissions that are significantly above the standards. As can be seen in Figure 6.16, data show that hydrocarbon emissions (HCs) increase approximately 77% as a car is driven more miles. This type of increase is seen for other pollutants as well, demonstrating that in-use emission increases are responsible for more than 50% of total mobile source emissions.

The 1990 Clean Air Act called for enhanced inspection and maintenance programs in the most polluted areas of the country and has effectively improved the quality of air that we breathe. A 1996 study in Colorado of roadside emissions showed some interesting tends that were not evident. Emissions from old cars, which were expected to be the most polluting group, were often not as great as some new cars. As can be seen in Figure 6.17, the worst 20% of new cars pollute more than the best 40% of 20-year-old cars. Later inspection of these new car polluters showed that 41% had deliberately tampered with their emission control equipment and that 25% had defective or missing emission control equipment.

The techniques and equipment used to measure automotive emissions must be able to provide accurate, rapid analyses at a relatively low cost. The monitoring equipment must be extremely durable because it is housed in automotive garages and must be easily calibrated and remain stable for long periods. Although the operators performing the tests are trained, it is the instrument that is calibrating itself, making the measurements, and writing the test reports.

Automobile Hydrocarbon Emissions

Internal combustion engines used in automobiles and trucks produce a complex mixture of organic molecules. The designs of the four-cycle internal combustion engine and the catalytic

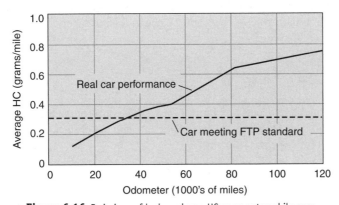

Figure 6.16 Emissions of hydrocarbons, HC as an automobile ages.

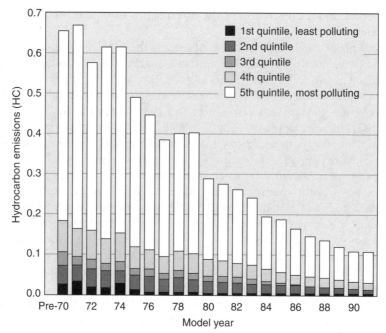

Figure 6.17 HC emissions from vehicles in California in 1991. Adapted from G.A. Bishop and D.H. Stedman, *Acc. Chem. Res.* 29 (1996): 489.

converter were discussed in Chapter 4. Depending on the air/fuel ratio of the engine and the compression ratio, more or less HCs are produced. The emissions contain saturated and unsaturated hydrocarbons, aromatic hydrocarbons, polynuclear aromatic hydrocarbons, as well as alcohols, aldehydes, ketones, and ethers.

A detector that will respond to almost all organic compounds that contain carbon–hydrogen bonds is the **flame ionization detector**, which uses a hydrogen/air flame to burn organic molecules in a sample, and as a result, it produces a current that is proportional to the carbon mass flow into the flame. A schematic diagram of the hydrogen flame detector can be seen in Figure 6.18. The dominant reaction that takes place in the hydrogen flame is as follows:

$$CH^\bullet + O^\bullet \rightarrow CHO^+ + e^-$$

$$CHO^+ + H_2O \rightarrow H_3O^+ + CO$$

The most stable ion in a hydrocarbon flame is H_3O^+. The current generated by the detector is carried from the burner nozzle to the ion collector plate. There is an opposing recombination reaction that reduces signal output:

$$H_3O^+ + e^- \rightarrow H_2O + H^\bullet$$

The applied voltage between the collector plate and the burner nozzle must be large enough so that the ions do not recombine before reaching the collector. When HC mole-

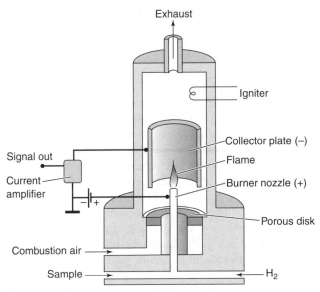

Figure 6.18 The flame ionization detector for measurement of hydrocarbons in auto tailpipe emissions. Courtesy of Horiba International Corp.

cules enter the flame ionization detector, they are burned, and the current between the burner nozzle and the ion collector plate increases proportionally to the amount of HC that is present.

Most HCs (saturated, unsaturated, and aromatic) give the same relative response. Alcohols do not respond as well as HC, and direct comparisons cannot be made for oxygenated fuels. A correction factor is incorporated for oxygenated fuels. Propane is used as a calibration gas. Because they do not burn, there is no direct interference from the other tailpipe gases: CO, CO_2, NO_x, and H_2O.

Automobile Nitric Oxide Emissions

The NO_x in automobile tailpipe emissions are measured by the use of **chemiluminescence**, chemical reactions that release energy produced by emitting light rather than heat. Fireflies use chemiluminescent reactions to produce light.

In this case, the NO in the tailpipe emissions is reacted with ozone. The energy produced in this reaction is released as light rather than as heat. The reaction is a two-step process. In the first step, the ozone reacts with the NO to produce an excited-state nitrogen dioxide molecule (NO_2^*). The electrons in the NO_2^* are not in the lowest energy state but are in an excited state. In the second step of the reaction, the NO_2^* loses excess energy as the excited electrons return to ground state and emit the excess energy as a photon of light (hv). The reactions can be written as follows:

$$NO + O_3 \rightarrow NO_2^* + O_2 \text{ (chemical activation)}$$

$$NO_2^* \rightarrow NO_2 + hv \text{ (emission of light)}$$

Not all of the NO_x in the tailpipe is NO. Significant amounts of NO_2 are usually present as well. If the tailpipe sample is passed through a heated-activated carbon catalyst, the NO_2 can be converted to NO. The reaction is as follows:

$$NO_2 + C \rightarrow NO + CO$$

An analysis of tailpipe emissions that have been treated this way gives the amount of NO_x, which is the sum of NO and NO_2 present. To measure the amount of NO in the tailpipe gas stream, the carbon catalyst is removed from the stream. The difference between the two results gives the amount of NO_2 present.

The NO_2^* loses energy and gives off electromagnetic radiation in the 600 to 2500 nm range, with a maximum absorption at 1200 nm. The emitted light is measured by a photo-multiplier; the intensity of light is proportional to the concentration of NO. An optical filter is placed between the reaction chamber and the photomultiplier to ensure that only light in the 600 to 2500 nm range reaches the detector. A diagram of the **chemiluminescent analyzer** (**CLA**) for NO_x is shown in Figure 6.19. The ozone used in the reaction is produced in a chamber just before the NO_x detector. In the ozone generator, air is irradiated with UV light to produce O_3.

The CLA is very dependable, has no moving parts, and is an ideal detector for continuous, rapid monitoring. There is no direct interference from the other tailpipe gases: CO, CO_2, HCs, or H_2O.

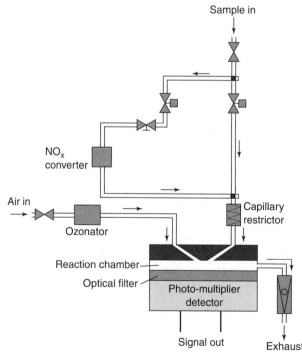

Figure 6.19 The CLA for NO_x emissions. Courtesy of Horiba International Corp.

Automobile Carbon Monoxide Emissions

Carbon monoxide in tailpipe emissions is monitored by IR spectrometry in the range of 400 to 4000 cm^{-1}. Some of the gases, such as O_2 and N_2, which are present in tailpipe emissions do not absorb in the IR region. The wavelength of maximum absorption of the other individual tailpipe gases that do absorb in the IR are listed in Table 6.1.

In a dispersive IR, the IR radiation from the source, which emits radiation across the 400- to 4000-cm^{-1} range, is dispersed by a grating into narrow bands of IR radiation. The instrument sequentially scans the IR recording absorption at each wavelength. After the data are recorded (approximately 20 minutes), it is presented as a *spectrum,* which is a graph of the absorbance versus wavelength. A dispersive IR is not used for measuring tailpipe emissions because it would take a dispersive IR too long to record the spectrum of a tailpipe sample.

CO emission monitoring uses a nondispersive infrared analyzer, NDIR. An NDIR analyzer differs from the dispersive IR spectrometer in that the radiation from the source is not dispersed by a grating. The NDIR analyzer is constructed for a specific analysis. A diagram of the NDIR instrument is shown in Figure 6.20.

The NDIR analyzer uses a heated tungsten filament, which is a broadband source that emits across the IR region. The reference cell is filled with nitrogen, and the sample to be measured flows through the sample cell. A flexible diaphragm separates the two sides of the detector. Optical filters are placed between the IR source and the sample and reference cell to remove IR wavelengths that would be absorbed by CO_2 in the sample.

When a sample containing CO is present in the sample cell, IR energy from the source will be absorbed, and less IR energy will reach the sample side of the detector cell. The difference in IR energy reaching the two halves of the detector causes a small pressure imbalance between the detector cells, which causes the diaphragm between the cells to bend. A constant voltage applied between the detector plate and the diaphragm causes a current to flow when the distance changes.

The voltage output of the NDIR is nonlinear because of the absorption characteristics of gases. The cell absorption is related to cell length by Beer's law. A 30-cm pathlength is needed

Table 6.1	
Wavelength of Maximum Absorbance of Tailpipe Gases	
Molecule	**λ (cm^{-1})**
CO	2170
NO	1940
H_2O	3756, 1595
CO_2	2349, 1343, 667
NO_2	1665, 1358, 757
CH_4	3019, 2916, 1534, 1306

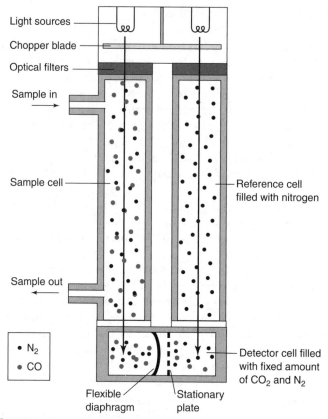

Light sources
Chopper blade
Optical filters
Sample in
Sample cell
Reference cell filled with nitrogen
Sample out

• N$_2$
• CO

Flexible diaphragm
Stationary plate
Detector cell filled with fixed amount of CO$_2$ and N$_2$

Figure 6.20 NDIR for the analysis of CO in automobile emissions. Courtesy Horiba International Corp.

for measurement of ppm levels of CO. The speed of the response is limited by the flow rate and sample cell volume. As the cell pathlength is increased to give more sensitivity, the cell volume increases. Because the flow rate is kept constant, this means that the time it takes to measure each car (cycle time) must increase.

Automobile Emissions: Sample Handling

Anyone who has had his or her car's emissions tested knows that tailpipe emissions testing must be completed rapidly. The results of the emissions test are only as reliable as the sample that is taken. Because tailpipe emissions are exiting from the automobile at an extremely high velocity and at an elevated temperature, manufacturers of testing equipment have developed a sampling system that would decrease the velocity of the emissions while maintaining an elevated temperature. Manufacturers, such as Horiba, have developed a **constant volume sampler** (Figure 6.21). The exhaust gas from the tailpipe is diluted with filtered ambient air. This system uses a critical venturi flowmeter to establish the flow rate. A constant flow is established through the venturi by a turbo compressor blower. The blower reduces the outlet pressure such that a choked condition occurs at the venturi throat. As long as this pressure is equal to or less than "critical" and a constant tempera-

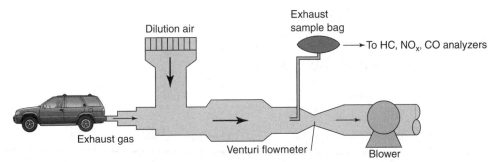

Figure 6.21 The Horiba constant volume sampler. Courtesy Horiba International Corp.

ture is maintained, the flow rate through the venturi is constant. Knowing the inlet pressure and temperature, flow at standard conditions can be computed. The data system uses this information to determine the dilution that has been made and to compute the final concentration of HC, CO, and NO_x.

■ Monitoring Emissions from Stationary Sources

The Clean Air Act of 1970 specified that the EPA design the National Ambient Air Quality Standards (NAAQSs) for criteria air pollutants. Criteria air pollutants are those whose permissible levels were based on health guidelines, or criteria, from air quality reports. NAAQSs were divided into primary standards designed to protect the public health, particularly the people who were most susceptible to respiratory problems, such as those with asthma, the elderly, and infants. Secondary standards were designed to protect the public welfare and included crops, visibility, and animals. The EPA was required to set primary standards based on health considerations alone, not on the cost of technology that would be required to meet the standard.

The six original criteria pollutants were CO, NO_2, SO_2, total suspended particulates (TSPs), HCs, and photochemical oxidants. Particulate lead was added to the list in 1976. O_3 replaced photochemical oxidants in 1979. HCs were removed from the list in 1983, and the TSP criteria were changed to include only particles with a diameter of 10 μm or less, which are called PM_{10}; a $PM_{2.5}$ standard (discussed later here and in Chapter 2) was added in 1997. Table 6.2 lists the current primary NAAQSs.

■ Monitoring Sulfur Dioxide Emissions

The release of SO_2 to the atmosphere is the primary cause of acid rain in the United States. Fossil fuel combustion at electric power-generating plants accounts for 59% of the SO_2 emissions; industrial sources contribute approximately 25%.

Coal frequently contains sulfur in the form of the mineral pyrite (FeS_2). Coal in the United States contains sulfur, typically between 1% and 4% by weight. When coal is burned, the sulfur is oxidized to SO_2.

Table 6.2

Current Primary Federal Air Quality Standards

Pollutant	NAAQSs
O_3	
1-Hour average	0.12 ppmv (235 µg/m³)
8-Hour average	0.08 ppmv (160 µg/m³)
CO_2	
1-Hour average	35 ppmv (40 mg/m³)
8-Hour average	9.5 ppmv (10.5 mg/m³)
NO_2	
Annual average	0.053 ppmv (100 µg/m³)
SO_2	
24-Hour average	0.14 ppmv (365 µg/m³)
PM_{10}	
24-Hour average	150 µg/m³
$PM_{2.5}$	
24-Hour average	65 µg/m³

The EPA method for analyzing the ambient air for SO_2 in determining compliance with the primary and secondary NAAQSs is described in the *Code of Federal Regulations*. Methods described in the *Code of Federal Regulations* have been shown to be useful for regulatory and legal purposes. When new methods are discovered, the EPA requires a careful examination of the performance of the new method by a carefully selected group of laboratories. A comparison of data obtained from the new method by each individual laboratory is made against the results from the older existing method. Once the new method has been shown to be superior by giving lower detection limits, more precise results, or faster analysis, then the EPA proposes a modification of the method.

Pararosaniline Spectrophotometric Method for the Determination of Sulfur Dioxide

The EPA pararosaniline method will measure SO_2 concentrations between 0.01 to 5.00 ppm in ambient air. Figure 6.22 shows that an air pump with a constant flow rate pulls the air sample into an impinger and bubbles the air to be measured through a solution containing potassium tetrachloromercurate. The SO_2 present in the air stream reacts with the tetrachloromercurate solution to form a stable disulfitonatomercurate complex as follows:

$$HgCl_4^{2-} + 2\,SO_2 + 2\,H_2O \rightarrow Hg(SO_3)_2^{2-} + 4\,H^+ + 4\,Cl^-$$

The disulfitonatomercurate complex resists air oxidation and is stable in the presence of strong oxidants such as ozone and NO_x. During subsequent analysis, the complex is reacted with

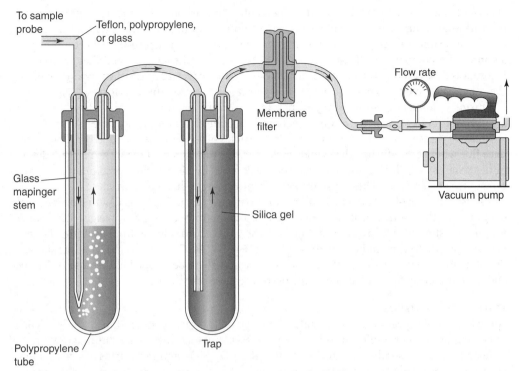

To sample probe

Teflon, polypropylene, or glass

Membrane filter

Flow rate

Vacuum pump

Glass mapinger stem

Silica gel

Polypropylene tube

Trap

Figure 6.22 Collecting a sample for the pararosaniline spectrophotometric SO_2 method. Adapted from EPA. *CFR* 40.

p-rosaniline dye and formaldehyde to form an intensely colored *p*-rosaniline methyl sulfonic acid. The concentration of this red-violet species is determined by measuring the absorbance at 548 nm. A Beer's law calibration curve is constructed by preparing standards with known concentrations of SO_2 and plotting absorbance versus concentration. The absorbance of an unknown sample is found by measuring its absorbance and by determining its corresponding concentration from the calibration curve. The absorbance is directly related to the amount of SO_2 collected. The total volume of air sampled, corrected to EPA reference conditions (25°C, 760 mm), is determined from the measured flow rate and the sampling time. The concentration of SO_2 in the ambient air is expressed in micrograms per standard cubic meter ($\mu g/m^3$).

■ Particulate Emissions

Fossil Fuel Combustion

Fossil fuels are an anthropogenic source of particulate matter (PM). Coal, oil, natural gas, gasoline, and diesel fuel all produce some PM. Combustion of gasoline in an automobile engine emits particles of organic material (soot), Si, Fe, Zn, and S. Diesel-powered trucks emit 10 to 100 times more particulate material than do gasoline-powered vehicles; most of the PM is organic material from unburned fuel.

The combustion of coal releases soot, sulfate, and fly ash. Coal is a solid fossil fuel that was formed millions of years ago by compaction of partially decomposed plant material. The classes of coal are distinguished by their percentage of carbon, which is directly related to their heating value. Seams of coal also form with other natural minerals such as iron pyrite (FeS). These minerals are often occluded in the pieces of coal and are responsible for the metals and sulfur compounds released on burning. Anthracite is the most sought-after of all coals; it contains 80% carbon, 4% moisture, and 1% sulfur and leaves 10% ash. Bituminous coal, on the other hand, contains 65% carbon, 9% moisture, and 2.8% sulfur and leaves 11% ash.

During combustion, some of the minerals in coal are converted into a solid that forms a glassy solid (slag), which does not become PM. Some fraction of the minerals produces particles of fly ash, so named because smaller particles of ash that are produced enter the furnace flue where they "fly up" the smokestack. The size (diameter) of the fly ash largely determines whether it will be removed from the stack gases by electrostatic precipitators or bag houses (see Chapter 4). These systems easily remove larger diameter particles.

Fly ash from coal-fired boilers produces approximately 1% to 2% fly ash, with a diameter of 0.1 μm. Even though this small diameter fraction contains only 1% to 2% by weight, it accounts for the largest number of particles and most of the surface area. Fly ash from coal-fired boilers has been shown to contain Fe, Zn, Pb, V, Mn, Cr, Cu, Ni, As, Co, Cd, Sb, and Hg. This smaller fly ash is an inhalation risk; smaller particles are less likely to be trapped in the nose or larynx and are more likely to be inhaled into the deep recesses of the lung.

Industrial Sources

Some industrial procedures burn metals with fossil fuels. Waste incineration and kiln drying of cement are two examples of industrial processes in which metals are emitted. PM from industrial process usually contains Fe_2O_3, Fe_3O_4, Al_2O_3, SiO_2, and carbonates of several metals. In these processes, heavy metals vaporize at high temperature and then recondense onto particles that are formed simultaneously.

Iron is released into the atmosphere far more than any other metal. Lead, an EPA criteria pollutant, is emitted from lead-ore smelters, lead-acid battery manufacturing, and solid waste disposal. Lead has not been used as a gasoline additive in the United States since 1987, and as a result, the ambient concentration of lead in the atmosphere has declined. Several countries still use leaded gasoline.

Miscellaneous Sources

Particulate matter can arise from many sources: Rubber particles from tires, pollen, viruses, and meteoric debris are all commonly found in the air. The constant rotation of tires on pavement causes tire wear at the interface between the tire and the road. Rubber particles usually larger than 2 μm in diameter are constantly emitted. Plants and biological organisms produce pollen and viruses. Meteorites that impact the stratosphere release Fe, Ti, and Al particles as they heat from the friction of entry into the atmosphere.

■ Monitoring Particulate Emissions

High-Volume Sampling

Airborne particles must be removed from the air sample before analysis. The high-volume method, which is described in the *Code of Federal Regulations,* measures the mass of total suspended particulate matter (TSP) in ambient air for determining compliance with the primary and secondary NAAQSs for PM.

An air sampler (Figure 6.23) is taken to the measurement site. It has an air pump that draws a measured quantity of ambient air into a covered housing and through a filter during a 24-hour sampling period. The sample flow rate and the geometry of the shelter favor the collection of particles 25 to 50 μm in diameter depending on the wind speed and direction. The filters used must have a collection efficiency of 99% for removing particles 0.3 μm and larger from the air stream.

The filter is weighed before and after use to determine the net weight of the collected particles. The total volume of air sampled, corrected to EPA standard conditions (25°C, 760 mm), is determined from the measured flow rate and the sampling time. The concentration of TSP matter in the ambient air is computed as the mass of collected particles divided by the volume of air sampled. The high-volume method is an appropriate test method for air, with TSP in the range of 2 to 750 μg/m^3.

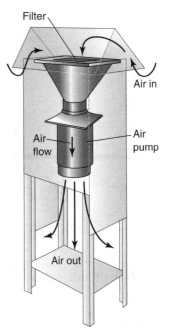

Figure 6.23 High-volume particulate sampler in shelter. Adapted from EPA. *CFR* 40.

Separation of Particles by Size

In 1997, the EPA proposed new NAAQSs for PM. In addition to revising the previous standard for PM$_{10}$, the EPA added a new standard for fine particles that are less than 2.5 μm in aerodynamic diameter (PM$_{2.5}$). To develop correlations between high PM$_{2.5}$ concentrations, especially those particles that are secondary components formed in the atmosphere through chemical reactions and condensation, it is necessary to sample not only for the mass of PM$_{2.5}$ in a fixed volume of air, but also for the chemical composition of the PM$_{2.5}$.

A variety of devices can separate particles by size and then measure the amount of each fraction collected. Several filters that remove successively smaller particles can be arranged together in a filter stack. Weighing the PM that each filter catches will give a quantitative analysis of the amount of PM in each range of particle sizes. Cyclone fractionators that use centrifugal force to separate particles by mass have been used to separate particles that are greater than 2.5 μm. Impactors are simple devices that direct a stream of gas at high velocity toward a collection plate. The gas stream makes a sharp turn, which causes PM$_{2.5}$ to be deposited on the collection plate. A cascade impactor separates particles by size as it directs the stream of air onto a series of collection plates through successively smaller orifices.

A schematic diagram of an ambient air sampler is shown in Figure 6.24. The air sample enters through a size-selective inlet followed by two cyclone fractionators that remove particles that are larger than 2.5 μm. The flow is split into four channels. The first channel is used to estimate atmospheric concentrations of particulate organic and elemental carbon. A quartz fiber filter that has been baked before the test is used to collect this fraction. A quartz filter is used because it contains no carbon. In the second channel, particulate material is collected on a Teflon filter for analysis of mass and metallic elements by **X-ray fluorescence (XRF)**. The third channel also uses a Teflon filter; however, this filter is

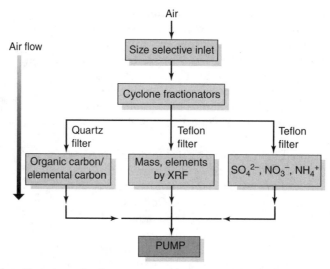

Figure 6.24 Ambient air sampler that separates particulate material by size for later chemical analysis.

extracted in water, and the extract is analyzed for sulfate and nitrate by ion chromatography. Teflon filter material is used for these samples because it does not contain the elements being measured. Ion chromatography is described in detail in Chapter 9.

XRF Spectrometry

XRF spectrometry is a spectrometric method that is based on the detection of X-ray radiation that is emitted from the sample being analyzed. XRF is a two-step process that begins with a focused X-ray beam striking the sample. The incident X-ray strikes an inner shell electron in the sample atoms, which causes the electron to be ejected like a pool ball being struck by the cue ball. The lowest electron shell is the "K" shell. The second lowest is the "L" shell, and the next is the "M" shell. The vacancy that is created by the loss of the emitted electron is filled by an outer shell electron that moves to fill the vacancy in the lower level (Figure 6.24). Because the outer shell electron is at higher energy, when it drops to the lower level, it loses excess energy by releasing a photon of electromagnetic radiation. The fluorescent photon has an energy that is equal to the difference between the two electron energy levels. The photon energies are designated as K, L, or M X-rays, depending on the energy level filled; for example, a K shell vacancy filled by an L level electron results in the emission of a K_α X-ray (Figure 6.25b). Because the difference in energy between the two electron levels is always the same, an element in a sample can be identified by measuring the energy of the emitted photon (or photons if there is more than one electron emitted). The intensity of the emitted photons is also directly proportional to the concentration of the element emitting the photon in the sample. The XRF instrument measures the photon energy to identify which element is present and the intensity of that photon to measure the amount of the element in the sample. Because this technique measures energy differences from inner shell electrons, it is insensitive to how the element (being measured) is bonded. The bonding shell electrons are not involved in the XRF process. XRF will not detect every element; the elemental range is

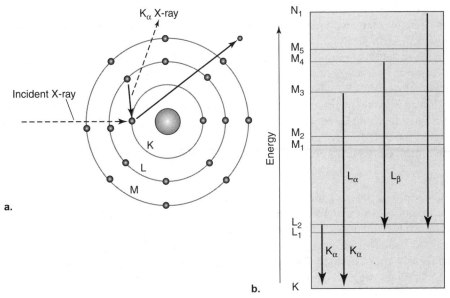

Figure 6.25 (a) The two-step XRF process. (b) The energy (or wavelength) of the photon released in XRF is determined by the energy difference between the two energy levels.

limited to elements that are larger than beryllium, and the detection of low atomic number (Z < 11, Na) elements is difficult.

A schematic diagram of the XRF instrument is shown in Figure 6.25. An X-ray tube produces X-rays that are directed at the surface of the sample. The incident X-rays cause the sample to release photons. The emitted photons released from the sample are observed at 90° to the incident X-ray beam. An energy-dispersive XRF detector collects all of the photons simultaneously. Each photon strikes a silicon wafer that has been treated with lithium and generates an electrical pulse that is proportional to the energy of the photon. The pulse height is proportional to the energy of the photon. The concentration of the element is determined by counting the number of pulses.

The XRF spectrometer (Figure 6.26) is very sensitive. It can determine transition metals at the ppm level. The lighter elements (Z < 19) have higher detection limits ranging

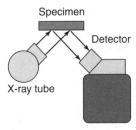

Figure 6.26 Energy dispersive XRF instrument.

from 10s to 100s of ppm. XRF is considered to be a bulk analysis technique; the X-ray source does not focus well into a narrow beam, and thus, it bombards the entire surface of the sample with X-rays.

XRF is a surface technique; it accurately reports the elements present on the surface of the PM. Because the surface contains metals that will directly interact with the lung if the particles are inhaled, the XRF analysis is most often used to asses the health risks that fly ash poses. If, on the other hand, an analysis of the total composition of the particle is needed, then the analysis must be made by another analytic technique. To determine the total metallic composition of the PM, dissolve in acid and measure the metals present by inductively coupled plasma-optical emission spectroscopy (described in detail in Chapter 13).

Fly Ash

U.S. power plants produce millions of tons of coal fly ash annually. The industry collects the vast majority with electrostatic precipitation or bag houses. More than 35% of the collected fly ash is used in a variety of applications, whereas the remainder is buried in landfills. During 2005, the U.S. electrical utility industry generated approximately 71.1 million tons of coal fly ash; approximately 29.1 million tons of the captured fly ash was used in industrial applications, which is more than twice times the average annual amount used prior to 1995.

The majority of fly ash products were used in construction-related applications, including cement production and concrete products, structural fills or embankments, soil stabilization, stabilization of waste materials, and mineral filler in asphalt paving. Fly ash products are used to supplement or replace Portland cement, a primary ingredient in concrete, to reduce raw material costs, and to strengthen the concrete. Industry experts estimate that using one ton of fly ash in concrete will eliminate approximately one ton of CO_2 that would have been emitted from cement production.

The Clean Air Act of 1990 requires power plants to reduce NO_x emissions, which complicates using fly ash in ready-mix concrete, the largest market for fly ash. Power plants often restrict oxygen during combustion to reduce NO_x emissions. This method of NO_x reduction unfortunately allows unburned carbon to remain in the fly ash and makes the fly ash unsuitable for use in concrete unless it is reprocessed.

■ Additional Sources of Information

Brasseur GP, Orlando JJ, Tyndall GS. *Atmospheric Chemistry and Global Change.* New York: Oxford University Press, 1999.

Harris DC. *Quantitative Chemical Analysis,* 7th ed. New York: W. H. Freeman and Company, 2006.

Rubinson KA, Rubinson JF. *Contemporary Instrumental Analysis.* Upper Saddle River, NJ: Prentice Hall, 2000.

Settle F. *Instrumental Techniques for Analytical Chemistry.* Upper Saddle River, NJ: Prentice Hall, 1997.

Skoog DA, Holler EJ, Crouch SR. *Principles of Instrumental Analysis,* 6th ed. Philadelphia, PA: Saunders, 2007.

■ Keywords

absorbance
atmospheric trace molecular spectroscopy (ATMS)
Beer's law
chemiluminescence
chemiluminescent analyzer (CLA)
constant volume sampler
emission experiment
flame ionization dector
ground state
in situ
irradiance
light detection and ranging (LIDAR)
limb infrared monitor of the stratosphere (LIMS)

limb paths
molar absorptivity
nadir
nondispersive, dual-detector, infrared detector (NDIR)
occultation experiment
PM_{10}
$PM_{2.5}$
remotely
remote measurements
scattering experiment
spectroscopic measurements
total ozone mapping spectrometer (TOMS)
transmittance (T)
X-ray fluorescence (XRF)

■ Questions and Problems

1. Describe the difference between an in situ measurement and a remote sensing measurement.
2. If the frequency of light is doubled, the energy of the light is _____.
3. If the wavelength of light is doubled, the energy of the light is _____.
4. Using Figure 6.1, arrange the following from highest to lowest energy: infrared, microwave, and X-ray radiation.
5. Which has higher energy, blue light with $\lambda = 400$ nm or purple light with $\lambda = 750$ nm?
6. Convert the following:
   a. A 22.6% transmittance into absorbance
   b. A 0.102 transmittance into absorbance
   c. A 31.3% transmittance into absorbance
7. Which of the following is directly proportional to concentration?
   a. Transmittance
   b. Molar absorptivity
   c. Absorbance
8. What are the units for absorptivity (in Beer's law) when the path length is in centimeters and the concentration is expressed as follows?
   a. ppm
   b. ppb
   c. $\mu g/m^3$
9. Convert the following:
   a. 0.055 absorbance into percent transmittance
   b. 0.245 absorbance into percent transmittance
   c. 0.020 absorbance into percent transmittance

10. What color would you expect to observe for the following solutions:
   a. Solution "A" that has an absorption maximum at 550 nm.
   b. Solution "B" that has an absorption maximum at 450 nm.
11. A sample of atmospheric gas is placed in a 10-cm cell, and an in situ absorption measurement is made. Will the transmittance of this sample be 10 times greater than the same sample in a 1-cm cell? What is the absorbance difference between the two?
12. Is electromagnetic radiation released by the sun monochromatic?
13. The UV absorption spectra of O_3 are shown in Figure 6.3. At what wavelength should the UV spectrometer be set to give the maximum sensitivity for the absorbance of O_3? Estimate how much less sensitive the technique would be if the measurement was made at 280 nm rather than at the optimum wavelength.
14. The UV absorption spectra of O_3 are shown in Figure 6.3.
   a. If the concentration of O_3 is expressed in ppm and the pathlength in meters, what are the units of ϵ?
   b. If the concentration of O_3 is expressed in mg/m^3 and the pathlength in centimeters, what are the units of ϵ?
15. Complete the following table:

A	% T	ϵ	b	c, M	c, ppm
0.420			1.0	1.25×10^{-4}	
	46%		1.0	8.2×10^{-5}	
		3.50×10^3	2.0		3.35
	49%		1.0		6.75

16. Explain what is meant by the following terms that are used to describe remote measuring techniques:
   a. nadir
   b. occultation
   c. limb
   d. scattering
17. Estimate where the following molecules will absorb in the IR region.
   a. NO
   b. HCl
   c. OH•
18. ATMOS
   a. Observes which region of the electromagnetic spectrum?
   b. Measures gases at what altitude?
   c. Detection limit?
   d. Uses this as the source.
   e. Takes measurements at this time of day.
19. LIMS
   a. Measures absorbance or emission?
   b. Observes which region of the electromagnetic spectrum?
   c. Uses a limb view. What is that?
   d. Is located where?

20. TOMS
   a. Measures which region of the electromagnetic spectrum?
   b. Measures what?
   c. Is an absorbance, emission, or backscatter technique?
21. The LIDAR
   a. Light sources are?
   b. Detector is?
   c. Uses differential absorption LIDAR. What is differential absorption LIDAR?
   d. Is used to make two different measurements. What are they?
22. List the three major automobile tailpipe pollutants. Why are these measured?
23. What detector is used to measure automobile HCs?
   a. Draw a schematic diagram of this detector.
   b. Write the most common reaction that takes place in the hydrogen flame.
   c. Why does CO_2 not interfere with the test for HC?
24. Does the automobile HC test give the same response for oxygenated fuels as for gasoline? Explain.
25. Does an automobile that is burning premium fuel give a higher HC response than one burning regular gasoline? Explain.
26. What is used to calibrate the flame ionization detector that measures HC emissions? Describe how the calibration is accomplished.
27. What detector is used to measure automobile NO_x emissions?
   a. Draw a schematic diagram of this detector.
   b. What is the reactant gas?
   c. What is chemiluminescence?
   d. Why does CO_2 not interfere with the test for NO_x?
28. For the CLA tailpipe NO_x monitor
   a. What reagents are added to the sample stream before measurement?
   b. Write the chemical reactions that take place.
29. For the CLA tailpipe NO_x monitor
   a. How does it differentiate between NO and NO_2?
   b. How is total NO_x measured?
30. How does the CLA detector differentiate the chemiluminescent signal from NO_x from other light that might leak into the system?
31. What detector is used to measure automobile CO emissions?
   a. Draw a schematic diagram of this detector.
   b. Why not use a dispersive IR instrument?
   c. Why does CO_2 not interfere with the test for CO?
32. Describe the NDIR detector.
   a. Source
   b. Detector
   c. Reference cell
33. How does the NDIR detect CO by IR if its detector is a diaphragm?
34. Is the response of the NDIR detector linear? Explain.
35. How are the NDIR cell pathlength and the time that it takes to measure the emissions from an automobile related?

36. How does the sample handling system of the automotive emissions test system dilute the tailpipe gases?
   a. Why are the gases diluted?
   b. Is the velocity of the tailpipe gas changed?
   c. Is the temperature of the tailpipe gas changed?
37. List the six original NAAQS criteria air pollutants.
38. List the current primary Federal Air Quality Standards.
39. For the Pararosaniline spectrophotometric method for the determination of SO_2
   a. Which is the limiting reagent?

$$HgCl_4{}^{2-} + 2\,SO_2 + 2\,H_2O \rightarrow Hg\left(SO_3\right)_2{}^{2-} + 4\,H^+ + 4\,Cl^-$$

   b. Why is the SO_2 made into the disulfitonatomercurate complex?
   c. If a smokestack sample has an SO_2 concentration of 1.5 $\mu g/m^3$, what is the concentration of the SO_2 in ppmv?
40. A smokestack effluent was measured for SO_2 concentration. A 500-m^3 volume of air at standard temperature and pressure (STP) was bubbled through 1.0 L of a 0.05-M potassium tetrachloromercurate solution. Subsequent analysis of a 10-mL aliquot at 548 nm showed the concentration of p-rosaniline methyl sulfonic acid to be 3.5×10^{-5} M. What was the concentration of SO_2 in the smokestack effluent?
41. List the following:
   a. Particulate matter released by gasoline engines
   b. Particulate matter released by diesel engines
   c. Particulate matter released by burning coal
42. For lignite and anthracite and bituminous coal, list the following:
   a. The percentage of carbon
   b. The percentage of moisture
   c. The percentage of sulfur
   d. The percentage of ash remaining
43. What fraction of the ash produced in a coal-fired boiler becomes fly ash?
   a. What is slag?
   b. What is the diameter of fly ash?
   c. What is the source of SO_2 that is released from burning coal?
44. List the following:
   a. Industrial sources of particulate matter
   b. Mobile sources of particulate matter
   c. Which metal is released into the atmosphere more than any other metal?
45. A high-volume TSP station has been constructed near the toll both of the Delaware Bay Bridge. During a busy summer day, the collection pump pulls a 1-L/min sample into the air sampler. An empty collection filter weighed 75.34 g before it was inserted into the air sampler. At the end of a 24-hour sampling period, the filter weighed 75.68 g. What was the TSP concentration of the air that was sampled?
46. Describe the following devices that collect PM:
   a. Cyclone fractionator
   b. Impactor
   c. Cascade impactor

47. One type of cascade impactor separates the PM into three fractions. Describe each fraction and the filter that is used to collect it.

48. What is XRF used to measure?

    a. What is XRF measuring in PM?

    b. Does XRF measure the total metallic composition of the PM?

    c. Why is XRF used rather than inductively coupled plasma-optical emission spectroscopy?

49. Draw a schematic diagram of the XRF spectrometer.

    a. What is used to generate the incident beam?

    b. Why are the released X-rays observed at 90° to the incident beam?

    c. Describe the detector.

    d. Can the XRF be used to determine whether carbon is present in the PM?

50. Describe the process that produces an X-ray for XRF analysis.

    a. Why does XRF not detect light elements?

    b. Would $PbCO_3$ give a different XRF signal for lead than would PbO?

    c. Can XRF determine elements at the ppb level?

51. How do power plants dispose of the huge amount of fly ash generated each year?

CHAPTER

7

Water Resources

WATER IS ONE OF THE WORLD'S MOST PRECIOUS RESOURCES; WITH-
OUT IT, THERE WOULD BE NO LIFE ON EARTH. We use water in the
home and for recreation, and water is crucial for agriculture
and industry. Although the total amount of water on Earth is
fixed and cannot be increased, we are in no danger of running
out. Water is constantly being recycled and replenished by
rainfall; there is plenty of fresh water to meet the needs of
everyone on Earth. Because of the uneven distribution of rainfall and the heavy use of water in certain areas, however, many
regions in the United States and other parts of the world are
experiencing severe water shortages.

Water in all of its forms—ice, liquid water, and water vapor—
is very familiar but has many unusual properties. For example,
water's exceptional ability to store heat modifies the Earth's climate, and ice's ability to float on water allows aquatic creatures
to survive in winter.

Water is an excellent solvent that can dissolve a wide variety
of ionic and polar substances. Thus, it is an effective medium
for carrying nutrients to plants and animals and is also a good
medium for carrying toxic substances and other pollutants.

In this chapter, we examine the composition of natural
waters, the way that water is distributed on the Earth, and how it
is recycled. We learn how the molecular structure of water
accounts for its extraordinary properties and consider the
implications of those properties for life on Earth. The ways that
water supplies are used and the measures that can be taken to
manage and conserve them better are discussed. Finally, we

study water pollution and examine its main sources, the effects of pollutants on human health and the environment, and the steps that can be taken to preserve water quality.

■ Distribution of Water on the Earth

Water is the most abundant compound on Earth, covering nearly three-quarters of the planet's surface. Although the total amount of water on Earth is enormous, only a small percentage is fresh water and much of that is not readily available for human use. Over 97% of the world's total supply of water is found in oceans and is too salty for drinking, irrigation, and most industrial and household needs. Of the remaining approximately 3%, approximately 2% is in the form of glaciers and polar ice caps. Most of the rest is found underground (groundwater), and much of that is too difficult or too expensive to tap. Lakes and rivers, which are the major sources of the world's **drinking water**, account for less than 0.01% of Earth's total water (Figure 7.1).

As can be seen in Table 7.1, water is the major component of all living things, constituting 70% of every adult human body and from 50% to 90% of all plants and animals. Water enters our bodies through the liquids that we drink and the foods that we eat; it leaves the body in the form of urine, feces, sweat, and exhaled air. Humans can survive only a few days without water.

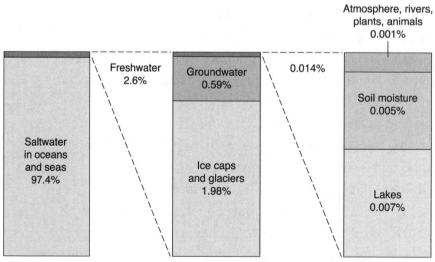

Figure 7.1 Most of the water on Earth is saltwater in the oceans. Less than 3% of freshwater and most of that exists as ice in glaciers and polar ice caps. The freshwater needs of humans must be supplied by the approximately 0.6% of the Earth's water that is found underground and in lakes and rivers. Data from D. Speidel and A. Agnew, "The World Water Budget," in D. Speidel et al., eds., *Perspectives on Water Uses and Abuses* (New York: Oxford University Press, 1988), p. 28.

Table 7.1

Water Content of Selected Organisms and Foods

	Percentage of Mass
Organisms	
Marine invertebrates	97
Human fetus (1 month)	93
Fish	82
Human adult	70
Foods	
Broccoli	90
Milk	88
Apples	85
Grapes	80
Eggs	75
Potatoes	75
Steak	73
Cheese	35

■ The Composition of Natural Waters

All bodies of natural water are solutions that contain varying concentrations of dissolved solids. Water containing up to 0.1% (1000 parts per million [ppm]) of dissolved solids is generally termed **freshwater**; however, such water is not necessarily suitable for drinking. For drinking water, the U.S. Public Health Service recommends a solid content of no more than 0.05% (500 ppm). The concentration of dissolved solids in seawater averages 3.5% (35,000 ppm). Water is termed brackish when the solid content lies somewhere between that of freshwater and seawater.

The ions present in seawater at a concentration of 1 ppm or more are shown in Table 7.2. At least 50 other elements are present at lower concentrations. The high concentration of salts in seawater comes primarily from the constant evaporation of water from ocean surfaces.

The concentration of ions in freshwater in streams, rivers, and lakes is much lower than that in seawater, and the distribution of ions is quite different (Table 7.3). In seawater, the main cation is sodium (Na^+), and the main anion is chloride (Cl^-). In fresh water, on the other hand, calcium (Ca^{2+}) and magnesium (Mg^{2+}) are the dominant cations, and bicarbonate (HCO_3^-) is the dominant anion. Ions in freshwater come from the weathering of rocks and soil. The main source of the Na^+ and Cl^- ions in rivers, particularly those near a seacoast, is salt spray that is thrown into the atmosphere from the ocean and then deposited on land and carried, by runoff, into the rivers.

Table 7.2	
Major Constituents of Seawater	
Ion	**Concentration (ppm)**
Chloride, Cl^-	19,000
Sodium, Na^+	10,600
Sulfate, SO_4^{2-}	2,600
Magnesium, Mg^{2+}	1,300
Calcium, Ca^{2+}	400
Potassium, K^+	380
Bicarbonate, HCO_3^-	140
Bromide, Br^-	65
Other substances	34
Total	34,519

■ The Hydrologic Cycle: Recycling and Purification

The Earth's supply of water is continually being purified and recycled in the **hydrologic cycle** (also **water cycle**; Figure 7.2). Through the processes of **evaporation, transpiration, condensation**, and **precipitation**, solar energy and gravity are responsible for the ceaseless redistribution of water among oceans, land, air, and living organisms. Heat of the sun warms the surface of the Earth, causing enormous quantities of water to evaporate from the oceans. On land, water vapor enters the atmosphere from plants as a result of transpiration, the process in which water escapes through the pores on leaf surfaces and evaporates. Additional water evaporates from lakes, rivers, and wet soil.

Table 7.3		
Comparison of the Concentrations of Major Ions in Freshwater and Seawater (Percentage of Total Ionic Concentration)		
Ion	**Freshwater**	**Seawater**
HCO_3^-	41.0	0.2
Ca^{2+}	16.0	0.9
Mg^{2+}	14.0	4.9
Na^+	11.0	41.0
Cl^-	8.5	49.0

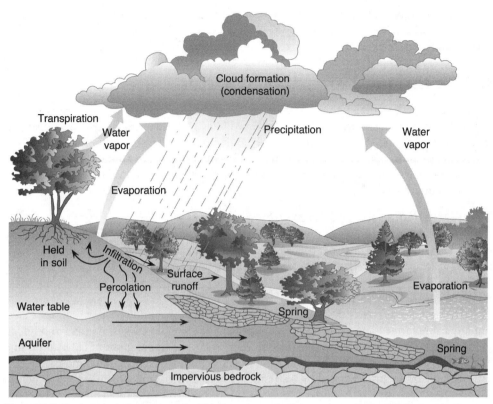

Figure 7.2 The hydrological cycle through which the water on Earth is continuously recycled. Powered by the sun, evaporation, condensation, and precipitation redistribute water among the oceans, the land, the atmosphere, and living organisms. Adapted from B.J. Nebel. *Environmental Science: The Way the World Works, Third edition*. Englewood Cliffs, NJ: Prentice Hall, 1990.

In the process of evaporation, water is purified. Dissolved substances are retained in the oceans, soil, and plants, and purified water vapor enters the atmosphere. As the water vapor rises, it cools and condenses into fine droplets that form into clouds. Prevailing winds carry moist air and clouds across the surface of the Earth. If the moist air cools sufficiently, water droplets or ice crystals fall to Earth as precipitation (rain, sleet, hail, or snow). Precipitation that falls through clean air is pure and contains only gases that are dissolved from the atmosphere and traces of dissolved salts. Some precipitation becomes locked in glaciers, but most of it either sinks into the soil or flows downhill, as runoff, into nearby streams and lakes, eventually making its way through rivers and wetlands to the ocean.

Precipitation that seeps into the soil is either taken by plant roots or continues to percolate into the ground until it reaches an impermeable layer of rock that stops its downward progress. Water collects in the porous rock above this impermeable layer, forming a reservoir of ground-water termed an **aquifer** (Figure 7.2). Groundwater is generally of high purity because the porous rock acts as a filter and retains suspended particles and bacteria. Groundwater is replenished much more slowly than is water at the surface of the Earth; it flows slowly (a speed of 15 meters or 50 feet per year is typical) through an aquifer until it reaches an exit to the surface, where it either emerges as a spring or seeps over a relatively wide area. These springs and seeps feed lakes, streams, rivers, and—ultimately—the oceans.

■ The Unique Properties of Water

Water is the only common pure compound on Earth that is a liquid and also possesses many unusual—even unique—properties. If it behaved like other chemical compounds of similar molecular weight and structure, life on Earth could not exist. The reason for most of water's extraordinary properties is hydrogen bonding. Although hydrogen bonds are much weaker than ionic or covalent bonds, they have a profound effect on the physical properties of water in both its liquid and solid states.

The Water Molecule and Hydrogen Bonding

A water molecule can form four hydrogen bonds (Figure 7.3). The oxygen atom in the molecule can bond to two hydrogen atoms in other molecules because it has two pairs of unshared electrons. Each hydrogen atom in a water molecule can bond to an oxygen atom in another molecule. These bonds are directional, which means that they can form only when the molecules are correctly oriented relative to each other.

In the liquid state, water molecules are in constant motion, and hydrogen bonds are continually being formed and broken. The arrangement of the molecules is random, and not all possible hydrogen bonds are formed. In solid water (*ice*), molecular motion is at a minimum, and molecules become oriented so that the maximum number of hydrogen bonds is formed. This results in an ordered, strong, extended, three-dimensional, open-lattice structure (Figure 7.4a). The size of the "holes" in the lattice is dictated by the bond angle in the water

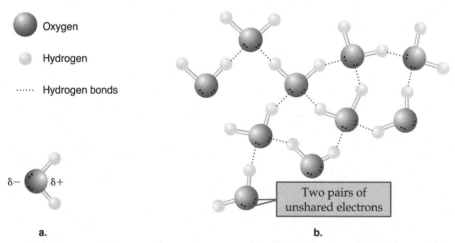

Figure 7.3 Hydrogen bonding in water: (a) Water is an angular polar molecule. Because oxygen is more electronegative that hydrogen, it has a partial negative charge (d–), and the two hydrogen atoms have partial positive charges (d+). (b) The slightly negatively charged oxygen atoms are attracted to the slightly positively charged hydrogen atoms, and hydrogen bonds form between adjacent water molecules. Each oxygen atom has two pairs of unshared electrons and can bond to two hydrogen atoms. Thus, each water molecule can form four hydrogen bonds.

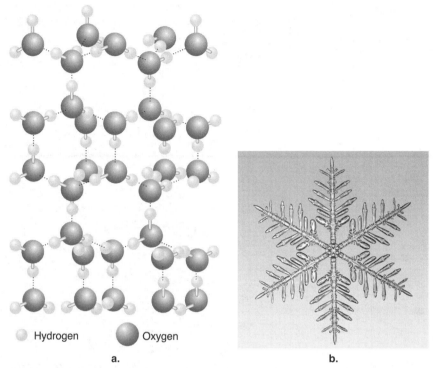

Hydrogen   Oxygen

a.   b.

Figure 7.4 (a) In the well-ordered three dimensional structure of ice, each water molecule is hydrogen bonded to four other water molecules. (b) The ordered arrangement of water molecules in ice accounts for the intricate hexagonal shapes of snowflakes.

molecule, which determines how close adjacent molecules can be. As a result, adjacent water molecules in ice are not as close to each other as they can be in liquid water. Thus, ice is less dense than water and floats. The ordered arrangement of atoms in ice accounts for the symmetry of ice crystals in snowflakes (Figure 7.4b).

Boiling Point and Melting Point

Compared with hydrogen compounds of other elements in group VIA (16) of the periodic table (H_2S, H_2Se, and H_2Te), water (H_2O) has an unexpectedly high boiling point (Figure 7.5). Normally, boiling points in a series of compounds of elements in the same group increase regularly with increasing molecular weight; this occurs for H_2S, H_2Se, and H_2Te. The hydrogen bonding that water molecules engage in causes the unexpectedly high boiling point of H_2O; the molecules of other hydrogen compounds in group VIA elements do not hydrogen bond to any significant extent. When water is converted to vapor, additional energy in the form of heat is required to break the hydrogen bonds; consequently, water's boiling point is higher than would be expected. If water boiled at the predicted temperature of −80°C (−112°F), it would be a gas at the temperatures found on Earth, and life as we know it would not be possible.

Water also has an exceptionally high melting point because of the large quantity of heat energy required to break its hydrogen bonds. When ice melts, approximately 15% of its hydrogen bonds are broken. The three-dimensional lattice structure collapses, and water forms.

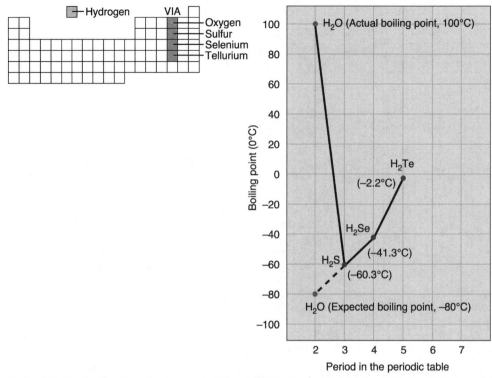

Figure 7.5 With the exception of oxygen, the boiling points of the hydrogen compounds of the Group VIA elements increase regularly going down the group. The unexpectedly high boiling point of water is due to hydrogen bonding, which occurs in H_2O but not to any extent in H_2S, H_2Se, and H_2Te.

Heat Capacity

Heat capacity is the quantity of heat that is required to raise the temperature of a given mass of substance by 1°C. It takes 1 calorie of heat to raise the temperature of 1 g of liquid water by 1°C. Water has the highest heat capacity of any common liquid or solid. From the definition of heat capacity, it follows that the higher the heat capacity of a substance, the less its temperature will rise when it absorbs a given amount of heat, and conversely, the less its temperature will fall when the same amount of heat is released from it.

The heat capacity (C) of a substance can be expressed as the ratio $q/\Delta T$, where q is the total heat flow into or out of an object, and ΔT is the temperature change produced. The heat flow equals the heat capacity times the temperature change:

$$q = (c)(\Delta T)$$

The heat capacity for 1 g of a substance is called the **specific heat**. The units of specific heat are Joules/gram C° (J/gC°). If we know the mass of the substance in grams, we can use the specific heat to determine the quantity of heat transferred when the temperature changes in ΔT.

$$q = (\text{grams of substance})(\text{specific heat})(\Delta T)$$

The specific heat of some selected substances is listed in Table 7.4. Water has one of the highest specific heats known. It is, for example, approximately five times as great as that of aluminum and approximately nine times as great as iron. That means that 4.184 J of heat will increase the temperature of 1 g of aluminum nearly 5°C and 1 g of iron more than 9°C. The temperature of 1 g of water, however, will be raised only one degree by this amount of heat. The high specific heat of water helps our body maintain the constant temperature of 37°C much easier. A human's body is approximately 70% water, and because of its high heat capacity, the body can release or absorb considerable amounts of energy with little effect on body temperature.

The high heat capacity of water has enormous implications for the Earth's climate. Oceans can absorb very large amounts of heat without showing a corresponding rise in temperature. The oceans absorb heat from the sun on warm days, primarily in the summer, and release it in the winter. If there were no liquid water to absorb and release heat, temperatures on the Earth would fluctuate as drastically as they do on the waterless moon and the planet Mercury, varying by hundreds of degrees during each light–dark cycle.

Heat of Fusion and Heat of Vaporization

Heat of fusion is the amount of heat that is required to convert 1 g of a solid to a liquid at its melting point; the same amount of heat is released when 1 g of the liquid is converted to the solid. **Heat of vaporization** is the amount of heat required to convert 1 g of a liquid to a vapor at its boiling point; the same amount of heat is released when 1 g of the vapor condenses to its liquid.

During the process of fusion (*melting*), heat energy is absorbed, but the temperature of the solid/liquid mixture does not begin to rise above the melting point until the melting is complete. During the reverse process of freezing, heat energy is released to the surroundings; the temperature does not begin to fall until freezing is complete. Similarly, there is no change in the temperature of a substance during vaporization and condensation until each process is complete. The changes in energy and temperature that occur when water transforms successively from a solid to a liquid and to a gas and then back again are shown diagrammatically in Figure 7.6.

Table 7.4

Specific Heat of Pure Substances

Substance	Specific Heat (J/gC°)
H_2O (l)	4.184
H_2O (s)	2.03
Al (s)	0.89
C	0.71
Fe	0.45
$CaCO_3$ (s)	0.85

l = liquid; s = solid.

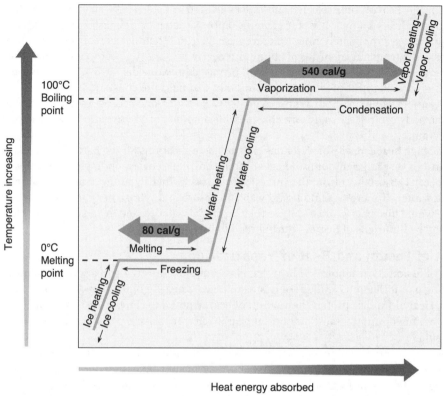

Figure 7.6 Changes in state as water is heated and cooled. There is no change in temperature as water melts, vaporizes, condenses, or freezes until the process has been completed. During melting and vaporization of water 80 and 540 cal/g, respectively, are absorbed; during condensation and freezing, 540 and 80 cal/g, respectively, are released.

Because heat of fusion and heat of vaporization are related to heat capacity, it is not surprising that their values are higher for water than for practically any other substance. Again, the explanation is hydrogen bonding: For ice to melt and water to vaporize, hydrogen bonds must be broken, and breaking them requires a considerable input of energy in the form of heat.

The fact that a relatively large amount of heat is required to evaporate a small volume of water has important consequences. It means that evaporation of a small amount of water (*perspiration*) from the skin can cool the human body efficiently. Extensive water loss from the body, which could upset the internal fluid balance, is thus kept at a minimum.

Anyone who has accidentally put his or her hand in the steam coming from a kettle of boiling water knows how painful the resulting burn can be. As the steam condenses, it releases heat, damaging the skin and causing pain. In contrast, a burn from water at the same temperature (100°C or 212°F) is far less severe.

Water's high heat of vaporization affects Earth's climate. In the summer, water evaporates from the surfaces of oceans and lakes. The heat energy needed for evaporation is drawn from the surroundings, and in consequence, nearby land masses are cooled. On a hot day, land close to a large body of water is always cooler than land farther away from the water. At night, when moist air cools, water vapor condenses, and heat is released; the temperature of the surroundings is raised. In this way, temperature variations between day and night are minimized. A similar modifying effect occurs in winter. When water freezes, heat energy is released, and the surroundings are warmed.

Temperature–Density Relationship

Density is defined as mass per unit volume. We often say that one substance is heavier or lighter than another. What we actually mean is that the two substances have different densities: A particular volume of one substance weighs more or less than the same volume of the other substance.

The density of most liquids increases with decreasing temperature and reaches a maximum at the freezing point, but the density of water does not. When water is cooled, its density reaches a maximum at 4°C—four degrees above the freezing point—and then decreases until the freezing point is reached at 0°C (32°F). The fortunate consequence of this property is that ice floats on the surface of water. This behavior is so familiar that we tend to forget that it is not typical of most liquids. For example, if a piece of solid paraffin is put into a container of liquid paraffin, it sinks to the bottom of the container because it is denser than the liquid (Figure 7.7).

The unusual behavior of ice is the result of the open-lattice structure of hydrogen-bonded molecules that forms when water freezes (see Figure 7.4). As noted earlier, molecules of water are farther apart in ice than in liquid water. As a result, when water starts to freeze, the number of molecules per unit volume (and thus the mass per unit volume or the density) decreases.

Because of the density of water, lakes develop a seasonal pattern. During the summer, lakes develop a three-layer profile. As can be seen in Figure 7.8, the **epilimnion layer** is near the surface. The **metalimnion layer** is the middle layer, and the **hypolimnion layer** is the bottom

Figure 7.7 Left: A solid piece of paraffin sinks in liquid paraffin. Right: Ice cubes float in water. This behavior of water is exceptional; most chemical compounds behave like paraffin.

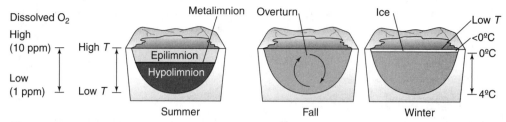

Figure 7.8 Changes in the temperature of air over a lake causes the layers of the lake to mix in a process that is called overturn.

layer. Photosynthesis takes place in the epilimnion as solar radiation penetrates, stimulating algae to produce oxygen. The hypolimnion is relatively dark but is rich in nutrients. Decomposition of organic matter on the bottom consumes oxygen and depletes the dissolved oxygen in the hypolimnion. When the warm season ends and cooling occurs, surface water (epilimnion) becomes denser than the hypolimnion, causing the surface water to sink, carrying with it dissolved oxygen. The deep water (hypolimnion) rises. This mixing process is called **overturn**, and it causes an increase in the chemical and microbiological activity of the lake. Water in the epilimnion, which has a large quantity of dissolved oxygen (10 ppm), drops oxygen to the creatures living deep in the lake. Conversely, water in the hypolimnion, in which the oxygen has been depleted, rises to the epilimnion where it can come in contact with atmospheric oxygen gas at the surface and start to replenish the dissolved oxygen.

The fact that ice is less dense than water has important consequences for aquatic life. When the air temperature falls below freezing in winter, water at the surface of lakes and ponds begins to freeze, and a layer of ice forms and floats on the water below it. The ice covering the surface acts as an insulating layer, reducing heat loss from the water under it. As a result, most lakes and large ponds in temperate climates never freeze to a depth of more than a few feet, and fish and other aquatic organisms are able to survive the winter in the water under the ice. If water behaved like most other liquids, ice would sink to the bottom as it formed, and lakes and ponds would freeze from the bottom up. Even deep lakes would freeze solid in winter, and aquatic life would be killed.

Water's unusual behavior when freezing has other consequences for the environment. When water trapped in cracks in rocks freezes, it expands, and the force of expansion is so powerful that the rock may split, an important factor in its weathering.

Acid–Base Properties

Water undergoes self-ionization, which is called **autoprotolysis**, in which it acts as both an acid and a base:

$$H_2O + H_2O \rightleftharpoons H_3O^+ + OH^-$$

$$H_2O \rightleftharpoons H^+ + OH^-$$

Water is a **protic solvent** because it can give up H^+. Diethyl ether ($CH_3CH_2OCH_2CH_3$), on the other hand, is an example of an aprotic solvent, which has no acidic protons.

The autoprotolysis constant for H_2O has the special symbol (Kw), where the "w" indicates the constant is for water.

$$H_2O + H_2O \rightleftharpoons H_3O^+ + OH^- \qquad Kw = [H^+][OH^-] = 1.0 \times 10^{-14}$$

The Kw is temperature dependent and has a value of 1.0×10^{-14} at $25°C$.

Acid–base reactions in water involve the loss or gain of H^+ by water. Many species act as acids in water by releasing H^+, and others act as bases by accepting H^+; water can act as either an acid or a base. For example, the dissociation of a weak acid, HB, in water is described in the following equation:

$$HB\,(aq) \rightleftharpoons H^+\,(aq) + B^-\,(aq)$$

The products are a proton, H^+, and the conjugate base, B^-, of the weak acid. An equilibrium expression can be written for this reaction:

$$Ka = \frac{[H^+][B^-]}{[HB]}$$

The equilibrium constant, Ka, is called the *acid ionization constant, Ka,* of the weak acid HB. The concentrations that appear in the brackets in this equation are always equilibrium concentrations in moles per liter. The Ka values of some common weak acids can be found in Appendix B. These constants are a measure of the extent to which the acid dissociates in water. The smaller the dissociation constant, the weaker the acid. Notice that in Appendix B the acids are arranged alphabetically and not in increasing or decreasing Ka. When comparing the strength of an acid, the term *pKa* is often used. By taking the log of Ka, a non-exponential number is produced that is easier to use.

$$pKa = -\log Ka$$

So, when comparing two weak acids such as hydrogen carbonate, HCO_3^-, and hydrogen sulfide, H_2S, for example, the Ka and pKa for each can be found in Appendix B.

Ka for $HCO_3^- = 4.7 \times 10^{-11}$, pka $= -\log (4.7 \times 10^{-11}) = -(-10.33) = 10.33$

Ka for $H_2S = 1.0 \times 10^{-7}$, pka $= -\log (1.0 \times 10^{-7}) = -(-7.00) = 7.00$

When the Kas of the two acids are compared, it is obvious that hydrogen carbonate is the weaker acid. So, in general, *the larger the value of pKa, the weaker the acid.*

Carbon Dioxide in Water

Carbon dioxide (CO_2) is the most important weak acid in water. It is dissolved in virtually all natural waters. The largest fraction of CO_2 found in natural waters comes as a result of the decomposition of organic material by bacteria. Atmospheric CO_2 also dissolves in natural waters. The concentration of CO_2 in the atmosphere, which is approximately 370 ppm or 0.037%, varies depending on the season and is increasingly caused by anthropogenic release.

The amount of gas that dissolves in water can be calculated using **Henry's law**, which states that the solubility of a gas in a liquid is proportional to the partial pressure of the gas that is in contact with the liquid.

$$[CO_2(aq)] = K_{CO_2} P_{CO_2}$$

Where

$[CO_2(aq)]$ = the molar concentration of CO_2 in water
K_{CO_2} = 3.38×10^{-2} mol/atm-L, the Henry's law constant for CO_2 (at 25°C)
P_{CO_2} = partial pressure of CO_2 (at 25°C)

To calculate P_{CO_2}, the partial pressure of water at 25°C must be known. *The Handbook of Chemistry and Physics* gives a value of P_{CO_2} = 0.0313 atm. Air has 370 ppm (0.037%) of CO_2.

$$P_{CO_2} = (1.00 - 0.0313 \text{ atm})(0.00037) = 3.58 \times 10^{-4} \text{ atm}$$

The concentration of CO_2 dissolved in water is calculated as follows:

$$[CO_2(aq)] = (3.38 \times 10^{-2} \text{ mol/atm-L})(3.58 \times 10^{-4}) = 1.18 \times 10^{-5} \text{ mol/L}$$

When CO_2 dissolves in water, HCO_3^- is formed:

$$CO_2 + H_2O \rightleftharpoons HCO_3^- + H^+$$

$$Ka_1 = \frac{[H^+][HCO_3^-]}{[CO_2(aq)]} = 4.45 \times 10^{-7} \quad pK_{a1} = 6.35$$

HCO_3^- can act as an acid by giving up a proton:

$$HCO_3^- \rightleftharpoons CO_3^{2-} + H^+$$

$$Ka_2 = \frac{[H^+][CO_3^{2-}]}{[HCO_3^-]} = 4.7 \times 10^{-11} \quad pK_{a2} = 10.33$$

To calculate the concentration of the various anions formed by CO_2 dissolving in water, begin by using the Kas and the total mass of these anions.

For mass balance, the total concentration (T) of all carbonate species is as follows:

$$T = [CO_2] + [HCO_3^-] + [CO_3^{2-}]$$

The fraction (α) of each carbonate containing anion can now be expressed in terms of only the pH and the Kas.

$$\alpha CO_2 = \frac{[CO_2]}{T} = \frac{[H^+]^2}{[H^+]^2 + [H^+]Ka_1 + Ka_1Ka_2}$$

$$\alpha HCO_3^- = \frac{[HCO_3^-]}{T} = \frac{Ka_1[H^+]}{[H^+]^2 + [H^+]Ka_1 + Ka_1Ka_2}$$

$$\alpha CO_3^{2-} = \frac{[CO_3^{2-}]}{T} = \frac{Ka_1Ka_2}{[H^+] + [H^+]Ka_1 + Ka_1Ka_2}$$

With these equations, the fraction of each anion can be calculated. The pH is all that needs to be known. Figure 7.9 shows the distribution of carbonate species as a function of pH. This figure shows the following:

- When the pH is 6.5 to 10, HCO_3^- is the predominant ion in natural waters.
- When pH = pK_{a1}, αCO_2 = αHCO_3^-.
- When pH = pK_{a2}, αHCO_3^- = αCO_3^{2-}.
- When the pH is low (less than 5.5), CO_2 is the predominant species.
- When pH is high (more than 10.5), CO_3^{2-} is the predominant species.

If the atmospheric concentration of CO_2 in the atmosphere is 370 ppm, we should be able to calculate the pH of rainwater that is in equilibrium with the atmosphere:

$$CO_2 + H_2O \rightleftharpoons HCO_3^- + H^+$$

$$Ka_1 = \frac{[H^+][HCO_3^-]}{[CO_2(aq)]} = 4.45 \times 10^{-7} \qquad pK_{a1} = 6.35$$

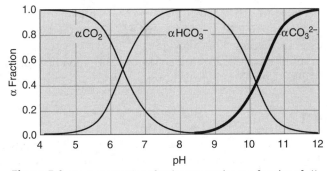

Figure 7.9 The concentration of carbonate species as a function of pH.

In this reaction, there is one HCO_3^- produced for every H^+. Thus, $[H^+] = [HCO_3^-]$ and

$$Ka_1 = \frac{[H^+]^2}{[CO_2]} = 4.45 \times 10^{-7}$$

Substituting this into the acid–base equation,

$$Ka_1 = \frac{[H^+]^2}{[CO_2]} = 4.45 \times 10^{-7} = \frac{[H^+]^2}{(1.18 \times 10^{-5}\ mol/L)}$$

$$(4.45 \times 10^{-7})(1.18 \times 10^{-5}\ mol/L) = [H^+]^2$$

$$5.25 \times 10^{-12} = [H^+]^2$$

$$2.29 \times 10^{-6} = [H]$$

$$5.64 = pH$$

The acid–base chemistry of atmospheric CO_2 dissolving in water explains why unpolluted rainwater has a pH of 5.64 rather than 7.0.

Alkalinity

Alkalinity is the capacity of water to accept H^+ to reach a pH of 4.5, which is the second equivalence point in the titration of carbonate (CO_3^{2-}) with H^+. Generally, alkalinity is controlled by the concentration of OH^-, CO_3^{2-}, and HCO_3^- in water. Because these ions can act as pH buffers and as a reservoir for inorganic carbon, the determination of the alkalinity of a natural water is a measure of its ability to support aquatic life.

$$Alkalinity = [OH^-] + 2[CO_3^{2-}] + [HCO_3^-]$$

When a water sample that has a pH of greater than 4.5 is titrated with acid to a pH 4.5 end point, all OH^-, CO_3^{2-}, and HCO_3^- will be neutralized. The end point of pH 4.5 can be measured using a pH meter with a H^+-sensitive glass electrode. Alkalinity is usually expressed as millimoles of H^+ needed to bring 1 L of water to pH 4.5. The method used to determine alkalinity in natural waters is discussed in Chapter 9.

Natural waters typically have an alkalinity of 1 mM. The individual contribution made by HCO_3^-, CO_3^{2-}, and OH^- depends on the pH of the water.

■ Acid Rain

Acid rain is a serious environmental problem. As we have seen, pure rainwater is naturally slightly acidic, with a pH of 5.6. Rainwater with a pH of less than 5.6 is characterized as **acid rain**. In recent years, the average pH of rainfall in many parts of northeastern North America has fallen below 4.6 (Figure 7.10), and rain with a pH as low as 2.9 has been recorded.

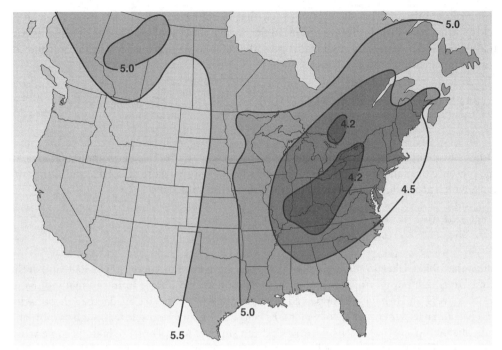

Figure 7.10 The average pH of precipitation in the United States and Canada. In many parts of North America, especially in the northeast, precipitation is abnormally acidic (pH below 5.6).

The Effects of Acid Rain

Weathering of rock is in large part caused by the acid that is naturally found in pure rainwater and soils. If rain is unusually acidic (pH less than 5.6), the normal weathering process is accelerated, and other chemical reactions also occur. The effects on the environment of both pure rainwater and acid precipitation depend on the type of soil and bedrock that the precipitation encounters.

In the normal process of weathering, when naturally acidic rain falls on soils derived from limestone ($CaCO_3$), some limestone dissolves, and the acid is neutralized:

$$CaCO_3 + H^+ \rightarrow Ca^{2+} + HCO_3^-$$

Although this reaction occurs very slowly, over thousands of years, great quantities of rock are dissolved. The extensive Carlsbad Caverns in New Mexico and many other caves were formed in this way (Figure 7.11).

When acid rain falls in regions where limestone is common, such as the Midwest and the Great Lakes area, the acid is neutralized according to the previously mentioned reaction, and adverse effects are minimized. In many parts of the United States and Canada, however, the underlying rock is

Figure 7.11 These cave formations are the result of the slow dissolution, over thousands of years, of underground deposits of limestone by naturally acidic rain.

primarily granite and basalt, igneous rock that is composed of silicate minerals such as feldspar that have little buffering action. In the normal process of weathering, feldspar is partially dissolved by naturally acidic rain and soils and is converted to kaolinite (a clay) and silica, with potassium ions going into solution:

$$2\,KAlSi_3O_8 \; + \; 2H^+ \; + \; H_2O \; \rightarrow \; Al_2Si_2O_5(OH)_4 \; + \; 4SiO_2 \; + \quad 2K^+$$

feldspar   rainwater   kaolinite   silica   potassium ions
                                     (clay)

Other silicate minerals weather in the same general way to produce clays and metal ions. In the presence of an increased concentration of acid, the clay partially dissolves, and aluminum ions, which are toxic to plants and aquatic life, go into solution:

$$Al_2Si_2O_5(OH)_4 \; + \; 6H^+ \; \rightarrow \; 2\,Al^{3+} \; + \; 2SiO_2 \; + \; 5H_2O$$

clay   acid   aluminum ions   silica   water

Since the 1950s, lakes in many parts of the world have become increasingly acid, and at the same time, fish populations have declined. Salmon are no longer found in many of Nova Scotia's rivers, and fish have disappeared from many lakes in eastern Canada, the Adirondacks, and Scandinavia. When the pH of water falls below 5.5, many desirable fish, such as trout and bass, die. At a pH of 5.0, few fish of any kind can survive. At a pH of 4.5, lakes become virtually sterile. The effects on aquatic life are caused by both increased acidity and the presence of toxic metal ions, particularly Al^{3+} ions. The dissolved Al^{3+} ions precipitate as a gel on contact with the less acidic gills of the fish and hinder the normal uptake of oxygen from the water, so the fish eventually dies from suffocation.

Since 1980, trees in many forests in the eastern United States and parts of Europe—including Germany's Black Forest—have suffered severe damage (Figure 7.12). Acid rain puts trees

Figure 7.12 In many parts of the eastern United States, the growth of forests at high elevations has declined and trees are dying. Acid deposition is blamed for much of this damage.

under stress, making them unusually susceptible to damage from disease, insects, and cold temperatures. Needed nutrients, such as K^+, Ca^{2+}, and Mg^{2+} ions, are leached from the soil by acid, often being replaced by toxic Al^{3+} ions. Unusually acidic soil damages fine root hairs and destroys beneficial microorganisms. Acid rain can damage surface structures on leaves and pine needles, causing them to wither and drop.

Building materials, particularly limestone and marble, which are composed of $CaCO_3$, are readily eroded by acid. As shown in Figure 7.13, many ancient statues in Europe, particularly those on historic cathedrals, have been severely and rapidly eroded over the last 50 years, and acid rain is considered to be responsible for their accelerated weathering.

The Causes of Acid Rain

The primary anthropogenic causes of acid rain are emissions of sulfur dioxide (SO_2) and nitrogen dioxide (NO_2) from coal- and oil-burning power plants and emissions of oxides of nitrogen from automobiles. In the atmosphere, these emissions are chemically converted into sulfuric acid (H_2SO_4) and nitric acid (HNO_3), which accumulate in droplets of water in clouds and fall to Earth as rain (Figure 7.14). Because these acids can be present in fog, sleet, snow, and fine particulates as well as in rain, this type of pollution is more accurately termed **acid deposition**.

Damage to lakes and forests in the northeastern United States and Canada has usually been blamed on emissions from industrial centers in the Midwest, which are carried from their source by the prevailing winds; however, recent research suggests that the natural process of soil formation may be an important factor in causing acidification of lakes and soil. Changes in land use (such as those that have occurred in the northeastern part of the United States and in Scandinavia) are known to acidify soils. Lowering acid-causing industrial emissions, therefore, may not be as effective as hoped in saving lakes and forests.

Figure 7.13 Over time, acid deposition has eroded many ancient limestone and marble monuments.

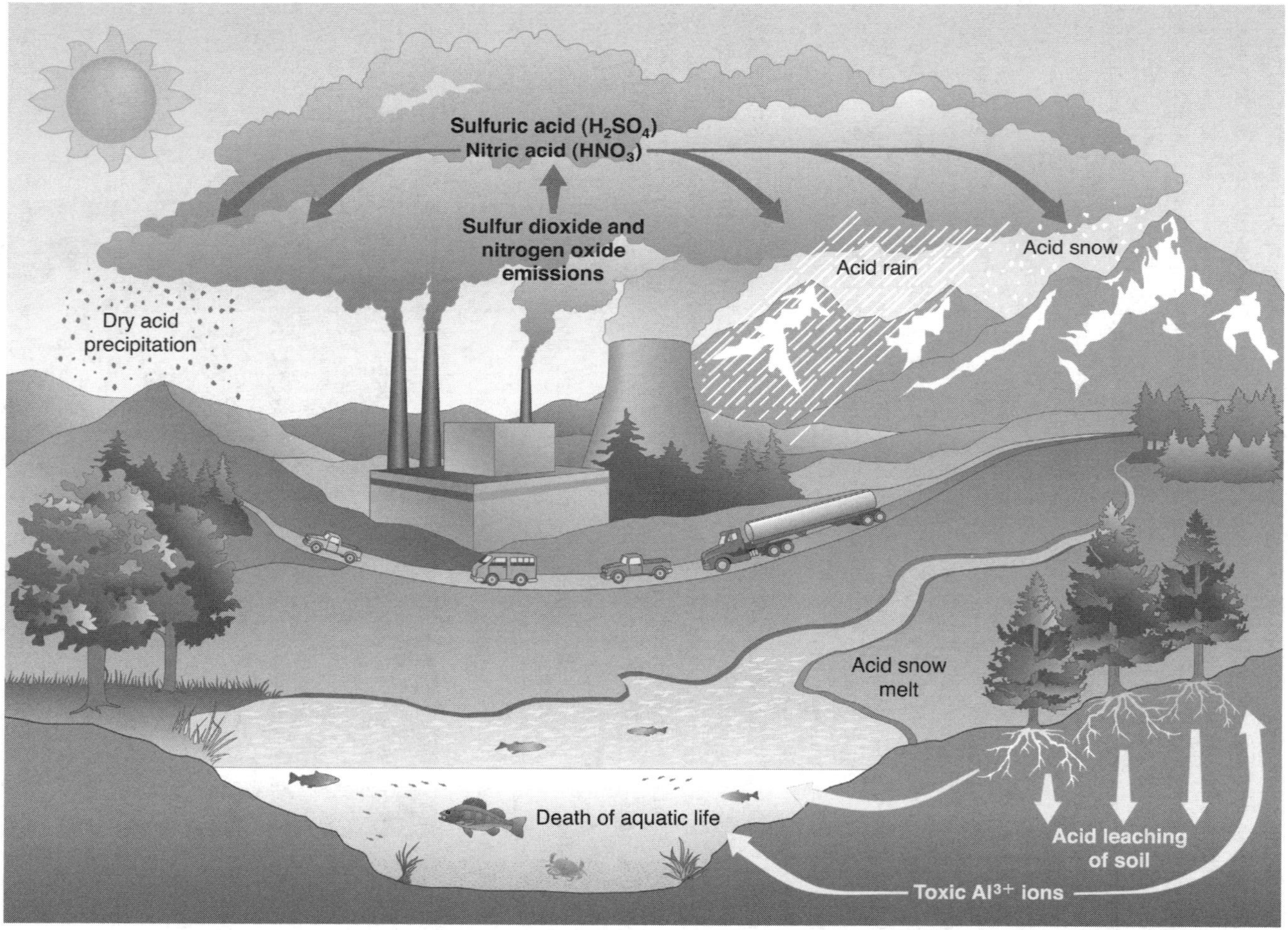
Sulfuric acid (H_2SO_4)
Nitric acid (HNO_3)
Sulfur dioxide and nitrogen oxide emissions
Dry acid precipitation
Acid rain
Acid snow
Acid snow melt
Death of aquatic life
Acid leaching of soil
Toxic Al^{3+} ions

Figure 7.14 The main sources of acid deposition are emissions from oil and coal-burning power plants and automobiles. Sulfur dioxide and nitrogen oxides from these sources combine with water vapor in the atmosphere to form sulfuric acid (H_2SO_4) and nitric acid (HNO_3). The acids collect in clouds and fall to Earth in rain and snow, causing many adverse effects. Adapted from B.J. Nebel. *Environmental Science: The Way the World Works, Third edition*. Englewood Cliffs, NJ: Prentice Hall, 1990.

■ Acid Mine Drainage

Acid released at certain types of mines can cause local increases in the acidity of soils, streams, and rivers. **Acid mine drainage** is associated primarily with the mining of coal deposits rich in the mineral pyrite (FeS_2). When underground sources of pyrite-rich coal are mined, the coal becomes exposed to water and the atmosphere. The pyrite reacts with oxygen and water, and acid is formed according to these equations:

$$2\,FeS_2 + 7O_2 + 2\,H_2O \rightarrow 2\,Fe^{2+} + 4\,SO_4^{\,2-} + 4\,H^+$$

$$4\,Fe^{2+} + O_2 + 10\,H_2O \rightarrow 4\,Fe(OH)_3 + 8\,H^+$$

Rainwater seeping through sulfur-rich wastes left at mine sites also produces acid. Although not widespread, the effects of acid runoff from mines into nearby streams can be severe and, in some instances, will have caused sizable fish kills.

Regulations now limit acid discharge from mine sites, and at most active mines, acid is neutralized with lime before it is released into natural water systems. Many abandoned mines have been sealed in an effort to prevent acid formation by excluding water and oxygen. These measures have not always been successful, and acid mine drainage remains a problem in some areas.

■ Water Use and Water Shortages

As human populations throughout the world have grown, the demand for water has increased tremendously. Water tables are falling on all continents, and many regions, including parts of the United States, are experiencing serious water shortages. In some countries, water scarcity is threatening food production; water pollution has added to the problem by reducing the proportion of water that is potable.

There are two sources of usable water: (1) **surface water** in lakes and rivers and (2) **groundwater** from wells drilled into aquifers (see Figure 7.2). Currently, only groundwater that lies within about 1000 m (approximately 3000 feet) of the Earth's surface can be tapped economically.

Worldwide, 70% of the water drawn from rivers, lakes, and aquifers is used for irrigation, and 20% is used by industry. In the United States, industry uses a much higher share, approximately 60%; irrigation accounts for 30%, and domestic and municipal use accounts for the remaining 10%. Oil refining, steel making, and paper manufacturing all require water; the major U.S. user, however, is the electric power industry, which uses large quantities of water for cooling purposes. This water is recycled at power plant sites and is reused. On the other hand, water used for irrigation, which accounts for 70% of all groundwater withdrawals, is consumed and must be continually resupplied.

The arid Plains States depend on the water in the vast Ogallala aquifer to irrigate over 10 million acres of cropland (Figure 7.15). Although the aquifer is enormous, the current rate at which water is being withdrawn greatly exceeds the rate of recharge, which is extremely slow because the aquifer underlies a region of low rainfall. It has been estimated

Figure 7.15 The vast Ogallala aquifer lies beneath the Great Plains, stretching from South Dakota to Texas. It was formed more than 2 million years ago from melting glaciers. Today, water is being withdrawn from the aquifer much more rapidly than it is being returned to it.

that at the present rate of withdrawal much of the Ogallala aquifer could be dry by 2035.

When withdrawal of water from an aquifer is excessive, the land above it may collapse, producing a large sinkhole (Figure 7.16). Depletion of groundwater near coastal areas can cause seawater to flow into an aquifer, making its water unfit for drinking. Regions of the United States where groundwater has been depleted are shown in Figure 7.17.

Potable water for domestic purposes, which is obtained almost equally from surface water and groundwater, comprises approximately 10% of all water used in the United States. Although each person needs to drink only approximately one-half gallon of water per day, the average per capita daily water use in the United States is 90 gallons, divided as shown in Table 7.5. This is nearly three times the average per capita daily use worldwide and up to 20 times the average per capita daily use in less developed countries. In some arid areas of the western United States, millions of gallons of water are used daily to water golf courses, and average per capita water consumption is approaching 400 gallons per day.

Such extravagant use of water has led to a serious water crisis. Although we cannot increase the Earth's supply of water, we can take steps to manage the water that we have more efficiently.

Figure 7.16 Excessive withdrawal of groundwater caused collapse of the land above it, forming this sink-hole in Winter Park, Florida.

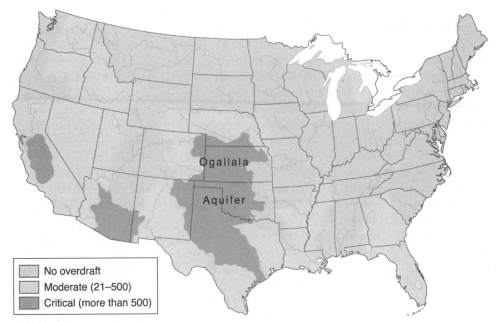

Figure 7.17 In many parts of the United States, groundwater is being withdrawn faster than the recharge rate, creating a groundwater overdraft. This map shows the groundwater overdraft in millions of gallons per day.

Table 7.5

Average per Capita Daily Water Use in the United States

Use	Rate (Gallons)
Flushing toilets	30
Bathing	23
Laundry	11
Drinking and cooking	10
Washing dishes	6
Miscellaneous	10
Total	90

■ Water Management and Conservation

Water Transfer

Limited water supplies in one region can often be increased by diverting river water to the depleted area or by building dams along rivers to collect water in large reservoirs to serve the area. These undertakings, however, are always controversial. When dams are built, agricultural land and wildlife habitat are lost, and nutrient levels fall in water downriver from the dam. Because it has been subjected to 10 major dams and many diversions, the Colorado River, for example, frequently runs dry before reaching the ocean.

River diversion often leads to an acrimonious debate between residents of neighboring regions, who are competing for the same water supplies. There have been frequent disputes about allocations from the Colorado River, which supplies water to seven states and is the main source of water for the arid Southwest. For years, California was able to withdraw more than its allotted share because less populous states were not taking their full shares. Now, as their populations have grown, Arizona and Nevada are claiming their allotments. Faced with a reduction in its water supply, California must decide how best to divide it between the needs of the cities and agriculture.

Worldwide, many nations that depend on the flow of water from other countries face similar problems, which in some cases could lead to hostilities. Egypt, which is entirely dependent on the Nile River for water, is concerned about diversions that are planned by Ethiopia, which controls the river's headwaters. In Turkey, dams on the Tigris and Euphrates Rivers will significantly reduce water flow into Syria and Iraq. The dry Middle East relies primarily on the Jordan River for water, and many believe that another war in that region could arise not over oil but over water, for which there is no substitute.

Conservation

Conservation of water is essential if serious shortages are to be avoided. Farmers could reduce their use of water substantially by lining irrigation canals with plastic to reduce loss by seepage and by burying pipes and irrigating from below ground. The commonly used method of spraying water over the land (lawn sprinklers) is inefficient because so much water evaporates before it can sink into the soil and reach plant roots.

In most industrialized countries, significant water use reductions could be made in the home. Before 1992, the average toilet in the United States converted 19 L (5 gallons) of drinkable water into wastewater with each flush. A 1992 Federal law mandated that only water-saving models (that work effectively with 1.6 gallons of water per flush) be sold. The excessive amount (almost 40 L or 10 gallons per minute) of water that older showerheads release was also reduced by switching to water-saving (2.5 gallons per minute) showerheads.

Water also can be saved by recycling wastewater from kitchen sinks and laundry tubs and using it for such purposes as flushing toilets and watering lawns (Figure 7.18). Another idea is for public water systems to provide two kinds of water: one of drinking quality and a second that would not meet all drinking water standards but would be free of bacteria and be suitable for many other household purposes.

In the United States and many other countries, there is currently little incentive to economize because water is so cheap. Farmers would be more likely to use efficient irrigation methods if their governments did not so heavily subsidize them. If water prices were higher,

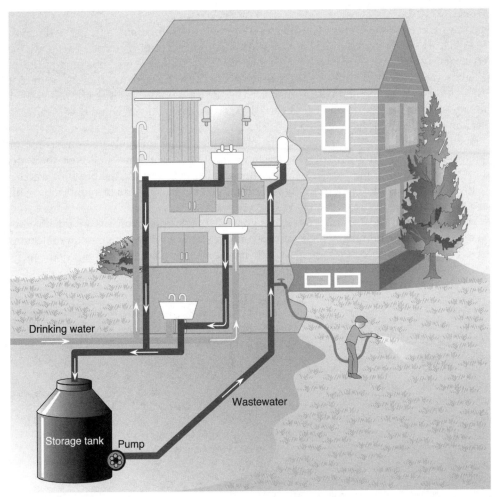

Figure 7.18 Water could be saved in homes if wastewater from sinks and bathtubs were piped into a storage tank and reused for flushing toilets, washing cars, and other activities that do not require drinking-quality water.

people in the dry Western states of the United States might exchange their green golf courses and lawns for more suitable desert vegetation.

Potable Water from Wastewater

In most urban areas of developed countries, wastewater from homes, businesses, and industrial sites flows through a network of sewer pipes to sewage treatment plants. After being treated, this water is discharged into rivers or oceans. Water could be conserved—and money saved—if the purified water from sewage plants was recycled into public water systems. Although the idea may not be attractive to many people, it is technologically feasible to obtain drinking-quality water from sewage effluent (wastewater treatment is discussed in Chapter 8).

Desalination

One way to increase the supply of freshwater is to employ **desalination**, the process of removing salts from **seawater** and **brackish water**. Although the technology exists, desalination is very expensive because it requires a large input of energy. Consequently, this process is used in only regions where there is no alternative source of freshwater. Another problem with desalination, besides its high cost, is disposing of the mountains of salt that it produces. If the salt is returned to the ocean near a coast, the increased salt content could have an adverse effect on wildlife in the coastal waters.

Distillation and reverse osmosis are the most widely used methods for converting salt water to freshwater. In **distillation**, salt water is heated to evaporate the water; the water vapor is condensed to pure water, leaving the salts behind. Because the need for water is greatest in hot, arid lands where sunlight is plentiful, energy costs can be reduced by using solar radiation to heat the salt water.

The process of **osmosis** is illustrated in Figure 7.19. Two compartments are separated by a **semipermeable membrane**, which contains minute pores through which water molecules, but not larger ions or molecules, can pass. When pure water is placed on one side of the mem-

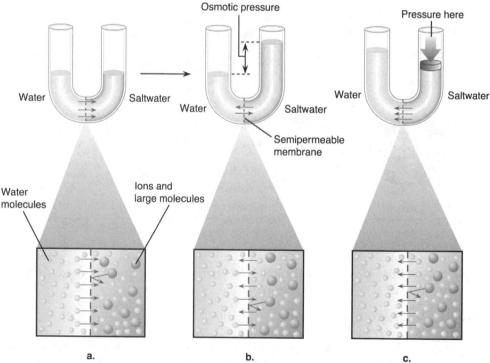

Figure 7.19 The process of osmosis and reverse osmosis: (a) Water and salt solution are separated by a semipermeable membrane. Osmosis causes water molecules to pass through the membrane into the salt solution, raising the level of liquid in that compartment. (b) The difference in levels between the two compartments represents the osmotic pressure. (c) In reverse osmosis, pressure in excess of the osmotic pressure is applied to force water molecules from the saltwater compartment into the other compartment.

brane and salt water on the other (Figure 7.19a), water molecules pass through the membrane from the water compartment to the salt compartment by the process of osmosis. The level of liquid in the salt compartment rises and reaches a maximum at equilibrium (when the movement of water molecules back and forth between the two compartments is equal; Figure 7.19b). The pressure that must be exerted to prevent this osmosis is termed the **osmotic pressure**. In the process of **reverse osmosis**, pressure in excess of the osmotic pressure is exerted on the liquid in the salt compartment, and pure water flows from the salt compartment through the membrane into the water compartment (Figure 7.19c). Reverse osmosis plants are a major source of freshwater in such arid places as Israel, Saudi Arabia, and the Mediterranean island of Malta, where freshwater is in very short supply.

■ Additional Sources of Information

Harris DC. *Quantitative Chemical Analysis,* 7th ed. New York: W. H. Freeman and Company, 2006.

Hemond HF, Fechner-Levy EJ. *Chemical Fate and Transport in the Environment,* 2nd ed. Academic Press, San Diego, CA, 2000.

Jacobson MZ. *Atmospheric Pollution.* Cambridge, UK: Cambridge University Press, 2002.

McKinney ML, Schoch RM. *Environmental Science Systems and Solutions,* 4th ed. Sudbury, MA: Jones and Bartlett Publishers, 2007.

Pinet PR. *Invitation to Oceanography,* 5th ed. Sudbury, MA: Jones and Bartlett Publishers, 2009.

■ Keywords

acid–base
acid deposition
acid mine drainage
acid rain
alkalinity
aquifer
autoprotolysis
brackish water
condensation
density
desalination
distillation

drinking water
epilimnion layer
evaporation
freshwater
groundwater
heat capacity
heat of fusion
heat of vaporization
Henry's law
hydrologic cycle (water
  cycle)
hypolimnion layer

metalimnion layer
osmosis
osmotic pressure
overturn
precipitation
protic solvent
reverse osmosis
seawater
semipermeable membrane
specific heat
surface water
transpiration

■ Questions and Problems

1. Explain why
   a. Water is a liquid at room temperature.
   b. Ice floats on liquid water.
   c. Water has such an exceptionally high boiling point.

2. Define heat of vaporization. Why is it cooler along the coast than inland? Is water, which has a large heat of vaporization, a more or less effective coolant for our bodies than another liquid with a smaller heat of vaporization?

3. What percentage of the Earth's water is seawater? List the four most concentrated metal ions present in seawater. What is the approximate concentration (in ppm) of dissolved solids in seawater?

4. For seawater,
   a. Which anion is present in the highest concentration and what is that concentration expressed in ppm?
   b. Which cation is present in the highest concentration, and what is that concentration expressed in ppm?

5. How much dissolved solids can each of the following contain?
   a. Brackish water
   b. Freshwater
   c. Drinking water

6. What is the dominant cation in each of the following:
   a. Seawater
   b. Hard water

7. Which is the dominant anion in freshwater?

8. In which of the following steps of the hydrologic cycle is water purified?
   a. Condensation
   b. Precipitation
   c. Evaporation
   d. Transpiration

9. If water molecules were linear rather than bent, would you expect ice to be less dense than liquid water? Explain.

10. At what temperature does water reach its maximum density? What are the implications of this property for life in a pond?

11. Why is the boiling point of H_2O so much higher that that of H_2S, H_2Se, and H_2Te? Why is the boiling point of water expected to be lower than the boiling point of all of these?

12. Define the following:
    a. Specific heat
    b. Heat capacity
    c. Joule

13. Calculate the following:
    a. The specific heat of water
    b. The heat capacity of 348 g of water
    c. The number of kilojoules needed to raise the temperature of 2.0 g of water from 25°C to 35°C

14. The specific heat of lead is 0.129 J/gC°. How many joules of heat are required to raise the temperature of 380 g of lead from 22°C to 37°C? Compare this with the amount of heat that is required to raise the temperature of the same amount of water the same temperature difference.

15. The specific heat of lead is 0.129 J/gC°.
    a. How does the specific heat of lead compare with other metals?
    b. How does the specific heat of lead compare with water?
    c. Would it be better to cool your car's engine with water, lead, or carbon?
16. Compare 1,000,000 L of ethanol to 1,000,000 L of water. The original temperature of each is 25°C, and quickly rises to 35°C. The specific heat of ethanol is 2.46 J/gC°. How much energy would each absorb? If lakes were filled with ethanol rather than water would the Earth's energy balance be affected?
17. Define the following:
    a. Heat of fusion
    b. Heat of vaporization
18. At what temperature is water at its maximum density? Why is solid water less dense than liquid water? Are the solid forms of most materials less dense than the liquid form?
19. Describe which of the following has the highest and lowest concentration of dissolved oxygen and which supports photosynthesis.
    a. Epilimnion
    b. Metalimnion
    c. Hypolimnion
20. Describe the seasonal process of overturn that occurs in lakes.
21. Describe what is meant by autoprotolysis. Write the autoprotolysis reaction and give the equilibrium constant for the reaction.
22. Using the autoprotolysis reaction, prove that the pH of pure water is 7.
23. Which of the following are protic solvents?
    a. Ethyl alcohol, CH_3CH_2OH
    b. Water, H_2O
    c. Diethyl ether, $CH_3CH_2OCH_2CH_3$
    d. Acetonitrile, CH_3CN
24. Suppose that the concentration of the greenhouse gas CO_2 continues to increase in our atmosphere until it reaches 500 ppm. What effect would this increase have on the pH of rainwater? How much would the pH increase or decrease?
25. What is the predominate carbonate species in natural waters with a pH of
    a. Greater than 11
    b. Less than 5
    c. Equal to 6.35
    d. Equal to 10.33
26. Air is 21% oxygen. Calculate the molar concentration of dissolved oxygen in lake water that is saturated with air. What is this concentration in ppm? The Henry's law constant for oxygen is 1.28×10^{-3} mol/atm-L.
27. Water with an alkalinity of 1.5 mM has a pH of 8.0. Calculate $[CO_2]$, $[HCO_3^-]$, $[CO_3^{2-}]$, and $[OH^-]$.
28. Water with an alkalinity of 2.5 mM has a pH of 7.0. Calculate $[CO_2]$, $[HCO_3^-]$, $[CO_3^{2-}]$, and $[OH^-]$.
29. Give equations to show how rainfall weathers limestone.

30. What is the pH of pure rain? Why is it acidic? Give equations to explain your answers.

31. Describe two natural sources of sulfuric acid.

32. Describe, giving equations, the natural acid weathering of the following:
  a. Limestone
  b. Feldspar

33. Define acid rain.
  a. What are the effects of acid rain on trees and other plant life?
  b. At what pH will a lake become sterile?

34. Why does acid rain create more serious problems in lakes in eastern North America than in the Great Lakes? Include chemical equations to illustrate your answer.

35. What is the main cause of acid mine drainage? Give an equation to show how this acid is formed.

36. What steps have been taken to control acid mine drainage? Have these methods been successful?

37. Describe the natural process that replenishes supplies of fresh water.

38. How close to the surface does groundwater need to be in order to be considered accessible?

39. Describe ways in which the average global temperature could affect the hydrologic cycle.

40. It has been proposed that water from the Great Lakes be diverted to raise the level of the Mississippi River. List possible benefits and problems that need to be considered before such a diversion begins.

41. How long does it typically take groundwater to travel 15 m (50 feet) through an aquifer?

42. Explain why land subsides when groundwater is depleted. If groundwater removal is discontinued, will the land rise to its previous level?

43. Many urban areas, such as Washington, D.C., draw their drinking water from a nearby river. Much of this water is then returned to the river in the form of treated sewage effluent. Does this affect the hydrologic cycle of the locality? How does this practice affect water quality?

44. San Diego, California, draws its water supply from the Colorado River, which is several hundred miles away. The water flows in open aqueducts from the river to the city. Would you expect this to have any affect on the quality of the water?

45. From where does the drinking water used in your community come? What alternative sources does your community have?

46. Assume that you are a homeowner and that you have a limited amount of water available from (1) water from your bath, (2) good well water, and (3) rain water drained from the roof. You want to cook dinner, give your little dog a bath, wash the car, and water your vegetable garden. Which source of water would you use for each task?

47. List five ways that you can conserve water. In your opinion, should the price of water for all uses be raised to encourage conservation in the United States?

48. During the Persian Gulf War, desalination plants in Saudi Arabia and other Middle Eastern countries were contaminated with crude oil intentionally released by the Iraqis.

Describe the following desalination processes. Do you think oil would interfere with these processes?

a. Distillation

b. Reverse osmosis

49. Two compartments are separated by a semipermeable membrane. There is pure water in one compartment and saltwater in the other. Which way will the flow proceed?

50. How is reverse osmosis different from osmosis? Explain how reverse osmosis is used to purify saltwater.

CHAPTER

8

Water Pollution and Water Treatment

- **Regulation of Water Quality**
  Rivers and Lakes
  Drinking Water

- **Additional Sources of Information**

- **Keywords**

- **Questions and Problems**

HUMAN WASTE WAS HISTORICALLY THE FIRST POLLUTION PROBLEM. In ancient times, people naturally settled near sources of water, and thus, communities grew beside lakes, along rivers, and in areas where spring or well water was available. People often drank water from the same river in which they disposed of their wastes and in which they washed themselves and their clothes. As a result, they often became sick because their drinking water was contaminated with disease-causing microorganisms from human and animal waste.

During the Industrial Revolution of the 19th century, cities in the United States and Western Europe grew at a tremendous rate. Refuse of all kinds—including great quantities of horse manure—ended up in the streets, open sewers, and nearby rivers. Devastating epidemics of water-borne diseases such as cholera, typhoid, and dysentery were common in large cities.

By the early part of the 20th century, the connection between diseases and sewage-borne microorganisms had been recognized, and safe water supplies were established in most industrialized nations. As a result, water-borne diseases have been virtually eliminated in those countries; however, they are still very common in less developed countries, where waste disposal systems are often inadequate or nonexistent. In industrialized nations, contamination with hazardous chemicals has become the main threat to water supplies.

■ Types of Water Pollutants

Water pollutants can be divided into the following broad categories: (1) disease-causing agents, (2) oxygen-consuming wastes, (3) plant nutrients, (4) suspended solids and sediments, (5) dissolved solids, (6) thermal pollution (heat), (7) toxic materials, (8) radioactive substances, (9) oil, and (10) acids. In this chapter, we consider the sources of the first six types of pollutants as well as their effects on the environment, and the steps that can be taken to control them. Oil pollution is discussed in Chapter 10 and radioactive substances are discussed in Chapter 11. Toxic materials are discussed in Chapters 13, 14, and 15.

Point and Nonpoint Sources of Water Pollutants

Pollutants enter waterways from both point and nonpoint sources (Figure 8.1). **Point sources** include sewage treatment plants, factories, electric power plants, mines, and off-shore oil-drilling rigs; in other words, these are sources that discharge pollutants at specific locations, usually through pipes. **Nonpoint sources** discharge pollutants over a wide area through runoff; these sources include feedlots, cultivated land, forests that have been

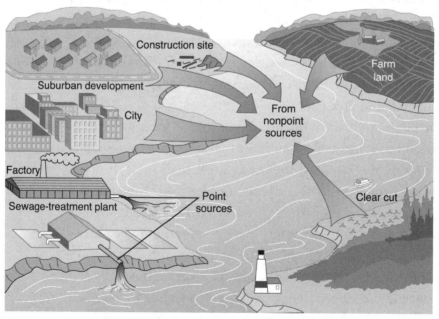

Figure 8.1 Point and nonpoint sources of water pollutants.

clear cut (that is, every tree has been cut), cities, and construction sites. Discharges from nonpoint sources are generally more dilute than those from point sources, but they are more difficult to identify and regulate.

■ Disease-Causing Agents

Outbreaks of disease, sometimes of epidemic proportions, can occur if feces from people that are infected with **pathogens** (disease-causing agents) contaminate water supplies. Diseases that are transmitted when people drink or swim in contaminated water include cholera, typhoid fever, dysentery, infectious hepatitis, and polio. According to the United Nations, in developing countries where clean water is scarce, as many as 10 million people, half of them children, die each year from drinking pathogen-contaminated water.

The coliform bacteria count is used to determine whether microorganisms contaminate water. **Coliform bacteria** live naturally in the human intestinal tract, and the average person excretes billions of them in feces each day. Coliform bacteria are harmless and cause no diseases, but their presence in water is an indication of fecal contamination. If none are found, the water is free from fecal contamination and can be assumed to be free from pathogens. Microbiological tests that are used to determine whether bacteria have contaminated water are described in detail in Chapter 9.

■ Oxygen-Consuming Wastes

Animals and plants that live in aquatic habitats depend on oxygen that is dissolved in the water for their survival. Oxygen is not very soluble in water, and the amount that does dissolve depends on both the temperature and the altitude of the water. As shown in Figure 8.2, the solubility decreases with increasing temperature and increasing altitude. At 20°C (68°F) at sea level, the concentration in oxygen-saturated water (water that contains the maximum amount of oxygen it can dissolve) is approximately 9 parts per million (ppm) compared with only 6 ppm at 30°C (86°F) at 2000 m (6600 feet).

At equilibrium, the concentration of a gas in water can be calculated using Henry's law. For dissolved oxygen, Henry's law can be written as follows:

$$[O_2] = K_H P_{O_2} \quad K_H \text{ is } O_2\text{'s Henry's law constant}$$

$$= 1.28 \times 10^{-3} \, \text{mol/L-atm} \, (25°C)$$

Where $[O_2]$ is the equilibrium concentration of oxygen in water (mol/L).

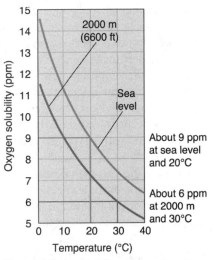

Figure 8.2 The solubility of oxygen in water at different temperatures and altitudes. The solubility decreases as temperature rises and as altitude increases.

The Henry's law constant varies with temperature, and its value for different temperatures can be looked up in a chemical handbook. Before using this equation, however, a correction must be made for the partial pressure of water by subtracting it from the total pressure of the gas. At 25°C, the partial pressure of water is 0.0313 atm (this also varies with the temperature and can be looked up in a handbook). Because dry air is 20.95% by volume oxygen, we can calculate the partial pressure of oxygen (P_{O_2}):

$$P_{O_2} = (1.00 \text{ atm} - 0.0313 \text{ atm})(0.2095) = 0.203 \text{ atm}$$

Substituting into Henry's law

$$[O_2] = K_H P_{O_2} = (1.28 \times 10^{-3} \text{ mol/L-atm})(0.203 \text{ atm})$$
$$[O_2] = 2.60 \times 10^{-4} \text{ mol/L}$$

The molecular weight of O_2 is 32 g/mol. The concentration of **dissolved oxygen (DO)** in water is as follows:

$$[O_2] = (2.60 \times 10^{-4} \text{ mol/L})(32 \text{ g/mol}) = 8.3 \text{ mg/L} = 8.3 \text{ ppm}$$

Dissolved oxygen in lakes and rivers is rapidly depleted if organic waste materials are released into the water. Typical examples of such wastes—collectively called **oxygen-consuming wastes**—are human and animal feces and industrial wastes from paper mills, tanneries, and food-processing plants. Wastes from slaughterhouses and meatpacking plants are a particularly concentrated source of oxygen-consuming wastes.

Organic detritus in aquatic ecosystems is ordinarily broken down by aerobic (oxygen-consuming) decomposers, primarily bacteria and fungi. If water is overloaded with organic wastes, the aerobic decomposers proliferate, and dissolved oxygen is consumed more rapidly than it can be replaced from the atmosphere. If the level of dissolved oxygen falls below 5 ppm, fish—particularly desirable game fish—start to die (Figure 8.3). If the concentration of dissolved oxygen continues to fall, invertebrates and aerobic bacteria will be unable to survive.

The **biological oxygen demand (BOD)** is the capacity of the organic material in a sample of water to consume the dissolved oxygen. The BOD is determined experimentally by adding heterotrophic microorganisms to a diluted effluent or water sample, saturating the sample with air, incubating the sample for five days, and then measuring the DO of the sample. The BOD is determined by comparing the initial DO to the final DO; the amount of oxygen used (mg/L) during the experiment is the BOD of the diluted sample. Unpolluted natural surface water in the United States has an average BOD value of 0.7 mg/L, whereas contaminated surface water has a BOD of greater than 5 mg/L, and municipal sewage effluent has a BOD of 50 mg/L (50 ppm).

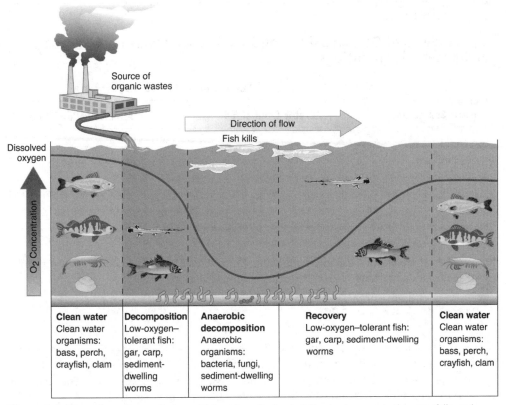

Figure 8.3 If organic wastes are discharged into a river, the level of dissolved oxygen in the water falls, and aquatic organisms begin to die. First to be affected by the decrease in dissolved oxygen are crayfish, clams and game fish such as bass and perch. If the level of dissolved oxygen continues to fall, aerobic bacteria disappear, other fish such as gar and carp die, and decomposition is taken over by anaerobic bacteria. Once the waste discharge ceases, the river recovers.

Another test that measures the contamination of natural waters is the **chemical oxygen demand (COD)**. This test, which is much faster than the BOD test, measures the concentration of organic substances that can be oxidized by acidified potassium dichromate at 100°C. A simple reaction that shows the oxidation of a generic organic pollutant, which is represented by the empirical formula of a sugar (CH_2O), follows:

$$3\,CH_2O \;+\; 16\,H_3O^+ \;+\; 2\,Cr_2O_7^{2-} \;\rightarrow\; 4\,Cr^{3+} \;+\; 3\,CO_2 \;+\; 27\,H_2O$$

The half-reaction for the oxidation by dichromate shows that six electrons are transferred:

$$Cr_2O_7^{2-} \;+\; 14\,H^+ \;+\; 6\,e^- \;\rightarrow\; 2\,H_2O$$

Table 8.1

End Products of Decomposition of Organic Compounds under Aerobic and Anaerobic Conditions

Element in Organic Compound	End Product(s) of Decomposition	
	Aerobic Conditions	Anaerobic Conditions
Carbon (C)	CO_2	CH_4
Nitrogen (N)	NO_3^-	NH_3 and amines
Sulfur (S)	SO_4^{2-}	H_2S
Phosphorus (P)	PO_4^{3-}	PH_3 and other phosphorus compounds

The number of moles of O_2 that the same sample would have consumed is 1.5 times the number of moles of dichromate because O_2 accepts only four electrons:

$$O_2 + 4H^+ + 4e^- \rightarrow 2H_2O$$

Because dichromate is such a strong oxidant, it sometimes oxidizes organic substances that would not undergo oxidation very quickly in surface water. Because of this, the COD value for the same sample is often greater than the BOD by a factor of two or more. The tests that are used to determine BOD and COD are described in detail in Chapter 9.

In the complete absence of dissolved oxygen, decomposition of organic detritus continues but it is taken over by anaerobic (non–oxygen-requiring) bacteria. If this stage is reached, the water begins to smell unpleasant because different decomposition pathways are being followed (Table 8.1). Many of the end products of anaerobic decomposition, including hydrogen sulfide, ammonia, amines, and phosphorus compounds, have disagreeable odors.

In a river, water downstream recovers quite quickly once the discharge of organic wastes ceases, particularly if the water is free flowing and turbulent. Organic material is diluted, and the water is reoxygenated as it is brought to the surface. Lakes, which have little flow of water, take much longer to recover.

■ Plant Nutrients

Water pollution can be caused by the use of synthetic fertilizers. If inorganic fertilizers in runoff water from agricultural land enter rivers and lakes, they can cause eutrophication (the excessive growth of plants and algae). Also, very soluble ions such as nitrate ions, which are toxic above certain levels, can percolate through the soil and contaminate groundwater.

Plants obtain carbon, hydrogen, and oxygen from air and water and, with energy from the sun, convert them to simple carbohydrates and oxygen through the process of photosynthesis:

$$6CO_2 + 6H_2O \rightarrow C_6H_{12}O_6 + 6O_2$$

The simple carbohydrates are then converted into the complex carbohydrates that form the basic structural material of all plants.

In addition to carbon, hydrogen, and oxygen, plants require at least 13 other elements that are called mineral nutrients because they are obtained from the soil. The **mineral nutrients** are divided into three groups according to the relative amounts of each that plants need to thrive: **primary nutrients** (nitrogen, phosphorus, and potassium), **secondary nutrients** (calcium, magnesium, and sulfur), and **micronutrients** (Table 8.2). Dissolved in water, mineral nutrients are absorbed from the soil through plant roots.

Nitrogen

Plants need nitrogen to manufacture amino acids and proteins that are necessary for the formation of leaves and stems. After carbon, hydrogen, and oxygen, nitrogen is the element that is needed in the greatest quantity. As a bushel (equal to 8 gallons) of corn ripens, it removes approximately 0.5 kg (1 pound) of nitrogen from the soil. If this nitrogen is not replaced, the soil's supply of nitrogen is gradually exhausted. Soil is more likely to be deficient in nitrogen than in any other nutrient.

Table 8.2

Chemical Nutrients Required by Plants

Name	Chemical Form Absorbed
Nonmineral	
Carbon (C)	CO_2
Hydrogen (H)	H_2O
Oxygen (O)	CO_2, H_2O, O_2
Primary nutrients	
Nitrogen (N)	NH_4^+, NO_3^-
Phosphorus (P)	$H_2PO_4^-$, HPO_4^{2-}
Potassium (K)	K^+
Secondary nutrients	
Calcium (Ca)	Ca^{2+}
Magnesium (Mg)	Mg^{2+}
Sulfur (S)	SO_4^{2-}
Micronutrients	
Boron (B)	$H_2BO_3^-$, $B(OH)_4^-$
Chlorine (Cl)	Cl^-
Copper (Cu)	Cu^{2+}
Iron (Fe)	Fe^{2+}, Fe^{3+}
Manganese (Mn)	Mn^{2+}
Molybdenum (Mo)	MoO_4^{2-}
Zinc (Zn)	Zn^{2+}

Although 78% of the atmosphere is nitrogen gas (N_2), very few plants can exploit this source directly. For most plants, **nitrogen fixation** must first change atmospheric nitrogen into a water-soluble form that plants can absorb through their roots. As we can see in Figure 8.4, nitrogen fixation is accomplished primarily by (1) soil bacteria, (2) bacteria in the soil or in nodules on the roots of leguminous plants, (2) blue-green algae (cyanobacteria) in soil and water, and (3) lightning. Nitrogen-fixing bacteria convert atmospheric nitrogen to ammonia (NH_3), which in the slightly acidic pH of most soils forms ammonium ions (NH_4^+). By the process of **nitrification**, other soil bacteria bring about the oxidation of ammonium ions to nitrate ions (NO_3^-), the preferred form for nitrogen for most plants.

During thunderstorms, electrical discharges in the atmosphere cause nitrogen to react with oxygen to produce nitric oxide (NO) and nitrogen dioxide (NO_2):

$$N_2 + O_2 \rightarrow 2\,NO$$
$$2\,NO + O_2 \rightarrow 2\,NO_2$$

Nitrogen dioxide dissolves readily in water forming nitrous acid (HNO_2) and nitric acid (HNO_3):

$$2\,NO_2 + H_2O \rightarrow HNO_2 + HNO_3$$

Nitric acid in rainfall replenishes the supply of nitrates in soil. The amount of nitric acid that thunderstorms produce is small and does not contribute significantly to the problems that acid rain created.

By the 19th century, in addition to fertilizing with manure, many farmers were practicing crop rotation, alternating a nitrogen-consuming crop, such as corn or wheat, with a nitrogen-fixing leguminous crop, such as beans, peas, alfalfa, or soybeans. The discovery of huge deposits of sodium nitrate (called Chile saltpeter) in the deserts of Chile provided a valuable additional source of nitrogen, but it was not until 1913 that a seemingly inexhaustible supply of nitrogen became available. The breakthrough came with the development by the German chemist Fritz Haber of a nitrogen-fixing process, which led directly to the production of synthetic nitrogen-containing fertilizers.

In the Haber process, nitrogen obtained from the atmosphere reacts directly with hydrogen (at a high temperature and pressure using a catalyst) to produce ammonia:

$$N_2 + 3\,H_2 \rightarrow 2\,NH_3$$

During World War I (1914–1918), the Germans used ammonia to make ammonium nitrate for the manufacture of explosives. It was not until the war was over that the process was used as the first step in the manufacture of fertilizers.

Today, ammonium salts, urea, and liquid ammonia are the major compounds that are used to restore nitrogen to the soil, and all have the Haber process as the first step in their production. The pathways by which nitrogen compounds reach the soil are shown in Figure 8.4.

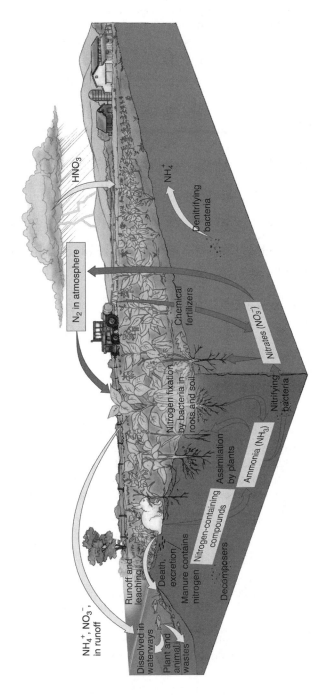

Figure 8.4 The nitrogen cycle. In nitrogen fixation, specialized bacteria convert atmospheric nitrogen to ammonia and nitrates, which absorb through their roots. Lightning fixes some nitrogen gas. Animals obtain nitrogen that they need to make tissues from plants. The wastes that animals produce and their dead bodies return nitrogen to the soil in the forms that plants can use. In denitrification, bacteria convert nitrates in soil back to nitrogen gas.

Anhydrous Ammonia

At normal temperatures and pressures, ammonia is a gas. At higher pressure, it is easily compressed into a liquid called *anhydrous ammonia,* which can be stored and transported in tanks. The liquid ammonia is injected directly into the soil, where it is converted into ammonium ions.

Ammonium Salts

The first step in the production of the fertilizer ammonium nitrate is oxidation of ammonia to nitric acid:

$$2\,NH_3 + 4\,O_2 \rightarrow 2\,HNO_3 + 2\,H_2O$$

The nitric acid then reacts with more ammonia to produce ammonium nitrate, the ammonium salt of nitric acid:

$$\underset{\text{ammonium nitrate}}{HNO_3 + NH_3 \rightarrow NH_4NO_3}$$

Ammonium sulfate and ammonium phosphate are prepared by treating ammonia with sulfuric acid and phosphoric acid, respectively.

Urea

Under high pressure, ammonia and carbon dioxide react to form urea:

$$2\,NH_3 + CO_2 \rightarrow \underset{\text{urea}}{NH_2 - \overset{\overset{\textstyle O}{\|}}{C} - NH_2} + H_2O$$

On contact with water in the soil, urea gradually decomposes, releasing ammonia as it does so. This slow release gives urea an advantage over the inorganic ammonium salts.

   The use of synthetic nitrogen fertilizers has been the main factor that has allowed food production to keep pace with population growth. The cost of manufacturing them, however, is high. Although there is an unlimited supply of nitrogen in the atmosphere, the hydrogen required to synthesize ammonia is obtained from petroleum. Small hydrocarbon molecules, such as propane (C_3H_8), are reacted with steam in the presence of suitable catalysts to yield carbon dioxide and hydrogen:

$$C_3H_8 + 6\,H_2O \rightarrow 3\,CO_2 + 10\,H_2$$

Thus, for now, the cost of synthetic fertilizers is tied to the cost and availability of petroleum. In the future, it may be possible to use solar energy to obtain hydrogen from the electrolysis of water.

   Scientists are working to understand how the unique enzymes in bacteria in legumes are able to fix atmospheric nitrogen. They hope to use genetic engineering to create nitrogen-fixing bacteria that can live on the roots of cereal plants as they do on legumes. Other approaches include improving the nitrogen-fixing abilities of legumes now being grown and treating soil with nitrogen-fixing blue-green algae (cyanobacteria).

Phosphorus

Plants need phosphorus for the synthesis of DNA and RNA and for the synthesis of adenosine 5′-triphosphate (ATP) and adenosine diphosphate (ADP), compounds that are involved in energy transfer in many metabolic processes, including photosynthesis. Phosphorus is particularly important for plants such as tomatoes that are cultivated for their fruit.

Plants absorb phosphorus from the soil in the form of phosphate ions. In natural soils at a near neutral pH, phosphates exist mainly as HPO_4^{2-} and $H_2PO_4^-$ ions, derived from the weathering of phosphate rock and from ancient deposits of skeletal remains of sea creatures and other animals. Phosphates are common in nearly all soils but are often present in low concentrations.

In water, phosphorus solubility is largely controlled by pH. The phosphate ion has three forms: $H_2PO_4^-$, HPO_4^{2-}, and PO_4^{3-}.

$$H_3PO_4 \rightarrow H_2PO_4^- + H^+ \quad pK_a \ 2.15$$

$$H_2PO_4^- \rightarrow HPO_4^{2-} + H^+ \quad pK_a \ 7.2$$

$$HPO_4^{2-} \rightarrow PO_4^{3-} + H^+ \quad pK_a \ 12.3$$

Figure 8.5 is a distribution diagram of the phosphate ion as a function of pH. In this case, the total concentration of phosphorus can be the sum of two different species, depending on pH. Notice that the distribution diagram points out that at each pK_a, the ratio of the acid and its conjugate base is one. The fraction of each aqueous species can be determined at any pH. For instance, at a pH of 4 to 5, $H_2PO_4^-$ is the predominant species, and there is a negligible amount of PO_4^{3-}.

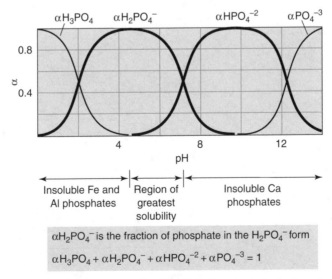

$\alpha H_2PO_4^-$ is the fraction of phosphate in the $H_2PO_4^-$ form

$\alpha H_3PO_4 + \alpha H_2PO_4^- + \alpha HPO_4^{-2} + \alpha PO_4^{-3} = 1$

Figure 8.5 Distribution of phosphate species as a function of pH. The value is the fraction of a particular species.

EXAMPLE 8.1

How were the fractions of phosphate determined for the distribution diagram shown in Figure 8.5?

Polyprotic acids dissociate stepwise (Remember that the brackets mean the "molar concentration of.")

$$H_3PO_4 \rightleftharpoons H_2PO_4^- + H^+ K_{a_1}$$

$$H_2PO_4^- \rightleftharpoons H_2PO_4^{2-} + H^+ \quad K_{a_2}$$

$$HPO_4^{2-} \rightleftharpoons PO_4^{3-} + H^+ \quad K_{a_3}$$

$$K_{a_1} = \frac{[H^+][H_2PO_4^-]}{[H_3PO_4^-]} \Rightarrow [H_2PO_4^-] = \frac{[H_3PO_4]K_{a_1}}{[H^+]}$$

$$K_{a_2} = \frac{[H^+][HPO_4^{2-}]}{[H_2PO_4]} \quad \begin{array}{l}\text{express } [H_2PO_4^-] \text{ in terms of } [H_3PO_4] \\ \text{from the previous equation}\end{array}$$

$$[HPO_4^{2-}] = \frac{[H_2PO_4^-]K_{a_2}}{[H^+]} = \frac{[H_3PO_4]K_{a_1}K_{a_2}}{[H^+]^2}$$

$$K_{a_3} = \frac{[H^+][PO_4^{3-}]}{[HPO_4^{2-}]}$$

$$[PO_4^{3-}] = \frac{[HPO_4^{2-}]K_{a_3}}{[H^+]} = \frac{[H_3PO_4]K_{a_1}K_{a_2}K_{a_3}}{[H^+]^3}$$

Mass balance: The total concentration (T) of all phosphate species is as follows:

$$T = [H_3PO_4] + [H_2PO_4^-] + [HPO_4^{2-}] + [PO_4^{3-}]$$

Substituting:

$$T = [H_3PO_4] + \frac{[H_3PO_4]K_{a_1}}{[H^+]} + \frac{[H_3PO_4]K_{a_1}K_{a_2}}{[H^+]^2} + \frac{[H_3PO_4]K_{a_1}K_{a_2}K_{a_3}}{[H^+]^3}$$

The fraction (α) of each phosphorous-containing anion can now be expressed in terms of only the pH and the K_as.

$$\alpha\, H_3PO_4 = \frac{\left[H_3PO_4\right]}{T} = \frac{\left[H^+\right]^3}{\left[H^+\right]^3 + \left[H^+\right]^2 K_{a_1} + \left[H^+\right]K_{a_1}K_{a_2} + K_{a_1}K_{a_2}K_{a_3}}$$

$$\alpha\, H_2PO_4^- = \frac{\left[H_2PO_4^-\right]}{T} = \frac{K_{a_1}\left[H^+\right]^2}{\left[H^+\right]^3 + \left[H^+\right]^2 K_{a_1} + \left[H^+\right]K_{a_1}K_{a_2} + K_{a_1}K_{a_2}K_{a_3}}$$

$$\alpha\, HPO_4^{2-} = \frac{\left[HPO_4^{2-}\right]}{T} = \frac{K_{a_1}K_{a_2}\left[H^+\right]}{\left[H^+\right]^3 + \left[H^+\right]^2 K_{a_1} + \left[H^+\right]K_{a_1}K_{a_2} + K_{a_1}K_{a_2}K_{a_3}}$$

$$\alpha\, PO_4^{3-} = \frac{\left[PO_4^{3-}\right]}{T} = \frac{K_{a_1}K_{a_2}K_{a_3}}{\left[H^+\right]^3 + \left[H^+\right]^2 K_{a_1} + \left[H^+\right]K_{a_1}K_{a_2} + K_{a_1}K_{a_2}K_{a_3}}$$

With these equations, the fraction of each anion can be calculated. The pH, K_{a_1}, K_{a_2}, and K_{a_3} are all that need be known.

Although it was not until the 19th century that scientific studies established a plant's need for phosphorus, records show that the practice of improving soil by adding ground bones was used in China as early as 2000 BC.

Because the phosphorus in bones and phosphate rock is tightly bound to calcium and is only slightly soluble in water, it is not readily available to plants. The phosphate can, however, be converted to a more soluble form, called **superphosphate**, by treatment with sulfuric acid.

$$\underset{\substack{\text{phosphate rock} \\ \text{or bone}}}{Ca_3\left(PO_4\right)_2} + 2\,H_2SO_4 \rightarrow \underset{\text{superphosphate}}{Ca\left(H_2PO_4\right)_2} + 2\,CaSO_4$$

Today, to replenish the phosphorus in soil, farmers mainly use superphosphate obtained by treating the phosphate mineral fluoroapatite $[Ca_5(PO_4)_3F]$ with sulfuric acid. An alternative fertilizer is ammonium phosphate $[(NH_4)_3PO_4]$, which supplies both phosphorus and nitrogen.

Although deposits of phosphate rock are widespread throughout the world, with major deposits in Morocco, China, and the United States, world reserves of high-grade phosphate may be depleted within the next 40 years unless new deposits are discovered. As existing supplies dwindle, more low-grade phosphate rock containing significant amounts of cadmium (Cd) must be used. Cd, which is toxic to many life forms and is subject to bioaccumulation in food chains, must be removed before the phosphate can be used; this step adds considerably to the cost of the fertilizer.

Potassium

Potassium, the third primary nutrient, is taken by plant roots as the ion K^+. Potassium takes part in the formation of starches and cellulose from simpler sugars and is essential for the normal functioning of many plant enzymes. Potassium may also be involved in the movement of carbohydrates within the plant and in the synthesis of proteins.

Potassium is distributed widely in the Earth's crust and is abundant in most soils. If crops are grown repeatedly on the same land without renewing potassium, however, the soil's supply of this nutrient will eventually be depleted. Commercial fertilizers usually contain potassium in the form of potash (K_2CO_3) or potassium chloride (KCl). Large deposits of these salts are found in the United States, Canada, and Germany. The KCl deposits under the Canadian prairies are enormous, but because they are 1.5 km (nearly 1 mile) underground, they are difficult and expensive to mine. Although there is no immediate shortage of potassium salts, it should be remembered that the deposits are nonrenewable.

■ Secondary Nutrients: Calcium, Magnesium, and Sulfur

Calcium is abundant in nearly all soils. Not only is it a component of many common minerals (limestone, dolomite, and many silicates), it is frequently spread on soil in the form of lime to neutralize acidity. Also, the application of superphosphate adds calcium ions as well as phosphate ions. Plants absorb calcium as the Ca^{2+} ion.

Magnesium, similar to calcium, is abundant in most soils. In some soils (particularly acidic ones), however, it becomes tightly bound, and there may be too few magnesium ions available in solution. The deficiency can be corrected by liming with crushed dolomite ($CaCO_3 \cdot MgCO_3$) or by adding magnesium sulfate ($MgSO_4$). Magnesium, which is absorbed through plant roots as Mg^{2+}, is a constituent of chlorophyll. A magnesium deficiency causes chlorosis (the yellowing of leaves).

Sulfur deficiencies occur most frequently in sandy, well-drained soils. The problem is treated by adding ammonium sulfate, potassium sulfate, or some other sulfate. Plants absorb sulfur as SO_4^{2-} ions. Sulfur is a constituent of amino acids and thus is essential for the synthesis of proteins.

Micronutrients

Although required in only trace amounts, micronutrients are just as essential for healthy plant growth; the effects of deficiencies can be severe. Onions grown in soil lacking zinc, for example, are stunted and develop abnormally.

It is often difficult to determine which, if any, micronutrient is in short supply in a particular soil and how much of a micronutrient should be applied to remedy a deficiency. Another problem is that the applied nutrient may bond to chemicals already in the soil and become unavailable to the plant. For example, agriculturists have found that it is preferable to add iron in the form of its ethylenediamine (EDTA) complex rather than as a simple inorganic salt. In the complexed form, the iron is protected from the soil chemicals that otherwise would bind it but is held loosely enough that it can be released gradually for uptake by plant roots.

Because plants vary considerably in their requirements, micronutrients must be applied with care. For example, the amount of boron that is ideal for sugar beets is toxic to soybeans and many other plants. Micronutrients are not always applied to the soil. Molybdenum compounds, for example, are often sprinkled on seeds before planting; other micronutrients are applied to leaves.

■ Mixed Fertilizers

Farmers and home gardeners usually correct soil deficiencies by applying fertilizers that contain a mixture of the three primary nutrients. Bags of mixed fertilizer are labeled with a set of three numbers that indicate, in the following order, the percentages of (1) nitrogen (N), (2) phosphorus as P_2O_5, and (3) potassium as K_2O (Figure 8.6). For example, if the numbers are 5–10–5, the fertilizer is composed of 5% N, 10% P_2O_5, and 10% K_2O.

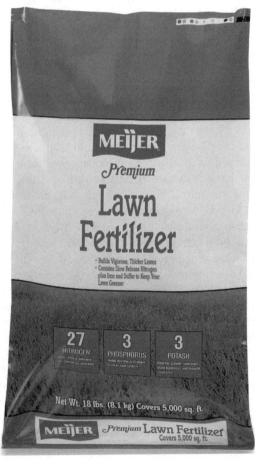

Figure 8.6 The three numbers on a bag of fertilizer refer to the percentage by weight of nitrogen (as N), phosphorus (as P_2O_5) and potassium (as K_2O).

The choice of fertilizer depends on the crop being grown. For lawns, the most important nutrient is nitrogen, and a 20–10–10 fertilizer is suitable. Fruits and vegetables require a fertilizer that is rich in phosphorus, and a 10–30–10 mix is a better choice.

■ Use of Synthetic Inorganic Fertilizers

Synthetic inorganic fertilizers were first manufactured on a large scale in the late 1930s; since then, farmers' dependence on them to return nutrients to the soil has grown dramatically. Between 1960 and 2006, fertilizer use in the United States more than tripled. The application of massive amounts of synthetic fertilizers undoubtedly has been a major factor in the huge increases in food production that were achieved worldwide in the second half of the 20th century. Unfortunately, however, the use of synthetic fertilizers can cause severe environmental problems.

If synthetic inorganic fertilizers are applied repeatedly without the simultaneous addition of sufficient organic materials to sustain the formation of humus, soil becomes compacted and loses its nutrient- and water-holding properties and its ability to fix nitrogen. Thus, increased amounts of the fertilizers must be added to maintain crop yields. Eventually, the soil becomes mineralized and increasingly susceptible to erosion. Another problem is that most commonly used fertilizers do not supply the needed micronutrients.

■ Plant Nutrients and Eutrophication

Aquatic plants need the elements nitrogen (as nitrates) and phosphorus (as phosphates) for proper growth. These are usually scarce in most natural bodies of water, thus keeping the spread of vegetation under control. The growth of primary producers is controlled by the limiting reagent, the element that is least available. Usually, the limiting nutrient is either nitrogen or phosphorus. If the amount of nitrogen or phosphorus in the water is increased, however, excessive plant growth occurs, and the algae population explodes.

As a result of this phenomenon, which is termed **eutrophication**, dense mats of rooted and floating plants are formed. Waterways become clogged and boat propellers are fouled. The cost of removing the unwanted vegetation is usually prohibitive. The carpets (also known as *blooms*) of green algae and cyanobacteria that appear on the water's surface release substances that are unpleasant smelling. The water becomes turbid, and once-attractive recreation areas are spoiled. As the vegetation and algae decay, they consume oxygen dissolved in the water, thereby producing the results that are described earlier. The effects of eutrophication are most severe in reservoirs, lakes, and estuaries or anywhere there is little or no flow of water.

A main cause of eutrophication is domestic sewage, which is made largely of nitrogen-containing human wastes and—except where the law prohibits their use—phosphorus-containing detergents. In the United States, 20 states now ban or limit phosphates in laundry detergents. In 2006, the State of Washington was the first in the nation to ban phosphates in dishwasher detergent. Other causes are animal wastes from feedlots, runoff from agricultural land that has been treated with nitrate and phosphate fertilizers, discharges from industries, and phosphate mining.

Excess nutrients in streams and rivers often become mixed with sediments (discussed in the next section). When these nutrient-rich sediments reach the ocean, they cause eutrophication.

The resulting blooms of ocean algae produce toxic substances that kill aquatic life, and as a result, many coastal regions, including up to 7000 square miles in the Gulf of Mexico, are now dead zones. Wastes from coastal fish farms add to the problem. Ocean algal blooms (often known as *red tides* because many of the algae are red) occur naturally and have been known since Biblical times, but they are occurring with increasing frequency and severity.

Control of Eutrophication

Action to control eutrophication has been focused on limiting discharges of phosphorus-containing wastes, which come mainly from point sources: sewage treatment plants, industrial plants that use large amounts of phosphorus-containing cleaning agents, and phosphate mines. Nitrogen-containing wastes are discharged primarily from nonpoint sources: agricultural land treated with manure or nitrate fertilizers, slaughterhouses, and stockyards. Nitrogen is usually present in wastes as nitrates, compounds that are very soluble in water and therefore difficult to remove. Phosphates, on the other hand, are much less soluble and can be removed by precipitation before the water is discharged (**sewage treatment** is discussed later in this chapter).

Limiting phosphate discharges is effective in controlling eutrophication. Without an adequate supply of phosphorus, excessive plant growth is greatly reduced even if sources of nitrogen remain abundant.

Anaerobic Decomposition of Organic Material

Organic material dissolved in water will decompose if anaerobic bacteria are present. Stagnant swamp water and the bottom of deep lakes are anaerobic. The bacteria cause a fermentation reaction to take place. This is a disproportionation reaction in which some of the organic carbon is oxidized to CO_2 and some is reduced to CH_4:

$$2\ CH_2O\ \rightarrow\ CH_4\ +\ CO_2$$

Because the methane is insoluble in water, it bubbles to the surface, where its aroma often is referred to as *swamp gas*. It is common to find that the tops of lakes are aerobic, whereas the bottoms of deep lakes are anaerobic.

Oxidation Reduction in Aqueous Systems

Many important **oxidation-reduction (redox)** reactions take place in aqueous systems. In many cases, bacteria are the catalysts for the reactions in natural waters. Bacteria act as catalysts in the reaction of molecular oxygen with organic material, the reduction of iron (III) to iron (II), and the oxidation of ammonia to nitrate ion.

Acid–base chemistry also influences redox reactions in natural waters. Often the transfer of electrons in a redox reaction is accompanied by the transfer of a H^+. It is a common practice to express the activity of the hydrogen ion generally in terms of acidity or basicity or specifically by the use of pH. By analogy, it is useful to have a similar unit that expresses the extent to which natural waters are oxidizing or reducing. The activity of the electron (e^-) is used to describe redox potentials in natural waters. Water with high electron activity, such as that in the (anaerobic) bottom of a deep lake, is considered to be reducing. On the other hand, water with a low electron activity, such as chlorinated drinking water, is considered oxidizing.

Analogous to **pH**, the measure of electron activity in aqueous solutions is defined as the negative logarithm of the electron activity (a_e) and is designated **pE**. (Both pH and pE are defined in terms of activity, but at a low concentration, activity and concentration can be used interchangeably.)

$$pE = -\log a_e$$

In natural waters, pEs range from −12 to 25. A large negative value of pE indicates that there are a lot of available electrons in solution, a condition that favors reduction. On the other hand, a large, positive pE indicates low electron activity, a condition that favors oxidation.

Electron Activity and pE

To determine the pE, we need to find the **standard reduction potential** (E°) for the half-reaction of interest. The standard reduction potential for most redox half-reactions has already been determined, and the E° values are listed in Appendix C. The E° is experimentally measured by connecting the electrochemical cell of interest to the standard hydrogen electrode (SHE). The concentration of all species in the cell of interest is set to be one. The potential produced is E°:

$$Fe^{3+} + e^- \rightleftharpoons Fe^{2+} \quad E° = +0.77 \text{ volts}$$

$$K_{eq} = \frac{[Fe^{2+}]}{[Fe^{3+}]a_e}$$

$$\frac{1}{a_e} = \frac{K_{eq}[Fe^{3+}]}{[Fe^{2+}]}$$

Where

a_e is the activity of the electron.

Take the log of both sides,

$$pE = -\log a_e = \log K_{eq} + \log\frac{[Fe^{2+}]}{[Fe^{3+}]}$$

Because $\Delta G° = -2.303RT \log K_{eq}$
$\Delta G° = -nFE°$

Where

E° = standard reduction potential for this half-cell
R = gas law constant [8.314 J/(Kmol) = 8.314 (VC/Kmol)]
T = temperature (K)
n = number of electrons transferred in the half-reaction
F = Faraday constant (9.648×10^4 C/mol)

$$\log K_{eq} = \frac{nFE°}{2.303RT}$$

Substituting $T = 298°K$ (25°C) gives the most useful form of this equation.

$$\log K_{eq} = \frac{nE°}{0.0591}$$

$$pE = \frac{nE°}{0.0591} + \log \frac{[Fe^{3+}]}{[Fe^{2+}]}$$

Thus, for this half-reaction at standard conditions, where $[Fe^{2+}] = [Fe^{3+}] = 1$

$$pE = \frac{(1)\,0.77}{0.0591} + \log \frac{1}{1}$$

$$pE = 13.0$$

If the concentration of $[Fe^{2+}] = 1.0 \times 10^{-5}$ and $[Fe^{3+}] = 1.0 \times 10^{-3}$ (nonstandard conditions), then

$$pE = \frac{nE°}{0.0591} + \log \frac{[Fe^{3+}]}{[Fe^{2+}]}$$

$$pE = \frac{(1)\,0.77}{0.592} + \log \frac{1.0 \times 10^{-3}}{1.0 \times 10^{-5}}$$

$$pE = 15$$

As the concentration of Fe^{3+} increases, the pE becomes more positive (13–15), and this indicates low electron activity, a condition that favors oxidation.

Generally, for a reaction

$$aA + ne^- \rightleftharpoons bB$$

where A and B are the oxidized and reduced forms of the species, and the reaction quotient (Q), which takes the structure of an equilibrium constant, is defined as follows:

$$Q = \frac{[B]^b}{[A]^a}$$

The general form of the equation is

$$pE = pE° - \frac{1}{n} \log Q$$

Table 8.3 lists the pE° values for important redox reactions in natural waters.

Table 8.3

pE⁰ Values of Redox Reactions Important in Natural Waters (at 25°C)

Reaction	pE⁰	pE⁰(W) *
(1) $1/4O_2(g) + H^+(W) + e^- \longleftrightarrow 1/2H_2O$	+20.75	+13.75
(2) $1/5NO_3^- + 6/5H^+ + e^- \longleftrightarrow 1/10N_2 + 3/5H_2O$	+21.05	+12.65
(3) $1/2MnO_2 + 1/2HCO_3^-(10^{-3}) + 3/2H^+(W) + e^- \longleftrightarrow 1/2MnCO_3(s) + H_2O$	—	+8.5†
(4) $1/2NO_3^- + H^+(W) + e^- \longleftrightarrow 1/2NO_2^- + 1/2H_2O$	+14.15	+7.15
(5) $1/8NO_3^- + 5/4H^+(W) + e^- \longleftrightarrow 1/8NH_4^+ + 3/8H_2O$	+14.90	+6.15
(6) $1/6NO_2^- + 4/3H^+(W) + e^- \longleftrightarrow 1/6NH_4^+ + 1/3H_2O$	+15.14	+5.82
(7) $1/2CH_3OH + H^+(W) + e^- \longleftrightarrow 1/2CH_4(g) + 1/2H_2O$	+9.88	+2.88
(8) $1/4CH_2O + H^+(W) + e^- \longleftrightarrow 1/4CH_4(g) + 1/4H_2O$	+6.94	−0.06
(9) $FeOOH(g) + HCO_3^-(10^{-3}) + 2H^+(W) + e^- \longleftrightarrow FeCO_3(s) + 2H_2O$	—	−1.67†
(10) $1/2CH_2O + H^+(W) + e^- \longleftrightarrow 1/2CH_3OH$	+3.99	−3.01
(11) $1/6SO_4^{2-} + 4/3H^+(W) + e^- \longleftrightarrow 1/6S(s) + 2/3H_2O$	+6.03	−3.30
(12) $1/8SO_4^{2-} + 5/4H^+(W) + e^- \longleftrightarrow 1/8H_2S(g) + 1/2H_2O$	+5.75	−3.50
(13) $1/8SO_4^{2-} + 5/4H^+(W) + e^- \longleftrightarrow 1/8H_2S(g) + 1/2H_2O$	+4.13	−3.75
(14) $1/2S + H^+(W) + e^- \longleftrightarrow 1/2H_2S$	+2.89	−4.11
(15) $1/8CO_2 + H^+ + e^- \longleftrightarrow 1/8CH_4 + 1/4H_2O$	+2.87	−4.13
(16) $1/6N_2 + 4/3H^+(W) + e^- \longleftrightarrow 1/3NH_4^+$	+4.68	−4.65
(17) $H^+(W) + e^- \longleftrightarrow 1/2H_2(g)$	0.00	−7.00
(18) $1/4CO_2(g) + H^+(W) + e^- \longleftrightarrow 1/4CH_2O + 1/4H_2O$	−1.20	−8.20

*(W) indicates $a_{H^+} = 1.00 \times 10^{-7}$ M and pE⁰(W) is a pE⁰ at $a_{H^+} = 1.00 \times 10^{-7}$ M.

†These data correspond to $a_{HCO_3^-} = 1.00 \times 10^{-3}$ M rather than unity and thus are not exactly pE⁰(W); they represent typical aquatic conditions more exactly than pE⁰ values do.

Source: Stumm WW, Morgan JJ. *Aquatic Chemistry.* New York: John Wiley, 1981, p. 449. Copyright 1981 by John Wiley & Sons, Inc. Reprinted with permission of John Wiley & Sons, Inc.

Measurement of pE

To measure pE in the environment, the sensing electrode must inert and develop a potential in response to redox reactions in the sample that are in equilibrium. A small platinum electrode is often used as the sensing electrode, with the saturated calomel electrode attached as the reference electrode. As can be seen in Figure 8.7, the platinum-sensing and calomel reference electrode are attached to a voltmeter (pH meter), which measures the voltage produced by the two half-cells.

If a sample of industrial wastewater is tested in the field using the apparatus in Figure 8.7 and a voltage of 750 mV is measured, the pE can be then determined. Our original definition of the Eo for a redox reaction involved measuring the half-cell against the SHE. By definition, the SHE's potential is 0.0 volts. Because the voltage produced by the calomel electrode is

+0.242 V, the E measured in the environmental sample can be determined by the following equation:

$$E \text{ vs. SHE} = E_{observed} + E_{sce}$$

$$E \text{ vs. SHE} = 0.750 \text{ V} + 0.242 \text{ V} = 0.992 \text{ V}$$

Thus, the pE can be calculated as follows:

$$pE = \frac{0.992}{0.0591} = 16.79$$

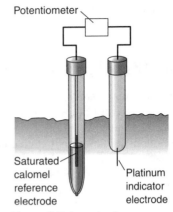

Figure 8.7 Apparatus for measuring pE in water or sediment.

This measurement indicates that the wastewater favors oxidization.

Many times the measurement of pE in an environmental sample is complicated by the fact that there may be several species present in the sample that are part of different redox systems. These samples produce potentials that are not one stable voltage, but rather drift to higher and higher potentials. Although it may not be possible to obtain a precise pE, these electrodes allow the measurement of an approximate pE that will classify a sample from a natural water as most likely to cause oxidization or a reduction.

■ Suspended Solids and Sediments

As a result of the natural erosion of rock and soil, all bodies of water contain undissolved particles termed **sediments**. Fine particles of clays remain suspended in water for months; coarser particles such as those of sand and silt settle out quite rapidly.

Many human activities increase the formation of sediments. For example, bulldozing for housing developments, clear cutting of timber, strip mining, overgrazing, and plowing are all practices that remove natural ground cover and accelerate soil erosion. Soil loss is greatest where the ground slopes steeply. It has been estimated that construction sites contribute 10 times as much sediment per unit area as farmland, 200 times as much as grassland, and 2000 times as much as undisturbed forest.

Increased loads of sediments in streams, rivers, and lakes can cause many problems. Suspended particles make water turbid, thereby reducing light penetration and slowing the rate of photosynthesis. As sediments settle, they can bury bottom-dwelling organisms and cover fish-spawning grounds, generally disrupting aquatic habitats. The effects of suspended solids can be especially damaging where rivers meet the sea in estuaries and bays, which are particularly important regions utilized as breeding grounds for fish and shellfish and as feeding grounds for birds and other wildlife.

Sediments cause problems by filling irrigation ditches and clogging harbors and lakes. Also, when toxic substances such as metals and pesticides are released into turbid water, the toxins adhere to the suspended particles and become concentrated in sediments. Then a subsequent disturbance in the water, such as an increase in acidity, may release the toxins. In clear water, where there are few suspended particles, toxic substances cause fewer problems because they are more likely to remain in solution and gradually become diluted to insignificant concentrations.

■ Dissolved Solids

Freshwater always contains some dissolved solids. The ions in solution are essential for the normal growth of most life forms, but at concentrations that are above certain levels, they become harmful and may destroy freshwater fish and other organisms. Freshwater that contains an abnormally high concentration of dissolved ionic solids is termed **saline water**. If water contains over 500 ppm of such salts, the U.S. Public Health Service does not recommend it for drinking. The salts in saline water may not be toxic but are considered pollutants if they are present in concentrations high enough to render the water unsuitable for normal purposes.

Irrigation is the main cause of increased salinity of natural waters. All over the world, food production has been increased dramatically by irrigating once-arid land. Unfortunately, this practice can lead to serious problems.

Agricultural land that is irrigated is located in regions where rainfall is sparse and the climate is generally hot and dry. Consequently, the rate of evaporation of water from the soil is high, and salts tend to accumulate on the soil surface. Runoff from the land and subsurface drainage carries an increased load of dissolved salts if returned to the water supply.

Because rainfall is scarce on irrigated land, the only way to remove excess salts is by further irrigation. Frequently, irrigation water is recycled and becomes increasingly saline with each recycling. It is estimated that 25% to 30% of the irrigated cropland in the United States suffers from increased salinity and that thousands of acres have ceased to be productive. Crops that are particularly susceptible to increased salinity include beans, carrots, and onions. Cabbage, broccoli, and tomatoes are more tolerant.

Other causes of increased salinity of natural waters are discharges from industrial and municipal waste-treatment plants and runoff from fertilized agricultural land and urban areas. Salt spread on roads to de-ice them can be a factor in regions where winters are severe.

By the natural processes of erosion, solids are continually added to a river as it flows to the sea. Thus, water near the mouth of a river always contains more dissolved solids than does water near the source. It has been estimated that in many rivers in the United States, including the Colorado, as much as half of the load of dissolved solids in the river water at the mouth is due to human activities.

■ Thermal Pollution

A different type of water pollution involves heat and is called **thermal pollution**. The electric power industry and many other industries draw huge volumes of water from rivers and lakes for cooling purposes. During operation, electric power plants produce enormous quantities of waste heat, which is removed in water that is circulated through the plant. As the water circulates, its temperature may rise by as much as 10°C to 20°C. If the warm water is returned directly into a waterway, the temperature of the water at the point of discharge increases.

A rise in water temperature can adversely affect aquatic life by increasing the body temperature of the organisms. High body temperatures can be fatal to some fish, particularly game fish. Warm water also raises respiration rates, which in turn increases oxygen consumption. At the same time, because the solubility of oxygen in water decreases with rising temperature (Figure 8.8), the dissolved oxygen content of the water decreases. Fish and other aquatic organisms may die from a lack of oxygen if they are not killed directly by increasing temperatures. The life cycles of many aquatic species are controlled by temperature. For example, the number of days that it takes for trout eggs to hatch depends on the temperature, and some fish migrate and spawn in response to slight changes in temperature. Any unusual changes in water temperature can disrupt their normal development.

Thermal pollution can be prevented in several ways. Heated water can be made to flow through a cooling pond or can be sprayed into a cooling tower and cooled by evaporation. More efficient, but more costly, is a dry tower, in which heated water transfers heat to the surrounding air as it circulates through pipes.

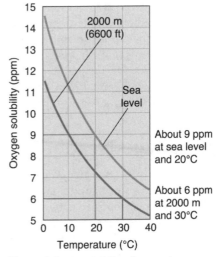

Figure 8.8 The solubility of oxygen in water at different temperatures and altitudes. The solubility decreases as temperature rises and as altitude increases.

■ Pollution of Groundwater

Thus far, we have concentrated on pollution of surface water—lakes, rivers, and streams—but of even greater concern is pollution of groundwater. Because groundwater flows and is renewed slowly, contaminants are not diluted and washed away as they are in a swiftly flowing river. Instead, they remain in the water for a very long time. Normally, groundwater is of such high quality that it meets safe drinking water standards without the need for purification or treatment and is usually pumped directly from the ground into homes. In a number of states, however, including New Jersey and Florida, formerly pure groundwater has been severely contaminated with hazardous substances, and hundreds of wells have had to be closed.

The main sources of hazardous materials that pollute groundwater are dump sites, where waste chemicals that industry produces are leaking from corroded metal drums. In the past, many industries disposed of their hazardous wastes carelessly—and often illegally—without considering the consequences. Chemical wastes generated in the production of paints, metals, textiles, fertilizers, pesticides, plastics, and petroleum products were often discarded in open dumps and landfills or were buried. The problem of cleaning up these dump sites is discussed in Chapter 18. Other types of groundwater pollutants include pesticides and fertilizers from farmland, sewage from septic tanks and leaking sewer pipes, and gasoline leaking from service station storage tanks.

As hazardous chemicals from dump sites and other sources seep through the ground, some pollutants are filtered by the soil and travel only short distances. Soluble substances such as nitrates, however, are not filtered and can percolate downward into groundwater from septic tanks, feedlots, and fertilized land. Consumption of very dilute solutions of nitrates can cause abortions in cattle and methemoglobinemia (blue baby disease) in human infants. Microorganisms in the human digestive tract convert nitrate ions to nitrite ions. Nitrite ions oxidize the iron in hemoglobin from Fe(II) to Fe(III), forming an abnormal hemoglobin called methemoglobin. Because methemoglobin is incapable of combining with oxygen, the individual's blood becomes oxygen deficient.

■ Sewage Treatment

Raw sewage is 99.9% water. The water comes primarily from flush toilets, showers and bathtubs, kitchen sinks, laundry facilities, car washes, and storm drains. In the United States, sewage is usually purified in two processes: primary and secondary treatments. Tertiary treatment, a process that is designed to take care of potentially harmful substances that are not removed in the first two stages, is currently used by few municipal treatment plants.

Primary Treatment

As the first step in **primary treatment** (Figure 8.9), the sewage is passed through a screen to remove large pieces of debris such as sticks, stones, rags, and plastic bags that have washed in through storm drains. The sewage then enters a grit chamber, where the flow rate is slowed just enough to allow coarse sand and gravel to settle on the bottom. The grit is collected and disposed of in landfills. As water enters the sedimentation tank, its flow rate is further decreased to permit suspended solids—which account for approximately 30% of the total organic wastes—to settle as **raw sludge**. Any oily material floats to the surface and is skimmed off.

Calcium hydroxide and aluminum sulfate are often added to speed up the sedimentation process. These two chemicals react to produce a gelatinous precipitate of aluminum hydroxide, which settles out slowly, carrying suspended material and bacteria with it.

$$3\,Ca(OH)_2 \;+\; Al_2(SO_4)_3 \;\rightarrow\; 2\,Al(OH)_3 \;+\; 3\,CaSO_4$$

In the past, the raw sludge that settled out was incinerated, put in landfills, or dumped at sea. Now, much of it is composted, or otherwise treated, to produce a product known as **biosolids**, a nutrient-rich, bacteria-free, humus-like material that is used as fertilizer. There is, however, concern that primary treatment does not remove toxic metals and non-biodegradable hazardous chemicals from this material. These pollutants get into sewage if metal-containing products, pesticides, and other dangerous compounds are carelessly flushed down drains.

In 1972—the year that the Clean Water Act was passed—one-third of all sewage treatment plants in the United States chlorinated the water remaining after primary treatment to kill pathogens and then discharged it into waterways without further treatment. This practice caused problems because the discharged water still contained a large amount of oxygen-consuming wastes, which often depleted dissolved oxygen in the waterways and caused eutrophication. Most sewage in the United States now receives both primary and secondary treatments, and as a result, water quality has improved in many waterways that were once seriously polluted.

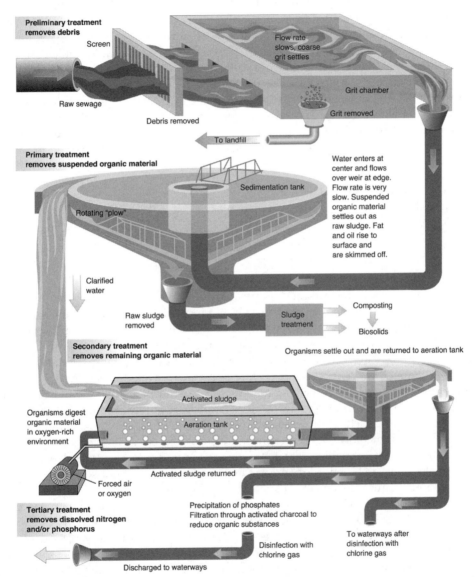

Preliminary treatment removes debris

Screen

Raw sewage

Debris removed

Flow rate slows, coarse grit settles

Grit chamber

Grit removed

To landfill

Primary treatment removes suspended organic material

Sedimentation tank

Rotating "plow"

Water enters at center and flows over weir at edge. Flow rate is very slow. Suspended organic material settles out as raw sludge. Fat and oil rise to surface and are skimmed off.

Clarified water

Raw sludge removed

Sludge treatment

Composting

Biosolids

Secondary treatment removes remaining organic material

Organisms settle out and are returned to aeration tank

Activated sludge

Organisms digest organic material in oxygen-rich environment

Aeration tank

Activated sludge returned

Forced air or oxygen

Tertiary treatment removes dissolved nitrogen and/or phosphorus

Precipitation of phosphates Filtration through activated charcoal to reduce organic substances

To waterways after disinfection with chlorine gas

Disinfection with chlorine gas

Discharged to waterways

Figure 8.9 The main steps in the processes used for sewage treatment. Adapted from B.J. Nebel. *Environmental Science: The Way the World Works, Third edition.* Englewood Cliffs, NJ: Prentice Hall, 1990.

Secondary Treatment

Secondary treatment (Figure 8.9) is a biological process that relies on aerobic bacteria and other detritus feeders to break down nearly all the oxygen-consuming wastes remaining in the water after primary treatment. In the most frequently used method, a mixture of organisms—termed **activated sludge**—is added to the sewage effluent. Air or oxygen is vigorously bubbled through pipes into the effluent as it moves slowly through a tank. In this oxygen-rich environment, the organisms digest the organic material and break it down into carbon dioxide and water. After settling in a sedimentation tank, the organisms, together with any remaining undercomposed material, are returned to the aeration tank.

Most municipal sewage treatment plants chlorinate the water after secondary treatment and then release it into waterways. Although approximately 90% of the original organic matter is removed by primary and secondary treatment, many other substances (including phosphates) are much less completely removed. Also, disinfection with chlorine can introduce hazardous chemicals.

Disinfection with Chlorine

Chlorine gas (Cl_2) reacts with water to form hypochlorous acid (HClO), which is a powerful oxidizing agent.

$$Cl_2 + H_2O \rightarrow HClO + H^+ + Cl^-$$

Hypochlorous acid kills disease-causing bacteria, destroys some (but not all) viruses, and removes color from the water. Disinfection with chlorine gas is relatively cheap and has almost completely eliminated the risk of contracting water-borne diseases from municipal water supplies. Chlorine can cause a number of environmental problems, however. It is very toxic, and there is always the danger of accidental release from tanks during transportation or at the plant. Furthermore, low concentrations of chlorine in water are toxic to fish. Perhaps the most serious problem is the finding that chlorine reacts with residual organic and inorganic substances in the water to form by-products known as **disinfectant by-products (DBPs)**, which include chloroform ($CHCl_3$), bromodichloromethane ($CHBrCl_2$), dibromochloromethane ($CHBr_2Cl$), and bromoform ($CHBr_3$), which are suspected of causing miscarriages, birth defects, and cancer. The bromine-containing compounds are produced as a result of the interaction of chlorine with bromine ions present in the water. The dangers associated with DBPs are minimized if tertiary treatment is used to remove organic substances.

Alternatives to chlorine that are sometimes used for disinfection include chloramine, chlorine dioxide, ozone, and ultraviolet (UV) radiation. Chloramines are formed by the reaction of ammonia with aqueous chlorine (that is, HOCl). Initially, chloramines were used for taste and odor control. However, it was soon recognized that chloramines were more stable than free chlorine and they were consequently found to be effective for controlling bacteria in the biolayer on the inside of the water pipes. As a result, chloramines were used regularly during the 1930s and 1940s for disinfection. Due to an ammonia shortage during World War II, however, the popularity of chloramination declined. Recent concern over chlorinated organics has renewed interest in chloramines because they form very few DBPs.

Chloramines are formed from the reaction of chlorine and ammonia. The mixture that results may contain monochloramine (NH_2Cl), dichloramine ($NHCl_2$), or nitrogen trichloride (NCl_3). When chlorine is dispersed in water, a rapid hydrolysis occurs according to the following reaction:

$$Cl_2 + H_2O \rightarrow HClO + H^+ + Cl^-$$

The equilibrium constant (K_{eq}) at 25°C is 3.94×10^4 M^{-1} for this reaction. In dilute solutions at pH greater than 3, the forward reaction is essentially complete. Hypochlorous acid (HOCl) is a weak acid that dissociates as follows:

$$HOCl \rightleftharpoons \rightleftharpoons OCl^- + H^+ \qquad pKa = 7.6$$

Relative proportions of HOCl and OCl^- are dependent upon pH. Both of the chlorine species in the above reaction are powerful oxidants, capable of reacting with many substances present in water. In aqueous solutions with pH 7.0 to 8.5, HOCl reacts rapidly with ammonia to form inorganic chloramines in a series of competing reactions. The simplified stoichiometry of chlorine-ammonia reactions are as follows:

$$NH_3 + HOCl \rightarrow NH_2Cl + H_2O \text{ (monochloramine)}$$

$$NH_2Cl + HOCl \rightarrow NHCl_2 + H_2O \text{ (dichloramine)}$$

$$NH_2Cl + HOCl \rightarrow NCl_3 + H_2O \text{ (nitrogen trichloride)}$$

These competing reactions, and several others, are primarily dependent on pH and are controlled to a large extent by the chlorine:ammonia nitrogen (Cl2:N) ratio. Temperature and contact time also play a role. Monochloramine is predominately formed when the Cl2:N ratio is less than 5:1 by weight. At Cl2:N ratios above 7.6:1, free chlorine and nitrogen trichloride are present.

Ozone (O_3) is a powerful oxidizing agent that is very effective in killing bacteria and, in the process, is converted to oxygen, which improves water quality. When ozone is used, it is manufactured on site by passing oxygen or air through an electric discharge. Ozone disinfection, which is relatively expensive, is used quite widely in Europe but less frequently in the United States. Unfortunately, ozone can also generate by-products that may be associated with health risks.

UV radiation is a relatively new disinfection technique and is effective in killing microorganisms; it is becoming increasingly popular in the United States as an alternative to chlorination. The UV source is a low-pressure mercury arc lamp.

At most sewage treatment plants in the United States, DBPs can be kept below the level allowed by the Safe Drinking Water Act by adjusting the amount of chlorine used and improving the sedimentation process. Where this is not possible, tertiary treatment may be necessary.

Tertiary Treatment

Tertiary treatment (Figure 8.9) removes DBPs by passing the chlorinated water that has undergone secondary treatment through granular charcoal that has been activated by heat. DBPs and other organic substances become attached to the surface of the carbon particles and are thus removed from the water. This procedure is effective but expensive. If necessary, reverse osmosis can be used to remove remaining dissolved organic and inorganic substances, including toxic metal ions.

Excess phosphates are removed in one of two ways. One method involves the addition of aluminum sulfate or lime (CaO), which results in the precipitation of phosphate as insoluble aluminum or calcium phosphates:

$$3\,PO_4^{3-} \;+\; Al^{3+} \;+\; 3\,Ca^{2+} \;\rightarrow\; AlPO_4 \;+\; Ca_3(PO_4)_2$$

The second method is similar to the use of activated sludge during secondary treatment. Microorganisms that absorb phosphate are added and then removed, taking the phosphates with them. The procedure also removes some remaining organic materials.

Because phosphorous is the limiting nutrient in fresh water eutrophication, it is usually the element of greatest concern to sewage treatment facilities. Nitrogen, however, is also very important because of its participation in eutrophication processes, and when it is in the ammonium form, it can react with oxygen and reduce the level of dissolved oxygen in the effluent from the treatment plant. Many areas are now being required to install nitrate removal technology in their sewage treatment plants.

Unlike phosphorous, it is not easy to precipitate nitrogen and it does not easily form insoluble precipitates. If nitrogen is present as an ammonium ion (NH_4^+), the pH of the water can be raised with calcium hydroxide, which will convert NH_4^+ to ammonia (NH_3). Subsequently, the water is passed into a stripping tower, where the water is dropped down a washboard-like device that aerates it with oxygen while simultaneously releasing the gaseous ammonia into the atmosphere. Another method of removing NH_4^+ involves passing the effluent through cation-exchange resins, which will absorb the NH_4^+.

Subsequent chlorination of the effluent also acts to remove NH_4^+ as it reacts to form mono-chloroamines, dichloroamines, and trichloroamines. As we have noted previously here, hypochlorous acid is the chlorinating species in water.

$$NH_4^+\,(aq) \;+\; HOCl\,(aq) \;\rightarrow\; NH_2Cl\,(aq) \;+\; H_3O^+\,(aq)$$

$$NH_2Cl\,(aq) \;+\; HOCl\,(aq) \;\rightarrow\; NHCl_2\,(aq) \;+\; H_2O$$

$$HOCl\,(aq) \;+\; NHCl_2\,(aq) \;\rightarrow\; NCl_3\,(aq) \;+\; H_2O$$

After treatment with chlorine, the effluent is passed through carbon filters to remove the chloramines. Because tertiary treatment is expensive, it is not used by all municipal sewage treatment plants.

A less expensive method to remove phosphorous and nitrogen may be to use a biological treatment. Nitrogen can be removed from wastewater by using nitrification bacteria in one step followed by a denitrification bacteria in the second step.

The effluent enters the first chamber where the atmosphere is aerobic and the bacterium, *Nitrosomonas nitrobacter*, which favors microbial nitrification is present. It converts aqueous ammonium to nitrate:

$$NH_4^+\,(aq) \;+\; 2\,O_2\,(aq) \;+\; H_2O \;\rightarrow\; NO_3^-\,(aq) \;+\; 2\,H_3O^+\,(aq)$$

The effluent then flows into a second chamber, where the atmosphere is anaerobic, and the bacterium, *Achromobacter* or *Pseudomonas,* which favors denitrification is present. These bacteria promote a reduction reaction where carbohydrates in the effluent act as electron donors and convert soluble nitrate into nitrogen gas:

$$4\,NO_3^-\,(aq)\,+\,5\,\{CH_2O\}\,\rightarrow\,2\,N_2(g)\,+\,4\,HCO_3^-\,(aq)\,+\,CO_2\,(aq)\,+\,3\,H_2O$$

The N_2 produced is then lost into the atmosphere.

Unfortunately, the facility required for this method must be designed specifically to remove one or both of the nutrients (P and/or N). One of the major disadvantages of the biological system is the large area that it takes to hold the reaction chambers as well as several large tanks that hold the biological system in the environment necessary so that it is ready for transfer to the reaction chambers.

■ Regulation of Water Quality

Through the 1960s, Congress passed various laws that gradually expanded federal responsibilities for maintaining water quality. Many of these laws were ineffective and were so widely ignored that water quality in the United States continued to deteriorate. Increased environmental awareness led to the passage of two important federal laws—the Clean Water Act and the Safe Drinking Water Act—in the early 1970s that were far more effective than earlier laws.

Rivers and Lakes

By the 1970s, many rivers and lakes in the United States had become so polluted that they were unsuitable for recreational activities, and it was often dangerous to eat fish taken from them. Some communities lacked safe drinking water. Since then, water quality has improved dramatically as a result of federal regulations, primarily the Clean Water Act of 1972 and its amendments of 1977 and 1978. The Act did the following: (1) provided for enforcement by the Environmental Protection Agency (EPA) with stiff penalties, (2) created a system for identifying new point sources, (3) established water-quality standards for discharged wastewaters, (4) set pretreatment standards for industrial wastes before discharge, and (5) provided federal funding to build wastewater treatment plants.

Although all five provisions of the Clean Water Act have been important, the funding of new waste treatment plants and the enforcement by EPA have been especially effective. Before the Act was passed, many localities could not afford wastewater treatment and often dumped raw sewage into local waters despite local and federal laws. In addition, the EPA was authorized to (1) establish specific water quality criteria that had to be met, (2) monitor surface waters to ensure compliance, and (3) punish violators. The EPA has established limits for more than 130 priority pollutants that are routinely being monitored.

Drinking Water

The Safe Drinking Water Act of 1974 and its amendments in 1986 focused on improving drinking water quality. It set more rigorous standards that applied to over 60,000 public water supply systems with 25 or more customers.

The Safe Drinking Water Act requires that all communities regularly monitor their drinking water supplies for toxic substances, including certain DBPs, to be sure that they do not exceed the EPA standards for drinking water. The Safe Drinking Water Act's drinking water standards fall into two categories: **Primary standards** specify contaminant levels based on health-related criteria, and **secondary standards** are based on nonhealth criteria. These improve the quality of water in aesthetic and other ways (odor, color, and taste). Because secondary standards do not directly affect life-threatening pollutants, they are generally not legally enforced. Another secondary standard is the hardness of drinking water. In Chapter 9, there is a discussion of hardness and how it is measured.

The EPA actively enforced primary standards. Contaminants are covered by primary standards and are arranged in four basic categories: inorganic chemicals, organic chemicals, radioactive matter, and microbes (pathogens). Table 8.4 lists the **maximum contaminant levels (MCLs)** for some inorganic chemicals. The MCLs are the highest concentration that the EPA allows. The EPA has similar listings for organic chemicals, radioactive material, and microbes, which are discussed in later chapters. You can easily find the list of contaminants by going to the EPA Web site (http://www.epa.gov) and by typing "MCL" into the search box.

Annual polls by the Gallup Organization show that approximately two-thirds of adult Americans have a "great deal" of concern about their drinking water. This has led to a rapid increase in the use (1) point-of-use water treatment and (2) bottled water. Point-of-use treatment refers to devices that can be used in the home to purify water; countertop carbon filters are the most common.

Table 8.4

MCLs for Certain Inorganic Chemicals

Contaminant	Principal Health Effects	MCLs (mg/L)
Arsenic	Dermal and nervous system toxicity effects	0.10
Barium	Circulatory system effects	2.0
Cadmium	Kidney effects	0.005
Chromium	Allergic dermatitis	0.1
Fluoride	Skeletal damage	4.0
Lead	Central and peripheral nervous system damage; kidney effects; highly toxic to infants and pregnant women	0.015
Mercury	Central nervous system disorders; kidney effects	0.002
Nitrate and nitrite	Methemoglobinemia (blue-baby syndrome)	10.0/1.0
Selenium	Gastrointestinal effects	0.05
Thallium	Liver/kidney effects	0.002

Source: Environmental Protection Agency, 2002.

The use of bottled water in the United States has quadrupled in the past 20 years. In 2007 the volume of bottled water sold in the U.S. surpassed 8.8 billion gallons. This means that on the average, every resident of the U.S. bought over 29 gallons. Some people buy bottled water for improved taste, smell, or color compared with tap water. Others buy it because they believe that it is healthier than tap water. This is not always true. Of the more than 700 brands of bottled water on the market, approximately 80% are not completely natural, and between 25% and 35% of the bottled-water companies use public water supplies. The discovery of benzene in bottled Perrier in 1990 caused consumers to realize that bottled water is not necessarily better than drinking water from public water supplies. Some cities including San Francisco, Albuquerque, Minneapolis, and Seattle have banned the sale of single-serve bottled water because their throwaway bottles create an unnecessary cost to taxpayers. In the past 10 years, the amount of polyethylene terephthalate (PET) bottles going into U.S. landfills rose from 1175 to 3900 million pounds. This increase would have been even bigger if at least some consumers tried to recycle their plastic bottles.

The use of point-of-use water treatment or bottled water, however, does not eliminate the risk of inhaling DBPs during showering. In 1998, because of growing concern about health problems, the EPA lowered the maximum allowable level for certain DBPs in drinking water by 20%.

■ Additional Sources of Information

Clesceri LS, Greenberg AE, Eaton AD, eds. *Standard Methods for the Examination of Water and Wastewater,* 20th ed. Washington, DC: American Public Health Association, 1998.

Environmental Protection Agency (EPA). Homepage. Accessed December 2008 from http://www.epa.gov.

Hemond HF, Fechner-Levy EJ. *Chemical Fate and Transport in the Environment,* 2nd ed. San Diego, CA: Academic Press, 2000.

McKinney ML, Schoch RM. *Environmental Science Systems and Solutions,* 4th ed. Sudbury, MA: Jones and Bartlett Publishers, 2007.

Pinet PR. *Invitation to Oceanography,* 5th ed. Sudbury, MA: Jones and Bartlett Publishers, 2009.

■ Keywords

activated sludge
biological oxygen demand (BOD)
biosolids
chemical oxygen demand (COD)
coliform bacteria
disease-causing wastes
disinfectant by-products (DBPs)
dissolved oxygen (DO)

eutrophication
maximum contaminant levels (MCLs)
micronutrients
mineral nutrients
nitrification
nitrogen fixation
nonpoint sources
oxidation-reduction (redox)

oxygen-consuming wastes

pathogens

pE

pH

point sources

primary nutrients

primary standards

primary treatment

raw sludge

saline water

secondary nutrients

secondary standards

secondary treatment

sediments

sewage treatment

standard reduction potential (E°)

superphosphate

tertiary treatment

thermal pollution

■ Questions and Problems

1. What is meant by point and nonpoint sources of pollution? Give an example of each.
2. List eight sources of water pollution.
3. In the middle of the 19th century, what was the main cause of water pollution in the United States and Europe? Explain briefly how this occurred.
4. Name three water-borne diseases that human waste causes. Are these diseases common in the United States today? Explain.
5. What are pathogens? Explain the test that is generally used to determine whether water is free from these pathogens.
6. How is the solubility of oxygen in water affected by each of the following?
   a. An increase in water temperature
   b. A decrease in pressure caused by increasing altitude
7. Calculate the concentration of CO_2 in water at 25°C using Henry's law. Assume that the water is saturated with air that contains 250 ppm of CO_2. The Henry's law constant for CO_2 is $K_H = 3.4 \times 10^{-2}$ mol/L-atm at 25°C. Express the result in molarity and ppm.
8. Calculate the concentration of CO_2 in rainwater at 25°C using Henry's law. Assume that the water is saturated with air that contains 350 ppm of CO_2. The Henry's law constant for CO_2 is $K_H = 3.4 \times 10^{-2}$ mol/L-atm at 15°C. Express the result in molarity and ppm.
9. Calculate the concentration of SO_2 in rainwater at 25°C that is in equilibrium with polluted air where the SO_2 concentration is 5.0 ppm. The Henry's law constant for SO_2 is $K_H = 1.2 \times 10^{-2}$ mol/L-atm at 25°C. Express the result in molarity and ppb.
10. Calculate the concentration of NO in rainwater at 25°C that is in equilibrium with polluted air where the NO concentration is 10.0 ppm. The Henry's law constant for NO is $K_H = 2.0 \times 10^{-4}$ mol/L-atm at 25°C. Express the result in molarity and ppb.
11. What are oxygen-consuming wastes? Name typical sources.
12. Why is coliform bacteria used to test for pathogens in drinking water?
13. What are the end products of the decomposition of organic materials containing the elements carbon, sulfur, and nitrogen under the following conditions:
    a. Aerobic conditions
    b. Anaerobic conditions
14. Explain:
    a. What type of pollutant the BOD measures in water?
    b. How to do a BOD test on a water sample.

c. Why a sample of water that you took from the lake that you visited on vacation gave a BOD measurement of 6 mg/L. Is the lake polluted? If so, what type of pollutant is present?

15. Explain:
   a. What type of pollutant the COD measures in water?
   b. How to do a COD test on a water sample.
   c. Why the COD result for a water sample twice as large as its BOD?

16. Compare the recovery of a lake to that of a river after a discharge of organic material.

17. What is eutrophication? How is it caused? Before regulations were introduced to limit its discharge, which consumer product was the primary cause of eutrophication?

18. Describe the effects of eutrophication on the aquatic life in an estuary.

19. Describe human activities that cause erosion and the deposition of sediments into waterways. List types of land use in the order in which they cause the most erosion.

20. Explain the following:
   a. Define pE.
   b. What is the range of pE in natural waters?
   c. A sample from a lake gave a pE = 10.5. Does the lake favor oxidation?
   d. A sample from a different lake gave a pE = −10.5. Does the lake favor oxidation?

21. At what depth in a lake would you expect to find high electron activity and low electron activity?

22. Determine the pE value for wastewater that contains 5.0×10^{-4} M S^{2-}. Does this wastewater favor oxidation or reduction?

$$S + 2e^- \rightarrow S^{2-} \qquad E^\circ = -0.50$$

23. Determine pE for wastewater that contains 5.0×10^{-7} M Cd^{2+}. Does this wastewater favor oxidation or reduction?

$$Cd^{2+} + 2e^- \rightarrow Cd \qquad E^\circ = -0.402$$

24. Determine pE for wastewater that contains 5.0×10^{-5} M H_3AsO_4 and 1.0×10^{-7} M H_3AsO_3. Does this wastewater favor oxidation or reduction?

$$H_3AsO_4 + 2H^+ + 2e^- \rightarrow H_3AsO_3 + H_2O \qquad E^\circ = +0.575 \text{ V}$$

25. A sample of industrial water is tested in the field using the apparatus in Figure 8.4; 550 mV are measured. Determine the pE for the wastewater and indicate whether it favors oxidation or reduction.

26. A sample of industrial water is tested in the field using the apparatus in Figure 8.4; −350 mV are measured. Determine the pE for the wastewater and indicate whether it favors oxidation or reduction.

27. A sample of industrial water is tested in the field using the apparatus in Figure 8.4; 350 mV are measured. Determine the pE for the wastewater. The wastewater is then passed through a remediation process to remove some of the pollutants. The treated water gives a reading of −200 mV. Calculate the pE of the treated water and report whether the remediation process succeeded.

28. Why does the electrode system in Figure 8.5 give voltages that are not stable?
29. The U.S. Public Health Service recommends drinking water have less than ___ ppm dissolved ionic solids.
30. Other than irrigation, list two causes of increased salinity of natural waters.
31. Are the salts in saline water classified as toxic?
32. What are the effects on aquatic life of increasing the amount of sediments in rivers? What are the economic consequences of increased sediments?
33. Water is drawn from a river to irrigate agricultural land. Explain how this can lead to increased salinity of the river water.
34. Explain how thermal pollution of natural waters has an adverse effect on fish. What effects does thermal pollution have on the metabolic rate and reproduction of aquatic organisms?
35. Using diagrams to show the flow of material, describe the following:
    a. Primary treatment
    b. Secondary treatment
    c. Tertiary treatment of sewage
36. Chlorine gas is used for disinfection of drinking water.
    a. What is the active oxidizing agent that is formed when chlorine is added to water?
    b. Describe any health risks associated with chlorination.
    c. What alternate disinfectant is used in Europe?
37. Draw the structure of three DBPs.
    a. Does treatment of water with chlorine dioxide form DBPs?
    b. How can some DBPs have bromine in their structure?
    c. DBPs are suspected of causing what medical problems?
    d. How can the levels of DBPs be lowered by sewage treatment plants?
38. What does primary treatment in a sewage treatment plant remove from the waste stream?
39. How do sewage plants deal with raw sludge obtained during the primary treatment of sewage?
40. What does secondary treatment in a sewage treatment plant remove from the waste stream?
41. What does tertiary treatment in a sewage treatment plant remove from the waste stream?
42. Describe two ways that phosphates are removed in tertiary treatment of sewage.
43. Describe two ways that nitrates are removed in tertiary treatment of sewage.
44. Why is it more difficult to remove nitrates than phosphates in tertiary treatment of sewage?
45. Describe how nitrogen can be removed from the wastewater by biological treatment.
46. For the biological treatment of wastewater containing NH_4^+, write the following:
    a. The nitrification reaction
    b. The denitrification reaction
47. Which U.S. government regulation improved the quality of water in lakes and rivers?
48. List the five provisions of the Clean Water Act. Which of these are considered the most important?
49. Which government regulation improved the quality of drinking water?
50. Define the following for the Safe Drinking Water Act:
    a. Primary standards
    b. Secondary standards

51. What are MCLs of the Safe Drinking Water Act?

52. Go to the EPA Web site and find the MCL for the following contaminants:
   a. Arsenic
   b. Cadmium
   c. Lead
   d. Benzene
   e. PCB (polychlorinated biphenyls)

CHAPTER

9

Analysis of Water and Wastewater

- Radioactive Substances

- Additional Sources of Information

- Keywords

- Questions and Problems

THE CONNECTION BETWEEN DISEASE- AND SEWAGE-BORNE MICRO-ORGANISMS WAS RECOGNIZED IN THE 1880S AND LED TO THE ORGA-NIZATION OF A SPECIAL COMMITTEE OF THE CHEMICAL SECTION OF THE AMERICAN ASSOCIATION FOR THE ADVANCEMENT OF SCIENCE, WHICH WAS TO PROPOSE "MORE UNIFORM AND EFFICIENT METHODS OF WATER ANALYSIS." A report that this committee issued in 1895 was entitled "A Method in Part, for the Sanitary Examination of Water, and for the Statement of Results."

In 1895, members of the American Public Health Association, recognizing the need for standard methods for the bacterio-logical examination of water, established a committee to draw up procedures for the study of bacteria in water. In 1905, this committee issued the first edition of *Standard Methods for the Examination of Water and Wastewater*. Revised and enlarged editions have been published since then, with the 21st edition being the latest published in 2006. *Standard Methods for the Examination of Water and Wastewater* presents **standard methods** that have been reviewed and validated by collaborative testing. The same sample is sent to many laboratories, and the proposed analytical method is performed by many different analysts to determine its reproducibility. Once a review committee has reviewed the results from all of the participating laboratories and has determined that the analytic measurements are consistent independent of laboratory, then the method is designated a standard method.

The Environmental Protection Agency (EPA) has also published methods of analysis. Some of its methods for water have incorporated the standard methods described in previous chapters. The EPA methods are written for specific tasks. A method for drinking water analyses, for instance, should not be used for hazardous waste analysis. Each EPA method has been given a number that allows it to be categorized. For example, EPA methods 1600, 1613, 1622, and 1623 are used to analyze water for bacterial contamination, and methods 601–613, 624, 625, 1624, and 1625 are used to analyze water for specific organic pollutants. To find the EPA method for a specific **analyte** (that which is being analyzed for) and/or **matrix** (what the analyte is contained in, such as water, solid waste, etc.), search the EPA Web site (http://www.epa.gov). Before using these methods to analyze drinking water samples, laboratories must be certified by EPA or the state. Once certified, laboratories must analyze performance evaluation samples, use approved methods, and undergo periodic onsite state audits.

■ Sampling Methods

One of the most important steps in an environmental analysis is obtaining a representative sample. Environmental sampling errors often greatly exceed other errors in the analytical procedure.

Many different water types are sampled and analyzed: surface waters, such as rivers, lakes and runoff water; groundwater and springwater; drinking water (**potable**); saline water; water from the atmosphere, such as fog, snow, and dew; and industrial water. Sampling methods depend on the type of water sample. Generally, samples can be retrieved from the surface of rivers or lakes and at different depths or at different points in municipal water systems. The most important factor is that the frequency and duration of sampling are sufficient to provide a representative and reproducible sample. **Grab samples** are taken at a single time and a specific location. In some cases, it may be necessary to use composite samples, where individual samples, which have been taken at frequent intervals, are combined. Generally, the average results from a large number of grab samples should give the same information as a composite sample.

For certain measurements, immediate analysis in the field is necessary to obtain reliable results because the composition of the sample may change before it arrives at the laboratory. The pH of water may change within a few minutes after collection; dissolved gases, such as carbon dioxide, hydrogen sulfide, and oxygen, may be lost, and other gases, such as oxygen and carbon dioxide, may be absorbed from the atmosphere. In certain cases, the gases can be fixed by reaction with a reagent, and then the analysis can be completed in the laboratory.

Containers that are used to collect water samples must be clean and must not contaminate or absorb analytes. Teflon containers are ideal for storing samples containing trace analytes. Generally, separate samples must be collected for chemical and biological analysis because the sampling and preservation methods are different for each. Samples should be analyzed as soon after collection as possible for maximum accuracy.

The most frequently used method for sample preservation is refrigeration to 4°C. Freezing is not recommended because it causes physical changes, such as the formation of precipitates and the loss of gas, which alter the composition of the sample. Instructions on specific preservation techniques that are used for each analyte are found in the *Standard Methods for the Examination of Water and Wastewater*.

■ Types of Water Pollutants

Chapter 8 states that water pollutants can be divided into the following broad categories: (1) disease-causing agents, (2) oxygen-consuming wastes, (3) plant nutrients, (4) suspended solids and sediments, (5) dissolved solids, (6) toxic substances, (7) heat (thermal pollution), (8) radioactive substances, (9) oil, and (10) acids. In this chapter, we examine the methods that are used to measure disease-causing agents, oxygen-consuming wastes, plant nutrients, suspended solids and sediments, dissolved solids, radioactive substances, and acids. The methods used to measure inorganic toxic substances are described in Chapter 13, and the methods that are used to measure organic toxic substances and oil are described in Chapter 14.

■ Disease-Causing Agents

Outbreaks of disease, often of epidemic proportions, can occur if human wastes that are infected with **pathogens** enter water supplies. Diseases that are transmitted when people drink contaminated water, swim in it, or eat contaminated food include cholera, typhoid fever, dysentery, infectious hepatitis, and polio.

Because it would be practically impossible to test water for each of the wide variety of pathogens that may be present, microbiological water-quality monitoring is primarily based on tests for indicator organisms. No single indicator organism can universally be used for all purposes of water-quality surveillance. Each of the wide variety of indicators available for this purpose has its own advantages and disadvantages, and the challenge is to select the appropriate indicator, or combination of indicators, for each particular purpose of water-quality assessment.

The coliform bacteria count is used to test water for contamination by microorganisms. **Coliform bacteria** (*Escherichia coli*) live naturally in the human intestinal tract, and the average person excretes billions of them in feces each day (Figure 9.1). The term *coliform bacteria* refers to a vaguely defined group of Gram-negative bacteria that have a long

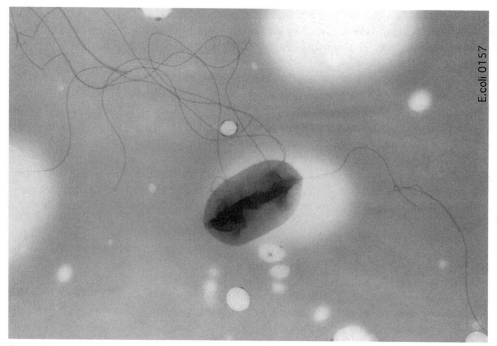

Figure 9.1 *E. coli* are extremely common symbiotic organisms in the human intestine.

history in water-quality assessment. Coliform bacteria are harmless and cause no diseases, but their presence in water is an indication of fecal contamination. If none are found, the water is free from fecal contamination and can be assumed to be free from pathogenic organisms. These bacteria, which can be determined by simple and inexpensive tests, are primarily used for assessment of the general sanitary quality of finally treated and disinfected drinking water.

In the United States, the EPA regulates the quality of public water supplies. These EPA regulations require municipal water districts to examine a minimum number of samples per month and establish the maximum number of coliform organisms allowable per 100 mL of drinking water, which is called the maximum contaminant level.

Microbiological Tests for Coliform

The methods that are used to test for coliform are the **multiple-tube fermentation technique** or the more sensitive **membrane filtration technique**, which the American Public Health Association describes in its publication *Standard Methods for the Examination of Water and Wastewater*.

Multiple-Tube Fermentation Technique

For the multiple-tube fermentation technique, results of the examination of replicate tubes and dilutions are reported in terms of the most probable number (MPN) of organisms present. This number, which is based on a probability formula, is an estimate of the mean density of **coliform bacteria** in the water sample. A set of five test tubes, filled with a lactose broth, are each inoculated with a 10-mL sample of drinking water. The test tubes are incubated at 35°C for 48 hours. Production of gas or acidic growth in the tubes is an indication of coliform contamination. Dilution of the water sample and a repeated inoculation of groups of five test tubes allow an experienced microbiologist to estimate the MPN for the water sample.

Membrane Filtration Technique
The membrane filtration technique passes a relatively large (200 mL) sample of drinking water through a sterile membrane filter. The membrane is carefully removed from the filter holder and placed on an agar (nutrient) Petri plate that is then incubated for 24 hours at 35°C to 37°C. The number of coliform colonies that have grown on the plate are counted using a stereoscopic microscope. The EPA action limit is four coliform colonies per 100 mL of water tested; if more than four colonies grow, then the sample is positive for coliform, and action has to be taken to reduce contamination of the water tested.

Ortho-nitrophenyl-β-d-galactopyranoside Test
An enzymatic assay can most rapidly identify coliform bacteria. Coliform possesses the enzyme β-d-galactosidase, which hydrolyses chromogenic substrates such as **ortho-nitrophenyl-β-d-galactopyranoside (ONPG)**, resulting in release of the chromogen and a color change to yellow in the liquid media. A kit for this ONPG test is commercially available. The inoculated media are placed in a 35°C water bath, and a yellow color after 30 minutes indicates that the sample is positive for coliform.

■ Oxygen-Consuming Wastes

Dissolved oxygen is vitally important in natural waters and can be consumed rapidly by the oxidation of organic material. In addition, oxygen in water can be consumed by the biooxidation of dissolved ammonium (NH_4^+) or by the chemical oxidation of chemical reducing agents such as Fe^{2+} or SO_3^{2-}. Analytical techniques are available that can measure the amount of dissolved oxygen directly or the organic content of the water or the O_2 required for biochemical degradation of organic materials by microorganisms.

Dissolved Oxygen
Although the concentration of dissolved oxygen in water (often abbreviated as DO) can be measured by titration or spectrophotometry, the most convenient method uses an oxygen sensor. The **Clark oxygen sensor** (Figure 9.2) has a thin, replaceable Teflon membrane that is stretched across the tip that allows diffusion of dissolved oxygen but is impermeable to ions in solution. It is an amperometric device that measures the current produced at a constant

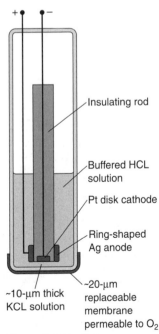

Figure 9.2 The Clark oxygen sensor.

Labels in figure:
Insulating rod
Buffered HCL solution
Pt disk cathode
Ring-shaped Ag anode
~20-μm replaceable membrane permeable to O_2
~10-μm thick KCL solution

potential. The cathode of the electrode is a platinum disk, and the anode is silver (in KCl). The electrochemical reactions are as follows:

Cathode (Pt)

$$\frac{1}{2}O_2(g) + H_2O + 2e^- \rightarrow 2OH^-(aq)$$

Anode (Ag)

$$2Cl^-(aq) + 2Ag(s) \rightarrow 2AgCl(s) + 2e^-$$

Overall

$$\frac{1}{2}O_2(g) + H_2O(l) + 2Ag(s) + Cl^- \rightarrow 2AgCl(s) + 2OH^-(aq)$$

When the oxygen sensor is immersed in water, oxygen diffuses through the membrane into the layer of KCl(aq) that is immediately adjacent to the Pt-disk electrode. Two diffusion processes take place: one through the membrane and the other through the KCl(aq). It is operated so that the rate of oxygen transport across the membrane determines the steady-state current. This rate is directly proportional to the dissolved oxygen concentration in solution (DO).

Total Organic Carbon

The organic carbon in water can be a complicated variety of organic compounds that are in various oxidation states. Figure 9.3 shows that the biological oxygen demand (BOD) decreases away from a sewage discharge point. The organic compounds that can be oxidized

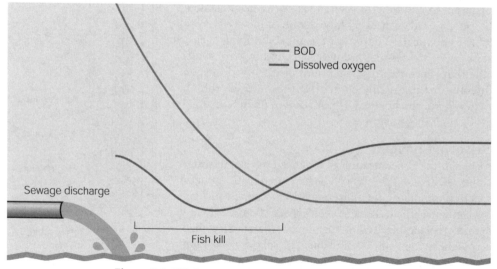

Figure 9.3 BOD deceases away from sewage discharge point.

farther by biological or chemical processes can be measured by microbiological tests that measure the BOD or the chemical oxygen demand. The measurement of **total organic carbon (TOC)**, on the other hand, is the best way of assessing the organic content of a water sample. The TOC is not dependent on the oxidation state of the organics in the water and does not measure other elements, such as nitrogen and inorganic elements that can contribute to the oxygen demand measured by other tests (BOD).

To determine the quantity of organic carbon in water samples, the organic molecules are oxidized. TOC instruments use a variety of methods, such as heat and oxygen, ultraviolet (UV) radiation, or chemical oxidants (discussed in Chapter 9), to convert the organic carbon to carbon dioxide. The CO_2 produced can be measured directly by a nondispersive infrared analyzer (described in Chapter 6); it can be reduced to methane and then measured with a flame ionization detector or can be determined by titration. Commercial instruments that measure TOC by thermal oxidation have detection limits of 4 to 50 parts per billion (ppb) (4 to 50 μg C/L). A 20-μL water sample can be analyzed in less than five minutes using a TOC instrument with a nondispersive infrared analyzer detector.

Biological Oxygen Demand

Biological oxygen demand is defined as the O_2 that is required for biochemical degradation of organic materials by microorganisms. The standard microbiological BOD method calls for incubating a sealed container of water for five days at 20°C in the dark. The incubating container must be completely filled with no extra space for air. If microbes are present, they metabolize organic compounds in the sample and in doing so consume O_2 dissolved in the solution. If there is no oxidizable organic material in the sample, no O_2 is consumed. The O_2 dissolved in the solution is measured before and after the incubation; the difference is the BOD.

Colorimetric Methods

Different chemical compounds absorb electromagnetic radiation at different wavelengths. The relationship between the concentration of the absorbing molecule and **absorbance** has been described in Beer's law (see Chapter 6):

$$A = \epsilon bc$$

Absorbance is dimensionless and, thus, all the constants and variables on the right side of the equation must have units that cancel. The concentration of the sample can be expressed in many ways: mol/L, parts per million (ppm), ppb, or mg/m³. For liquid samples, the pathlength, b, is usually 1 cm. For visible and UV spectroscopy, a liquid sample is usually placed in a cell called a *cuvet*, which has flat, fused silica faces. Glass cells can be used for measurements in the visible region, but not for UV spectroscopy, because glass absorbs UV radiation. The most common cuvets have a 1-cm pathlength and are sold in matched sets for sample and reference. The quantity ϵ (epsilon) called the molar absorptivity (or extinction coefficient) is the characteristic of a molecule that indicates how much light it will absorb at a particular wavelength. The ϵ can be expressed by using many different units. Depending on how concentration and pathlength are expressed, ϵ must have units that make the products of ϵbc dimensionless.

In most spectrophotometric analyses, it is important to prepare a reagent blank containing all reagents, but with the analyte replaced by distilled water. Any absorbance by the reagent blank must be subtracted from the absorbance of the sample.

A series of standards is usually prepared using the same procedure that will be used on the unknown sample. The absorbance of the unknown sample should fall within the region that the standards cover. If the unknown sample and the standards are prepared and measured in the same way, then the concentration of the unknown sample can be calculated from a least-squares equation for the calibration. This calculation is usually completed by spreadsheet program such as Excel.

UV-Visible Spectrometer

In Chapter 6, Figure 6.2 shows a schematic diagram for a single-beam spectrometer. Light from a lamp is separated into narrow wavelength bands by a monochromator, passed through a sample, and then measured by a detector. In making an absorbance measurement, one must be certain that only the sample analyte is absorbing. This means a reference, or blank, sample must be prepared that contains the solvent being used and any other chemical that will be present in the sample being measured. Although a single-beam instrument is inexpensive, it can be inconvenient because the sample and reference must be alternately placed in the beam. Every time that the wavelength of the spectrometer is changed, the reference sample has to be measured again.

A more convenient, but more expensive, alternative is the double-beam spectrometer. In this spectrometer, light from the source lamp alternately passes through the sample and reference (blank) cells directed by a rotating mirror (chopper). When light passes though the sample cell, the detector measures **irradiance (P)**. When the chopper directs the light through the reference cell, the detector measures P_o. The chopper divides the beam several times per second, and the instruments circuits automatically compare P and P_o to obtain absorbance and **transmittance**. This instrument automatically corrects for any changes in the source lamp intensity or detector response with time.

The UV-visible spectrometer uses a tungsten lamp as an inexpensive source of visible light. The tungsten lamp produces radiation in the range 320 to 2500 nm. The tungsten lamp covers the visible region, but a deuterium arc lamp is needed for the UV region. Figure 9.4 shows the intensity as a function of wavelengths for the two lamps. In a typical UV-visible spectrometer, a change in source lamps is made in the 300- to 350-nm region.

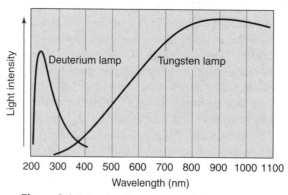

Figure 9.4 Intensity of a tungsten and deuterium lamp.

■ Plant Nutrients

Aquatic plants need the elements nitrogen (as nitrates) and phosphorus (as phosphates) for proper growth. These elements are usually scarce in most natural bodies of water, normally keeping the spread of vegetation under control; however, if a new source of nitrogen or phosphorus is introduced into the water, excessive plant growth occurs, and the algae population explodes (see Chapter 8).

Spectrophotometric Determination of Phosphorus

There is a standard method for measuring the concentration of phosphate in natural waters. Aqueous phosphate reacts with ammonium molybdate in acidic conditions to form molybdophosphoric acid. When vanadium is added to the solution, vanadomolybdophosphoric acid, which has a yellow color, is formed. The method is so designed that the intensity of the yellow color is directly proportional to the phosphate concentration.

The wavelength at which the Beer's law measurement is made depends on the concentration level that needs to be measured. The intensity of the absorbance of the vanadomolybdophosphoric acid varies 10-fold over the 400- to 490-nm wavelength range. The response for phosphate can be seen in Table 9.1.

Once a wavelength has been chosen, a calibration curve is constructed by preparing at least four standards with known concentrations of phosphate and measuring the absorbance of each. The calibration curve (Figure 9.5) produces a straight line and shows that Beer's law is valid for this system. Each unknown sample is prepared in the same way, and its absorbance is measured. Once the absorbance of the unknown is measured, its concentration is interpolated from the calibration curve.

Spectrophotometric Determination of Nitrogen

The concentration of nitrate in water samples can also be determined spectrophotometrically. First, nitrate in the solution is reduced to nitrite, which is then reacted with sulfanilamide and coupled with N-(1-naphthyl)-ethylenediamine dihydrochloride to produce a highly

---

Table 9.1	
Phosphate Response as a Function of Wavelength	
P Range (ppm, mg/L)	**λ nm**
1.0–5.0	400
2.0–10	420
4.0–18	470

Source: Standard Methods for the Examination of Water and Wastewater. (Washington, D.C.: American Public Health Association, 2004).

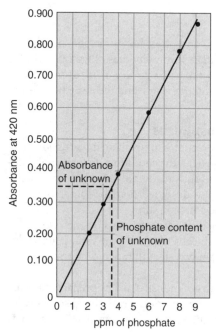

Figure 9.5 Calibration curve showing the relationship of concentration and absorbance for phosphate.

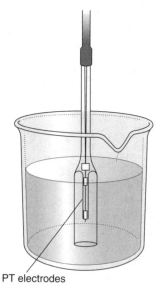

PT electrodes
Figure 9.6 The conductivity cell.

colored azo-compound that absorbs at 540 nm. It takes the azo-compound 10 minutes to form once the water sample is added to the reagents and mixed. The analyst must be careful to wait 10 minutes for the reaction to go to completion; otherwise, the results obtained will give a nitrate concentration that is arbitrarily low. This colorimetric method can determine the concentration of NO_3^- in natural waters from 5 to 1000 ppb µg/L.

■ Dissolved Solids

Fresh water always contains some dissolved solids. The ions in solution (see Table 7.3) are essential for the normal growth of most life forms, but if present at concentrations above certain levels, they become harmful and may destroy freshwater fish and other organisms.

Conductivity Measurements

The easiest way to estimate the total dissolved solids (TDS), total ionized solids (TIS), or salinity (g/kg) in a water sample is to make a conductivity measurement. Ionic solutions conduct electrical current because of the movement of ions in solution. If two electrodes are placed into the solution, they will measure its resistance. The resistance measured, however, depends on the size of the electrodes and the distance between them. To standardize the measurement, a cell is used that has a volume of 1 cm³ between two electrodes, which are each 1 cm² and 1 cm apart. This measurement of resistance is *call resistivity.*

Conductivity is the reciprocal of the resistivity and is expressed in units of ohms⁻¹ or mhos. Conductivity measurements are usually made with a conductance cell (Figure 9.6) that has platinum electrodes placed 1 cm apart and are reported in micromhos per centimeter (µmhos/cm). Conductivity reported in SI units use the term *siemens (S)* for the reciprocal of the ohm. Conductivity is reported as millisiemens per meter (mS/M): 1 mS/M = 10 µmhos/cm. To convert to mS/m, divide µmhos/cm by 10.

The conductivity of a water sample depends on the ions that are present and their concentration. All of the ions present in the solution participate in the conduction process, but not all ions conduct with the same efficiency. H^+, which is the most mobile cation in solution, is approximately five times more efficient in conducting charge in water than are the other cations, and OH⁻, the most mobile anion in solution, is approximately three times better than other anions. Conductivity depends on the concentration of ions in solution: as the concentration of an electrolyte increases, the conductivity increases because more ions are available to conduct electricity. Conductivity also varies with temperature; measurements are usually taken at 25°C.

In environmental analyses, the conductivity of a sample is often compared with a known standard. Geochemists, for instance, measure the salinity of natural waters by comparing the conductivity of a sample to that of a standard seawater sample (35 g/kg).

Electrochemical Sensors

There are several useful methods for the analysis of water that use electrochemical sensors. These methods make potentiometric measurements, where the relationship between the potential produced by the measuring electrode is related to the analyte by the logarithm of the analyte concentration.

Glass pH Electrode

The glass electrode used to measure pH is the most common ion-selective electrode. A typical pH combination electrode, which incorporates both indicator and reference electrode in one body, is shown in Figure 9.7. The glass bulb at the bottom of the electrode is the pH-sensitive part.

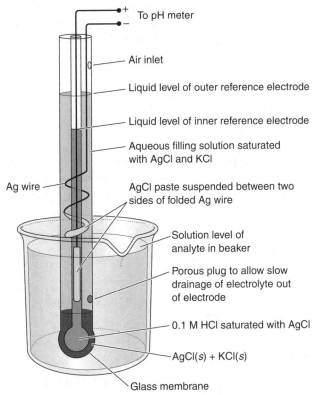

To pH meter

Air inlet

Liquid level of outer reference electrode

Liquid level of inner reference electrode

Aqueous filling solution saturated with AgCl and KCl

Ag wire

AgCl paste suspended between two sides of folded Ag wire

Solution level of analyte in beaker

Porous plug to allow slow drainage of electrolyte out of electrode

0.1 M HCl saturated with AgCl

AgCl(s) + KCl(s)

Glass membrane

Figure 9.7 Diagram of the glass combination electrode.

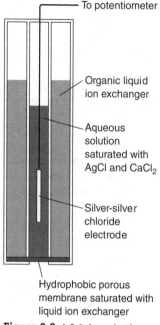

To potentiometer

Organic liquid
ion exchanger

Aqueous
solution
saturated with
AgCl and $CaCl_2$

Silver-silver
chloride
electrode

Hydrophobic porous
membrane saturated with
liquid ion exchanger

Figure 9.8 A Ca^{2+} ion-selective
electrode.

The response of the glass electrode is described by the Nernst equation:

$$E = constant + (0.059)log[H^+]$$

Where

E is the potential produced

The voltage produced is a function of the $[H^+]$, and the voltage of an ideal pH electrode changes 59 mV for every pH-unit change of $[H^+]$. A pH electrode must be calibrated before it can be used. The calibrating standards should bracket the pH of the unknown sample that is being measured.

Ca^{2+}-Sensing Ion-Selective Electrodes

The Ca^{2+}-sensing electrode is a liquid-based, ion-selective electrode. It uses a mobile carrier to transport the Ca^{2+} ion across a membrane impregnated with a solution of the carrier. A schematic diagram of the Ca^{2+} ion-selective electrode can be seen in Figure 9.8.

The carrier in the Ca^{2+} electrode is calcium didecylphosphate dissolved in dioctyl phenylphosphonate. The didecylphosphate anion binds Ca^{2+} and transports it across the membrane.

$$\left[\left(C_{10}H_{21}O\right)_2 PO_2\right]_2 Ca \rightarrow 2\left(C_{10}H_{21}O\right)_2 PO_2^- + Ca^{2+}$$
Didecylphosphate

A solution of the carrier is in the reservoir surrounding the internal Ag, AgCl electrode. Equilibration of with ion exchanger in the membrane establishes a voltage at the boundary between the membrane and analyte solution. The response of the electrode can be described with this equation:

$$E = constant + \frac{(0.059)}{2} log[Ca^{2+}]$$

The 2+ charge of the Ca^{2+} requires the factor of 2 in the denominator before the logarithm. The voltage produced is a function of the $[Ca^{2+}]$.

■ Alkalinity

In Chapter 7, the term *alkalinity* was used to describe the ability of a body of water to neutralize acids. Alkalinity takes into account the proton-accepting components of natural waters and is a very important indicator of the extent of acidification of lakes and rivers. Alkalinity is usually defined as follows:

$$\text{Alkalinity} = \underbrace{[OH^-] + [HCO_3^-] + 2[CO_3^{2-}]}_{\text{proton acceptors}} - \underbrace{[H_3O^+]}_{\text{proton donors}}$$

Typical values for the alkalinity of natural waters fall in the range of less than 50 μM to more than 2000 μM.

Another broader measure of water quality is the acid-neutralizing capacity (ANC) of natural waters. The ANC is defined as follows:

$$\text{ANC} = \underbrace{[OH^-] + [HCO_3^-] + 2[CO_3^{2-}] + [B(OH)_4^-] + [H_3SiO_4^-] + [HPO_4^{2-}] + [HS^-]}_{\text{all proton acceptors}}$$

$$+ \underbrace{[NOM^-] - [H_3O^+] - 3[Al^{3+}]}_{\text{all proton donors}}$$

Where NOM^- is natural organic material that is able to accept a proton. (The total number of proton acceptors and proton donors may be greater for some polluted natural waters than the total represented here.)

The silicon-, phosphorous-, sulfur-, and boron-containing anions that are able to contribute to the ANC are those whose pKa constants are in the pH range of natural waters. Appendix B lists the pKa constants for these acids. In almost all cases, the contribution of these species to neutralization is minimal because the concentration of these species is too small to have a substantial effect. For most unpolluted natural waters, the alkalinity is equal to the ANC, which means the only buffering species present in natural waters are carbonate species and hydroxide ions.

Measuring Acidity of Natural Waters

Acidity of natural waters refers to the total acid content that can be titrated to pH 8.3 with NaOH. The titration can be carried out using a pH-sensitive glass electrode attached to a pH meter to monitor the pH of the solution as the NaOH solution is added. The pH 8.3 endpoint is the first equivalence point for the titration of carbonic acid (H_2CO_3) with OH^-.

$$H_2CO_3 \rightleftharpoons HCO_3^- + H^+ \qquad Ka_1 = 4.45 \times 10^{-7}$$

$$HCO_3^- \rightleftharpoons CO_3^{2-} + H^+ \qquad Ka_2 = 4.69 \times 10^{-11}$$

Acidity is expressed as mg/L of $CaCO_3$ needed to bring 1 L of water to pH 8.3.

Measuring Alkalinity of Natural Waters

When water in which the pH is greater than 4.5 is titrated with acid to a pH 4.5 endpoint (which is measured with a pH electrode), all of the OH^-, CO_3^{2-}, and HCO_3^- will have been neutralized. Alkalinity is normally expressed as millimoles of H^+ needed to bring 1 L of water to a pH 4.5 endpoint.

Comparing Acidity, Alkalinity, and Hardness

The measurement of the pH of natural waters gives an intensity factor, which is a measure of the concentration of acid or base that is immediately available for reaction. Alkalinity, on the other hand, is a capacity factor that measures the ability of a body of natural water to sustain reaction with acids or base. Acidification of a lake can be considered to be a gigantic slow titration of the natural water with acid. Lakes are described as well buffered, transitional, or acidic, depending on their alkalinity. A large alkalinity value means that the lake is buffered and can accept a large amount of H^+ before the pH begins to change.

The hardness of water is defined as the sum of the total concentrations of dissolved Ca^{2+} and Mg^{2+}. Hardness is usually expressed as the equivalent number of milligrams of $CaCO_3$ per liter. If $[Ca^{2+}] + [Mg^{2+}] = 1$ mM, the hardness would be expressed as 100 mg of $CaCO_3$ per liter because the molecular weight of $CaCO_3$ is 100. If the amount of $CaCO_3$ per liter is above 270 mg, the water is considered to be "hard."

Alkalinity and hardness are important factors to measure for irrigation water. Alkalinity that is in excess of the hardness is called *residual sodium carbonate*. Water with a residual sodium carbonate content of greater than 2.5-mmol H^+/L is not suitable for irrigation. Residual sodium carbonate between 1.25- and 2.5-mmol H^+/L is marginally acceptable, but water that is less than 1.25-mmol H^+/L is ideal for irrigation.

■ Ion Chromatography

The analysis of water for common anions such as bromide, chloride, fluoride, nitrate, phosphate, and sulfate is often needed to assess its quality and to determine the need for a specific treatment. As we have seen, there are standard methods for analyzing each individual anion separately; however, only **ion chromatography** provides a single instrumental technique that can be used for the simultaneous analysis of all of these ions. Ion chromatography can also be used for the analyses of some cations.

Ion Exchange

Ion chromatography is a high-performance technique that separates ions by ion-change chromatography. In ion-exchange chromatography, anions in the sample are passed over an ion-exchange resin. The ion-exchange resins are polystyrene beads that have been treated to attach ion-exchange sites to the surface of the beads. Anion-exchange resins have a quaternary ammonium cation that is attached to the surface. Figure 9.9 shows the structure of a typical anion-exchange resin. The counter ion in this figure is the Cl^- ion, and the anion-exchange resin is said to be "in the chloride form."

All anions in the sample are attracted to the quaternary ammonium cation on the surface of the resin. Each anion competes with others in the sample for these bound-cation sites. This competition is the basis of the separation technique. Each anion has a selectivity that can be described as follows:

$$Resin\text{-}N(CH_3)_3^+Cl^- + NO_3^- \rightarrow Resin\text{-}N(CH_3)_3^+NO_3^- + Cl^-$$

Selectivity coefficient
$$K = \frac{\left[Resin\text{-}N(CH_3)_3^+NO_3^-\right]\left[Cl^-\right]}{\left[Resin\text{-}N(CH_3)_3^+Cl^-\right]\left[NO_3^-\right]}$$

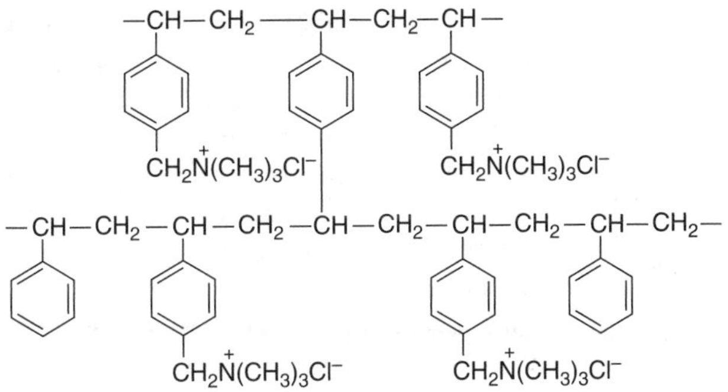

Strongly basic anion-exchange resin

Figure 9.9 Structure of an anion-exchange resin.

Relative selectivity coefficients have been experimentally measured (some are in Table 9.2). The larger the selectivity coefficient, the more efficiently the ion attaches to the ion-exchange site on the resin and the more strongly is the anion retained by the resin. NO_3^- would be expected to be held tighter than Cl^- but not as hard as I^-. Generally, the order of elution of anions from the IC will be the same as their order of increasing selectivity coefficients. In this manner, using the information, a separation of anions in a sample can be done.

The ion exchange resin is packed into a tube that is made of plastic that can withstand pressure of 2000 pounds per square inch (psi) with plumbing fittings on each end that allow it to be attached to the instrument. This device (Figure 9.10) is called the *separation column* and is generally 5 to 30 cm long and 1 to 2 cm in diameter.

Table 9.2

Relative Selectivity Coefficients of Quaternary Ammonium Anion-Exchange Resin

Anion	Relative Selectivity
F^-	0.09
OH^-	0.09
Cl^-	1.0
Br^-	2.8
NO_3^-	3.8
I^-	8.7

Source: Rohm and Hass Co.

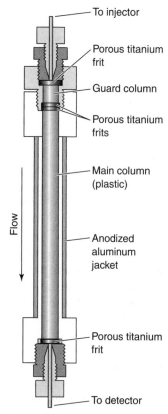

To injector

Porous titanium frit

Guard column

Porous titanium frits

Main column (plastic)

Flow

Anodized aluminum jacket

Porous titanium frit

To detector

Figure 9.10 Ion chromatography separation column. Courtesy of Upchurch Scientific, Oak Harbor, WA.

Ion Chromatography Instrument

The ion chromatography instrument consists of a pump that is capable of pumping solutions at constant flow (0.1 to 10 mL/min) and relatively high pressure (more than 1000 psi), a sample injection valve, a separation column, a suppressor, and a conductivity detector. Figure 9.11 shows a commercial ion chromatography instrument. The one common feature of all anions is that they have a negative charge. This means that a relatively simple and inexpensive conductivity detector will detect them all.

Separation and Detection of Anions by Ion Chromatography

A water sample containing low concentrations of NaF, NaCl, and Na_2SO_4 is injected into an ion chromatograph that has an anion-exchange separator column in the OH^- form. A dilute aqueous solution of KOH is pumping at a constant flow through the instrument. The OH^- is pumped through the system because it is a "driving ion" that competes with the sample anions, F^-, Cl^-, and SO_4^{2-}, for the exchange site on the resin in the column. The F^-, Cl^-, and SO_4^{2-} in the sample will repeatedly attach to different exchange sites as they move through the column. Each is then displaced by OH^- until the ions elute in narrow bands from the end of the separator column, as can be seen in the upper part of Figure 9.12. The sample anions cannot be easily detected at this point by the conductivity detector because the KOH solution that is pumped through the column has a high background conductivity, making it difficult to detect the sample anions above it. The Na^+ ions do not interact with the anion-exchange resin and are not retained and quickly pass through the column; the Na^+ ions are in the large "counter ion" peak that elutes from the column first.

To reduce the background conductivity, the solution travels into a suppressor, in which the Na^+ counter ions from the sample and K^+ counter ions from the KOH eluent are exchanged with H^+. H^+ exchanges with Na^+ through a cation-exchange membrane in the suppressor. H^+, which is at high concentration on the other side of the membrane, diffuses through the membrane to a low concentration solution. Na^+ and K^+ diffuses out of the sample stream through the membrane, as can be seen in the lower part of Figure 9.12. The net result is that the KOH solution, which has high conductivity, is converted to H_2O, which has low conductivity, and the sample anions are converted to HNO_3 and H_2SO_4, which have high conductivity. The strong acids ionize to H^+, which is the most conductive of all cations. The suppressor lowers the background conductivity while increasing the conductivity of the sample anions.

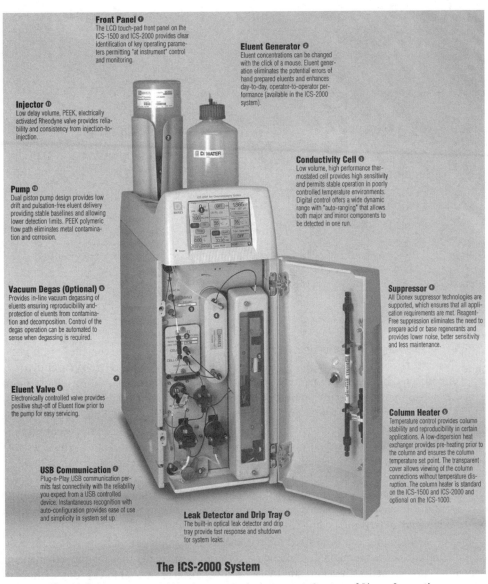

Front Panel ❶
The LCD touch-pad front panel on the ICS-1500 and ICS-2000 provides clear identification of key operating parameters permitting "at instrument" control and monitoring.

Eluent Generator ❷
Eluent concentrations can be changed with the click of a mouse. Eluent generation eliminates the potential errors of hand prepared eluents and enhances day-to-day, operator-to-operator performance (available in the ICS-2000 system).

Injector ⑪
Low delay volume, PEEK, electrically activated Rheodyne valve provides reliability and consistency from injection-to-injection.

Conductivity Cell ❸
Low volume, high performance thermostated cell provides high sensitivity and permits stable operation in poorly controlled temperature environments. Digital control offers a wide dynamic range with "auto-ranging" that allows both major and minor components to be detected in one run.

Pump ⑩
Dual piston pump design provides low drift and pulsation-free eluent delivery providing stable baselines and allowing lower detection limits. PEEK polymeric flow path eliminates metal contamination and corrosion.

Vacuum Degas (Optional) ❺
Provides in-line vacuum degassing of eluents ensuring reproducibility and protection of eluents from contamination and decomposition. Control of the degas operation can be automated to sense when degassing is required.

Suppressor ❹
All Dionex suppressor technologies are supported, which ensures that all application requirements are met. Reagent-Free suppression eliminates the need to prepare acid or base regenerants and provides lower noise, better sensitivity and less maintenance.

Eluent Valve ❾
Electronically controlled valve provides positive shut-off of Eluent flow prior to the pump for easy servicing.

Column Heater ❻
Temperature control provides column stability and reproducibility in certain applications. A low-dispersion heat exchanger provides pre-heating prior to the column and ensures the column temperature set point. The transparent cover allows viewing of the column connections without temperature disruption. The column heater is standard on the ICS-1500 and ICS-2000 and optional on the ICS-1000.

USB Communication ❼
Plug-n-Play USB communication permits fast connectivity with the reliability you expect from a USB controlled device. Instantaneous recognition with auto-configuration provides ease of use and simplicity in system set up.

Leak Detector and Drip Tray ❽
The built-in optical leak detector and drip tray provide fast response and shutdown for system leaks.

The ICS-2000 System

Figure 9.11 A commercial ion chromatography instrument. Courtesy of Dionex Corporation.

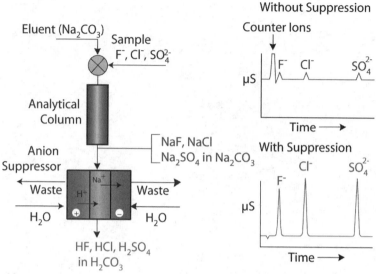

Figure 9.12 The separation and detection process in ion chromatography. Courtesy of Dionex Corporation.

Figure 9.13 shows the results of a student experiment to measure anions present in pond water. A mixture of standard anions with concentrations in the ppm range is carefully prepared and injected into the ion chromatograph at standard operating conditions. Each anion exits the instrument sequentially, giving a Gaussian peak. The time that it takes for a peak to elute is called the "retention time," and it is characteristic for each individual anion. In this analysis, NO_3^- enters the detector (elutes from the suppressor) at a six-minute retention time and SO_4^{2-} at a ten-minute retention time. Using standard conditions, we would expect NO_3^- in the pond sample to also elute at six minutes. The area under the peak is directly proportional to the concentration of the anion present. By injecting standards of known concentration and recording the area obtained allows us to construct a calibration curve of concentration versus area. The concentration of the anions in the pond water (shown in Figure 9.13) were determined using this method.

Ion chromatography has been used for the analyses of anions in many types of environmental samples. As was mentioned in Chapter 6, airborne particulates that are trapped on membrane filters are washed with water, and the solution is injected into an ion chromatograph to determine the anions that are present on the particle. The measurements of the concentration of anions in Antarctic snow provide an indication of global atmospheric pollution because there is no local source of pollution. The results of one study of Antarctic snow provided the results in Table 9.3.

■ Radioactive Substances

Uranium, thorium, and radium are naturally occurring radioactive elements that emit α, β, or γ radiation. These naturally occurring elements produce radioactive radon and thorium

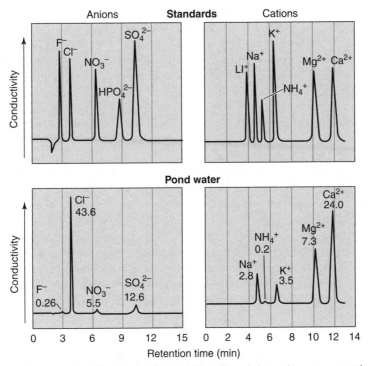

Figure 9.13 Ion chromatography of pond water. Concentrations of ions in lower chromatogram are in ppm (μg/ml). Adapted from K. Sinniah and K. Piers, *J. Chem. Ed.* 78 (2000): 358.

Table 9.3		

Anions in Antarctic Snow

Anion	Minimum ppb	Maximum ppb
F^-	0.10	6.2
Cl^-	25	40,100
Br^-	0.8	49.2
NO_3^-	8.6	354
SO_4^{2-}	10.6	4,020
$H_2PO_4^-$	1.8	49

Source: R. Udisti, S. Bewllandi, and G. Piccardi, 1994. Analysis of Snow from Antarctica. Fresenius J. Anal. Chem. 349: 289.

that contribute to the radioactivity of some groundwater. Additional naturally radioactive elements include ^{40}K, ^{87}Rb, ^{147}Sm, ^{176}Lu, and ^{187}Re. With the development of nuclear reactors, large quantities of radioactive elements such as ^{137}Cs and ^{60}Co are being produced.

The EPA has established maximum contaminant levels for natural radioactive isotopes in water. Radioactive contamination of water is normally determined by measuring the gross β (total β emission from all isotopes present) and gross α emission (total α emission from all isotopes present). This measurement is much easier to make than one that detects each individual isotope. The water to be measured is evaporated on a small pan to produce a very thin layer. The pan is then inserted inside an internal proportional counter, an instrument with a very thin window separating the components of the instrument from the sample. The thin window allows both the α and β particles to penetrate into the instrument, which is similar in construction to the Geiger counter that is described in Chapter 11. The sample is considered safe if it contains less than 50 pCi/L of β emitters and 5 pCi/L of α emitters.

If the radioisotope emits γ radiation, such as ^{137}Cs, the concentration can be determined directly by counting a large liquid sample (4 L) with a γ spectrometer. The γ spectrometer has multiple solid-state detectors that are able to measure γ rays of different wavelength, and this allows the identification of the isotope that emitted the γ ray. Using this method, it is possible to identify a number of radionuclides in the same sample without prior chemical separation.

■ Additional Sources of Information

Clesceri LS, Greenberg AE, Eaton AD, ed. *Standard Methods for the Examination of Water and Wastewater,* 20th ed. Washington, DC: American Public Health Association, 1998.

Environmental Protection Agency (EPA). Homepage. Accessed December 2008 at http://www.epa.gov.

Harris DC. *Quantitative Chemical Analysis,* 6th ed. New York: W. H. Freeman, 2003.

Hemond HF, Fechner-Levy EJ. *Chemical Fate and Transport in the Environment,* 2nd ed. San Diego, CA: Academic Press, 2000.

Rubison KA, Rubison JF. *Contemporary Instrumental Analysis.* Upper Saddle River, NJ: Prentice Hall, 2000.

Skoog DA, Holler FJ, Crouch SR. *Instrumental Analysis,* 6th ed. Philadelphia, PA: Saunders College Publishing, Harcourt Brace College Publishers, 2007.

■ Keywords

acidity	grab samples
absorbance	ion chromatography
analyte	irradiance (P)
Clark oxygen sensor	matrix
coliform bacteria	membrane filtration technique

multiple-tube fermentation technique
ortho-nitrophenyl-β-D-galactopyranoside
  (ONPG)
pathogens

potable
standard methods
total organic carbon (TOC)
transmittance

■ Questions and Problems

1. List two different agencies that issue analytic methods that are used for the analysis of water and wastewater.
2. Describe the process by which a standard method of analysis is certified.
3. How are the EPA methods of analysis categorized?
4. Define the following:
   a. Analyte
   b. Matrix
5. What is meant by "obtaining a representative sample?"
   a. What is a grab sample?
   b. What is a composite sample?
   c. What is an analyte?
6. What is the most common way to preserve a sample?
7. The coliform bacteria count is used to test water for what type of contamination?
   a. Do coliform bacteria cause disease?
   b. What is the EPA maximum contaminant level?
   c. How do you treat water that contains pathogens?
8. Describe the multiple-tube fermentation technique for coliform bacteria.
9. Describe the membrane filtration technique for coliform bacteria.
10. Describe the ortho-nitrophenyl-β-D-galactopyranoside test for coliform bacteria.
11. Draw a schematic of the Clark oxygen sensor and label all parts. Write the reaction that:
    a. Occurs at the cathode
    b. Occurs at the anode
    c. Shows the overall reaction
12. The TOC measurement assesses what water-quality parameter?
    a. How are TOC measurements made?
    b. How can the results of the TOC test differ from the BOD?
    c. What detectors are used on the TOC instrument? How do they work?
13. Describe how to measure the BOD.
14. Arrange the following in order of increasing energy: X-rays, infrared light, UV light, visible light, and microwaves.
15. Convert the following absorbance (A) measurements into percent transmittance (%T).
    a. 0.065
    b. 0.320
    c. 0.425

16. Convert the following transmittance measurements into absorbance (A).
   a. 22.6%
   b. 31.4%
   c. 0.102

17. Complete the following table:

A	% T	ϵ	b	c, M	c, ppm
0.420			1.0	1.25×10^{-4}	
	46%		1.0	8.2×10^{-5}	
		3.50×10^3	2.0		3.35
	49%		1.0		6.75

18. A solution containing 4.5 ppm of anthracene has a transmittance of 0.360 in a 1-cm cell at a wavelength of 280 nm. Calculate the molar absorptivity, ϵ. Indicate ϵ's units.

19. Iodine deficiency can cause a goiter, which is an enlargement of the thyroid gland. You are sent to Pakistan to make field measurements of the concentration of iodide (I^-) that is present in groundwater.

Procedure:
   a. Oxidize I^- in sample to I_2 with potassium peroxymonosulfate.
   b. Add the dye leuco crystal violet, which reacts with I_2 to form a violet color, which absorbs at 590 nm.

Results:
   a. A standard solution containing 3.5×10^{-6} M solution of I^- displayed an absorbance of 0.240 at 590 nm.
   b. A sample of groundwater from Pakistan gave an absorbance of 0.140. What is the concentration of I^- in the groundwater?

20. A colored substance Z has an absorption maximum at 350 nm. A solution containing 2 mg of Z per liter had an absorbance of 0.850 using a 1-cm cell. The formula weight of Z is 148.
   a. Calculate the absorptivity of Z at 350 nm.
   b. A solution containing Z gave an absorbance of 0.450. What is the concentration of Z in this sample in ppm?

21. The unionized form of the weak acid HA does not absorb in the visible region. When it ionizes, its conjugate base (A^-) absorbs strongly at 510 nm. The following data were obtained by measuring the absorbance at 510 nm of solutions with a concentration of 1.0×10^{-4} M HA in different pH buffers.

pH	Absorbance
2.0	0.00
3.0	0.00
4.0	0.00

pH	Absorbance
5.0	0.025
6.0	0.180
7.0	0.480
8.0	0.570
9.0	0.595
10.0	0.595
11.0	0.595
12.0	0.595

What is the approximate pKa of HA?

22. Draw a schematic of a single beam UV-visible spectrometer.
   a. What lamp is used for the visible region?
   b. What lamp is used for the UV region?
   c. Can glass cells be used to hold samples for UV and visible absorbance measurements?

23. Draw a schematic of a double beam UV-visible spectrometer.
   a. What is the function of the chopper?
   b. How does this instrument correct for changes in source lamp intensity?

24. Describe the spectrophotometric method that is used to determine the plant nutrient phosphorous in water samples.

25. a. A 3.42×10^{-5}-M solution of phosphate when reacted with ammonium molybdate in the presence of vanadium has a maximum absorbance of 0.434 at 420 nm in a 1-cm cell. Find the molar absorptivity of vanadomolybdophosphoric acid.
   b. A 10-mL aliquot (sample withdrawn) of groundwater from a farm was mixed with ammonium molybdate in the presence of vanadium and diluted to a final volume of 250.0 mL to give an absorbance of 0.780 at 420 nm in a 1-cm cell.
Find the concentration of phosphate in the unknown sample.

26. Draw a schematic of an ion chromatograph.
   a. How is the sample introduced into the instrument?
   b. Describe the ion-exchange resin that is used for anion separations.
   c. What type of detector is used? Why?

27. Predict the order of elution of the following ions from a quaternary ammonium anion-exchange resin: I^-, F^-, Br^-, NO_3^-, and SO_4^{2-}.

28. Describe the function of the suppressor in ion chromatographic separations of anions.

29. Referring to Figure 9.13, assume that the standard contains all five anions at 50 ppm each. Estimate the concentration of Cl^-, F^-, NO_3^-, and SO_4^{2-} in the pond water sample from the sample chromatogram. Estimate the error in your measurement.

30. Describe the easiest way to measure total dissolved solids, total ionized solids, and salinity.

31. Define the following:
   a. Resistivity
   b. Conductivity

32. A solution has a conductivity of 200 mS/m. Covert to µmhos/cm.

33. Which is the most mobile cation in solution? How much more efficient is this cation?
34. Which is the most mobile anion in solution? How much more efficient is this anion?
35. How is the salinity of natural waters measured?
36. Describe the construction of the pH electrode.
37. Describe the construction of the calcium ion electrode.
38. A Ca^{2+} ion-selective electrode was calibrated with a $[Ca^{2+}]$ standard with a concentration of 1.00×10^{-6}. This standard gave a potential of -54 mV.
    a. Construct a calibration curve.
    b. A groundwater sample from an area rich in limestone gave a potential of $+16$ mV. What was the $[Ca^{2+}]$ in ppm for this sample?
    c. Another groundwater sample gave a potential of -38 mV. What was the $[Ca^{2+}]$ in ppm for this sample?
39. Describe how you would calibrate a pH electrode that was going to be used to measure the pH of rainwater.
40. The sensitivity of the Ca^{2+} electrode is 50 times as great for Ca^{2+} than for Mg^{2+} and 1000 times greater than for Na^+. If you needed to measure Ca^{2+} at the 1-ppm level in an environmental water sample,
    a. What concentration of Mg^{2+} would cause a 3% error in the Ca^{2+} measurement?
    b. What concentration of Na^+ would cause a 2% error in the Ca^{2+} measurement?
41. Define the following for a water sample:
    a. Alkalinity
    b. ANC
    c. Acidity
    d. Hardness
42. Why is the measurement of pH in natural waters an intensity factor and the measurement of alkalinity a capacity factor?
43. What is NOM?
    a. Typically, how many proton-accepting sites are on NOM from natural waters?
    b. What is a typical NOM compound?
44. Using equations, show how alkalinity differs from pH.
45. What are the pKas for the following:
    a. $B(OH)_4^-$
    b. HPO_4^{2-}
    c. HS^-
46. Two water samples are collected.

    Sample #1 pH $= 8.8$, $[HCO_3^-] = 0.01$ M

    Sample #2 pH $= 7.0$, $[HCO_3^-] = 0.00$ M

    a. Compare the alkalinity of the two samples.
    b. Which of these two samples was taken from a well-buffered lake?
47. Sea water has a pH of approximately 8. It has a $[HCO_3^-] + [CO_3^{2-}] = 2$ mM. What is its alkalinity? Is it well buffered?

48. The hardness of water can be determined by cation ion chromatography. Describe an ion chromatography IC system that would make the measurement. In your description include the following:

    a. What cation(s) are you going to measure? At what concentration?

    b. Describe the ion-exchange resin that will be placed in the separation column.

    c. Describe how a suppressor would work, including chemical equations.

    d. What detector would be used?

49. How are radioactive substances that may be present in water detected?

50. Describe the following instruments that are used to detect radioisotopes in water:

    a. γ Spectrometer

    b. Proportional counter

CHAPTER

10

Fossil Fuels:
Our Major Source of Energy

ENERGY EXISTS IN MANY FORMS, INCLUDING SOLAR ENERGY, HEAT ENERGY, MECHANICAL ENERGY, ELECTRICAL ENERGY, AND CHEMICAL ENERGY, WHICH ARE CONTINUALLY BEING CONVERTED FROM ONE FORM TO ANOTHER. For example, when animals run, their bodies convert chemical energy stored in carbohydrates into mechanical energy to contract muscles. In a power plant, combustion of coal produces heat energy that is used to change water into steam. The steam turns a turbine, and the mechanical energy of the turning turbines produces electricity (electrical energy).

In accordance with the second law of thermodynamics, some energy is wasted in these and all other energy conversions. For example, when you drive a car, only approximately 10% of the chemical energy in the gasoline is converted to mechanical energy of motion. As much as 90% is converted to heat, which is lost to the surroundings.

All life forms and all societies require a constant input of energy. The more industrialized and technologically advanced the society is, the greater its need for energy. Today, the United States and other industrialized nations obtain most of their energy from fossil fuels: coal, petroleum, and natural gas. **Fossil fuels** are the remains of plants, animals, and microorganisms that lived millions of years ago and, over time, were buried and converted through compression and/or heat to the three fossil fuels. Thus, for energy, modern societies are almost entirely dependent on photosynthesis that occurred billions of years ago.

Fossil fuels, like all living and once-living materials, are composed of carbon compounds. The majority of these compounds are **hydrocarbons,** which contain only carbon and hydrogen. Differences in properties of the three fossil fuels are due primarily to differences in their hydrocarbon composition. Petroleum, natural gas, and coal are all excellent fuels because when they are burned some of the chemical energy stored in their hydrocarbon bonds is released as high-intensity heat energy, which can perform useful work.

This chapter reviews the development of energy sources and discusses why fossil fuels are such a good source of energy. We study the formation, extraction, composition, and uses of each of the three fossil fuels and examine the environmental impact of modern society's reliance on fossil fuels.

■ Energy Use: A Historical Overview

Throughout human history, as societies have become more advanced, their energy use has increased. In primitive cultures, people relied on their own physical labor to supply their energy requirements. They obtained the food that they needed by hunting animals and gathering wild plants. With the discovery of fire, humans began to use energy that was not obtained as food. Heat obtained from the burning of wood was used to heat dwellings and cook food.

As civilization advanced, the animals such as the horse and ox were domesticated and put into service to supply the energy needed for farming, transportation, and other activities. Water power and wind power were harnessed to do other useful work. As early as the first century AD, Egyptians were using waterwheels to grind grain. By the Middle Ages, windmill technology had spread from Persia throughout Europe, England, and China, furnishing the energy needed to grind grain and pump water.

As late as the first quarter of the 19th century, the main sources of energy—apart from human and animal muscle—were wood fires, windmills, and waterwheels (which could be used to power lumber mills, textile manufacturing mills, and other small industries). A great surge in the use of energy came with the development of the steam engine in the late 18th century, which for the first time enabled heat energy from fuel to be converted directly into useful work. This technology was the beginning of the Industrial Revolution, the period when machines began to replace human and animal labor. In the United States, an essentially agrarian economy changed to one that was based on manufacturing. Because steam engines could be located almost anywhere and provided unprecedented amounts of energy, the availability of water no longer dictated where factories could be located. Steam engines could be mounted on wheels, and the development of the steam locomotive marked the beginning of mass transportation. Enormous quantities of fuel were required to supply energy to operate the new machinery; in the 200 years between 1725 and 1925, per capita energy consumption in industrialized nations increased 10-fold.

The changing pattern of energy use in the United States since 1860 is shown in Figure 10.1. Wood initially fueled steam engines, including those in railroad locomotives and riverboats; however, in the 1830s, wood became scarce, and coal was gradually substituted. By 1910, coal was the source of 70% of the energy used in the United States. Petroleum did not become a significant energy source until the end of the 19th century. It was first discovered in the United States, near Titusville, Pennsylvania, in 1859, and in the Middle East, in Iran, in 1908. For the first 60 years after its discovery in the United States, petroleum was used principally to produce kerosene for lighting and heating.

The development of petroleum as a major energy source paralleled the development of the internal combustion engine and the growth of the automobile industry. As a fuel for transportation, petroleum has a tremendous advantage over coal: Gasoline-powered engines are many times smaller and lighter than coal-powered steam engines.

Natural gas, which is often found with petroleum, was not used as a fuel until an economical means for transporting it from the well to the consumer was established. In the early part of the 20th century, as much as 90% of the natural gas found with petroleum was

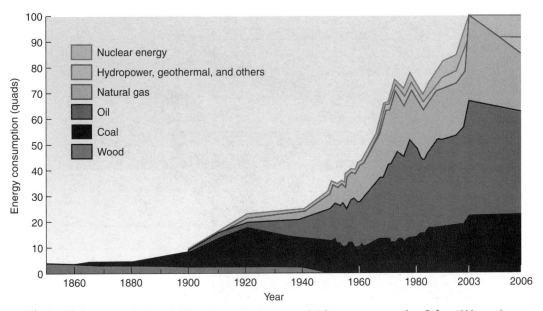

Figure 10.1 The contribution of different energy sources to total U.S. energy consumption. Before 1880, wood was the major source of energy, but then coal consumption began to increase sharply. Petroleum consumption rose dramatically throughout the first three-quarters of the 20th century as the number of automobiles and trucks on U.S. highways multiplied. At the same time, the use of natural gas for home heating increased. The brief decline in fuel consumption in the 1970s was the result of an oil shortage engineered by the major oil-producing nations. Nuclear power continues to be a relatively minor source of energy.

burned ("flared") at the well site as a waste product. By 1945, a network of pipelines had been constructed throughout the United States to link the gas fields, located mainly in the Southwest, with the states in the Midwest and Northeast, where gas for heating was in greatest demand.

Current Use of Energy

In the industrialized nations of the world, energy is used for four main purposes: (1) transportation, (2) industrial processes, (3) heating and cooling buildings, and (4) generating electricity. Figure 10.2 summarizes the major sources of energy in the United States and the uses to which they are put.

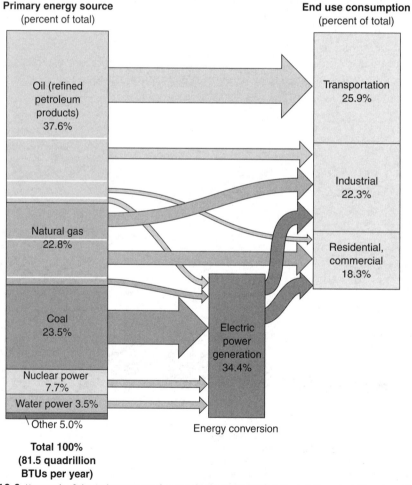

Figure 10.2 How each of the major sources of energy in the United States is used. Transportation depends almost entirely on petroleum products for energy; coal is used almost exclusively for generating electricity; natural gas is used about equally for commercial and residential heating and for industrial purposes. BTU = British thermal unit.

Transportation, which includes road vehicles, trains, boats, and airplanes, consumes approximately 29% of all of the energy used in the United States. Energy for transportation comes almost exclusively from gasoline and diesel oil, both of which are derived from crude oil (*oil* is frequently used as a synonym for petroleum). Huge amounts of energy are required to manufacture the goods that Americans depend on for a high standard of living; industries producing these goods account for approximately one-third of the U.S. energy consumption. The U.S. Department of Energy last reported that in 2007 personal residences consumed 21% while other commercial uses accounted for another 18%. The energy for industry comes almost equally from petroleum, natural gas, and electricity; mainly natural gas and electricity supply the energy for heating and cooling. Almost 40% of the energy that is consumed in the United States is used to produce electricity. **Electricity**, unlike fossil fuels, nuclear power, and water power, is not a primary source of energy. It is called a secondary source because it is produced from a primary source.

Figure 10.2 shows that fossil fuels—petroleum, natural gas, and coal—are used to meet almost all of the energy requirements of the United States. Nuclear energy, water power, solar energy, and other less conventional sources of energy fulfill a minute fraction of U.S. energy needs.

■ Energy and Power

It is important to understand the distinction between energy and power. **Energy** is the ability to do work and is measured in units such as calories, joules, and British thermal units. **Power** is a measure of the rate at which energy is used. It is expressed in terms of units of energy per unit of time, such as calories per second or joules per second (watts). Units in which energy and power are measured are explained in Table 10.1.

■ Energy from Fuels

Fuels are defined as substances that burn readily in air and release significant amounts of heat energy. The process of burning (or **combustion**) is an oxidation process and is **exothermic**. Most materials on the Earth—minerals in rocks and soil and water, for example—are already in an oxidized state so they do not burn and cannot be used as a source of heat energy.

What makes petroleum, natural gas, and coal such excellent fuels? The answer lies in their chemical composition. Fossil fuels are composed primarily of hydrocarbons, which when burned in an adequate supply of air are oxidized, forming carbon dioxide and water. Chemical bonds are broken. New bonds are formed, and as a result, energy is released as heat.

Equations for the combustion of methane (CH_4, the major component of natural gas) and octane (C_8H_{18}, a component of gasoline) are shown:

$$CH_4 + 2\,O_2 \rightarrow CO_2 + 2\,H_2O + heat$$

$$2\,C_8H_{18} + 25\,O_2 \rightarrow 16\,CO_2 + 18\,H_2O + heat$$

When these hydrocarbons are burned in air, their carbon–hydrogen bonds are broken, as are bonds between oxygen atoms in oxygen molecules. New bonds form between carbon and

Table 10.1

Glossary of Energy Terms

Term	Meaning
Calorie (cal)	The amount of heat needed to raise the temperature of 1 g of water 1°C
Kilocalorie (kcal)	One thousand calories (1000 × 1 cal)
Joule (J)	The joule is the standard unit of energy: 4.18 J = 1 cal
British thermal unit (BTU)	The amount of heat needed to raise the temperature of 1 lb of water 1°F
Quad	One quadrillion (1000 trillion) British thermal units (BTU): one quad = 10^{15} BTU
Horsepower	One horsepower = 33,000 ft lb of work/min
Watt (W)	An amount of power available from an electric current of 1 ampere (A) at the potential difference of one volt (V): 1 W = 1 V x 1 A; 1 W = 1 J/s
Kilowatt (kW)	One thousand watts (1000 x 1 W)
Megawatt (MW)	One million watts, or 1000 kW
Kilowatt-hour (kWh)	Unit of electrical energy equivalent to the energy delivered by the flow of 1 kW of electrical power for 1 hour

oxygen atoms and between hydrogen and oxygen atoms, as CO_2 and H_2O molecules are produced. Heat is produced in these reactions because the energy released in making the new bonds in carbon dioxide and water is greater than the energy required to break the bonds in the hydrocarbons and in oxygen. (If the energy released in making the new bonds were less than the energy required to break the old bonds, heat would be consumed, and the reaction would be **endothermic**.)

The amount of energy required to break 1 mole of a particular kind of bond is the **bond energy**. Average values for bond energies are given in Table 10.2 in kilocalories, which can be converted to kilojoules by multiplying by 4.18. The same amount of energy is released if the bond is reformed from individual atoms.

We can use the values given in Table 10.2 to calculate the heat energy that will be released when 1 mole of CH_4 is oxidized to yield carbon dioxide and water, as shown in the previous equation.

Thus, more energy is released in making new bonds than is consumed in breaking the bonds in the hydrocarbon and oxygen molecules. The excess energy ($828 - 632 = 196$ kcal) is released as heat. Because 1 mole of CH_4 releases 196 kcal/mol and a mole of CH_4 weighs 16 g, CH_4 releases approximately 12.25 kcal/g. When you compare the heat released per gram of CH_4 to the value given in Table 10.3 for natural gas, you will notice that it is larger (12.25 vs. 11.7 kcal/g). As you will see later, natural gas is not pure CH_4, and the impurities that it contains lower the amount of heat that it releases per gram.

Table 10.2

Approximate Bond Energies for Selected Bonds

Type of Bond	Bond Energy (kcal/mole)
$C-H$	99
$C-C$	83
$C=C$	147
$C-O$	85
$C=O$	173
$C=O$ in CO_2	192
$C-Cl$	78
$H-H$	104
$H-Cl$	103
$O-O$	35
$O-O$ in O_2	118
$O-H$	111
$Cl-Cl$	58

Table 10.3

Fuel Values of Some Common Fuels

Fuel	Fuel Value (kcal/g)
Wood (pine)	4.3
Bituminous coal	7.4
Anthracite	7.6
Crude oil	10.8
Gasoline	11.5
Natural gas	11.7
Hydrogen	33.9

EXAMPLE 10.1

How much energy is released when 1 mole of propane (C_3H_8) is burned?

Solution

1. Write the balanced equation for the reaction:

$$C_3H_8 + 5 O_2 \rightarrow 3 CO_2 + 4 H_2O$$

2. Determine the type and number of bonds broken by drawing the structural formula of C_3H_8:

```
    H  H  H
    |  |  |
H — C — C — C — H
    |  |  |
    H  H  H
```

Clearly, 8 C—H and 2 C—C bonds are broken; 5 O—O bonds are also broken.

3. Use Table 10.2 to obtain the bond energy of each type of bond. Calculate the total energy required to break all the bonds:

Bond	Bond Energy (kcal)	Number of Bonds Broken	Bond energy/mole (kcal)
C—H	99	8	$8 \times 99 = 792$
C—C	83	2	$2 \times 83 = 166$
O—O	118	5	$5 \times 118 = \underline{590}$
			Energy required to break bonds = 1548

4. Similarly, calculate the energy released when the new bonds are formed:

Bond	Bond Energy (kcal)	Number Formed	Bond energy/mole (kcal)
C=O	192	6	$6 \times 192 = 1152$
O—H	111	8	$8 \times 111 = \underline{888}$
			Energy released in forming bonds = 2040

5. Find the difference between the energy consumed and the energy released: $(2040 - 1548) = 492$

Thus, 492 kcal/mol of energy are released when 1 mole of C_3H_8 is burned. Because 1 mole of C_3H_8 weighs 44 g/mol, C_3H_8 releases 11.18 kcal/g when it is burned. If you compare this with CH_4, you will see that it is slightly lower (12.25 vs. 11.18 kcal/g). The larger the H/C ratio, the more energy will be liberated.

Petroleum is primarily a mixture of saturated hydrocarbons. For the combustion of petroleum, the calculation can be simplified to be the reaction between a CH_2 group and oxygen. When accounting for the bonds that are broken, one C–C is counted for every CH_2:

$$2\left(-CH_2-\right) + 3\,O_2 \rightarrow 2\,CO_2 + 2\,H_2O$$

This reaction, as written, releases 292 kcal (1220 kJ) or 10.4 kcal/g of petroleum. As hydrocarbon chains become longer, the H/C ratio declines.

EXAMPLE 10.2

How much energy is released when 1 mole of ethane (C_2H_6) is burned?

Answer to Practice Example 10.2

344 kcal/mol

These calculations explain why fossil fuels are valuable as energy sources. When they are burned, numerous carbon–hydrogen bonds are broken, and as carbon dioxide and water are formed, a relatively large quantity of energy is produced. The energy output (fuel value) of various common fuels is given in Table 10.3.

■ Petroleum

The Formation of Oil Fields

Petroleum probably originated from microscopic marine organisms that once lived in great numbers in shallow coastal waters. When the organisms died, their remains collected in bottom sediments, where the supply of oxygen was insufficient to oxidize all of the organic material. Over millions of years, sediments buried the organic matter, and in some locations, the seas dried, leaving behind a new land mass. The buried material was subjected to high temperature and pressure; chemical reactions converted some of the organic matter to liquid hydrocarbons (crude oil or petroleum) and gaseous hydrocarbons (natural gas).

Based on their understanding of the Earth's history, geologists are able to recognize the major geologic features that indicate the presence of large accumulations of oil (Figure 10.3). They then assess the probable size of the oil-bearing formation, its location, and other factors to determine the economic feasibility of extracting the petroleum.

Petroleum and natural gas are generally found together. They accumulate in porous, permeable rock called **reservoir rock**. Rock is described as porous if, like a sponge, it contains

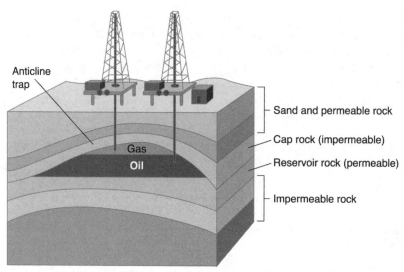

Labels in figure:
Anticline trap
Gas
Oil
Sand and permeable rock
Cap rock (impermeable)
Reservoir rock (permeable)
Impermeable rock

Figure 10.3 Geological features associated with petroleum and natural gas deposits: As a result of folding and faulting of rock layers, oil and gas become trapped under a dome of impermeable cap rock.

open spaces in its structure where gases and liquids can collect. It is permeable if the spaces are interconnected so that fluids—gases, water, and oil—can migrate through the pores. The porosity of reservoir rock, which is usually sandstone or limestone, ranges from 10% to 30%; usually one-half of the pore space is occupied by water.

For petroleum to accumulate in quantity there must be a layer of impermeable rock (known as **cap rock**) above the reservoir rock to prevent the fluids in the oil-bearing rock from escaping by moving upward. There also must be a folding or faulting of the rock strata to create a dome, or **anticline trap**, under which the petroleum and natural gas are held. The trap, which may extend over many miles, prevents lateral movement of fluids in the reservoir rock and allows oil and gas to collect. In time, the fluids in the reservoir rock separate according to density. Because petroleum and natural gas are less dense than water, they rise above the water and collect directly under the anticline trap, with the gas above the oil (Figure 10.3).

Oil cannot be withdrawn from an oil field easily because it does not accumulate in liquid pools. Instead, it is dispersed within the pores of the reservoir rock and must be forced out. The oil and gas are under pressure. Drilling into the reservoir rock releases the pressure, and the crude oil is forced up the bore hole. If the pressure is sufficiently high, the oil is released as a gusher. As the oil flows out, the pressure gradually decreases, and the rate of flow declines. The average oil field ceases to be productive after only approximately one-third of the oil in the formation has been recovered. That is, two barrels of oil are left in the ground for every barrel extracted. A further one-third can be obtained by repressurizing the oil field. In repressurizing, auxiliary wells are drilled, and water, steam, or a gas such as carbon dioxide or nitrogen is injected into the reservoir rock to force out more oil. Repressurizing is costly; whether it is worthwhile depends primarily on the prevailing price of oil.

The Composition of Petroleum

Crude petroleum is a complex mixture of thousands of organic compounds. The majority of these compounds are hydrocarbons, including alkanes, cycloalkanes, alkenes, and aromatic hydrocarbons. The proportions of different hydrocarbons in a sample of crude petroleum vary with location. For example, the percentage of straight-chain and branched-chain hydrocarbons is much higher in crude oil from Pennsylvania than in crude oil from California, which has a high content of aromatic hydrocarbons.

The sulfur content of crude oil ranges from 0.1 wt% for low-sulfur to as high as 3.7% for high-sulfur petroleum. The sulfur is incorporated into thiophene or thiophene derivatives.

Thiophene

2,3-Benzthiophene

Nitrogen compounds are also present, but constitute less than 1% by weight. Nitrogen, like sulfur, also is present in organic compounds that form heterocyclic ring systems such as thiazole or quinolines.

Thiazole

Quinoline

Petroleum Refining

Crude petroleum in the form in which it is recovered has few immediate uses. To be useful, it must be sent to a refinery and separated into fractions by **fractional distillation** (Figure 10.4a), a process that separates the hydrocarbons according to their boiling points.

Figure 10.4b shows the essential parts of a fractional distillation tower. The tower may be as tall as 30 m (100 ft). Crude oil at the bottom of the tower is heated to approximately 500°C (930°F), and the majority of the hydrocarbons vaporize. The mixture of hot vapors rises in the fractionating tower; components in the mixture condense at various levels in the tower and are collected. The temperature in the tower decreases with increasing height. The more volatile fractions with lower boiling points condense near the top of the tower; high-boiling, less volatile fractions condense near the bottom. Any uncondensed gases are drawn off at the top of the tower. Residual hydrocarbons that do not vaporize collect at the bottom of the tower and are transferred to a vacuum distillation tower, where they are vaporized under reduced pressure to yield further fractions. This process makes the full range of hydrocarbons in crude oil available for useful purposes. The properties and uses of typical petroleum fractions are listed in Table 10.4.

The relative amount of each fraction produced by fractional distillation varies with the composition of the original crude petroleum and rarely reflects consumer needs. Before the

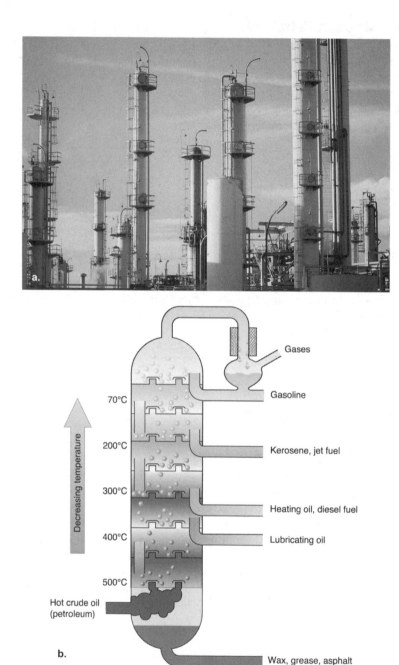

Figure 10.4 (a) Distillation towers at a refinery. (b) Crude petroleum is heated at the bottom of a distillation tower, and most of the hydrocarbon constituents vaporize. Inside the tower, temperature decreases with increasing height. High-boiling components, or fractions, condense in the lower part of the tower; lower-boiling components (including the gasoline fraction) condense near the top. Fractions are removed at various heights as they condense.

Table 10.4

Typical Petroleum Fractions

Fraction	Boiling Point (°C)	Composition	Uses
Gas	Up to 20	Alkanes from CH_4 to C_4H_{10}	Synthesis of other carbon compounds; fuel
Petroleum ether	20–70	C_5H_{12}, C_6H_{14}	Solvent; gasoline additive for cold weather
Gasoline	70–180	Alkanes from C_6H_{14} to $C_{10}H_{22}$	Fuel for gasoline engines
Kerosene	180–230	$C_{11}H_{24}$, $C_{12}H_{26}$	Fuel for jet engines
Light gas oil	230–305	$C_{13}H_{28}$ to $C_{17}H_{36}$	Fuel for furnaces and diesel engines
Heavy gas oil and light lubricating distillate	305–405	$C_{18}H_{38}$ to $C_{25}H_{52}$	Fuel for generating stations; lubricating oil
Lubricants	405–515	Higher alkanes	Thick oils, greases, and waxy solids; lubricating grease; petroleum jelly
Solid residue	–	–	Pitch or asphalt for roofing and road material

advent of the automobile, the fraction in greatest demand was **kerosene**, which was used for heating and lighting. Today, the greatest need is for the more volatile fraction known as **straight-run gasoline**, which provides the fuel for automobiles (Table 10.4).

To meet the increasing need for gasoline, refineries must convert higher boiling fractions to gasoline by the process of **cracking**. The higher boiling fractions contain long-chain hydrocarbons, for which there is little demand. When heated under pressure in the absence of air, the long-chain hydrocarbons break (or crack) into shorter chain hydrocarbons, including both alkanes and alkenes, some of which are in the desired gasoline range (their molecules have 6 to 10 carbons).

$$C_{14}H_{30} \xrightarrow[\substack{\text{catalyst} \\ \text{absence of air}}]{\substack{400°C\text{–}600°C \\ \text{and pressure}}} \underset{\text{alkene}}{C_7H_{16}} + \underset{\text{alkene}}{C_7H_{14}}$$

If an aluminosilicate, which is impregnated with potassium, is used as the catalyst, the reaction proceeds at lower pressure, and the yield of hydrocarbons in the gasoline range is increased.

Another way to increase the gasoline fraction is to react smaller hydrocarbons together to produce a larger one. This is the alkylation process.

$$\underset{\text{light hydrocarbons}}{\begin{array}{c} CH_3 \\ \diagdown \\ \end{array} C=C \begin{array}{c} H \\ \diagup \\ \end{array}} \; + \; C_4H_{10} \xrightarrow{\;\; H^+ \;\;} \underset{\substack{\text{octane}\\(\text{gasoline})}}{C_8H_{18}}$$

By the use of cracking and alkylation reactions, the gasoline fraction of crude oil can be increased from approximately 20% by volume to 40% to 45%.

To meet the increased demand for heating oil in the winter months, a refinery can switch to the reverse process, in which smaller hydrocarbons are combined to produce larger ones.

Octane Rating

Straight-run gasoline consists primarily of straight-chain hydrocarbons and is of limited value as an automobile fuel. A high content of straight-chain hydrocarbons causes a fuel to begin burning before the spark plug ignites it. This premature ignition produces a knocking sound and leads to a loss of engine power and eventual damage to the engine. The tendency of gasoline to cause knocking is rated according to an arbitrary scale known as the **octane rating**.

The octane rating was established in 1927, when a large number of straight- and branched-chain hydrocarbons in the gasoline range were tested separately for performance in a standard engine. The branched hydrocarbon 2,2,4-trimethylpentane (isooctane) was found to be a superior fuel that burned without causing knocking; it was assigned an octane rating of 100. The straight-chain hydrocarbon n-heptane, in contrast, caused serious knocking and was given a rating of 0.

$$\underset{\text{2,2,4-trimethylpentane (isooctane)}}{\begin{array}{c} \quad\; CH_3 \qquad CH_3 \\ \quad\; | \qquad\quad | \\ CH_3{-}CH{-}CH_2{-}C{-}CH_3 \\ \quad\qquad\qquad\;\; | \\ \quad\qquad\qquad\; CH_3 \end{array}} \qquad \underset{n\text{-heptane}}{CH_3{-}CH_2{-}CH_2{-}CH_2{-}CH_2{-}CH_2{-}CH_3}$$

The octane rating of a particular gasoline is determined by burning it in a standard engine and comparing its knocking properties with those of standard mixtures of isooctane and n-heptane. A gasoline that performs in the same way as a mixture containing 90% isooctane and 10% n-heptane is assigned an octane rating of 90. Table 10.5 lists the octane ratings for pure samples of some of the hydrocarbons that are present in gasoline.

The straight-run gasoline fraction from the distillation tower has an octane rating of between 50 and 55, much too low for today's automobile engines, which require gasoline with a rating of between 87 and 93. The octane rating of gasoline can be increased in

Table 10.5

Octane Ratings of Selected Hydrocarbons

Name	Formula	Octane Rating
Straight-chain and branched hydrocarbons		
n-Butane	C_4H_{10}	94
n-Pentane	C_5H_{12}	62
2-Methylbutane		94
n-Hexane	C_6H_{14}	25
2-Methylpentane		73
2,2-Dimethylbutane		92
n-Heptane	C_7H_{16}	0
2-Methylhexane		42
2,3-Dimethylpentane		90
2-Methylheptane	C_8H_{18}	22
2,3-Dimethylhexane		71
2,2,4-Trimethylpentane (isooctane)		100
Aromatic hydrocarbons		
Benzene		106
Toluene		118
o-Xylene		107
p-Xylene		116

three main ways: (1) cracking, (2) catalytic reforming, and (3) the addition of octane enhancers.

Cracking increases the octane rating by increasing the percentage of short-chain hydrocarbons, which in general have higher octane numbers than longer-chain hydrocarbons (Table 10.5). For example, as the length of the hydrocarbon chain decreases from heptane (C_7H_{16}) to butane (C_4H_{10}), the octane rating increases from 0 to 94.

Branched-chain hydrocarbons have higher octane numbers than the corresponding straight-chain isomers. It can be seen in Table 10.5 that as the degree of branching increases in the three isomeric hexanes (C_6H_{14})—*n*-hexane, 2-methylpentane, and 2,2-dimethylbutane—the octane rating increases from 25 to 92. Table 10.5 also shows that aromatic hydrocarbons have higher octane ratings than do nonaromatic hydrocarbons. Benzene, toluene, and the xylenes, for example, have octane ratings that are above 100, which are obtained by comparing a fuel with reference samples containing isooctane (octane number, 100) and known amounts of an octane enhancer. A conversion chart is used to obtain the octane number.

It therefore follows that if the percentages of branched-chain hydrocarbons and aromatic hydrocarbons can be increased, the octane rating of a gasoline will be improved. This improvement can be achieved by **catalytic reforming**, a process in which hydrocarbon vapors from straight-run gasoline are heated in the presence of suitable catalysts such as platinum. By this means, *n*-hexane, for example, is converted to 2,2-dimethylbutane:

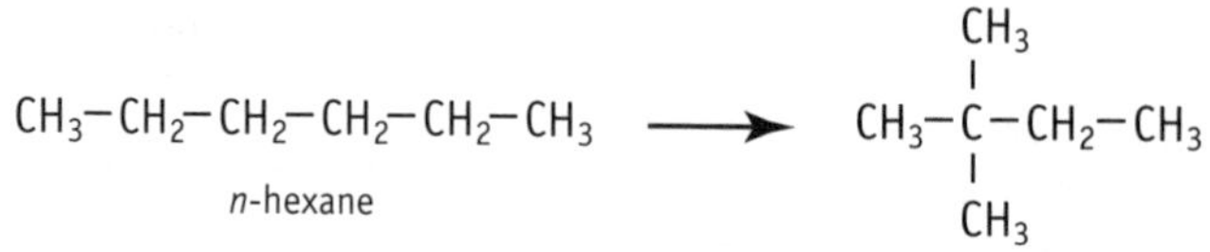

$$CH_3-CH_2-CH_2-CH_2-CH_2-CH_3 \longrightarrow CH_3-\underset{\underset{CH_3}{|}}{\overset{\overset{CH_3}{|}}{C}}-CH_2-CH_3$$

$$\text{\textit{n}-hexane} \qquad\qquad\qquad\qquad \text{2,2-dimethylbutane}$$

Also, *n*-heptane can be converted to toluene by heating heptane to 500°C to 600°C at high pressure in the presence of an Re-Pt-Al$_2$O$_3$ catalyst:

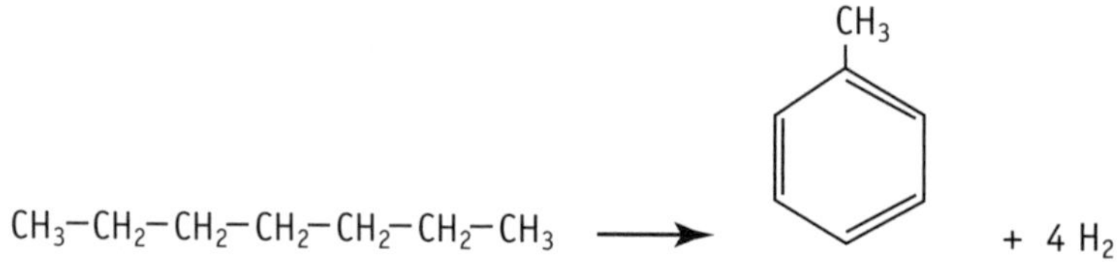

$$CH_3-CH_2-CH_2-CH_2-CH_2-CH_2-CH_3 \longrightarrow$$

(toluene structure) + 4 H$_2$

n-heptane toluene

The octane rating of gasoline can also be increased by adding antiknock agents, or **octane enhancers**. Before 1975, the most widely used octane enhancer was tetraethyl lead (TEL), $(C_2H_5)_4Pb$, which was both cheap and effective.

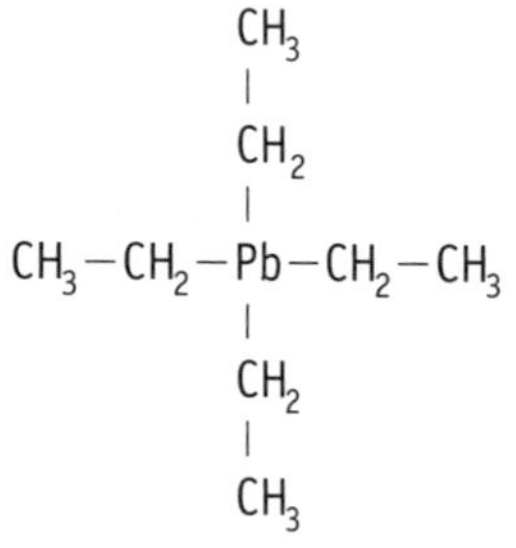

$$CH_3-CH_2-\underset{\underset{CH_2}{\underset{|}{\underset{CH_3}{|}}}}{\overset{\overset{CH_3}{|}}{\overset{\overset{|}{CH_2}}{Pb}}}-CH_2-CH_3$$

TEL

Adding as little as 0.1% of TEL to gasoline can increase the octane rating by 10 to 15 points; however, lead is toxic, and the recognition of the health hazards associated with its release into the atmosphere from automobile exhausts led to the mandatory phasing out of TEL as a gasoline additive. The 1975 requirement that all new cars be fitted with a catalytic converter to control pollutants was a further factor in reducing the use of TEL. Only unleaded gasoline can be used in automobiles fitted with catalytic converters because lead inactivates the catalysts.

Octane enhancers that have replaced TEL include methyl-t-butyl ether (MTBE), ethyl-*t*-butyl ether, methanol, and ethanol, all of which have high octane ratings (Table 10.6). The most

Table 10.6

Octane Number of Gasoline Additives

Additive	Octane Number
Methanol	107
Ethanol	109
MTBE	116
Ethyl-*t*-butyl ether	118

popular octane enhancer is MTBE, which is made by reacting methanol and isobutene in the presence of an acid catalyst.

MTBE

MTBE is made at refineries in which isobutene is generated in the catalytic cracking reaction. Because of leaking underground storage tanks and MTBE detected in water supplies, for health reasons, the Environmental Protection Agency recommended a reduction in its use. By early 2007, MTBE's use as a gasoline oxygenate decreased and gasoline retailers replaced MTBE with ethanol. Methanol and ethanol, which are both fuels, have been added to gasoline. Ethanol added in amounts up to 10% to produce gasoline blends known as **gasohol** or **E10**. This mixture can be used as a fuel without modification of the automobile. E85, which is a mixture of 85% gasoline and 15% ethanol can be used in flexible fuel vehicles (FFVs) that are designed to run on gasoline, E85, or any mixture of the two. The use of methanol as an alternative fuel is considered in Chapter 12.

Oil Shale and Tar Sands

A largely untapped source of petroleum is **oil shale** (Figure 10.5a), a common sedimentary rock that contains a solid organic material called **kerogen**, which consists primarily of heavy hydrocarbons together with small quantities of sulfur-, nitrogen-, and oxygen-containing compounds. Oil shale is found close to the Earth's surface and can be mined like coal. When oil shale is crushed and heated in the absence of air at approximately 450°C to 500°C, a thick brown liquid called **shale oil** is produced. After it has been treated to remove sulfur and nitrogen impurities, shale oil can be refined like petroleum. Fractional distillation yields mainly high molecular weight hydrocarbons, which can then be cracked to yield desirable hydrocarbons in the gasoline range.

The United States has immense deposits of oil shale. It is estimated that the Green River Formation, which is located in Wyoming, Colorado and Utah, contains 1.2 to 1.8 trillion

Figure 10.5 Oil can be extracted from (a) oil shale, a sedimentary rock rich in the hydrocarbon material called kerogen, and from (b) tar sands, a mixture of sandstone and a black, high-sulfur, tarry substance called bitumen.

barrels of oil, which is at least three times greater than the proven reserves of Saudi Arabia. Although the technology is available to exploit this resource, environmental and economic problems must first be overcome. For example, very large quantities of waste rock are produced when oil shale is mined, and because this waste is alkaline and poor in nutrients, very little vegetation will grow on it. Furthermore, for every gallon of shale oil that is produced, 2 to 4 gallons of water are required. When oil prices are relatively low, mining of oil shale for fuel is not economically attractive.

Tar sands are another potential source of petroleum (Figure 10.5b) and consist of sandstone mixed with a black, high-sulfur, tar-like oil known as **bitumen.** The amount of bitumen in tar sands tends to be less than 15% of the total weight. Large deposits exist in the United States, particularly in Utah, but the most extensive known deposits are found in Northern Alberta, Canada, where they are nearer the surface than in the United States and much easier to mine. The extraction process involves injecting steam under pressure into the tar sand to liquefy the bitumen so that it can float to the surface and be collected. After it is purified, the oil can be refined in the usual manner.

Producing oil from tar sands is expensive and requires a great deal of energy. As with oil shale, environmental and economic considerations will determine the future of tar sands as a viable energy source. As petroleum reserves are depleted, interest in both oil shale and tar sands will undoubtedly increase.

The Petrochemical Industry

Petroleum is the primary source of the hydrocarbons that is needed as starting materials for the synthesis of organic compounds used in the manufacture of plastics, synthetic fibers, synthetic rubber, pesticides, pharmaceuticals, and numerous other consumer products (Figure 10.6). Approximately 3% of the petroleum that is refined today is used for this purpose. The organic

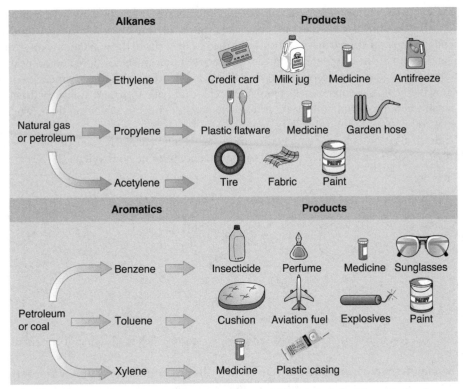

Figure 10.6 Some products made from petrochemicals (chemicals derived from petroleum).

compounds synthesized from petroleum hydrocarbons are called **petrochemicals**. In the future, as petroleum reserves decline, more organic compounds are expected to be produced using synthesis gas derived from coal (discussed later in this chapter) as the starting material.

Oil Pollution

The environmental cost of producing, transporting, and using oil is considerable. The worst oil spill in U.S. history occurred in March of 1989, when the supertanker *Exxon Valdez* ran aground 28 miles south of the trans-Alaska pipeline terminal at Valdez and spilled 260,000 barrels (11 million gallons) of crude oil into Prince William Sound. More than 1000 miles of shoreline were contaminated with oil, and tens of thousands of oil-coated seabirds and at least a thousand sea otters died. While 20 years later the surface oil has largely been cleaned, over 26,600 gallons (100 tons) are still immediately below the surface, leaching into the Sound and negatively affecting digging birds and animals as well as the local economy that depends on fishing and tourism.

When crude oil is spilled into the ocean, it is dispersed and changed by many physical and chemical processes. Nearly all of the components in crude oil are less dense than water and are insoluble in it; they float on the ocean's surface as a layer that gradually thins as it spreads outward. The most volatile components in the layer (usually approximately 25% of the oil) evaporate into the atmosphere. Less volatile components are broken by wave action into fine droplets that become dispersed in the water. Near shore, oil droplets tend to adsorb onto any suspended sand and silt particles and sink. At the ocean surface, the wind and wave actions whip oil into an oil-in-water emulsion (or mousse). Oil that washes ashore and oil near the ocean surface is gradually decomposed by bacteria and sunlight, but these processes occur very slowly in cold seawater. As the emulsion breaks, tar balls form and are washed ashore. The small fraction of crude oil that is heavier than water sinks to the ocean floor, where it can coat and destroy bottom-dwelling organisms.

Crude oil also contains aromatic hydrocarbons in which benzene has one or more methyl groups attached. This fraction is referred to as **benzene, toluene, and xylene (BTX)**.

benzene   toluene   p-xylene

It is the BTX fraction that is most toxic to marine creatures when an oil spill occurs in the ocean.

The long-term environmental effects of oil spills are difficult to assess. Where strong waves pound exposed rocky shores, cleanup is rapid, but on sheltered sandy beaches and in salt marshes and tidal flats, oil may become buried and remain without decomposing for two to three years. Hydrocarbons are insoluble in water; when ingested by marine creatures, they dissolve in the animals' fatty tissues. Fish metabolize and excrete hydrocarbons rapidly, but shellfish may remain contaminated for a long time, thus smelling and tasting unpleasant.

Accidental oil spills from supertankers receive the most publicity, but nearly 70% of oil pollution in the ocean comes from the normal routine operation of these tankers. Tankers often leak. Oil is frequently spilled during unloading and refueling, and most tankers intentionally flush oil wastes directly into the sea.

Offshore oil drilling operations are a minor cause of oil pollution, mainly from the occasional well blowout. The main environmental concern is the wastewater that is brought up with the oil from below the ocean floor. This ancient water contains barium, zinc, lead, and trace amounts of radioactive materials, and when it is returned to the ocean, it can destroy nearby natural habitats. Another concern is the clay mud that is used as a drill lubricant; it contains lignite, barium, and other chemicals and is usually dumped at sea after use. The clay can bury and destroy bottom-dwelling organisms, which form an essential part of the marine food chain. The most serious effects of offshore drilling probably occur along the shore, where facilities built to receive the oil destroy wetlands.

Motorboats are a source of oil pollution on recreational waters. Many outboard two-cycle engines run on a mixture of oil and gasoline and emit this fuel half burned through their exhaust systems. For many years, motorboat emissions were not regulated, but starting in 1998 the Environmental Protection Agency enacted strict emission controls for new outboard motors.

A significant cause of oil pollution is improper disposal of lubricating oil from machines and automobile crankcases. Many car owners who change their own motor oil pour the used oil into storm sewers and other areas that drain into creeks, rivers, and groundwater. Few realize that as little as 1 quart of oil can contaminate 2-million gallons of drinking water; 1 gallon of oil can form an oil slick that measures nearly 8 acres. Because of this danger, many communities have set up collection centers where used oil can be brought for recycling.

■ Natural Gas

The geologic conditions that are necessary for the formation of **natural gas** are similar to those required for the formation of oil (see Figure 10.3). As we saw earlier, oil and gas are often found together, but natural gas is also found by itself. In the formation of natural gas, part of the original, buried organic material was converted to light gaseous hydrocarbons; in the formation of oil, heavier liquid hydrocarbons were formed.

Natural gas is composed chiefly of CH_4 and small amounts of C_2H_6, C_3H_8, and butane. Typically, it contains 60% to 80% CH_4, but the exact composition varies with the source. Crude natural gas also contains small quantities of larger alkanes, carbon dioxide, nitrogen, hydrogen sulfide (H_2S), and helium. The major natural gas-producing states are Texas, Wyoming, Oklahoma, New Mexico, and Louisiana.

Before crude natural gas can be used as a commercial fuel, it must be treated to remove carbon dioxide, sulfur compounds, water vapor, and most of the hydrocarbons with molecular weights that are greater than that of C_2H_6. Carbon dioxide is undesirable because it reduces the heat output of the gas, and H_2S causes corrosion. Water vapor can cause problems if it condenses in a pipeline. After the natural gas has been purified, mixtures of propyl ($CH_3CH_2CH_2SH$) and butyl mercaptan ($CH_3CH_2CH_2CH_2SH$), which have foul, skunk-like odors, are added at the low parts per million level to natural gas. They do not alter the combustion properties of natural gas but make it smell bad so that leaks are more easily detected.

C_3H_8 and butane (C_4H_{10}) are useful by-products. After they have been removed from crude natural gas, these by-products are converted to a liquid known as **liquefied petroleum gas** and are marketed as heating fuel. Helium, which is often present in crude natural gas, is another important by-product. It is used to provide an inert atmosphere for welding, as a lighter-than-air gas for balloons, and when condensed into a liquid ($-269°C$), it is used to cool the superconducting magnets in magnetic resonance imaging instruments (MRIs).

Natural gas is nontoxic and safe, but many people mistakenly believe that it is dangerous. This misconception dates from the early part of the 20th century, when coal gas (discussed in the next section), a toxic mixture of carbon monoxide and hydrogen, was piped into homes. Of course, if natural gas accumulates in a confined space, a spark will ignite it explosively.

Natural gas is relatively inexpensive, is a superior fuel, and has many advantages over coal and petroleum. It burns cleanly and leaves no residue, and weight for weight, it provides a higher heat output than any other common fuel (see Table 10.3). It emits less carbon dioxide per unit of energy than other fossil fuels and generally produces no sulfur oxides. The extensive system of pipelines that crisscross the United States, taking gas directly to the consumer, reduces the need for expensive storage facilities.

■ Coal

The Formation of Coal Deposits

Three hundred million years ago, when the Earth was much warmer than it is now, plants grew in great abundance in the widespread tropical freshwater swamps and bogs. When the luxuriant growth died, some of the plant material sank under water before it could be oxidized by atmospheric oxygen in the usual way. In the absence of air, little decomposition occurred. The material accumulated and became buried under sediments, falling leaves, and other vegetation. In time, it was compressed and converted to a porous brown organic material that is now known as **peat**. In some locations, because of geologic changes, the peat became more deeply buried; increasing pressure compressed and changed it to a harder material called **lignite**.

Over thousands of years, deeper burial and the resulting increase in temperature and pressure transformed the lignite into various grades of **bituminous coal** (also known as *soft coal*). In areas where mountains formed as a result of deformation and uplifting of rock formations, further changes occurred. The very high pressure and temperature associated with these geologic processes converted bituminous coal to **anthracite coal** (or *hard coal*).

With each step in the transformation from peat to anthracite, chemical reactions occur. Volatile compounds containing carbon, oxygen, and hydrogen are released; the water content of the material decreases, and the carbon content increases (Table 10.7). At the same time, the material becomes hard and shiny. The quality of coal as a fuel increases with its carbon content. Anthracite, which is approximately 90%, is the most desirable form of coal. Peat is not a true coal and is considered a low-grade fuel because of its very high water content (90%) and low carbon content (5%). Dried peat, however, is used as a fuel in some parts of the world. The locations of the principal coal deposits in the United States are shown in Figure 10.7. Most of the anthracite has already been mined.

The Composition of Coal

Coal is a complex mixture of organic compounds and is composed primarily of carbon, hydrocarbons, and small amounts of oxygen-, nitrogen-, and sulfur-containing compounds.

Table 10.7

Characteristics of Different Types of Coal

Type of Coal	Carbon (%)	Water (%)	Fuel Value
Peat	5	90	Very low
Lignite	30	40	Low
Sub-bituminous coal	40	9	Medium
Bituminous coal	65	3	High
Anthracite	90	3	High

Values may vary considerably with the source of coal.

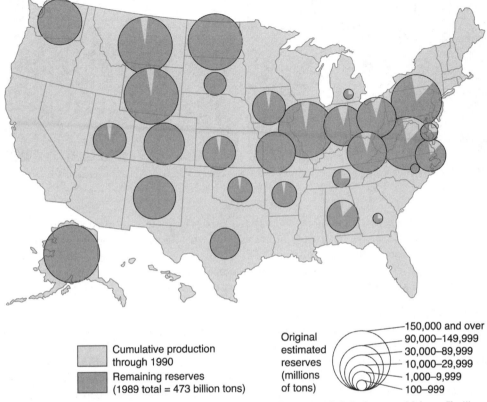

Figure 10.7 Major coal deposits in the United States. Adapted from B.J. Nebel. *Environmental Science: The Way the World Works, Sixth edition.* Englewood Cliffs, NJ: Prentice Hall, 1990.

Compared with petroleum, coal contains a much higher percentage of aromatic hydrocarbons, including many polycyclic aromatic compounds and, as a result, a much lower percentage of hydrogen. The composition of coal varies with its type and the location of the deposit.

Problems with Coal

Most coal is extracted from the Earth by strip mining, which devastates the land unless adequate steps are taken for reclamation (Figure 10.8). Both strip mining and underground mining produce large quantities of waste rock, and acid leaching from the wastes can contaminate nearby streams and rivers. Coal is dirty, bulky, and expensive to transport. When coal is burned, large quantities of ash are produced, presenting a disposal problem. Some ash is used to make cement and some for fill. All coal contains some sulfur, and sulfur dioxide emitted during the combustion of coal is the primary cause of acid rain. Also, the burning of coal produces carbon dioxide and, as we saw earlier, the constantly increasing concentration of carbon dioxide in the atmosphere has the potential to change the Earth's climate drastically. In 2006, coal accounted for 20% of annual global carbon emissions (in the form of carbon dioxide); it produces 29% more carbon per unit of energy than oil and 80% more than natural gas. The environmental effects of coal use are pictured in Figure 10.9.

Figure 10.8 (a) Strip mining of coal can devastate the land, leaving it bare and subject to erosion. (b) Land destroyed by strip mining can be reclaimed by regrading the surface, followed by planting trees and grass.

Apart from the environmental problems associated with its use, solid coal is not a versatile fuel and is not suitable for today's home and office heating systems or as a fuel for automobiles or airplanes. The main consumer of coal today is the electric power industry. Many of the disadvantages of coal as a fuel can be overcome if it is converted to a gas or liquid before it is burned. In a gaseous or liquid form, coal is easier to transport and burns cleaner.

Coal Gasification

Coal gasification (the production of a gas from coal) is not a new idea. As early as 1807, a municipal system was used to light streets in London. Coal was heated in the absence of air—by a process known as *pyrolysis*—and yielded coal gas, which is a mixture of hydrocarbons, hydrogen, and carbon monoxide that could be burned to produce light. As late as 1932, gas supplied to homes in the eastern United States was derived from coal, but with the construction of the pipeline network in the 1940s, less expensive and safer natural gas replaced it. Coal makes up the largest reserve of fossil fuel in the United States, and as concerns over our dependence on imported oil have grown, interest in both coal gasification and coal liquefaction has been renewed.

In the first step in a modern coal gasification process, heated, crushed coal is treated with superheated steam at high temperature (900°C) and pressure. This reaction, which is called the "steam-reforming" reaction, produces carbon monoxide and hydrogen:

$$C + H_2O \rightarrow CO + H_2 - 131.4\,kJ$$

coal steam

Carbon monoxide then reacts with hydrogen to produce CH_4 and water and with steam to produce carbon dioxide and more hydrogen.

$$CO + 3H_2 \rightarrow CH_4 + H_2O$$

$$CO + H_2O \rightarrow CO_2 + H_2$$

Conditions are adjusted to give the maximum possible yield of CH_4. Unreacted carbon monoxide and hydrogen are recycled through the system, and sulfur compounds and other impurities are removed. The final product, which is predominantly CH_4, is called **synthesis gas** (or *syn gas*). The process, which is relatively inexpensive, also can be used to produce methanol, another useful fuel.

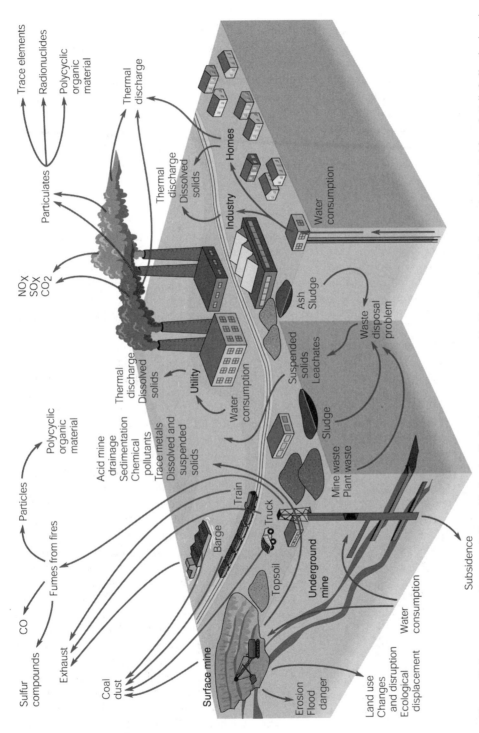

Figure 10.9 Environmental impacts of the coal fuel cycle. Similar impacts are caused by the oil and natural gas cycles. Adapted from C. Hall, C. Cleveland, and R. Kaufman. *Energy and Resource Quality*. Niwot, CO: University Press of Colorado, 1986. Used with permission of the publisher.

Coal Liquefaction

Petroleum-like liquids can be obtained from coal if the complex organic molecules in coal are broken into smaller molecules and if the hydrogen content of the molecules is increased. This result can be achieved by subjecting a mixture of coal and hydrogen to high pressure in the presence of a metal catalyst. A material similar to crude oil is produced and can be fractionally distilled to yield products that are similar to those obtained from petroleum: kerosene, diesel fuel, gasoline, and lubricants.

The future of coal liquefaction depends on the world price of oil and the degree of U.S. reliance on imported oil. It should be remembered that the conversion of coal to a liquid or a gas in no way reduces the amount of carbon dioxide released to the atmosphere during combustion.

■ Electricity

Electricity, the most important secondary source of energy in the United States, plays a critical role in modern society. When the first electrical generating plants began operating at the end of the 19th century, electricity was used only for lighting. It was not until the beginning of the 20th century, when the electrification of industry began, that electrical appliances were introduced in the marketplace. The first clothes washers and vacuum cleaners were produced in 1907. Today, 40% of all of the energy consumed in the United States is used to generate electricity. Approximately two-thirds of this production powers electric motors, which are vital to the industrial, domestic, and military sectors of the U.S. economy. Elevators, air conditioners, communications systems, television sets, microwave ovens, toasters, and numerous other systems and appliances that depend on electricity have become commonplace, taken for granted as necessities in the American way of life. In the United States, coal provides 49% of the energy used to generate electricity. Nuclear power provides nearly 20%. Natural gas supplies 20%, and hydropower, oil, and renewable sources supply around 11% combined (Figure 10.10).

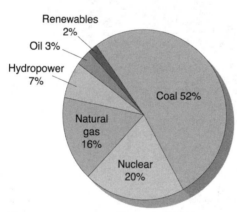

Figure 10.10 In 2006, fossil fuels (oil, natural gas, and coal) compose more than 87% of the fuel sources for electric power generation in the United States. U.S. Department of Energy, Energy Information Administration, 2006.

Electricity is produced whenever a coil of wire is moved through a magnetic field, or whenever a magnetic field is moved within a coil of wire. The movement induces a flow of electrons (electricity) in the wire (Figure 10.11). A rotating coil of wire within a circular arrangement of magnets operates most modern electric generators. Energy, of course, is needed to rotate the coil. In the United States, this energy is obtained primarily from coal. Heat from the combustion of coal converts water in a boiler into high-pressure steam. The steam strikes the blades of a turbine, which is coupled to a generator. The turbine turns, causing the coil of wire to rotate in a magnetic field, and electricity is generated (Figure 10.12a). After the steam has passed through the turbine, it is cooled, condensed back to water, and returned to the boiler to be reused.

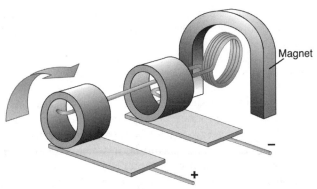

Figure 10.11 In an electric generator, when a coil of wire is rotated in a magnetic field, an electric current flows in the wire. Adapted from B.J. Nebel. *Environmental Science: The Way the World Works, Sixth edition*. Englewood Cliffs, NJ: Prentice Hall, 1998.

Generation of electricity is not an efficient process. Inevitably, as a consequence of the second law of thermodynamics, energy is wasted. Thermodynamics defines **efficiency** as follows:

$$E = 1 - (Tc/Th)$$

Where

Th is the temperature of the heat source (boiler), and Tc is the temperature of the condenser (both temperatures are in K degrees).

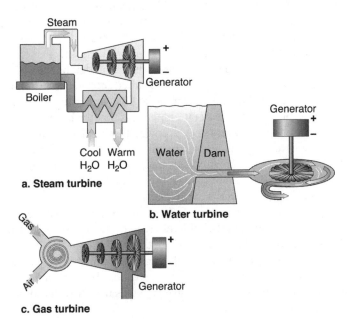

Figure 10.12 (a) Steam turbines, (b) water turbines, or (c) gas turbines can be used to generate electricity.

In a modern coal-fired plant, the boiler temperature reaches 550°C (820°K). The condenser is cooled by water that is at an ambient temperature, 25°C (398°K). According to the equation, the efficiency is as follows:

$$E = 1 - (298/820) = 0.64 \text{ or } 64\%$$

Thus, 64% is the maximum theoretic efficiency. In the average coal-fired electric power plant, 60% to 70% of the energy used to produce steam is lost in the form of heat as the steam is cooled. Only 30% to 40% of the energy derived from coal is converted into electricity. The waste heat is removed by circulating large quantities of water, usually taken directly from a nearby river or lake, around the condenser. Chapter 8 pointed out that the increased temperature of the water (thermal pollution) as it is returned to its natural source can adversely affect fish and other marine organisms.

Steam turbines generate most of the electricity produced in the United States. Coal is the major heat source used to convert water to steam, but oil and nuclear energy are also used. Approximately 7% of the electricity produced in the United States is generated with hydroelectric power (Figure 10.12b). In this process, the kinetic energy released when water held behind a dam or at the top of a waterfall drops to a lower level is used to turn the blades of a turbine.

Another 10% of U.S. electricity is generated by gas turbines (Figure 10.12c), which can be run at much higher temperatures than a boiler (1250°C), and because of this, their overall efficiency (approximately 50%) is higher than that of a boiler. High-pressure gases produced in the combustion of natural gas turn the turbine blades. Increased use of gas turbines in the future is expected to increase the total fraction of electricity generated by gas turbines to 20% by 2010.

An old technology, which has found new proponents and sometimes can more than double the usable energy produced by a power plant, is **cogeneration**. The basic principle behind cogeneration is that a power plant that can simultaneously produce several types of energy, such as electricity and heat, is much more efficient. An electrical power plant may heat water to produce steam; the steam is used to turn a turbine, which is attached to a generator that produces electricity. In conventional power plants, the steam that has passed though the turbine is treated as waste, and its heat content is discarded into the environment. In a cogeneration plant, the spent steam (hot water) is used for some other purpose, such as heating a building. The major problem with cogeneration plants is that the farther they are from where the heat is to be used, the less efficient they are; ideally, the plant should be located in the community that they serve. For this reason, relatively small-scale cogeneration plants have more potential for community acceptance and energy efficiency than do large-scale generators. Cogeneration plants are often built adjacent to chemical plants where the spent steam is used to heat chemical reactions and run the plant.

Electricity is a very clean and convenient form of energy at the consumer level, but its production creates many environmental problems, which are related to the primary energy source. In the United States, coal provides 49% of the energy used to generate electricity; its combustion by power plants produces the adverse environmental effects described earlier here. As demand for electricity has escalated, the world has become increasingly dependent on coal and nuclear energy to meet energy needs. Nuclear energy and other nonfossil fuel energy sources are considered in Chapter 11.

Sequestering CO$_2$

The single biggest problem with continued fossil fuel consumption is the release of CO$_2$ into the atmosphere. Rather than limiting CO$_2$ release, why not capture CO$_2$ from smoke stacks and store it in places where it will not be released into the atmosphere? Until recently, the idea of sequestering CO$_2$ was not seriously considered because the cost to do so would be prohibitive. Today, at the Sleipner oil field off the coast of Norway in the North Sea, 1 million tons a year of CO$_2$ is injected into the pores of a sandstone layer 1000 m below the sea. Norway imposes a carbon dioxide tax on generators, and the owners of the Sleipner field determined that sequestering the CO$_2$ and injecting it into the deep sandstone layer would be less expensive than releasing the CO$_2$ into the atmosphere and paying the tax. In another project in Canada, the Weyburn oil field injects 13,000 tons of waste CO$_2$ per day underground into the oil field, which prior to this project had become unproductive. When CO$_2$ is pumped at high pressure into the reservoir, the CO$_2$ mixes with the oil, causing it to swell and become less viscous. The swelling forces oil out of the pores in the rocks, so that it can flow more easily. It is predicted that the CO$_2$ injection will enable an additional 130 million barrels of oil to be produced, extending Weyburn's commercial life by approximately 25 years.

In 2007, 39% of U.S. CO$_2$ emissions came from power plants. These plants would be suitable for CO$_2$ capture except that flue gas emitted from coal-fired power plants is diluted with N$_2$ and contains only 10% to 12% CO$_2$ by volume. The CO$_2$ released from fossil fuel combustion would have to be selectively removed from the smoke stack and concentrated. The CO$_2$ must be concentrated so that it is greater than 90% by volume so that it can be liquefied. The trapped CO$_2$ must then be placed in long-term storage or reacted to form a material that cannot harm the atmosphere.

The U.S. Department of Energy (DoE) is funding research on finding an efficient way to trap and concentrate CO$_2$ in flue gas. One DoE project is developing high-temperature gas separation membranes that would selectively pass CO$_2$ and concentrate it while allowing the N$_2$ to continue up the smoke stack. Another DoE trapping project reacts the CO$_2$ with alkali carbonates, which converts them to alkali bicarbonates. The alkali bicarbonate is removed, and in a regeneration process, it releases CO$_2$ and water. Two options that are being considered for long-term storage of CO$_2$ are shown in Figure 10.13. Deep ocean storage is attractive because the low temperature and high pressure would keep the CO$_2$ liquefied. At a depth of more than 2 miles, the density of the liquid CO$_2$ would be greater than that of water, and it would stay in a pool on the ocean's floor. There is worry that the pool of CO$_2$ may harm marine life, and it is well known that CO$_2$ dissolves in water to form bicarbonate. Another option is underground storage of CO$_2$ in salt domes or other deep geologic formations. The biggest problem with this option is the cost of capturing CO$_2$ and transporting it from the power plant to the deep ocean.

Chapter 1 pointed out that weathering is the natural process in which CO$_2$ reacts with basic minerals and is converted into a solid carbonate. If this extremely slow reaction could be accelerated, then reaction of CO$_2$ with basic minerals would produce solid carbonates that could be buried.

If any of these processes could be made economically feasible, we could continue to rely on fossil fuels until our supply was depleted, without adding more CO$_2$ to the atmosphere.

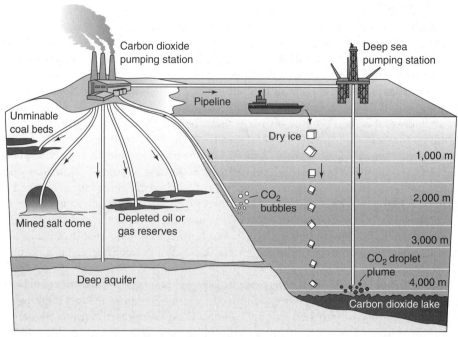

Figure 10.13 Storage sites for CO_2 in the ground and deep under the sea. Adapted from H. Howard, B. Eliasson, and O. Kaarstad, *Sci. Am.* (2000): 72–79.

■ Additional Sources of Information

Aubrecht GJ. *Energy*, 3rd ed. Upper Saddle River, NJ: Prentice Hall, 2004.

Bisio A, Boots S. *The Wiley Encyclopedia of Energy and the Environment.* New York: John Wiley and Sons, 1996.

Brasseur GP, Orlando JJ, Tyndall GS. *Atmospheric Chemistry and Global Change.* New York: Oxford University Press, 1999.

Jacobson MZ. *Atmospheric Pollution.* Cambridge, UK: Cambridge University Press, 2002.

McKinney ML, Schoch RM. *Environmental Science Systems and Solutions,* 4th ed. Sudbury, MA: Jones and Bartlett Publishers, 2007.

■ Keywords

anthracite coal
anticline trap
bitumen
bituminous coal
bond energy

benzene, toluene, and
  xylene (BTX)
cap rock
catalytic reforming
cogeneration

combustion
cracking
efficiency (thermo-
  dynamics)
E10

electricity	kerogen	peat
endothermic	kerosine	petrochemicals
energy	lignite	power
exothermic	liquefied petroleum gas	reservoir rock
fossil fuel	natural gas	shale oil
fractional distillation	octane enhancers	straight-run gasoline
gasohol	octane rating	synthesis gas (syn gas)
hydrocarbons	oil shale	tar sands

■ Questions and Problems

1. Describe two ways in which the steam engine changed society.
2. Before the discovery of petroleum, what was the major source of energy in the United States?
3. Describe the difference between energy and power. What units are used to measure each?
4. What is the origin of petroleum?
5. What are the geologic features that are required for the formation of petroleum in amounts that can be extracted from the Earth?
6. Besides crude oil and natural gas, what other substance is often found in the reservoir rock? What problem does this substance create?
7. Why was natural gas once treated as a waste product of petroleum production?
8. Give two reasons why natural gas is considered to be a good fuel.
9. Why is the fuel value for crude oil higher than the value for coal or wood?
10. Write a chemical equation for the complete combustion of butane.
    a. How much heat is released when 580 g of butane are burned?
    b. How many grams of CO_2 are produced when 1060 g of butane are burned?
11. Assume that natural gas is pure CH_4 and that the fuel used in gas barbecue grills is pure C_3H_8.
    a. Calculate and compare the energy released by burning 1000 g of each.
    b. Calculate and compare the energy released per mole of CO_2 released.
    c. Which fuel causes less damage to the atmosphere?
12. Calculate the amount of energy released by burning octane (gasoline). Compare this with the amount of energy released by burning the same amount of ethanol. Would you expect automobiles that used ethanol as a fuel to get lower mileage? How much less?
13. Assume that diesel fuel is $C_{14}H_{30}$ and that gasoline is octane.
    a. Calculate the amount of energy each releases per gram.
    b. Compare each as a fuel for trucks.
    c. Why do so many large trucks have diesel engines?
14. Why does sulfur pose a problem when it is a component of crude oil?
    a. How much sulfur does high-sulfur oil contain?
    b. How much sulfur does low-sulfur oil contain?
    c. Is the sulfur in crude oil elemental sulfur? Explain.
15. a. How much nitrogen is present in crude oil?
    b. Draw the structure of a characteristic nitrogen-containing component.
    c. Do these nitrogen-containing molecules pose a problem for the refinery?

16. Describe what is meant by fractional distillation.
    a. What physical property describes a "fraction?"
    b. What is asphalt?
    c. What physical properties describe the gasoline fraction?
    d. What is the difference between gasoline and kerosene?
17. Describe the following:
    a. Straight-run gasoline
    b. The octane rating of heptane
    c. The octane rating of regular gasoline
18. Answer the following for the cracking of petroleum:
    a. Write a chemical equation that shows the cracking of $C_{16}H_{34}$.
    b. The catalyst used for cracking petroleum.
    c. Why must air be excluded from the reaction?
19. Describe three methods that are used to increase the fraction of gasoline that is produced from each barrel of crude oil.
20. Describe what is meant by catalytic reforming.
21. Draw all hydrocarbon isomers with the formula C_5H_{12}. Using Table 10.5, list them according to increasing octane number.
22. Which has a higher octane rating, toluene (C_7H_8) or heptane (C_7H_{16})?
23. For each of the following pairs of hydrocarbons, indicate which has the higher octane rating.
    a. Pentane or octane
    b. 2-Methylpentane or 2,2-dimethylbutane
    c. Octane or 2-methylheptane
24. List four compounds that are used as octane enhancers.
25. Give two reasons why TEL was banned as a gasoline additive.
26. What is MTBE?
    a. What is it made from?
    b. Why is its use being reduced?
27. List the two main components of gasohol with their concentration.
28. What is the BTX fraction of crude oil? Draw structures of the molecules that comprise this fraction.
29. What is oil shale? What organic material does it contain?
30. Describe the problems associated with recovering oil from oil shale.
31. How does the current price of petroleum affect the production of shale oil?
32. What are tar sands? What organic material do they contain?
33. What is the difference in chemical composition between commercial natural gas and crude natural gas?
34. What is the difference between bituminous and anthracite coal? Why is anthracite coal a better fuel?
35. Why is peat not considered a good fuel?
36. What are the two main methods for mining coal, and which one is used more extensively? What is the environmental impact of each method?
37. Identify two environmental problems that are associated with strip mining.
38. Write a chemical equation for the complete combustion of coal.

39. Why is coal, which was a major fuel for residential heating in the 1930s, considered unsuitable for home heating in the 2000s?

40. Why is coal gasification considered to be a way to convert coal to a more suitable fuel for home use?

41. Steam reforming is used in coal gasification. Write a chemical reaction that shows the reactants and products.

42. What is synthesis gas? How is it formed in coal gasification? Is it a high-energy fuel?

43. The generation of electricity produces significant quantities of various pollutants. List the types of fuels that are used to generate electricity, including the percentage that each contributes to total U.S. production of electricity and the pollutants that each produces.

44. Producing electric power from fossil fuels causes two different environmental problems. What are they?

45. What is the average efficiency of a coal-burning electric-generating power plant?

46. What is the maximum efficiency of a gas turbine that operates with a combustor temperature of 1250°C and a condenser temperature of 25°C?

47. What is the maximum efficiency of a gas turbine that operates with a combustor temperature of 1250°C and a condenser temperature of 25°C?

48. Describe the relationship between the demand for a fossil fuel and growth in the following U.S. industries (indicate increasing or decreasing demand and which fossil fuel).
   a. Farming
   b. Food processing
   c. Manufacturing
   d. Transportation

49. Describe how an electric generator makes electricity.

50. Describe how energy produced by burning coal is used to turn a turbine and generate electricity.

51. Why are electric power plants often located near large bodies of water?

52. Every day 100 railroad cars of coal enter a medium-sized coal-burning electric power plant. By the end of the day, 30 railroad cars of ash are removed from the plant. Describe what happened to the contents of the railroad cars.

53. Describe how a gas turbine makes electricity.

54. What fraction of the U.S. CO_2 emissions comes from electric power plants?
   a. What is the concentration of CO_2 in the flue gas?
   b. Why do we not capture CO_2 before it reaches the atmosphere?

55. List two ways of concentrating the CO_2 in flue gas emissions, and state what concentration of CO_2 is needed to make sequestering practical.

56. List four places where sequestered CO_2 could be stored. List advantages and disadvantages for each.

57. Propose a chemical reaction that would sequester CO_2 by converting gaseous CO_2 into a solid.

CHAPTER

11

Nuclear Power

THE EXPLOSION OF THE FIRST ATOMIC BOMB IN 1945 DEMONSTRATED THE ENORMOUS DESTRUCTIVE POWER OF THE ATOMIC NUCLEUS AND THE DANGERS OF NUCLEAR RADIATION. With the end of the Cold War and the collapse of the Soviet Union in 1992, the threat of nuclear war has diminished but many concerns remain. Radioactive wastes that nuclear weapons plants and the nuclear power industry have produced threaten the environment, and safe disposal of these wastes is a major problem. There is also fear that another catastrophic accident, such as the one at Chernobyl in the Ukraine in 1985, could occur at one of Eastern Europe's aging nuclear power plants.

■ The Nature of Natural Radioactivity

Ernest Rutherford determined that the radiation emitted by naturally occurring radioactive materials is of three distinct types. Rutherford passed a beam of such radiation between electrically charged plates onto a photographic plate (Figure 11.1). He found that the beam of radiation split into three components. One component was attracted to the negatively charged plate and was therefore positively charged. A second component was attracted to the positively charged plate and was therefore negatively charged, and the third component was not deflected from its original path and could be assumed to have no charge. Rutherford named these components alpha (α), beta (β), and gamma (γ) **rays** after the first three letters in the Greek alphabet. It was not until some 30 years later that these three kinds of radiation were completely characterized.

α Rays in fact are made of particles that each consist of two protons and two neutrons. An α particle, therefore, has a mass of 4 amu and a +2 charge and is identical to a helium nucleus (a helium atom minus its two electrons). α Particles are represented as $_2^4\alpha$ or $_2^4\text{He}$.

β Rays are made of particles also. A β particle is identical to an electron and therefore has negligible mass and a charge of -1. It is usually represented

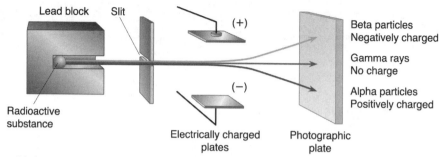

Figure 11.1 In an electric field, radiation from a radioactive source splits into three components: alpha particles, beta particles, and gamma rays. The positively charged alpha particles are attracted to the negatively charged plate; the negatively charged beta particles are attracted to the positively charged plate; and the uncharged gamma rays are not deflected from their original path. Adapted from T.L. Brown, E. LeMay, and B. Bursten. *Chemistry: The Central Science: Fifth edition.* Englewood Cliffs, NJ: Prentice Hall, 1991.

as $_{-1}^{0}\beta$ (or $_{-1}^{0}e$). Although a β particle is identical to an electron, it does not come from the electron cloud surrounding an atomic nucleus, as might be expected. It is produced from inside the atomic nucleus and then ejected.

γ Rays, unlike α and β rays, are not made up of particles and therefore have no mass. Similar to X-rays, they are a form of high-energy electromagnetic radiation with a very short wavelength and are usually represented as $_{0}^{0}\gamma$. All three types of radiation have sufficient energy to break chemical bonds and thus disrupt living organisms.

Penetrating Power and Speed of the Types of Radiation

α Particles, β particles, and γ rays are emitted from radioactive nuclei at different speeds and have different penetrating powers. α Particles are the slowest and are emitted at speeds that are approximately equal to one-tenth the speed of light and can be stopped by a sheet of paper or by the outer layer of a person's skin (Figure 11.2). β Particles are emitted at speeds of almost equal to the speed of light, and because of their greater velocity and smaller size, their penetrating power is approximately 100 times greater than that of α particles. β Particles can pass through paper and several millimeters of skin but are stopped by aluminum foil. γ Rays are released from nuclei at the speed of light and are even more penetrating than X-rays. They pass easily into the human body and can be stopped only by several centimeters of lead or several meters of concrete. The properties of the three types of natural radiation are summarized in Table 11.1.

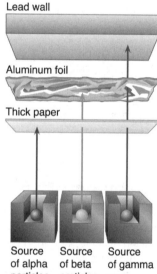

Figure 11.2 The penetrating abilities of α particles, β particles, and γ rays differ: α particles are the least penetrating and are stopped by a thick sheet of paper or the outer layer of the skin. β particles pass through paper but are stopped by aluminum foil or a block of wood. γ Rays can be stopped only by a lead wall several centimeters thick or a concrete wall several meters thick. Adapted from J.P. Sevenair and A.R. Burkett. *Introductory Chemistry: Investigating the Molecular Nature of Matter.* Dubuque, IA: Wm. C. Brown, 1997.

Table 11.1

Properties of the Three Types of Radiation Emitted by Radioactive Elements

Name	Symbol	Identity	Charge	Mass (amu)	Velocity	Penetrating Power
α	$^4_2\alpha$, ^4_2He	Helium nucleus	2+	4	1/10 the speed of light	Low, stopped by paper
β	$^0_{-1}\beta$	Electron	1−	0	Close to the speed of light	Moderate, stopped by aluminum foil
γ	$^0_0\gamma$	High-energy electro-magnetic radiation	0	0	Speed of light (3×10^{10} cm/s)	High, stopped by several centimeters of lead

Nuclear Stability

The nuclei of the atoms of most—but not all—naturally occurring elements are very stable, despite the electrical repulsions between the positively charged protons that tend to pull them apart. This stability is achieved by a powerful, although not well understood, localized force within each nucleus that overcomes the repulsions and holds the nuclei firmly together. If we plot the number of neutrons (N) in an isotope against the number of protons (Z) for all stable nuclei, we find that they all fall in a narrow band called the **band of stability** (Figure 11.3). At low atomic numbers, stable nuclei have an equal number of protons and neutrons. As the atomic number increases for stable nuclei, the number of neutrons exceeds the number of protons, and the ratio becomes approximately 1.5 for the heaviest stable nuclei.

Nuclei that lie outside of the band of stability are unstable and spontaneously emit high-energy radiation as a means of achieving greater stability. Nuclei that lie outside the band of stability are unstable and decompose to give a nucleus with a more stable neutron-to-proton ratio. For example, a nucleus that lies above the band of stability must either gain protons or lose neutrons to become more stable. The isotope ^{14}C, which lies above the band of stability, decays by the emission of an electron, because the process converts a neutron into a proton:

$$^1_0\text{n} \rightarrow {}^1_1\text{p} + {}^0_{-1}\text{e}$$

$$^{14}_6\text{C} \rightarrow {}^{14}_7\text{N} + {}^0_{-1}\text{e}$$

Nuclei located below the band of stability must increase their neutron-to-proton ratio to achieve stability. They can do this by either positron emission (^0_1e) or by electron capture. Position emission converts a proton into a neutron:

$$^{11}_6\text{C} \rightarrow {}^{11}_5\text{B} + {}^0_1\text{e}$$

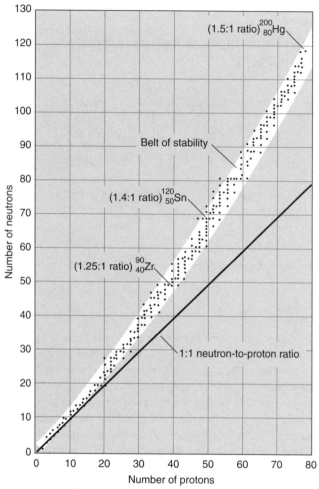

Figure 11.3 A plot of the number of protons versus the number of neutrons presents the band of stability for stable nuclei. As the atomic number increases, the neutron-to-proton ratio of the stable nuclei increases. The majority of radioactive nuclei lay outside the belt of stability.

Electron capture converts a proton to a neutron:

$$^{7}_{4}Be + ^{\ 0}_{-1}e \rightarrow ^{7}_{3}Li$$

The electron is captured from an inner shell of the atom, which leaves the atom in an electronically excited state. An electron from the valence shell quickly drops to fill the vacant orbital, and an X-ray is emitted. The wavelength of the X-ray corresponds to the energy of the transition and can be used to identify the presence of the isotope.

$$^{40}_{19}K + ^{\ 0}_{-1}e \rightarrow ^{40}_{18}Ar + X\text{-ray}$$

Nuclei with atomic numbers that are greater than 83 cannot achieve stability by electron or positron emission or electron capture. Unstable isotopes are called **radioisotopes** because they are **radioactive** and decompose to a more stable isotope by emitting an α, β, or γ. Approximately 25 naturally occurring elements have one or more radioisotopes.

Nuclear Reactions

When a radioactive isotope of an element emits an α or β particle, a nuclear reaction occurs, and the nucleus of that isotope is changed.

As examples, the element uranium has several radioactive isotopes, including $^{238}_{92}U$ (also written uranium-238), which spontaneously emits α particles. Thus, the spontaneous emission of an α particle from a uranium atom results in the formation of a completely different element. This **transmutation** of uranium to thorium is represented by the following nuclear equation:

$$^{238}_{92}U \rightarrow {}^{234}_{90}Th + {}^{4}_{2}\alpha$$

Thorium-234, the main product of this reaction, is always found in uranium-238 ore deposits. It also is radioactive but emits β instead of α particles. It may seem improbable that a negatively charged β particle ($^{0}_{-1}\beta$) identical to an electron can be released from a nucleus made of positively charged protons ($^{1}_{1}p$) and uncharged neutrons ($^{1}_{0}n$), but this does occur. The accepted explanation is that a neutron in the nucleus changes into a proton and a β particle according to the following equation:

$$^{1}_{0}n \rightarrow {}^{1}_{1}p + {}^{0}_{-1}\beta + \text{energy}$$

Here is the balanced nuclear equation showing the emission of a β particle from the atom of thorium-234 and the formation of a new element:

$$^{234}_{90}Th \rightarrow {}^{234}_{91}Pa + {}^{0}_{-1}\beta$$

After emitting an α or β particle, the nuclei of naturally occurring radioisotopes still have excess energy, which they release by emitting γ rays. Because γ rays are waves, not particles, and have no mass and no charge, the identity of the emitting element does not change when a γ ray is emitted. γ Rays are not therefore included in nuclear equations. The energy of the γ ray released is dependent on the reaction; a γ spectrometer can measure their energy and identify from which radioisotope they were emitted.

Radioactive Decay Series

The transformation of one element into another as a result of radioactive emissions is termed **radioactive decay**. The spontaneous decay of uranium-238 to form thorium-234 and that of thorium-234 to form proactinium-234 is the first two steps in the **uranium decay series**, which continues through a total of 14 steps to finally yield the stable isotope lead-206 (Figure 11.4). Of particular interest in this decay series is radon-222 ($^{222}_{86}Rn$), the radioactive gas that (as discussed later in this chapter) accounts for most of the potentially harmful radiation in the environment.

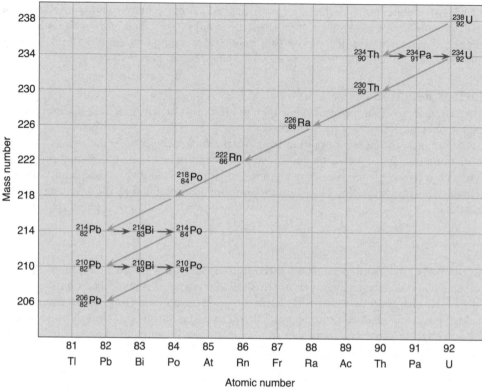

Figure 11.4 The uranium decay series: $^{238}_{92}$U spontaneously decays in a series of steps until the stable element $^{206}_{82}$Pb is formed. Each time an alpha particle is emitted, the mass number decreases by 4 and the atomic number decreases by 2 (blue arrows). Each time a beta particle is emitted, the mass number remains the same but the atomic number increases by 1 (red arrows). Note that the gas radon (Rn), which accounts for most of our everyday exposure to radiation, is formed in the series a result of the disintegration of $^{226}_{88}$Ra. Adapted from T.L. Brown, E. LeMay, and B. Bursten. *Chemistry: The Central Science: Fifth edition.* Englewood Cliffs, NJ: Prentice Hall, 1991.

Besides the uranium decay series, two other naturally occurring decay series exist: the thorium series, which starts with thorium-232 and ends with lead-208, and the actinium series, which starts with uranium-235 and ends with lead-207. All of the naturally occurring radioisotopes with high atomic numbers belong to one of these three decay series.

■ The Half-Life of Radioisotopes

Different radioisotopes—whether naturally occurring or artificially produced—decay at characteristic rates. The more unstable the isotope is the more rapidly it will emit α or β particles and change into a different element. The rate at which a particular radioisotope decays is expressed in terms of its **half-life**, the time required for one-half of any given quantity of the isotope to decay. There is no way of knowing when a particular nucleus will decay, but after one half-life, half the nuclei in the original sample will have done so.

Half-lives range from billionths of a second to billions of years. For example, the half-life of boron-9 is only 8×10^{-19} seconds. That of thorium-234 is 24 days, and that of uranium-238 is 4.5 billion years. Half-lives of a number of radioisotopes are given in Table 11.2.

We can construct a **decay curve** (Figure 11.5) to show graphically the quantity of a given radioisotope that will remain after a given length of time. Assume that we start with 16 g of $^{32}_{15}P$, a radioisotope that has a half-life of 14 days and decays by emitting a β particle to form $^{32}_{16}S$. After one half-life (14 days), one-half of the original 16 g of $^{32}_{15}P$ will have decayed and been converted to $^{32}_{16}S$; 8 g of $^{32}_{15}P$ will therefore remain (and 8 g of $^{32}_{16}S$ will have been formed). At the end of two half-lives (28 days), one-half of the 8 g of $^{32}_{15}P$ that was present at the end of one half-life will have decayed; 4 g will remain. At the end of three half-lives (42 days), 2 g will remain, and so on. After six half-lives (84 days), just 0.25 g of the original amount of $^{32}_{15}P$ will remain; the rest of the sample will be $^{32}_{16}S$. Even after many half-lives, a minute fraction of the original radioisotope will remain.

Radioactive decay is an example of a first-order rate process. If there are no nuclei of an isotope at time (t) zero and N at time t, then $\Delta N = No - N$ nuclei have disintegrated in the time interval $\Delta t = t - to$

$$\text{Rate of decay} = \frac{\Delta N}{\Delta t} = kt$$

The number of nuclei N remaining after time t can be calculated from this equation:

$$\ln \frac{N_0}{N} = kt$$

Table 11.2

Half-Lives of Some Radioactive Isotopes

Radioactive Isotope	Half-Life
Oxygen-13	8.7×10^{-3} seconds
Bromine-80	17.6 minutes
Iodine-132	2.4 hours
Technetium-99	6.0 hours
Radon-222	3.8 days
Barium-140	12.8 days
Hydrogen-3 (tritium)	12.3 years
Strontium-90	28.1 years
Radium-226	1620 years
Carbon-14	5730 years
Plutonium-239	24,400 years
Beryllium-10	4.5 million years
Potassium-40	1.3 billion years
Uranium-238	4.5 billion years

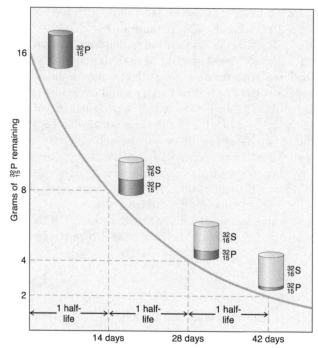

Figure 11.5 Decay curve for a 16-g sample of $^{32}_{15}P$ (half-life, 14 days) to $^{32}_{16}S$ by emission of β particles. After one half-life, one-half of the original 16 g of $^{32}_{15}P$ will have been converted to $^{32}_{16}S$; that is, 8 g will remain. After another 14 days, half of that 8 g remain. At the end of 42 days (three half-lives), 2 g of $^{32}_{16}S$ will have been formed and so on.

To find the half-life, $t_{1/2}$ (that is, when one-half of the radioactive nuclei in any sample has decayed)

$$N = 1/2N_0$$

$$\ln \frac{N_0}{1/2N_0} = k\,t_{1/2} \ \text{ or}$$

$$k\,t_{1/2} = \ln 2$$

$$t_{1/2} = \frac{0.693}{k}$$

EXAMPLE   11.1

The half-life of radon, $^{222}_{86}Rn$, is 3.8 days. If your basement contains 45 g of $^{222}_{86}Rn$ today, how much $^{222}_{86}Rn$ will remain after 8.5 days (assuming that only radioactive decay is occurring)?

Determine the rate constant k:

$$k = \frac{0.693}{t_{1/2}} = \frac{0.693}{3.8 \text{ days}} = 0.18 \text{ day}^{-1}$$

$$\ln \frac{No}{N} = \ln \frac{\text{Initial mass of Rn}}{\text{Mass Rn after 8 days}} = kt = \left(0.18 \text{ day}^{-1}\right)\left(8.5 \text{ days}\right) = 1.5$$

Therefore

$$\ln \frac{45 \text{ g Rn}}{\text{Mass Rn after 8 days}} = 1.5$$

$$\frac{45 \text{ g Rn}}{\text{Mass Rn after 8 days}} = 4.5$$

$$\text{Mass Rn after 8 days} = \frac{45 \text{ g Rn}}{4.5} = 10 \text{ g Rn}$$

Thus, there will be 10 g of radon remaining after 8.5 days.

Many of the radioisotopes in the wastes generated in the production of nuclear weapons and the operation of nuclear power plants have long half-lives. This is a matter of concern. If long-lived radioisotopes escape into the environment, they persist for many years, creating the possibility that they will accumulate in food webs. At least 10 half-lives must elapse before a radioisotope has decayed to the point at which it is no longer considered to be a radiation hazard.

EXAMPLE 11.2

A sample of metal pipe was removed from a nuclear power plant. An analysis of the pipe showed that it contained 0.30 g of ^{60}Co. Another measurement was made 1.4 years later. This measurement showed 0.25 g of ^{60}Co to be remaining. What is the half-life of ^{60}Co? How long should we wait until this radioisotope has decayed to a point at which it is no longer considered a radiation hazard?

$$\ln \frac{\text{Initial mass of Co}}{\text{Mass Co after 1.4 years}} = \ln \frac{0.30 \text{ g}}{0.25 \text{ g}} = kt = \left(k\right)\left(1.4 \text{ years}\right)$$

$$0.182 = \left(k\right)\left(1.4 \text{ years}\right)$$

$$0.13 \text{ year}^{-1} = k$$

We know that

$$t_{1/2} = \frac{0.693}{k}$$

so

$$t_{1/2} = \frac{0.693}{0.13 \text{ year}^{-1}} = 5.3 \text{ years}$$

We must wait at least 10 half-lives, or 53 years, until the ^{60}Co is no longer considered a radiation hazard.

■ The Harmful Effects of Radiation on Humans

When radioactivity was discovered at the beginning of the 20th century, its harmful effects were not recognized. Marie Curie, who worked for many years with radioactive materials, suffered from anemia and died of leukemia; she was probably one of the first victims of radiation poisoning. Other early sufferers were women who, in the 1920s, worked in factories painting radium on watch dials to make them glow in the dark. The women frequently licked their brushes to obtain a fine point and often developed cancer of the lips; as a result of ingesting the radioactive material, many of them developed bone cancer or leukemia and died at an early age.

Why Is Radiation Harmful?

Radiation emitted by radioisotopes is harmful because it has sufficient energy to knock electrons from atoms and thus form positively charged ions. For this reason, it is called **ionizing radiation**. Some of these ions are highly reactive and, by disrupting the normal workings of cells in living tissues, can produce abnormalities in the genetic material DNA and increase the risk of cancer.

Factors Influencing Radiation Damage

The degree of damage caused by ionizing radiation depends on many factors, including (1) the type and penetrating power of the radiation, (2) the location of the source of the radiation (inside or outside the body), (3) the type of tissue exposed, and (4) the amount and frequency of exposure.

α Particles from a source outside of the body constitute the least dangerous type of ionizing radiation because α particles cannot penetrate the skin and enter the body; however, if an α emitter is ingested, for example, in contaminated food, or is inhaled by breathing air containing radon gas, damage to internal body tissues can be severe. Once inside the body, α particles are more damaging than β particles or γ rays because they are more effective in forming ions in the surrounding tissues. They travel only short distances through tissues and rapidly transfer the bulk of their energy to a small area. β Particles and γ rays travel farther than α particles and transfer their energy over a wider area of tissue. As a result, the radiation received per unit area is lower for β and γ radiation, and the ability to form ions is reduced. The greater the penetrating power of a particular source of radiation, the weaker is its ionizing power.

β Particles, unlike α particles, can penetrate the outer layers of skin and clothing and produce severe burns and can also cause skin cancer and cataracts. Inside the body, β particles are less disruptive to individual cells than are α particles; outside the body, however, they are more damaging than α particles. γ Rays, because of their great penetrating power (Figure 11.2), are more dangerous than α or β particles outside of the body; they are less dangerous inside the body than the other two.

X-rays are a fourth type of ionizing radiation. Their penetrating power, and thus their ability to cause tissue damage, lies between that of β particles and γ rays.

Body tissues vary widely in their sensitivity to ionizing radiation. Rapidly dividing cells are very vulnerable to radiation. Such cells are found in bone marrow, the lining of the gastrointestinal tract, the reproductive organs, the spleen, and the lymph glands. Embryonic tissue is particularly easily damaged. Therefore, unless there are compelling medical reasons, pregnant women should avoid exposure to all radiation, including X-rays. Cancer cells, which divide very rapidly, are more easily killed by radiation than are healthy cells; this fact explains the success of radiation treatment for some types of cancer.

Detection of Radiation

In order to measure radiation exposure, it is necessary to have a means of detecting it. The most common instrument for detecting and measuring radioactivity is the **Geiger counter** (Figure 11.6), which is essentially a modified cathode-ray tube. Argon gas is contained in a metal cylinder, which acts as the cathode; a wire anode runs down the axis of the tube. Radiation from a radioactive source—for example, contaminated soil—enters the tube through a window of thin mica; when it does so, it causes ionization of the argon gas. As a result of the ionization, pulses of electric current flow between the electrodes and are amplified and converted into a series of clicks that are counted automatically.

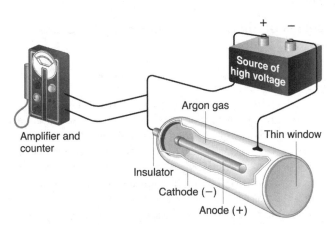

Figure 11.6 In a Geiger counter, the metal cylinder, which is filled with a gas (usually argon), acts as the cathode; the metal wire projecting into the cylinder acts as the anode. The window in the cylinder is permeable to alpha, beta, and gamma radiation. When radiation enters the cylinder, the gas is ionized and small pulses of electric current flow between the wire and the metal cylinder. The electrical pulses are amplified and counted. The number of pulses per unit of time is a measure of the amount of radiation. Adapted from T.L. Brown, E. LeMay, and B. Bursten. *Chemistry: The Central Science: Fifth edition.* Englewood Cliffs, NJ: Prentice Hall, 1991.

Units of Radiation

Nuclear disintegrations are measured in **curies** (Ci); 1 Ci is equal to 3.7×10^{10} disintegrations per second. A curie represents a very high dose of radiation. Natural background radiation amounts to only approximately two disintegrations per second. Damage that radiation causes depends not only on the number of disintegrations per second but also on the radiation's energy and penetrating power. Another unit of radioactivity, the **rad**, measures the amount of energy released in tissue (or another medium) when it is struck by radiation. A single diagnostic X-ray is equivalent to 1 rad. The rad is being replaced by a new international unit, the **gray** (1 gray = 100 rad).

A more useful unit for measuring radiation is the **rem**, which takes into account the potential damage to living tissues caused by the different types of ionizing radiation. For X-rays, γ rays, and β particles, 1 rad is essentially equivalent to 1 rem, but for α particles, because of their greater ionizing ability, 1 rad is equivalent to 10 to 20 rems. The new international unit to replace the rem is the **sievert** (1 sievert = 100 rem).

How Much Radiation Is Harmful?

Scientists still do not know for certain how much radiation is harmful. The study of Japanese survivors of the World War II bombings of Hiroshima and Nagasaki and of people exposed to fallout from nuclear power plant accidents—primarily those at Three Mile Island near Harrisburg, Pennsylvania in 1979 and at Chernobyl, Soviet Union in 1986—has revealed the probable effects of high short-term doses of radiation given to the entire body. However, there is no agreement on how the effects of these high short-term doses can be used to predict the effects of low doses. A summary of the effects of high doses is given in Table 11.3.

Table 11.3

Effects on Humans of Short-Term Whole-Body Exposure to Various Doses of Radiation

Dose (rem)	Effects
<50	Effects inconsistent and difficult to demonstrate.
50–250	Fatigue, nausea, decreased production of white cells and platelets in blood; increased probability of leukemia.
250–500	Same as for 50–250 rem, but more severe; vomiting, diarrhea, damage to intestinal lining; very susceptible to infections because of low white cell count; hemorrhaging because of impaired clotting mechanism; 50% die within months.
500–1000	Damage to cardiovascular system, intestinal tract, and brain; death within weeks.
1000–10,000	Same as for 500–1000 rem, but more severe; coma; death within hours at 10,000 rem.
100,000	Immediate death.

Some scientists believe that because the body has the ability to repair radiation damage, many small doses of radiation over a long period of time produce no lasting effects. Others believe that there is no safe level of radiation exposure. Recently, a study of survivors of the atomic bomb blasts in Japan found that although large numbers of those closest to the explosions died of cancers, those with limited exposure to the radiation from the bombs were actually living longer than Japanese people who had been far enough from the bombs to avoid exposure. Another study has shown that U.S. soldiers who participated in the nuclear weapons testing in the early 1960s have no more health problems than veterans who were not involved in the tests. Currently, there is no agreement on what constitutes a "safe" annual dose of radiation.

■ Everyday Exposure to Radiation

It is impossible to avoid exposure to every source of radiation in daily life. Low-level radiation is all around us in the environment. It has been estimated that the average exposure of a member of the U.S. population to ionizing radiation amounts to approximately 360 millirem (mrem) per year. The largest contribution by far (82%) comes from natural sources. The remaining 18% comes primarily from medical procedures, consumer products, and occupational activities (Table 11.4). Medical personnel and other people who work near radioactive substances wear badges containing film that is sensitive to radiation. Any radiation affects the film in the same way that light affects a photographic film. The amount of darkening of the film negative gives a measure of the person's exposure.

Natural Sources of Radiation

The average dose from all natural sources of radiation is approximately 300 mrem/year. It is now recognized that the major natural source is **radon**, a gas that on average contributes 200 mrem to the total annual exposure (Table 11.4). Radon-222 (a half-life of 3.8 days), an α emitter, is a naturally occurring product of the decay of uranium-238 (see Figure 11.4), a radioisotope of uranium that is present in widely varying concentrations in most soils and rocks. Radon-222 is an odorless, tasteless, inert gas, which as it escapes from soil and rocks enters the surrounding water and air. The Environmental Protection Agency has conducted a nationwide survey of radon in U.S. homes to determine where problems are likely to exist (Figure 11.7). In areas of the country where crustal rocks and soil have high concentrations of uranium-238, radon can seep into homes through basements and in well-sealed houses may reach potentially dangerous levels. Inhalation of radon gas can increase the risk of cancer. This effect is due less to the radon itself than to its α- and β-emitting decay products, which when the gas is inhaled become deposited in the respiratory tract.

Cosmic rays—a form of short-wavelength electromagnetic radiation that reaches the Earth's surface from outer space—account for 8% of natural radiation, or an average dose of 27 mrem/year. Because the dose increases with altitude, residents of mile-high Denver receive approximately 50 mrem more radiation each year from this source than do people living at sea level in New Orleans.

Terrestrial radiation in rocks and soil, originating from radioisotopes other than radon, accounts for another 8% of natural radiation (28 mrem/year). Naturally occurring carbon-14

Table 11.4

Average Annual Exposure (1990) to Radiation for U.S. Residents

Source of Radiation	Dose (mrem)	Percentage of Total Dose
Natural		
Radon gas	200	55
Cosmic rays	27	8
Terrestrial (radiation from rocks and soil other than radon)	28	8
Inside the body (naturally occurring radioisotopes in food and water)	39	11
Total natural	294	82
Artificial		
Medical		
X-rays	39	11
Nuclear medicine	14	4
Consumer products (building materials, water)	10	3
Other		
Occupational (underground miners, X-ray technicians, nuclear plant workers)	<1	<0.03
Nuclear fuel cycle	<1	<0.03
Fallout from nuclear weapons testing	<1	<0.03
Miscellaneous	<1	<0.03
Total artificial	64	18
Total natural plus artificial	358	100

Source: National Council on Radiation Protection and Measurement (NCRP87b), Washington, DC: National Academy Press, 1990.

and traces of potassium-40, thorium-223, and uranium-238 are present in food, water, and air and enter the body when ingested or inhaled. This **inside-the-body** source makes approximately 11% of natural radiation (39 mrem/year).

Radiation from Human Activities

The average annual dose of radiation arising from everyday human activities amounts to approximately 64 mrem (Table 11.4). Diagnostic **X-rays** account for 39 mrem/year, and other medical procedures and treatments add 14 mrem/year (together accounting for 15% of the total average annual dose). **Consumer products** add 10 mrem/year.

Smokers receive an additional dose of radiation from polonium-210, a naturally occurring α emitter present in tobacco. Workers who are potentially at risk of higher exposure to

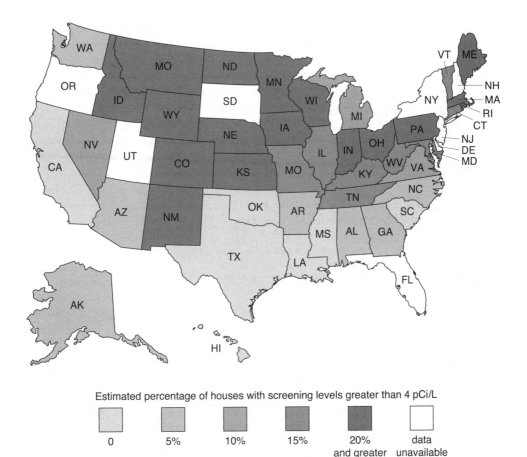

Figure 11.7 Results of an Environmental Protection Agency 1993 survey of radon exposure in U.S. residential dwellings. Radon levels below 4 pCi/L are considered to be average, and generally no action is recommended. if the level is much greater than 4 pCi/L, action to reduce the radon exposure is recommended. (pCi/L = picocuries per liter of air; 1 pCi = 1×10^{-12} Ci.) Courtesy of EPA.

radiation than the average for the U.S. population include underground miners, radiologists, X-ray technicians, and nuclear power plant workers. The average additional dose for these workers is less than 1 mrem/year. Although residents downwind from nuclear weapons testing sites were exposed to radioactive fallout in the past, nuclear weapons production and nuclear power plant operations together contribute negligible amounts of radiation today; however, nuclear wastes from both of these activities pose a serious hazard.

■ Uses of Radioisotopes

Although it may present potential health hazards, radiation from radioisotopes has useful applications in fields as diverse as archeology, medical diagnosis and treatment, agriculture, scientific research, and the power industry.

Determining the Dates of Archeological and Geological Events

An ingenious and reliable method based on the half-lives of certain naturally occurring radioisotopes such as carbon-14 can be used to estimate the age of ancient objects. A radioisotope decays at a constant rate that is defined by its half-life, and as it decays, it changes into a different isotope. By measuring the relative amounts of the original and the product radioisotopes in an object, the age of the object can be estimated.

Uranium-238, a natural constituent of most rocks, has a half-life of 4.5-billion years. It decays to form the stable isotope lead-206 and can be used to date very ancient rocks. To take a simple example, suppose that equal amounts of uranium-238 and lead-206 are found in a rock that we wish to date. This finding tells us that the uranium in the rock must have gone through one half-life (4.5 billion years); therefore, the rock is 4.5 billion years old. In coming to this conclusion, we make the assumption that the rate of decay of uranium-238 has always been constant and that no lead-206 was initially present in the sample; that is, all of it came from uranium-238.

EXAMPLE 11.3

A sample of rock is found to contain 13.2 mg of ^{238}U and 3.42 mg of ^{206}Pb. If the half-life of $^{238}_{92}U$ is 4.51×10^9 years, what is the age of the rock?

Solution

We have to estimate how much ^{238}U was present in the rock when it was formed. It must equal 13.2 mg plus the amount that decayed to produce ^{206}Pb. We can calculate that amount by multiplying the current amount of ^{206}Pb by the ratio of the atomic mass of uranium to that of lead. The original amount of $^{238}_{92}U$ is as follows:

$$\left(3.42 \, \text{mg of } ^{206}Pb\right) \frac{\left(238 \, \text{g U mol}^{-1}\right)}{\left(206 \, \text{g Pb mol}^{-1}\right)} = 3.95 \, \text{mg } ^{238}U$$

The initial amount of ^{238}U in the ore $= 13.2 \, \text{mg} + 3.95 \, \text{mg} = 17.2 \, \text{mg}$

$$\ln \frac{N_0}{N} = kt = \ln \frac{N_0}{N} = \frac{0.693\,(t)}{t_{1/2}}$$

$$\ln \frac{17.2 \, \text{mg}}{13.2 \, \text{mg}} = \frac{0.693\,(t)}{4.51 \times 10^9 \, \text{years}}$$

$$t = 1.7 \times 10^9 \, \text{years}$$

The $^{238}_{92}U$ in the rock was formed 1.7-billion years ago.

Using the decay rate of uranium-238, scientists have dated the oldest rocks on Earth at between 3 and 3.5 billion years. Meteorites and moon rocks have been found to be approximately 4.5 billion years old, and on the basis of this information, geologists conclude that the planets, including the Earth, were formed 4.5 to 5.0 billion years ago.

To obtain an accurate date, the half-life of the radioisotope used for dating must be fairly close to the age of the object being studied. Uranium-238 is appropriate for dating rocks that are billions of years old, but for objects that are several thousand years old, carbon-14, which has a half-life of 5730 years, is more suitable. Other radioisotopes besides uranium-238 and carbon-14 are used to estimate the age of objects; the choice depends on the nature and the age of the object.

Carbon-14 is formed naturally at a fairly constant rate in the upper atmosphere when neutrons ejected from stable nuclei by cosmic rays interact with ordinary nitrogen nuclei.

$$^{14}_{7}N + ^{1}_{0}n \rightarrow ^{14}_{6}C + ^{1}_{1}p$$

The radioactive carbon-14 reacts with oxygen in the atmosphere to form radioactive carbon dioxide ($^{14}_{6}CO_2$). Both radioactive carbon dioxide and carbon dioxide ($^{12}_{6}CO_2$) are incorporated into plants through photosynthesis and into animals through normal food chains and respiration. Although an organism is alive, the ratio of carbon-14 to carbon-12 in its tissues remains constant, but when the organism dies and incorporation of carbon dioxide ceases, the ratio begins to change. The carbon-14 in the tissues slowly disappears as it emits β particles and becomes nitrogen-14, but the nonradioactive carbon-12 does not change. The ratio of carbon-14 to carbon-12 decreases.

$$^{14}_{6}C \rightarrow ^{14}_{7}N + ^{0}_{-1}\beta$$

The carbon-14 emits β particles at a constant rate given by its half-life. Thus, measuring the half-life of carbon-14 in the plant or animal artifact gives the age of the artifact.

Carbon-14 dating, also known as **radiocarbon dating**, is used to date ancient artifacts such as cloth, wood, leather, and bone, which were derived from once-living matter and thus contain carbon. The method assumes that the flow of carbon-14 into the environment has always been constant. Studies of the annual growth rings in trees have shown this assumption to be justified over the past 7000 years. There were some fluctuations, however, in earlier times. Even so, other dating methods generally agree very well with carbon-14 dating.

The amount of carbon-14 in living matter is very small and is difficult to measure with any accuracy. It is much easier to measure the rate at which the carbon-14 disintegrates by counting the disintegrations per second per gram of material with a Geiger or scintillation counter. In all living organisms, there are 15.3 disintegrations of carbon-14 per minute per gram of carbon. When the organism dies, this rate of disintegration decreases with a half-life of 5730. The rate of disintegration, R, at time t is proportional to N, the number of radioactive nuclei at time t.

$$t = \frac{t_{1/2}}{0.693} \ln \left(\frac{R_o}{R} \right)$$

For carbon, $R_o = 15.3$ disintegrations per minute per gram, and $t_{1/2}$ is 5730 years.

$$t = \frac{5730}{0.693} \ln\left(\frac{15.3}{R}\right) = 8.27 \times 10^3 \ln\left(\frac{15.3}{R}\right)$$

The time that has elapsed since any living organism died can be determined by using this equation.

EXAMPLE 11.4

A sample of wood from an ancient settlement was placed in front of a Geiger counter tube. It gave a reading of 13.6 disintegrations per minute per gram. What is the age of the wood?

Solution

$$t = 8.27 \times 10^3 \ln\left(\frac{15.3}{R}\right) = 8.27 \times 10^3 \ln\left(\frac{15.3}{13.6}\right) = 974 \text{ years}$$

The wood is 974 years old.

■ Nuclear Fission

Until approximately 70 years ago, scientists believed that the only way an unstable nucleus such as uranium-238 could change into one that was more stable was by a series of successive steps. In each of these steps, a small nuclear particle, such as an α or β particle, a proton, or a neutron, was emitted. As a result of such emissions, atomic numbers changed by no more than one or two units at a time.

In 1938, on the eve of World War II, these beliefs had to be completely revised. Two German chemists, Otto Hahn and Fritz Strassman, found that if they bombarded uranium-238 with neutrons, they obtained not only the expected products—uranium-239 (atomic number 92) and neptunium-239 (atomic number 93)—but also small amounts of barium (atomic number 56), lanthanum (atomic number 57), and cerium (atomic number 58). Hahn and Strassman were astounded to find elements with such low atomic numbers among the products. They realized that the original uranium nucleus must have been split almost in half, a process that they named **nuclear fission.**

The implications of nuclear fission were immediately evident to Hahn's colleague, Lise Meitner, who because she was Jewish had fled to Sweden in 1938 when her homeland, Austria, was annexed by Hitler. Meitner and her nephew, Otto Frisch, used calculations based on Albert Einstein's famous equation, $E = mc^2$, to determine the tremendous amount of energy that would be released if an atomic nucleus was split. They realized that if this energy could be harnessed, it could be used to construct very powerful bombs.

The Energy Source in Nuclear Fission

As you learned earlier, despite repulsions between protons, a powerful force within the nucleus holds it together. This **binding energy** comes from the conversion of a very small amount of mass into energy that occurs when protons and neutrons are packed together to form nuclei. Einstein had discovered that energy and mass are two aspects of the same thing. In his equation, E represents energy, m represents mass, and c is the speed of light (300,000 km/s or 186,000 mi/s). Because c^2 is a very large number, the equation means that even if the mass (m) converted is very small, the amount of energy produced (m $\times$ c^2) will be enormous.

The binding energy in the nuclei of various elements is related to the mass of their nuclei (Figure 11.8). Nuclei of iron atoms—and other nuclei of similar mass—are more stable and need less energy to hold them together than either lighter or heavier nuclei. When a heavy nucleus such as a uranium nucleus is split to form lighter, more stable nuclei, some of the binding energy is released. It has been calculated that fission of 1 g of uranium releases approximately 10 million times more energy than burning 1 g of coal.

Figure 11.8 also shows that energy is released when nuclei of very light elements such as hydrogen and helium join, or fuse, together. Fusion reactions, which can only occur at extremely high temperatures, occur continually in the sun and provide the energy that sustains life on earth. Nuclear fusion also occurs in the explosion of a hydrogen bomb.

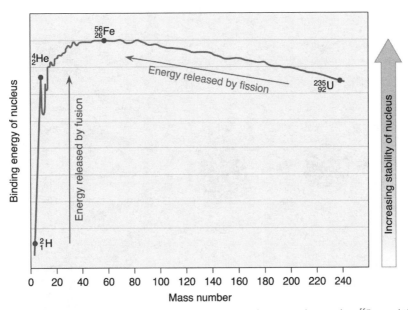

Figure 11.8 Intermediate-size nuclei with mass numbers between about 30 and 65, such as $^{56}_{26}$Fe, need the least amount of energy to hold them together and are the most stable. During fission, nuclei of elements with high mass numbers, such as $^{235}_{92}$U, split apart to form smaller, more stable nuclei, and energy is released. Energy is also released during fusion when small nuclei such as $^{2}_{1}$H and $^{4}_{2}$He join to form larger, more stable nuclei. Adapted from T.L. Brown, E. LeMay, and B. Bursten. *Chemistry: The Central Science: Fifth edition.* Englewood Cliffs, NJ: Prentice Hall, 1991.

Fission Reactions

If a fissionable isotope such as uranium-235 or plutonium-239 is bombarded with neutrons, the nuclei can split in more than one way to produce a variety of lighter elements. Whatever way a nucleus splits, the fission reaction releases between two and four neutrons and a great deal of energy. One way a uranium-235 nucleus splits is shown in the following equation:

$$^{235}_{92}U + ^{1}_{0}n \rightarrow ^{146}_{57}La + ^{87}_{35}Br + 3\,^{1}_{0}n + energy$$

A slow-moving neutron enters the uranium-235 nucleus and causes fission (Figure 11.9). If a certain minimum amount of fissionable nuclei, called the **critical mass**, is present, the neutrons emitted by the first reaction can cause the fission of more uranium-235 nuclei and thus set off a **chain reaction** (Figure 11.10). Such a reaction is self-perpetuating and can continue with the release of ever-increasing amounts of energy until all of the uranium nuclei have been split.

The Atomic Bomb

In August 1939, after President Roosevelt received a letter signed from Einstein warning him that the Germans were already working on a bomb based on nuclear fission, the United States launched the secret Manhattan Project to develop an **atomic bomb.**

To obtain a critical mass of uranium-235 (which makes up only 0.7% of natural uranium), the U.S. government built a top-secret facility in Oak Ridge, Tennessee, for enriching uranium.

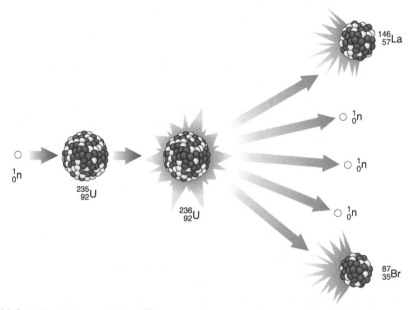

Figure 11.9 Nuclear fission occurs when $^{235}_{92}U$ atoms are bombarded with slow-moving neutrons. During fission, the uranium-235 atoms split apart to form two lighter atoms and from two to four neutrons. Because the new elements formed are more stable than uranium, a large amount of energy is released. Uranium atoms split in many different ways. Up to 35 elements have been identified among the fission products, including $^{146}_{57}La$ and $^{87}_{35}Br$.

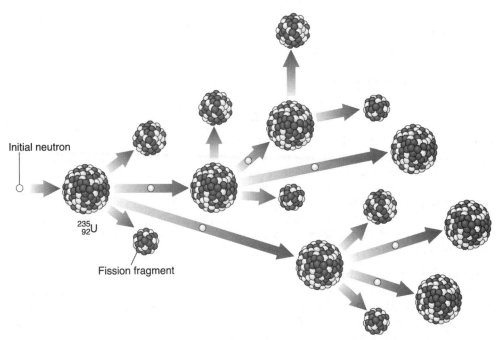

Figure 11.10 A nuclear chain reaction can begin when fission of a $^{235}_{92}$U atom by a single neutron releases two to four neutrons from the uranium nucleus. Each of these neutrons can trigger fission of another uranium atom, and the release of more neutrons. As a result, an uncontrollable chain reaction occurs, releasing enormous amounts of energy.

At the same time, a facility was established in Hanford, Washington, to produce plutonium-239, a fissionable radioisotope that is not naturally occurring.

In 1942, the first sustained chain reaction, using uranium-235 as the starting material, was achieved in a laboratory under the football stadium at the University of Chicago. The buildup of neutrons was carefully controlled to prevent a dangerous explosion. By early 1945, sufficient plutonium-239 had been prepared for the construction of a bomb. In an atomic bomb, two separate masses of fissionable material, each incapable of sustaining a chain reaction separately, are brought together to form a critical mass at the moment of detonation.

In July 1945, the first atomic bomb was successfully tested in the desert near Alamogordo, New Mexico. Those who witnessed the explosion were awed by the magnitude of the blast; the brilliance of the light, the tremendous heat generated, and the huge mushroom cloud that formed were overwhelming. On August 6, 1945, President Harry Truman, compelled by his desire to avoid the millions of American and Allied casualties expected in an invasion of Japan, ordered a uranium bomb to be dropped on Hiroshima. Three days later, a plutonium bomb was dropped on Nagasaki. The two bombs killed or injured approximately 200,000 Japanese. The war ended with the surrender of Japan on August 14, 1945.

Peaceful Uses of Nuclear Fission

Nuclear fission is used to create bombs and other weapons of war but also can be harnessed for peaceful purposes. If fission is controlled so that the concentration of neutrons produced is sufficient to maintain the fission reaction but not great enough to allow an uncontrolled chain reaction, the process can be used as a source of energy for electrical power generation. Nuclear power plants generate much of the electricity used in Europe, but the United States is still heavily dependent on fossil fuels to meet its energy needs.

■ Nuclear Energy

Few issues have generated as much controversy as the future of nuclear energy. When controlled, nuclear fission reactions became a reality in the early 1940s; nuclear energy appeared to hold great promise as a cheap, clean, and safe source of energy. In the 1950s, it was predicted that by the year 2000, nuclear power plants would supply up to 20% of the world's energy; however, because of fears about safety and the relatively low cost of other sources of energy, the actual figure was less than 5%. In 2008 the United States had 104 operating nuclear reactors that were supplying approximately 20% of total electrical energy (Figure 11.11). Other industrialized countries are much more dependent on nuclear power; France, which is not well endowed with coal, relies on nuclear power for 77% of its electrical power. Table 11.5 lists some of the leading countries in the use of nuclear power.

Nuclear Fission Reactors

Nuclear power is used primarily for the production of electricity. A diagram of a **pressurized water reactor (PWR)**, the most common type of reactor in use, is presented in Figure 11.12. The basic design is essentially the same as that for a coal-fired power plant (see Figure 10.12), except that nuclear fuel is used instead of coal as the source of heat for converting water to steam.

At the **core** of a **nuclear fission reactor** are several hundred steel **fuel rods** (Figure 11.12) containing the fissionable uranium fuel. Interspersed between the fuel rods are **control rods** made of a material—usually cadmium or boron—that absorbs neutrons. When the rods are withdrawn from the core, the rate of fission increases; when the rods are inserted, more neu-

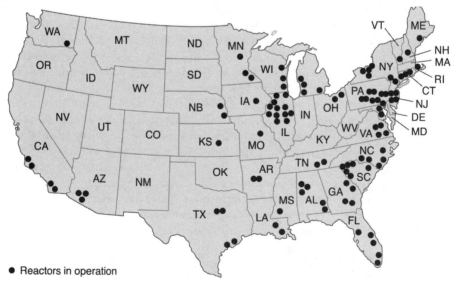

● Reactors in operation

Figure 11.11 Nuclear power plants in operation in the United States in 2008. Based on information from the U.S. Department of Energy (www.eia.doe.gov/cnaf/nuclear/page/nuc_reactors/reactsum.html).

Table 11.5

Some of the Leading Countries in the Use of Nuclear Power, 2008

Country	Number of Reactors	Electricity from Nuclear Generators (% of Total)	Nuclear Energy Supplied in 2007 (Billions of Kilowatt-Hours)
Belgium	7	54	46
Bulgaria	2	32	13.7
Canada	18	14.7	88.2
China	11	1.9	59.3
Czech Republic	6	30.3	24.6
Finland	4	29	22.5
France	59	77	420.1
Germany	17	26	133.2
Hungary	4	37	13.9
India	17	2.5	15.8
Japan	55	27.5	267
Korea (South)	20	35.3	136.6
Lithuania	1	64.4	9.1
Russia	31	16	148
Slovakia	5	54	14.2
Slovenia	1	42	5.4
Spain	8	17.4	52.7
Sweden	10	46	64.3
Switzerland	5	43	26.5
Taiwan	6	19.3	39
Ukraine	15	48	87.2
United Kingdom	19	15	57.5
United States	104	19.4	806.6

trons are absorbed, and the rate slows. If there is an accident or a need to make repairs or remove spent fuel, operators can stop the reaction by inserting the rods to their limit. Water circulating around the fuel rods and control rods acts as a **moderator**, slowing the neutrons to speeds that are optimal for splitting uranium-235 atoms. The water, which is circulated through a heat exchanger, also serves to keep the fuel rods cool and to prevent the reactor from overheating. The entire reactor is housed in a thick-walled containment building.

The Nuclear Fuel Cycle

Because most uranium ore that is mined contains no more than 0.2% uranium, producing the fuel suitable for a nuclear power plant is expensive and energy intensive. In the first steps, milling

a.

b.

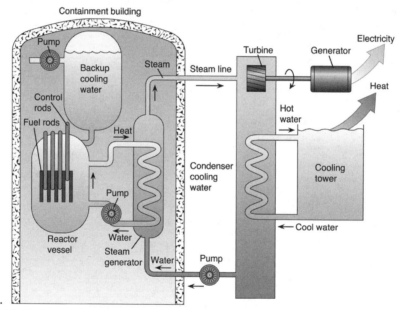

c.

Figure 11.12 (a) Nuclear power plant at Three Mile Island in Pennsylvania. (b) A technician inspecting fuel rods. (c) Schematic diagram of a pressurized water reactor.

and chemical treatment convert the ore to a product that is approximately 80% uranium oxide (U_3O_8). The uranium in this oxide is 99% nonfissionable uranium-238 and only 0.7% fissionable uranium-235. The uranium content must be **enriched** to approximately 3% uranium-235; then the enriched material is fabricated into small pellets, which are packed into the fuel rods.

The ^{235}U and ^{238}U isotopes exhibit essentially identical reaction chemistry, and thus, reaction rate differences cannot be used as a method of separation. Physical methods that are able to differentiate the mass of each have been used successfully to separate the two. During

World War II, gaseous diffusion chambers that were able to capitalize on Graham's Law of Effusion were used for the separation. The rate at which a gas penetrates a thin membrane is its rate of effusion. **Graham's law** states that the effusion rate of a gas is inversely proportional to the square root of its molar mass. The uranium isotopes were reacted with hydrofluoric acid HF to form UF_6, which is a gas at room temperature. The rate of effusion for two gases can be calculated from the following equation:

$$\frac{r_1}{r_2} = \sqrt{\frac{M_1}{M_2}}$$

Where

r_1 and r_2 are the rates of effusion of gases 1 and 2.
M_1 and M_2 are the masses of gases 1 and 2.

For $^{235}UF_6$ and $^{238}UF_6$, the equation becomes as follows:

$$\frac{r_{238}}{r_{235}} = \sqrt{\frac{352}{349}} = 1.004$$

Thus, for every pass through a membrane the lighter isotope, $^{235}UF_6$ is enriched by only a factor of 1.004. Uranium ore has a minimal concentration of the ^{235}U isotope of approximately 0.7% to 0.9%. To enrich it to 3%, ^{235}U requires approximately 350 passes through the membrane (enriching it to weapons grade [more than 95% ^{235}U] requires approximately 2000 passes). After the gaseous isotopes are separated, the uranium is converted back into a solid form. The isotope enrichment plants that complete this separation are expensive and use huge amounts of energy to push the $^{235}UF_6$ through so many membranes.

The **gas diffusion process** was replaced with a **gas centrifuge process**. Centrifugal force is a more efficient separation technique. It is proportional to mass, not the square root of mass, as in Graham's law. Centrifugal separation is giving way to laser isotope enrichment. Lasers produce monochromatic electromagnetic radiation. If the energy the laser emits is of a narrow enough wavelength, it is possible that that energy would be absorbed by only one of the isotopes of a molecule but not the others. The molecule that absorbed the energy would undergo a transition to an excited state. If the excited state molecule then reacted with a reagent to produce anther product, isotopic enrichment would be achieved. It may be possible for laser enrichment to lower the cost of ^{235}U enrichment.

As fission occurs in the reactor, the uranium-235 concentration decreases, and after approximately three years, fuel rods must be removed and replaced. Spent rods remain radioactive. They can be either permanently stored as wastes or reprocessed to recover unreacted uranium-235 and the fissionable plutonium-239 produced during the nuclear reactions. Because of the danger that the recovered plutonium could be stolen from a reprocessing plant and used to make nuclear weapons, the United States currently does not reprocess nuclear fuel; however, even though the cost of reprocessing is very high, Japan, Britain, and France all reprocess spent fuel rods.

Problems with Nuclear Energy

Most people consider the risk of a catastrophic explosion to be the greatest problem with the use of nuclear energy, but a power plant could never blow up like a nuclear bomb because the fuel is not sufficiently enriched in uranium-235. However, as was demonstrated in the Three Mile Island incident and Chernobyl disaster, human error and mechanical failure can lead to serious accidents. In both incidents, a loss of cooling water caused overheating of the reactor and **core meltdown**. At Three Mile Island, the amount of radiation that escaped into the atmosphere was small; in the Chernobyl accident, however, enormous quantities of radioactive gases and particles were released. Huge tracts of contaminated land will remain uninhabitable for hundreds of years.

Some risk of radiation exposure exists at every step of the nuclear fuel cycle. The huge quantities of crushed waste rock left from the processing of uranium ores are a source of low-level radioactivity, which may leach into groundwater or be dispersed to the atmosphere in wind-blown dust. A major problem (studied in Chapter 18) is the disposal of the high-level radioactive wastes produced during the operation of a nuclear power plant. These wastes, which include fission products and spent fuel rods, must be stored so that they are completely isolated from the atmosphere.

Nuclear power plants have a limited life expectancy. After approximately 30 years of operation, continual bombardment of plant components with neutrons makes the metals brittle. The chances of cracking and leakage of radiation are increased, and for safety reasons, the plant must be shut down (decommissioned). Even after spent fuel rods and circulating water have been removed, a plant is still radioactive.

There are several ways to deal with a decommissioned plant. The plant could be closed, fenced, and put under guard. Such mothballing is perhaps the cheapest solution in the short run. The plant can be dismantled, or refurbished, in the future when the radiation has decreased. A second option is to entomb the plant (that is, seal it in concrete). Entombment is currently being erected at the damaged Chernobyl reactor after the first containment structure was found to be unsound. Finally, a decommissioned plant could be totally dismantled. All components are taken apart and cut into smaller pieces. All materials contaminated with radioactivity would be removed to a safe facility for the permanent storage of radioactive waste. The estimated cost to decommission and dismantle a nuclear power plant completely is $3 billion or more. By 2012, 19 of the nuclear reactors operating in the United States will have come due for decommissioning. By 2030, all will be due for retirement.

Nuclear Breeder Reactors

The world's supply of uranium ores is not abundant; in the 1960s, when rapid expansion of the use of nuclear energy was expected, there were fears that shortages of uranium-235 would develop. The solution appeared to be the development of **nuclear breeder reactors**, which not only produce heat from fission but also yield a new supply of fissionable fuel.

If uranium-238 is bombarded with fast-moving neutrons, the following series of reactions occurs, and fissionable plutonium-239 is formed:

$$^{238}_{92}U + {}^{1}_{0}n \rightarrow {}^{239}_{92}U$$
nonfissionable   unstable

$$^{239}_{92}U \rightarrow {}^{239}_{93}Np + {}^{0}_{-1}\beta$$
unstable

$$^{239}_{93}Np \rightarrow {}^{239}_{94}Pu + {}^{0}_{-1}\beta$$
fissionable

Fission of uranium-235 provides the neutrons that are needed to start the reaction sequence. Because two or in some cases three neutrons are produced in every uranium-235 fission, two (or three) plutonium-239 atoms may be formed from each uranium-235 atom (Figure 11.13). The amount of fuel produced thus exceeds the amount consumed. Water cannot be used as the moderator in a breeder reactor because it slows the neutrons needed to produce the plutonium-239. Instead, liquid sodium is used.

The future of breeder reactors is uncertain. In addition to the problems associated with conventional nuclear reactors, there are several other drawbacks. Plutonium-239 has an extremely long half-life—24,000 years—and is one of the most toxic substances known. Inhalation of even a minute quantity can cause lung cancer. Another problem is that plutonium-239 can be used more easily than uranium-235 to make nuclear weapons, thus increasing the need for security. The conversion of uranium-238 to fissionable plutonium-239 is difficult to control, and the sodium used as the moderator reacts explosively if it comes in contact with water. Currently, no breeder reactors are operating in the United States.

Nuclear Fusion

When two very light atomic nuclei are combined, or fused, a heavier nucleus is formed. There is a loss of mass, and an enormous amount of energy is released. Fusion of hydrogen atoms to form helium is the primary source of the energy emitted by the sun. Theoretically, for each gram of fuel, **nuclear fusion** releases four times as much energy as the fission of uranium-235 does and approximately a million times as much as the combustion of fossil fuels does. Many scientists believe that if controlled fusion could be achieved on Earth, it would solve the

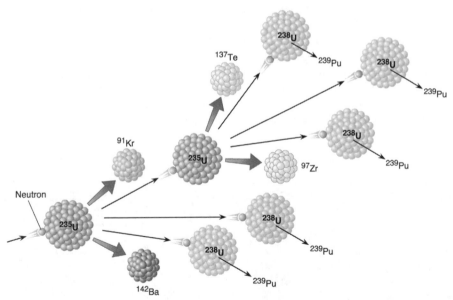

Figure 11.13 Typical reactions in a breeder reactor: When Uranium-235 atoms are bombarded with neutrons, fission occurs. In each fission reaction, a uranium-235 atom produces two atoms of lighter elements and two or three neutrons. The neutrons bombard other uranium-235 atoms or uranium-238 atoms, converted the latter to atoms of fissionable plutonium-239. Fission of one uranium-235 atom can result in the formation of more than one fissionable plutonium-239 atom. Thus, the amount of fuel produced (plutonium-239) exceeds the amount of fuel consumed (uranium-235).

world's energy problems; however, enormous technical difficulties must be overcome before this source of energy can be exploited.

A temperature approaching that in the sun is required for hydrogen atoms to combine; even if this temperature could be attained, the problem of finding a container that could withstand the heat without vaporizing remains. A fission (or atomic) bomb was used to produce the high temperature needed for the hydrogen bomb, but this is hardly an option for controlled energy production.

Fusion of the two isotopes of hydrogen, deuterium ($^{2}_{1}$H) and tritium ($^{3}_{1}$H), is receiving the most attention because fusion of these two atoms requires a lower temperature (approximately 100,000,000°C) than other fusion reactions.

$$^{2}_{1}\text{H} + {}^{3}_{1}\text{H} \rightarrow {}^{4}_{2}\text{He} + {}^{1}_{0}\text{n} + \text{energy}$$
$$\text{deuterium} \quad \text{tritium} \qquad \text{helium}$$

Deuterium, a naturally occurring isotope, can be obtained in unlimited quantity from seawater. Tritium, however, is an unstable radioactive isotope and must be produced by bombardment of lithium with neutrons.

$$^{1}_{0}\text{n} + {}^{6}_{3}\text{Li} \rightarrow {}^{3}_{1}\text{H} + {}^{4}_{2}\text{He}$$

Although nuclear fusion produces little radioactive waste, other drawbacks are the danger of leakage of tritium (half-life of 12.3 years) and thermal pollution. Also, known reserves of lithium ores are very limited.

Two approaches to achieving nuclear fusion are being tested. One is the Tokamak reactor, which Soviet physicists pioneered. In this device, very high temperatures strip electrons from deuterium and tritium atoms, creating a gas-like plasma of energetic nuclei and free electrons. The greater the energy, the more likely the nuclei are to fuse. The hot plasma is contained within a powerful magnetic field (Figure 11.14). In the second approach, laser beams are focused narrowly on a minute pellet containing frozen deuterium and tritium (Figure 11.15). A rapid increase in pressure and temperature causes the nuclei to fuse. So far, neither method has achieved sustained, controlled fusion, and nuclear fusion is unlikely to become a practical source of energy in the near future.

 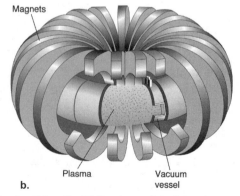

Figure 11.14 (a) The Tokamak Fusion Test Reactor at Princeton University. (b) The plasma of the Tokamak reactor is contained in the tunnel inside the doughnut-shaped magnet. Courtesy of Princeton Plasma Physics Laboratory.

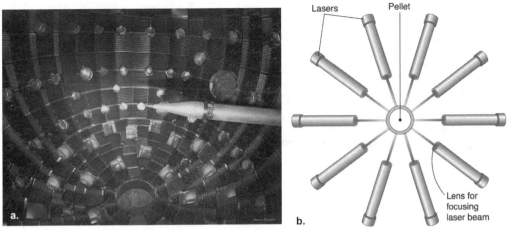

Figure 11.15 (a) The experimental laser fusion reaction at the Lawrence Livermore Laboratory. (b) Powerful laser beams are focused on a tiny frozen pellet of a mixture of deuterium and tritium: the high pressure and temperature cause the hydrogen to fuse. Adapted from B.J. Nebel. *Environmental Science: The Way the World Works, Third edition.* Englewood Cliffs, NJ: Prentice Hall, 1990.

■ Additional Sources of Information

Aubrecht GJ. *Energy,* 3rd ed. Upper Saddle River, NJ: Prentice Hall, 2004.

Bisio A, Boots S. *The Wiley Encyclopedia of Energy and the Environment.* New York: John Wiley and Sons, 1996.

Duffy RJ. *Nuclear Politics in America: A History and Theory of Government Regulation.* Lawrence, KS: University of Kansas Press, 1997.

Nuclear Energy in the Twenty-first Century: Examination of a Contentious Subject. *Ann Rev Energy Environ* 1999;24:113–137.

Wong SSM. *Introductory Nuclear Physics,* 2nd ed. New York: John Wiley and Sons, 1998.

■ Keywords

α rays
atomic bomb
band of stability
β rays
binding energy
chain reaction
consumer products
control rods
core
core meltdown
cosmic rays
critical mass
curies

decay curve
enriched uranium
fuel rods
γ rays
gas centrifuge process
gas diffusion process
Geiger counter
Graham's Law of
    Effusion
gray
half-life
Inside-the-body
ionizing radiation

moderator
nuclear breeder reactors
nuclear fission
nuclear fission reactor
nuclear fusion
pressurized water reactor
    (pwr)
rad
radiation poisoning
radioactive
radioactive decay
radiocarbon dating
    (carbon-14 dating)

radioisotopes   sievert   uranium decay series
radon   terrestrial   X-rays
rem   transmutation

■ Questions and Problems

1. Write balanced nuclear equations for the following:
   a. Emission of a β particle by a magnesium-28 nucleus
   b. Emission of an α particle by a by lawrencium-255 nucleus
   c. Emission of a β particle by a nickel-65 nucleus
2. Which of the following has the greatest penetrating ability: an α particle, a β particle, or an γ ray?
3. What type of shield is necessary to stop the following:
   a. X-rays
   b. α Particles
   c. β Particles
   d. γ Rays
4. Why are α particles very dangerous when the source of the particles is inside your body and not very dangerous when it is outside your body?
5. Thorium-231 is the product when a certain radioisotope emits an α particle. Thorium-231 is radioactive and emits a β particle. Write a nuclear equation for the reaction in which isotope thorium-231 is produced. Write a second reaction that shows the product of thorium-231 decay.

$$\underline{\hspace{2cm}} \rightarrow \ ^{231}_{90}\text{Th} + \ ^{4}_{2}\alpha$$

$$^{231}_{90}\text{Th} \rightarrow \underline{\hspace{2cm}} + \ ^{0}_{-1}\beta$$

6. Fill in the missing symbol in each of the following nuclear equations.
   a. $^{210}_{83}\text{Bi} \rightarrow \ ^{4}_{2}\alpha + \underline{\hspace{2cm}}$
   b. $^{15}_{8}\text{O} \rightarrow \ ^{15}_{7}\text{N} + \underline{\hspace{2cm}}$
   c. $\underline{\hspace{2cm}} \rightarrow \ ^{4}_{2}\alpha + \ ^{222}_{86}\text{Rn}$
7. Fill in the blank in each of the following:
   a. $^{9}_{4}\text{Be} + \underline{\hspace{2cm}} \rightarrow \ ^{12}_{6}\text{C} + \ ^{1}_{0}\text{n}$
   b. $^{27}_{13}\text{Al} + \ ^{2}_{1}\text{H} \rightarrow \underline{\hspace{2cm}} + \ ^{4}_{2}\alpha$
8. What is the effect on the mass number and atomic number of the reacting isotope when the following transmutations occur?
   a. A β particle is emitted.
   b. An α particle is emitted.
   c. A γ ray is emitted.
9. Write balanced nuclear equations for the following:
   a. Nickel-65 emits a β particle.
   b. Neodymium-150 emits an α particle.
   c. Magnesium-28 emits a β particle.
10. A sample of sulfur-35 is radioactive and emits β particles. Sulfur-32 is not radioactive. What differences in chemical reactivity would you expect between sulfur-35 and non-radioactive sulfur-32? Explain your answer.

11. When cobalt-60 emits a γ ray, what is the other product of the reaction?

12. Which of the following nuclei would you expect to be radioactive?

    a. $^{32}_{16}S$

    b. $^{115}_{52}Te$

    c. $^{208}_{84}Po$

13. Which of the following nuclei would you expect to be stable?

    a. $^{108}_{50}Sn$

    b. $^{92}_{40}Zr$

    c. $^{18}_{9}F$

14. The following nuclei are radioactive and undergo either β or positron emission. Which of the following is most likely to be a positron emitter?

    a. $^{137}_{53}I$

    b. $^{105}_{45}Rh$

    c. $^{51}_{25}Mn$

15. Which of the following radioactive nuclei are expected to be β emitters?

    a. $^{137}_{53}I$

    b. $^{103}_{45}Rh$

    c. $^{51}_{25}Mn$

16. The half-life of barium-131 is 12.0 days. How many grams of barium-131 will remain after 30 days if you begin with 100 g?

17. The radioisotope americium-241 is used commercially in home smoke detectors. This isotope has a half-life of 433 years.

    a. How long will it take for the activity of this material to drop to less than 1% of its original activity?

    b. On the basis of your answer, should a 20-year-old smoke detector be disposed of by placing it in the municipal trash?

18. Tritium ($^{3}_{1}H$) has a half-life of 12.3 years. If 68 mg of tritium is accidentally released from a nuclear power plant, how many milligrams of the radioactive material remain after 12.3 years? After 60 years?

19. Pu-239 is produced in the nuclear fuel cycle in nuclear power plants. The half-life of Pu-239 is 24,000 years. What fraction of the Pu-239 generated today will still exist in the year 4000?

20. Describe how a Geiger counter works and how radioactivity is detected.

21. A sample of barium-140 has an initial activity of 6600 counts per minute on a Geiger counter. After 30 days, the activity of the sample gives 490 counts on the same Geiger counter. What is the half-life of barium-140?

22. A sample of curium-243 had an initial activity of 6024 disintegrations per second. After 1 year, the radioactivity of the sample declined to 5842 disintegrations per second. What is the half-life of curium-243?

23. Plutonium-240, which is produced in nuclear power plants, has a half-life of 6.58×10^3 years. The used nuclear fuel rods are removed from the nuclear reactor and placed in storage. Calculate the fraction of plutonium-240 that will remain after 100 years.

24. If we assume that the Earth is 4.5×10^9 years old, what fraction of uranium-235 that was present during the formation of the Earth is present today? The half-life of uranium-235 is 7.1×10^8 years.

25. The synthetic radioisotope technetium-99 (^{99}Tc), which is a β emitter, is the most widely used isotope in nuclear medicine. The following data were obtained using a Geiger counter.

Disintegrations per Minute	Time (hours)
180	0.0
130	2.5
104	5.0
77	7.5
59	10.0
46	12.5
24	17.5

    a. Make a graph of this information similar to Figure 11.5 and determine the half-life of technetium-99.

    b. How long would it take for 90% of the technetium-99 to decay?

26. Chemical reactions can often be used to change a toxic compound into another compound that is not as toxic. Why can chemical treatment technology not be applied to make nuclear waste harmless?

27. A so-called expert suggested that strontium-90 deposited in the Nevada desert during nuclear testing would undergo radioactive decay more rapidly than strontium-90 elsewhere because of the high desert temperatures. Do you agree with this assessment? Explain.

28. Describe the sources of everyday radiation to which the general public is exposed and indicate the percentage that each source contributes to average daily exposure.

29. What is the naturally occurring radioisotope that people are most likely to encounter?

30. Why is the rem the most useful unit of radiation to use when making an assessment of the impact of nuclear radiation on people?

31. Before the Nuclear Test Ban Treaty of 1963, the average exposure to radioactive fallout from nuclear weapons testing amounted to 30 mrem/year per person. In 1990, the average exposure was only 1 mrem/year per person. Estimate the additional exposure for an average individual for the period from 1990 to 2000, if open-air nuclear bomb testing had not been banned (assume that the average exposure per year would continue to be 30 mrem per person).

32. An oil painting alleged to be painted by Rembrandt (1606–1669) is subjected to radiocarbon-14 dating. The carbon-14 content of the canvas used is 0.96 times that of a living tree. If the half-life of carbon-14 is 5720 years, could this painting be an original Rembrandt?

33. Would the radioactive decay of uranium-238 to lead-206 provide an accurate method for determining the age of a material thought to be approximately 500 years old?

34. A mineral sample contains 100 mg of uranium-238 and 28 mg of lead-206. The half-life for the process uranium-238 → lead-206 is 4.5×10^9 years. What is the age of the mineral?

35. A wooden artifact from a temple has a carbon-14 activity of 52 counts per minute. A wood sample from a living tree has an activity of 63 counts per minute. What is the age of the artifact?

36. Radioactive fallout from a nuclear explosion contains ^{90}Sr with a half-life of 28.1 years. How long will it take this radioisotope to decay?

    a. 90%

    b. 99%

37. A wooden artifact from a temple contained 50 g of carbon and had an activity of 260 disintegrations per minute. What is the age of the artifact?
38. What percentage of uranium-235 does naturally occurring uranium ore contain?
39. One gram of uranium-235 produces as much energy as 14 barrels of crude oil. How many barrels of crude oil can be saved for every kilogram of uranium-235 used in a nuclear power plant?
40. Why does limiting the number of neutrons limit the rate of a nuclear reaction in a nuclear electrical power plant that uses uranium-235 as a fuel?
41. Which countries of the world place a high reliance on nuclear power?
42. Draw a schematic diagram of a pressurized water reactor and label all parts.
43. What percentage of uranium-235 must the fuel pellets that are used in a nuclear power plant contain?
44. Are spent nuclear fuel rods recycled in the United States? Are they recycled in other countries?
45. Can nuclear bombs be made from reactor-grade uranium? Explain.
46. Describe the nuclear fuel cycle for the production of reactor-grade uranium.
47. How often do individual fuel rods in a nuclear reactor need to be replaced? What happens to spent fuel rods?
48. For how many years can a nuclear power plant be expected to generate electricity before it reaches the end of its useful life?
49. More nuclear power plants should be built in the United States to reduce dependence on imported petroleum and to slow global warming. Do you agree with this statement?
50. Show that $^{235}UF_6$ is enriched by a factor of 1.004 for each pass through a membrane.
51. List two techniques that are used to enrich nuclear fuel.
52. List three ways to decommission nuclear plants. Arrange them in order of cost.
53. How much radiation would workers be exposed to in each of the three methods listed in Question 52.
54. How does a nuclear breeder reactor produce more fuel than it uses?
55. List two reasons why a nuclear breeder reactor may be more dangerous than a nuclear fission reactor.
56. The U.S. government should develop the nuclear breeder reactor to conserve uranium resources and avoid becoming dependent on other countries for uranium. Do you agree? Why or why not?
57. Describe the function of the following in a nuclear power plant:
    a. Control rods
    b. Fuel rods
    c. Moderator
    d. Containment building
58. What roles do deuterium (^2H) and tritium (^3H) play in nuclear fusion?
59. In a fusion reaction, a plasma (called the fourth state of matter) is formed. Describe the composition of a plasma.
60. How much more energy can be obtained from a pound of fuel using nuclear fusion than using nuclear fission?
61. How are the very high temperatures necessary for a fusion reaction going to be contained?

CHAPTER

12

Energy Sources for the Future

ENERGY CONSERVATION IS AN IMPORTANT OPTION IN PLANNING AN
ENERGY POLICY FOR THE FUTURE. Measures such as turning off
lights, using less hot water, and turning down thermostats are
important, but for the long term, conserving energy by improving the **efficiency** of energy use is far more effective. Improving
efficiency means finding a way to decrease the amount of energy
that is needed to perform a particular task. Already, industry
has made great strides: Home appliances such as refrigerators,
washers, and dryers are much more efficient. Insulation in new
homes and buildings has improved, and there has been an
increase in the use of fluorescent light bulbs. In the 1970s, new
passenger cars averaged 13 miles per gallon (mpg); by 1985,
they were required by law to average 27.5 mpg. In late 2007
new Corporate Average Fuel Economy (CAFÉ) standards were
signed into law requiring automakers to increase gas mileage to
35 mpg by 2020. If the automobile industry makes use of available technology, new cars could average 60 mpg or more by
2100. Already, European manufacturers have produced cars
that achieve over 80 mpg. There are a number of existing technologies that automobile manufactures could implement today
to improve fuel efficiency. The Union of Concerned Scientist
suggest that the use of cylinder deactivation, which shuts down
half the cylinders when full power is not needed, flexible fuel
components, more efficient transmissions, and better aero-
dynamics could almost double the mileage of most cars.

Greater use of mass transportation, less intensive use of
energy for agriculture, the manufacture of goods that are more

durable, greater use of energy-efficient appliances, and more recycling are all measures that can be taken to save energy. The initial costs are often high, but savings will be considerable in the long run.

Many industries have improved efficiency by using waste heat that previously was dissipated to supply energy for a second process. At some power plants, waste heat is being used to heat nearby buildings.

Energy conservation through increased efficiency is not an answer to the energy problem but is an essential step if we are to avoid severe shortages while alternative sources of energy are being developed to replace dwindling fossil fuel supplies.

■ Energy-Efficient Lighting

According to the last Residential Energy Consumption Survey, U.S. households have, on average, 5.4 indoor lights that are on 1 or more hours per day and 8.9 lights that are on 15 minutes or more per day. They use 940 kWh of electricity for lighting and spend $83 on electricity for lighting each year. Their expenditures on electricity for lighting are approximately 10% of their total electric bill. Most of the lights used for residential purposes are inefficient incandescent bulbs. Changing to more efficient lighting would result in significant reductions in electrical demand.

In an incandescent lamp, an electric current heats a metal filament in the bulb, making it glow white hot and give off light. The problem is that only 10% of the electricity is used to make light—the rest ends as heat. Compact fluorescent bulbs, one type of energy-efficient lights, are much more efficient at turning electricity into light.

A typical compact fluorescent is a one-piece light that holds both the fluorescent tube and the electronic ballast that controls the electric current. It is designed to screw easily into a standard incandescent fixture without modification and can be used outdoors as long as it is protected from the weather; thus, it is suitable for use in barns and storage sheds. In very cold weather, a slight delay exists before the fluorescent reaches full brightness.

The compact fluorescent lights are more expensive than standard incandescent lights, but if they are in use for more than six hours a day, they will pay for themselves usually within two years. Because they last eight to ten times longer, there is a further savings in the labor needed to replace the lights. Compact fluorescent bulbs need approximately one-third of the power required by incandescent bulbs to emit the same amount of light.

Larger incandescent fixtures, such as pole lights or floodlights, should be replaced with more efficient lights such as sodium or metal halide lamps. These are designed specifically to cast a lot of light over a wide area but with significantly less energy consumption. These lights, which require unique fixtures, are typically seen in streetlights, modern warehouses, and large stores.

The environmental benefits of making the change are considerable. Replacing a 75-W incandescent light with a 20-W compact fluorescent saves approximately 550 kWh over its lifetime. If the electricity comes from a coal-fired generating plant, the savings represents approximately 1300 pounds of carbon dioxide and 20 pounds of sulfur dioxide that would have otherwise been released into the atmosphere.

■ Solar Energy

The most attractive source of energy is probably the sun. During daylight hours in sunny locations, huge quantities of **solar energy** reach the Earth's surface. This energy comes to us free and is nonpolluting and, for all practical purposes, infinitely renewable; however, it is widely dispersed, and concentrating it and converting it to a usable form are both difficult and costly.

Solar Heating for Homes and Other Buildings

Solar energy can be used very simply to heat buildings and water. A building made of appropriate materials and suitably constructed and oriented can capture the sun's heat. In a typical **passive solar heating** system, solar energy enters through windows facing the sun, and convection currents distribute passively the heat around the building. Some of the heat is stored in rock below the building for release when the sun is not shining.

In an **active solar heating system**, heat gathered in solar collectors located on the roof of a building is circulated by means of pumps (Figure 12.1). A typical flat-plate collector consists of a black surface covered with a glass or plastic plate. As anyone who has walked barefoot on asphalt in summer knows, sunlight is absorbed by a black surface and converted to heat. In a solar collector, the glass plate allows sunlight to enter but traps the heat that is produced. Circulating water is heated as it passes between the glass and the black surface.

Depending on climate and the availability of sunshine, passive and active solar systems can provide from 50% to 100% of home heating requirements. Although initial construction costs are high and a back-up heating system may be needed, savings on energy bills are substantial. As oil becomes scarce and solar technology advances, solar heating should become increasingly attractive and affordable.

Electricity from Solar Thermal Power Collectors

Heat from solar collectors can be used to generate electricity. Three **solar thermal power** (STP) designs are currently under development: the parabolic trough, the central receiver, and the **parabolic dish**. One of the most successful developments is the parabolic trough collector (Figure 12.2a). The trough is a reflector that focuses sunlight onto a pipe running down its center. Oil or other fluid in the pipe is heated to a temperature of up to 400°C (750°F). The heat is used to boil water and produce steam that can be used to turn a turbine and generate electricity. The most up-to-date of these facilities converts approximately 20% of the sunlight reaching the troughs into electricity. There are nine solar trough facilities in the Mojave Desert in California that are expected to increase production in the next few years.

The central receiver, or power tower (Figure 12.2b), uses a circular array of **heliostats** (large individually tracking mirrors) to concentrate sunlight onto a central receiver mounted at the top of a tower. Although a number of designs exist, one unit in the desert of Southern California pumps molten nitrate salts as a heat transfer media into the central receiver. The salt is heated to approximately 550°C and is then pumped either to a "hot" tank for storage or through heat exchangers to produce superheated steam. The superheated steam is used in a conventional turbine generator to produce electricity. If more heat is captured by this system than is needed to make electricity, the molten salt is placed in the "hot" tank for use in making steam in times of peak electricity demand.

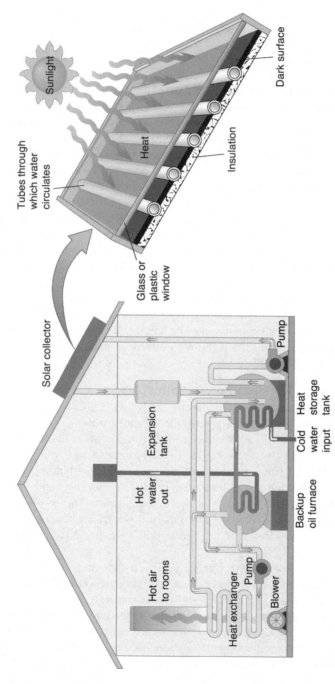

Figure 12.1 In an active solar heating system, water circulating through flat-plate collectors on the roof is heated by the sun and conveyed to a storage tank; the heat from the stored hot water is used to provide hot water, as well as hot air for space heating. The system is termed "active" because energy other than solar energy is required to pump the water through the house. Adapted from B.J. Nebel. *Environmental Science: The Way the World Works, Third edition.* Englewood Cliffs, NJ: Prentice Hall, 1990.

Figure 12.2 STP plants generate electricity from solar thermal energy. (a) A parabolic trough, (b) a central receiver or power tower, (c) a parabolic dish.

Parabolic dishes (Figure 12.2c) use an array of parabolic dish-shaped mirrors to concentrate sunlight onto a receiver located at the focal point of the dish. The receiver absorbs energy reflected by the concentrators, and fluid in the receiver is heated to approximately 750°C. Because this unit heats the transfer fluid to a higher temperature, this design achieves a higher efficiency than the power tower or parabolic trough. The simpler design of the parabolic dishes make them attractive choices as power sources in remote locations.

Electricity generated from STP plants is currently more costly than electricity from fossil fuel plants. It is, however, becoming more attractive as a source of electricity during peak demand.

Because the hot days of summer, when air conditioning demand is at its peak, are optimal for STP plants, they produce the maximum output just when it is needed. As the price of fossil fuels increases, the cost of electricity from STP plants will become more competitive.

Electricity from Photovoltaic Cells

Another approach to converting solar energy directly into electricity is the **photovoltaic cell** (or **solar cell**). Most solar cells consist of two layers of almost pure silicon. The top, very thin layer contains a trace of arsenic; the lower, thicker layer contains a trace of gallium (the addition of a trace of another element to silicon is known as *doping*). Recall that silicon has four valence electrons. Arsenic has five, and gallium has three. In a pure silicon crystal, each silicon atom is covalently bonded to four other silicon atoms (Figure 12.3). When arsenic atoms are included in the silicon structure, four of each of the arsenic atom's electrons form bonds with silicon atoms. The fifth electron is relatively free to move. **Doping silicon** with arsenic produces a semiconductor, which is termed an n-type semiconductor, because negative charge carriers (electrons) are created. When gallium (Ga) atoms are included, there is a shortage of one bonding electron around each Ga atom, and a positive hole is created. Doping silicon with Ga produces a p-type semiconductor that has positive charge carriers.

When a p-type semiconductor is joined with an n-type semiconductor, a p–n junction is formed. As a result, there is a tendency for free electrons in the arsenic-doped layer to migrate to the gallium-doped layer to fill the holes. Once the holes near the interface between the two layers are filled, the flow of electrons ceases. If the mobile electrons are sufficiently energized by exposure to sunlight and the layers are connected by an external circuit, however, electrons will flow through the external circuit, producing an electric current that can do useful work (Figure 12.4).

Solar cells were first used in the 1950s to provide power to space satellites. Today, they are widely used in calculators, watches, and other small devices. In the last 25 years, the cost of producing electricity from solar cells has fallen dramatically. At the same time, the efficiency of solar cells has increased from approximately 10% to more than 20%. As new micromorphous silicon materials and possible substitutes for silicon, such as germanium (Ge),

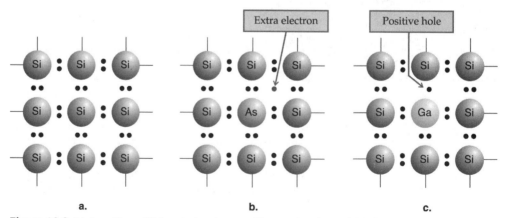

Figure 12.3 (a) Pure silicon. (b) Arsenic doped n-type silicon semiconductor. (c) Gallium doped p-type silicon semiconductor.

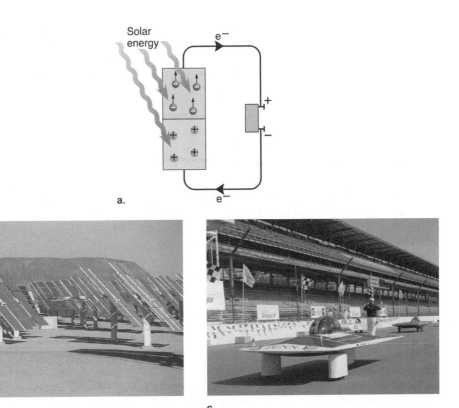

Figure 12.4 (a) In a photovoltaic cell, electrons are energized by sunlight and flow through an external circuit. (b) Solar panels. (c) The Sunraycer, an experimental car built by General Motors, is powered by solar cells. Silver-zinc batteries provide energy on cloudy days or if needed on steep hills.

gallium arsenide (GaAs), copper indium gallium selenide (CIGS), and cadmium telluride (CdTe) are developed, efficiency is expected to increase further.

As technology continues to advance and prices continue to fall as projected, rooftop arrays of solar cells are being installed to provide electricity to many new homes. Duke Power is proposing to regulators that it install solar panels on 850 North Carolina sites including homes, schools, stores, and factories. Already, experimental automobiles that run on solar energy have been built. An even more significant breakthrough is the development in California of an electric-generating plant powered by banks of solar cells that can provide enough electricity to meet the needs of approximately 10,000 people. Although solar-powered plants are not yet competitive with conventional fossil-fueled plants, they have many advantages: They are relatively inexpensive to construct, produce no pollution, require little maintenance, and have no fuel costs.

Solar cells hold enormous promise for the future, particularly in developing countries such as India where installing photovoltaic systems to supply electricity to remote, rural communities is much cheaper than extending existing power lines. The solar cells charge batteries with excess electricity produced during the day so that electricity is available during the night.

As a large number of aging nuclear-, petroleum-, and coal-fueled power plants in the United States near the end of their useful lives, legislators are urging power companies to seriously consider replacing them with solar-powered plants. In 2008 only 0.1% of U.S. electricity was generated by solar; on the other hand, Japan and several European countries are investing heavily in solar technology.

■ Energy from Biomass

Biomass, which is defined as any accumulation of biological materials, can be used as a source of energy. Examples of biomass are wood and crop residues. Like fossil fuels, biomass can be burned directly to provide energy in the form of heat. It can also be converted to methane (natural gas) and the liquid fuels methanol and ethanol.

Burning biomass or products made from it releases carbon dioxide into the atmosphere. One way to look at biomass is to consider the CO_2 that was recently in the air. Growing plants take up CO_2 from the air and convert it to biomass through photosynthesis. As biomass is burned, CO_2 is returned to the air, where it can quickly be absorbed by more growing plants that make more biomass. As long as the rate of biomass growth is larger than the rate of biomass burning, there is no net change in the CO_2 level in the atmosphere. The CO_2 that is released is recent CO_2 rather than ancient carbon from fossil fuels.

Burning Biomass

Wood, a form of biomass, is the major source of energy for cooking and heating for approximately 80% of people in the less developed nations of the world. In many areas of these countries, the constant search for firewood has led to excessive deforestation, with resulting erosion and degradation of soil. If trees continue to be cut faster than they are replaced, severe shortages of firewood are anticipated in the very near future in many places in the world.

Many sawmills obtain most of their power from burning wood waste, and wood-burning stoves are popular in homes; however, it is unlikely that burning wood or wood wastes will ever be more than a very minor source of energy in the United States or other industrialized countries. Cutting timber on a scale large enough to provide a significant amount of energy would have an adverse effect on the environment. Furthermore, wood stoves, unless they include control devices, are a source of both indoor and outdoor air pollution and are often dangerous. For this reason, some cities, including London and cities in South Korea, among others, have banned wood burning.

A number of cities in the United States now burn municipal trash, which is generally approximately 40% wastepaper, as a source of energy. The heat evolved by a typical modern waste incinerator can generate enough electricity to supply the needs of 50,000 homes.

Crop residues are another source of energy that can be exploited. In Hawaii, the fibrous residue from sugar cane is burned to produce electricity. Another possibility is to raise crops of fast-growing, high-energy–yielding plants that are specifically to be used as fuel. Trees and grasses that are native to a region are the best crops for energy. Some trees grow very fast and some trees will grow back after being cut off close to the ground, which is a technique called *coppicing*. Coppicing trees can be harvested every three to eight years for 20 or 30 years before replanting. These trees grow as high as 40 feet high between harvests. In the cooler, wetter regions of the northern United States, poplar, maple, black locust, and willow are coppicing candidates. In the warmer Southeast, sycamore and sweetgum are candidates, while in the warmest parts of Florida and California, eucalyptus is a candidate. These "energy plantations," however, have several disadvantages. Large areas of land are needed, and because of the energy that must be expended to raise, harvest, and dry the crop, the net energy yield is very low.

Production of Biogas

Plants, organic wastes, manure, and other forms of biomass can be used as sources of methane. The process of converting biomass to methane (or other fuels) is called **bioconversion**. In the absence of air, digestion of the organic materials in biomass by anaerobic bacteria produces **biogas**, which is approximately 60% to 70% methane. The most suitable starting materials are sewage sludge or manure. In India, as many as 2-million digesters are producing biogas from cow dung. The biogas is used for cooking, lighting, and heating, or it is used to generate electricity. The nitrogen-rich slurry that is produced as a by-product provides an excellent fertilizer. Dung from two cows can supply sufficient fuel to meet the needs for cooking and lighting for the average rural family. China has over 5 million digesters, which use human waste as well as animal dung.

Alcohols from Biomass

The methane in biogas can be converted to methanol as shown in the following equations:

$$CH_4 + H_2O \rightarrow CO + 3H_2$$
$$CO + 2H_2 \rightarrow CH_3OH$$

Natural gas or methane produced by coal gasification (discussed in Chapter 10) can be converted to methanol in the same way. Methanol also can be produced by destructive distillation of wood. As solid pieces of wood are heated, the wooden structure breaks down, and methanol is released. Fermentation of sugars and starches in plants produces ethanol.

$$C_6H_{12}O_6 \rightarrow 2CO_2 + 2C_2H_5OH$$

Suitable plant materials for bioconversion to alcohol are sugar cane, sugar beets, cassava, corn, and sorghum, all of which have a high content of sugars and starches. The alcohol can be concentrated by distillation. A major problem with producing alcohol in this way is that large-scale use of land to grow fermentable crops could infringe on land used to produce food crops. Both methanol and ethanol have high octane ratings (Table 10.6) and can be used directly as automobile fuels. Methanol requires major modifications in the conventional engine design, but ethanol can be used in today's automobiles with only minor changes. Using sugar cane as the starting material, Brazil has pioneered the development of alcohol as fuel; most of its cars run on ethanol or gasohol, a mixture of gasoline and 10% to 20% (by volume) ethanol.

In the United States, as a result of the Clean Air Act of 1992, most major metropolitan areas are required to sell oxygenated gasoline during the winter months to reduce carbon monoxide emissions. The oxygen content of gasoline can be increased by adding ethyl alcohol, but most U.S. petroleum companies increased it by adding methyl-t-butyl ether, which has a higher octane rating than ethanol. Because the Environmental Protection Agency recommended a reduction in the use of methyl-*t*-butyl ether due to health considerations, more ethanol is being used in gasoline today than has been in the last 20 years.

Although it is unlikely that bioconversion will ever become a major source of energy, it is a useful way to supplement other sources. It is particularly valuable as a means of converting plant and animal agricultural wastes and wastepaper in municipal garbage into usable energy.

■ Wind Power

Wind power was one of the earliest forms of energy humans harnessed to do useful work. Until the 1930s, rural America relied heavily on windmills for pumping water, grinding corn, and generating electricity. Then, as rural electrification schemes brought cheap power to farming communities, most windmills fell into disuse. Windmills, which when used to generate electricity are called **wind turbines** (Figure 12.5), are now making a dramatic comeback.

Figure 12.5 Five different wind turbine designs.

A modern wind turbine consists of three fiberglass blades mounted on a steel tower. The blades are much stronger and lighter than older, conventional ones, and the entire system is automated and very reliable. The turbines are usually arranged in wind farms, which consist of several hundred turbines, each with blades measuring about 15 m (45 ft) and capable of producing 10 to 15 kW of electricity (1 kW = 0.001 MW). Single units can be used for individual homes and farms, but arrays of units are more efficient. The land between the wind turbines can still be cultivated or used for grazing. California already has three large wind farms, which together can generate sufficient electricity to meet the needs of San Francisco, and wind power is expanding in west Texas, on the Great Plains and in coastal areas of the east. Proposed wind farms in beautiful vacation spots, such as Nantucket Sound and off shore in Rehobeth, Delaware, have pitted wind power advocates against vacation homeowners who argue that wind farms will spoil their ocean views. For wind power to be cost-effective, winds must blow fairly steadily at approximately 10 mph (16 km per hour). Many areas in the world meet this criterion, and wind power holds great promise for developing nations, especially for those that lack large reserves of fossil fuels. Today, over 50,000 wind turbines are in operation worldwide. The United States is second, behind Germany, in having the most wind capacity in the world. In 2008, the United States had a wind capacity to provide electricity to 4.9 million people. The largest U.S. wind farm, and in the world, is the Horse Hollow Center in Taylor county Texas, which has 421 wind turbines. Other large wind farms are located in Oregon and California, and many other states are planning investments in even more farms.

Because of dramatic decreases in the cost of the technology, in many areas, electricity can now be generated more cheaply from wind than from coal. It is projected that wind power will supply over 10% of the world's electrical energy by 2050. Already Denmark, which is the third leading country in the production of wind turbines, generates nearly 20% of its electricity from wind power. Unlike the U.S. government, many European governments, including those of Spain and Germany, are investing heavily in research and development of wind technology.

Wind is free and abundant, and wind power produces no pollutants and no carbon dioxide. Wind, however, does have a number of disadvantages: It is intermittent, making it necessary to have a storage system or an alternative source of energy when the wind is not blowing; wind turbines are also noisy and pose a danger to birds if they are located along migratory routes.

■ Geothermal Energy

Geothermal energy is heat energy that is generated deep within the Earth's interior by the decay of radioactive elements. Geothermal energy is abundant, but like solar energy, it is widely dispersed. It becomes accessible in only certain unstable regions of the world, where, as a result of geological activity, magma rises from great depths to near the Earth's surface. The hot magma, at temperatures of between 900°C and 1000°C (1600°F to 1800°C), heats rock and groundwater that comes in contact with the rock. Heated groundwater may emerge as a geyser or hot spring or may remain as a reservoir below the surface, sealed by a layer of impermeable cap rock.

Reservoirs may contain hot water or steam. Geothermal wells, basically similar to oil wells, can be drilled into the reservoirs to release the hot water or steam. Hot water deposits are the most common and are used primarily to heat buildings. Steam is used to generate electricity. The largest known geothermal field is The Geysers, located 145 km (90 miles) north of San Francisco. Between 1960 and 1988, this field was producing electricity more cheaply than fossil-fueled or nuclear-powered plants. Since then, however, power output has fallen because the underground reservoir is drying up. The largest electric-generating plant based on steam is in New Zealand.

Both water and steam deposits are abundant in Iceland. In Reykjavik, the capital, nearly all of the buildings and greenhouses for producing fruits and vegetables are heated with hot water obtained from geothermal wells. Hot water reservoirs also heat homes in Boise, Idaho, and Klamath Falls, Oregon.

Hot rock can also be exploited to obtain steam. Cold water is injected into wells drilled into the hot rock. The water returns to the surface as steam, which can be used to generate electricity.

The future of geothermal energy is uncertain. Many geothermal fields occur in rugged, inaccessible terrain; many are located in scenic areas, such as Yellowstone National Park, and cannot be developed. Most steam deposits contain hydrogen sulfide, small amounts of which are released during plant operation, and pollute the atmosphere. Salts dissolved in the water released with steam from reservoirs can corrode pipes and other fixtures and, if not prevented from reaching streams or rivers, can cause severe ecological damage. Despite these problems, the use of geothermal energy is likely to increase in areas where it can be exploited as oil becomes scarce and more expensive.

■ Water Power

To produce large quantities of electricity from water, huge dams are built across rivers. As water in the reservoir behind the dam is released, the flow is used to drive turbines that produce electricity. **Hydropower** is a very efficient (80% efficiency) means of producing electricity and is essentially renewable and nonpolluting; however, the dams that must be built to produce hydroelectric power create many environmental problems. Water impounded behind a dam may flood scenic stretches of river, valuable cropland, places of historical or geological interest, and/or people's homes. Downstream from a dam, as water flow is adjusted to meet electrical demand, the constantly changing water level alters the natural ecosystem. Silt becomes trapped behind the dam, and the amounts of sediments and nutrients downstream are reduced. Dams disrupt the migration and spawning of fish. For example, although fish ladders have been installed around dams on the Columbia River in the state of Washington, the ladders have not been effective in preventing drastic reductions in the salmon population.

Worldwide, many new dams are planned, and great concern exists about the environmental impact of two of the largest projects: the Three Gorges Dam on the Yangtze River in China and the James Bay project in Quebec in Canada. In the United States, because most of the best dam sites have already been used and because of environmental concerns, no new dams are expected to be constructed; one in Maine has already been removed, and four on the Columbia River may be removed in the future. Approximately 7% of the electricity in the United States and approximately 24% worldwide are supplied by hydropower.

◼ Tidal Power

The twice-daily rise and fall of ocean tides represents an enormous potential source of energy. This **tidal power** can be exploited by building a dam across the mouth of a bay or inlet. The incoming tide generates electricity as it flows through turbines constructed in the dam. The turbine blades are then reversed so that the outgoing tide also produces electricity. To be practical, the difference in level between low and high tide must be 6 m (20 ft) or more; this difference ranges from 1 to 10 m (3 to 30 ft) around the world. Three tidal power plants have been built: one in Canada, one in France, and the other in Russia.

Tidal power plants alter the normal flow of water and disrupt and disfigure the natural environment. Because of the adverse environmental effects and lack of suitable sites, tidal power is never likely to be more than a minor source of energy in the United States.

◼ Energy from Hydrogen Gas

Hydrogen gas may someday replace oil and natural gas. Hydrogen could be transported through pipes like natural gas. It could be used for heating homes and for heating water to produce steam for electric power generation. With minor changes to the carburetor, today's cars could run on hydrogen. Hydrogen burns cleanly, combining with oxygen in the air to produce water vapor, and it releases more energy gram for gram than coal, gasoline, or natural gas. Transporting hydrogen through pipelines is cheaper than transmitting electricity over power lines.

A major problem is that practically no free hydrogen exists on Earth. Technically, hydrogen is not a "fuel" but rather a means for transporting energy obtained from other sources, just like electricity. Hydrogen (along with oxygen) can be produced by the electrolysis of water, but this process requires the input of energy. In fact, because of the second law of thermodynamics, it requires more energy to break the bonds between hydrogen and oxygen atoms in water molecules than is released when hydrogen is burned; however, it is projected that the decreasing cost of solar cells will make it economical to use solar-generated electricity to produce hydrogen by electrolysis. Seawater could provide an almost inexhaustible source of water.

If solar energy could be used directly to decompose water into hydrogen and oxygen, hydrogen could be produced much more economically. In the initial stages of photosynthesis, water is decomposed to hydrogen and oxygen. This photodissociation has been duplicated with some success in the laboratory by exposing blue-green algae to sunlight. The hydrogen and oxygen produced were separated and collected.

One problem is that hydrogen is even more flammable than gasoline, but one advantage is that because hydrogen is much lighter than air it dissipates rapidly. Being heavier than air, gasoline vapor tends to accumulate at the site of a leak. There is concern that releases of hydrogen may be adding another greenhouse gas into the atmosphere. Another problem is how to store hydrogen in the vehicle. Tanks that can hold hydrogen gas under very high pressure are available but have limited capacity and some danger of rupture. A promising technique that is being investigated is absorption of hydrogen by certain metals to form hydrides, which then readily release the hydrogen as it is needed. This technology is discussed later.

■ Fuel Cells*

Fuel cells are a more promising way of using hydrogen as an energy source for automobiles and for generating electricity. Instead of burning hydrogen to power the vehicle or generating station, electricity generated by the oxidation of hydrogen in a fuel cell is used as a source of power. Fuel cells are lightweight and are well suited for producing electricity and have been used for years in spacecraft.

Fuel cells depend on an oxidation-reduction reaction that converts chemical energy directly into electrical energy. The oxidation of hydrogen by oxygen is the basis of the fuel cells used in space.

$$2\,H_2 + O_2 \rightarrow 2\,H_2O + \text{electrical energy}$$

The electrochemical cell is two to three times more efficient than an internal combustion engine in converting fuel into power.

EXAMPLE 12.1

What is the theoretic maximum voltage that could be generated by a fuel cell? If we know the Gibbs free energy change (ΔG) for the reaction, we should be able to calculate the maximum cell voltage (ΔE) from the following:

$$\Delta E = -\Delta G/nF$$

Where

n is the number of moles of electrons involved in the reaction per mole of H_2, and F is Faradays constant 96,487 coulombs (joules/volt).

At a constant pressure of 1 atmosphere and at 25°C, the ΔH is −285,000 J, and ΔS is −163.2 J/K. To calculate ΔG, use the following:

$$\Delta G = \Delta H - T\Delta S$$

$$\Delta G = -285,000\,J - (298\,K)(-163.2\,J/K)$$

$$\Delta G = -237,200\,J$$

Substituting ΔG

$$\Delta E = -\Delta G/nF$$

$$\Delta E = -(-237,200\,J)/2\,(96,487\,J/V)$$

$$\Delta E = 1.23\,V$$

---

*Portions of this section are from http://www.eere.energy.gov/hydrogenandfuelcells/fuelcells/.

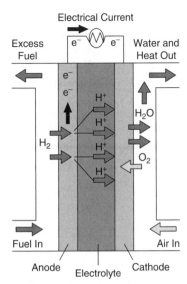

Electrical Current

Excess Fuel

Water and Heat Out

e^- e^-

H_2

e^-
e^-
H^+
H_2O
H^+
H^+
O_2
H^+

Fuel In

Air In

Anode

Electrolyte

Cathode

Figure 12.6 A PEM fuel cell. Department of Energy: www.eere.energy.gov/hydrogenandfuelcells/fuelcells/how.html.

The temperature of the fuel cell increases as it is operating, and humidified air is usually used rather than pure oxygen, both of which decrease the optimum cell voltage to 1.16 V.

The fuel cell would generate 1.16 volts if it were operating at 100% efficiency. A good measure of fuel cell efficiency is the ratio of the actual cell voltage to the theoretic maximum voltage for the cell. Fuel cells that generate 0.7 volts are generating approximately 60% of the maximum useful energy available from the fuel in the form of electricity. If the same fuel cell is operating at 0.9 volts, approximately 78% of the maximum useful energy is delivered as electricity. The remaining energy is lost in the form of heat.

The voltage from one single fuel cell is approximately 0.7 volts, too small to power a car. When the fuel cells are stacked in series, the operating voltage increases to 0.7 volts multiplied by the number of cells stacked.

Fuel cells are classified primarily by the kind of electrolyte used. The electrolyte limits the chemical reactions that can take place in the cell, the type of catalyst required and the temperature at which the cell operates. We describe six promising types of fuel cells that are currently undergoing engineering development.

Polymer Electrolyte Membrane Fuel Cell

Polymer electrolyte membrane (PEM) fuel cells—also called proton exchange membrane fuel cells—deliver high-power density and offer the advantages of low weight and volume compared with other fuel cells (Figure 12.6). PEM fuel cells use a solid polymer as an electrolyte and porous carbon electrodes containing a platinum catalyst. They need only hydrogen, oxygen from the air, and water to operate and do not contain corrosive fluids as do some fuel cells. They are typically fueled with pure hydrogen supplied from storage tanks.

The PEM is usually made from the polymer polyperfluorosulfonic acid, which is manufactured by DuPont under the trade name of Nafion.

$$
\begin{array}{c}
| \\
CF_2 \\
| \\
CF-O-CF_2-CF-O-CF_2-CF_2-\overset{\displaystyle O}{\underset{\displaystyle O}{\overset{\displaystyle \|}{S}}}-O^-H^+ \\
| \qquad\qquad | \qquad\qquad | \\
CF_2 \qquad\quad CF_3 \qquad\quad O \\
|
\end{array}
$$

The Nafion membrane material has three regions:

1. A Teflon-like fluorocarbon backbone, with hundreds of repeating units ($-CF_2-CF-CF_2-$)
2. Side chains, which connect the inert backbone with the sulfonate functional group ($-O-CF_2-CFCF_3-O-CF_2-CF_2-$)
3. The sulfonic acid functionality, $SO_3^-H^+$

A negative charge is permanently attached to the sulfonate and cannot move. When the membrane becomes hydrated by absorbing water, the H^+ ions become mobile. Ion movement occurs by H_3O^+ hopping from SO_3^- site to SO_3^- site within the membrane. The membrane can be considered to be a H^+ conductor. Because Nafion is expensive, new patents describe the replacement of Nafion with Solupor. Solupor is a porous polyethylene that is claimed to be more durable and less expensive than Nafion.

The PEM fuel cell uses platinum as the electrode material. Platinum (Pt) is a unique material because it is sufficiently reactive in binding H and O intermediates as required to facilitate the electrode processes and is also capable of effectively releasing the intermediate to form the final product. For example, the anode process requires Pt sites to bond H atoms when the H_2 molecule reacts, and these Pt sites next release the H atoms as H^+:

$$H_2 + 2\,Pt \rightarrow 2\,Pt\text{--}H$$

$$2\,Pt\text{--}H \rightarrow 2\,Pt + 2\,H^+ + 2\,e^-$$

This reaction requires a material that will bind H atoms not too weak and not too strong. Pt is the best catalyst for this purpose, but it costs three times the price of gold. Most PEM development is focused on lowering the amount of Pt necessary for each cell. Electrodes are being constructed from porous carbon to which a very small amount of Pt is bonded. The electrode is porous so that the H_2 and O_2 can diffuse through the electrode to reach the catalyst. The electrochemical reactions are as follows:

$$\text{Anode:} \quad H_2 \rightarrow 2\,H^+ + 2\,e^-$$

$$\text{Cathode:} \quad \frac{1}{2}\,O_2 + 2\,H^+ + 2\,e^- \rightarrow H_2O$$

$$\text{Cell:} \quad H_2 + \frac{1}{2}\,O_2 \rightarrow 2\,H_2O$$

PEM fuel cells operate at relatively low temperatures, approximately 80°C (176°F). The low temperature operation allows them to start quickly (less warm-up time) and results in less wear on system components, resulting in better durability. The platinum catalyst is also extremely sensitive to CO poisoning, making it necessary to employ an additional reactor to reduce CO in the fuel gas if the hydrogen is derived from an alcohol or hydrocarbon fuel. This also adds cost. Developers are currently exploring platinum/ruthenium catalysts that are more resistant to CO.

PEM fuel cells are used primarily for transportation applications and some stationary applications. Because of their fast startup time, low sensitivity to orientation, and favorable power-to-weight ratio, PEM fuel cells are particularly suitable for use in passenger vehicles, such as cars and buses. Hydrogen is a noncondensible gas; its critical temperature (−240°C, −400°F) is extremely low. This means that hydrogen gas cannot be liquefied above this temperature, no matter how high the pressure applied. Because it is not practical to lower the temperature of the fuel tank to its critical temperature, a hydrogen fuel tank at ambient temperature will hold a much smaller number of moles of fuel than it would if the fuel was a liquid. It is difficult to store enough hydrogen onboard to allow vehicles to travel the same distance as gasoline-powered vehicles before refueling, typically 300 to 400 miles. Higher density liquid fuels such as methanol, ethanol, natural gas, liquefied petroleum gas, and gasoline can be used for fuel, but the vehicles must have an onboard fuel processor to reform the methane to hydrogen. The reforming reaction passes steam with the carbon-based fuel over a selective catalyst. The reaction for methane is as follows:

$$CH_4 + H_2O \rightarrow 3\,H_2 + CO$$

$$H_2O + CO \rightarrow CO_2 + H_2$$

The gas from this reaction is passed into a reactor where air is injected so that oxygen reacts with the remaining CO over a Pt catalyst to convert CO to CO_2. The final gas mixture from this reaction is approximately 70% H_2, 24% CO_2, 6% N_2, and traces of CO. This step increases costs and maintenance requirements. The reformer also releases carbon dioxide (a greenhouse gas), although less than that emitted from current gasoline-powered engines.

Phosphoric Acid Fuel Cell

Phosphoric acid fuel cells (PAFCs) use liquid phosphoric acid as an electrolyte—the acid is contained in a Teflon-bonded silicon carbide matrix—and porous carbon electrodes containing a platinum catalyst (Figure 12.7).

The electrochemical reactions occurring in the PAFCs are as follows:

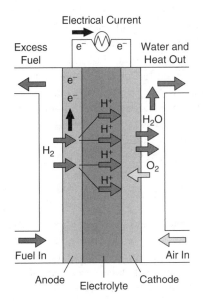

Figure 12.7 A phosphoric acid fuel cell. Courtesy of the U.S. Department of Energy.

Anode: $H_2 \rightarrow 2\,H^+ + 2\,e^-$

Cathode: $\frac{1}{2}\,O_2 + 2\,H^+ + 2\,e^- \rightarrow H_2O$

Cell: $H_2 + \frac{1}{2}\,O_2 \rightarrow 2\,H_2O$

The PAFC is considered the "first generation" of modern fuel cells. It is one of the most thoroughly developed cell types and is the first to be used commercially, with over 200 units currently in use. This type of fuel cell is typically used for stationary power generation, but some PAFCs have been used to power large vehicles such as city buses.

PAFCs are more tolerant of impurities in the reformed H_2 than PEM cells, which are easily "poisoned" by carbon monoxide; carbon monoxide binds to the platinum catalyst at the anode, decreasing the fuel cell's efficiency. PAFCs are 90% efficient when used for the co-generation of electricity and heat, but less efficient at generating electricity alone (37% to 42%). This is only slightly more efficient than combustion-based power plants, which typically operate at 33% to 35% efficiency. PAFCs are also less powerful than other fuel cells, given the same weight and volume. As a result, these fuel cells are typically large and heavy. PAFCs are also expensive. Like PEM fuel cells, PAFCs require an expensive platinum catalyst, which raises the cost of the fuel cell. A typical PAFC costs between $4000 and $4500 per kilowatt to operate.

Direct Methanol Fuel Cell

Most fuel cells are powered by hydrogen, which can be fed to the fuel cell system directly or can be generated within the fuel cell system by reforming hydrogen-rich fuels such as methanol, ethanol, and hydrocarbon fuels. **Direct methanol fuel cells (DMFCs)**, however, are powered by pure methanol, which is mixed with steam and fed directly to the fuel cell anode.

The electrochemical reactions occurring in a DMFC are as follows:

$$\text{Anode:} \quad CH_3OH + H_2O \rightarrow CO_2 + 6\,H^+ + 6\,e^-$$

$$\text{Cathode:} \quad \frac{3}{2}O_2 + 6\,H^+ + 6\,e^- \rightarrow 3\,H_2O$$

$$\text{Cell:} \quad CH_3OH + \frac{3}{2}O_2 \rightarrow CO_2 + 2\,H_2O$$

The DMFC has some technological problems. The amount of platinum necessary to achieve a high current is much greater than the amount used in the PEM fuel cell, making the DMFC very expensive. Methanol has permeated some membrane material and crossed from anode to cathode, decreasing the performance of the air cathode and wasting fuel.

DMFCs do not have many of the fuel storage problems that are typical of some fuel cells because methanol has a higher energy density than hydrogen, although less than gasoline or diesel fuel. Methanol is also easier to transport and supply to the public using our current infrastructure because it is a liquid, such as gasoline. This advantage in terms of simplicity and cost means that DMFC systems present an attractive alternative to hydrogen.

DMFC technology is relatively new compared with that of fuel cells powered by pure hydrogen, and research and development are roughly three to four years behind that of other fuel cell types.

Alkaline Fuel Cell

Alkaline fuel cells (AFCs) were one of the first fuel cell technologies developed and were the first type widely used in the U.S. space program to produce electrical energy and water onboard spacecraft. As can be seen in Figure 12.8, AFCs use a solution of potassium hydroxide in water as the electrolyte and can use a variety of nonprecious metals as a catalyst at the anode and cathode. High-temperature AFCs operate at temperatures between 100°C and 250°C (212°F and 48°F); however, more recent AFC designs operate at lower temperatures of roughly 23°C to 70°C (74°F to 158°F).

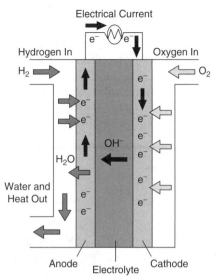

Electrical Current

Hydrogen In   Oxygen In

Anode   Electrolyte   Cathode

Figure 12.8 An alkaline fuel cell. Courtesy of the U.S. Department of Energy.

The electrochemical reactions are as follows:

Anode: $H_2 + 2\,OH^- \rightarrow 2\,H_2O + 2\,e^-$

Cathode: $\frac{1}{2}\,O_2 + 2\,H_2O + 2\,e^- \rightarrow 2\,OH^-$

Cell: $H_2 + \frac{1}{2}\,O_2 \rightarrow 2\,H_2O$

AFCs are high-performance fuel cells because of the rate at which chemical reactions take place in the cell. They are also very efficient, reaching efficiencies of 70% in space applications. The disadvantage of this fuel cell type is that it is easily poisoned by CO_2. In fact, even the small amount of CO_2 in the air can affect the cell's operation, making it necessary to purify both the hydrogen and oxygen used in the cell. This purification process is costly. Susceptibility to poisoning also affects the cell's lifetime, further adding to cost.

Cost is less of a factor for remote locations such as space or under the sea. To compete effectively in most commercial markets, however, these fuel cells will have to become more cost-effective. AFC stacks have been shown to maintain sufficiently stable operation for more than 8000 operating hours. To be economically viable in large-scale utility applications, these fuel cells need to reach operating times exceeding 40,000 hours. This is possibly the most significant obstacle in commercializing this fuel cell technology.

Molten Carbonate Fuel Cell

Molten carbonate fuel cells (MCFCs) are currently being developed for natural gas- and coal-based power plants for electrical utility, industrial, and military applications. As can be seen in Figure 12.9, MCFCs are high-temperature fuel cells that use an electrolyte composed of a molten carbonate salt mixture suspended in a porous, chemically inert ceramic lithium aluminum oxide ($LiAlO_2$) matrix. Because they operate at extremely high temperatures of 650°C (roughly 1200°F) and above, nonprecious metals can be used as catalysts at the anode and cathode, reducing costs.

The electrochemical reactions occurring in a molten carbonate fuel cell are as follows:

Anode: $CO_3^{2-} + H_2 \rightarrow CO_2 + H_2O + 2\,e^-$

Cathode: $\frac{1}{2}O_2 + CO_2 + 2\,e^- \rightarrow CO_3^{2-}$

Cell: $H_2 + \frac{1}{2}O_2 + CO_2 \rightarrow H_2O + CO_2$

Improved efficiency is another reason that MCFCs offer significant cost reductions over PAFCs. MCFC plants can reach efficiencies approaching 60%, considerably higher than the 37% to 42% efficiencies of a PAFC plant. When the waste heat is captured and used, overall fuel efficiencies can be as high as 85%.

Unlike alkaline, phosphoric acid, and PEM fuel cells, MCFCs do not require an external reformer to convert more energy-dense fuels to hydrogen. Because of the high temperatures at which they operate, these fuels are converted to hydrogen within the fuel cell itself by a process called internal reforming, which also reduces cost. MCFCs are not prone to carbon monoxide or carbon dioxide "poisoning"—they can even use carbon oxides as fuel—making them more attractive for fueling with gases made from coal. Although they are more resistant to impurities

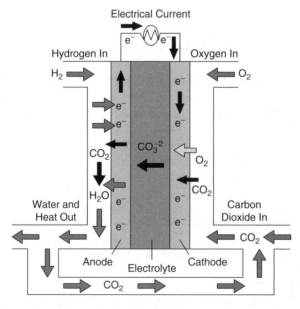

Figure 12.9 A molten carbonate fuel cell (MCFC). Courtesy of the U.S. Department of Energy.

than other fuel cell types, scientists are looking for ways to make MCFCs resistant enough to impurities from coal, such as sulfur and particulates.

The primary disadvantage of the current MCFC technology is durability. The high temperatures at which these cells operate and the corrosive electrolyte used accelerate component breakdown and corrosion, decreasing cell life.

Solid Oxide Fuel Cell

Solid oxide fuel cells (SOFCs) use a hard, nonporous ceramic compound as the electrolyte. Because the electrolyte is a solid, the cells do not have to be constructed in the plate-like configuration that is typical of other fuel cell types (Figure 12.10). SOFCs are expected to be approximately 50% to 60% efficient at converting fuel to electricity. In applications designed to capture and use the system's waste heat (co-generation), overall fuel use efficiencies could top 80% to 85%.

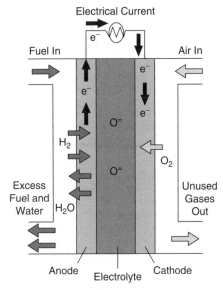

Figure 12.10 A solid oxide fuel cell (SOFC). Courtesy of the U.S. Department of Energy.

The electrochemical reactions for the SOFC are as follows:

$$\text{Anode:} \quad H_2 + O^{2-} \rightarrow H_2O + 2\,e^-$$

$$\text{Cathode:} \quad \frac{1}{2}O_2 + 2\,e^- \rightarrow O^{2-}$$

$$\text{Cell:} \quad H_2 + \frac{1}{2}O_2 \rightarrow 2\,H_2O$$

SOFCs operate at very high temperatures, approximately 1000°C (1830°F). High-temperature operation removes the need for precious-metal catalyst, thereby reducing cost. It also allows SOFCs to reform fuels internally, which enables the use of a variety of fuels and reduces the cost associated with adding a reformer to the system.

SOFCs are also the most sulfur-resistant fuel cell type; they can tolerate several orders of magnitude more sulfur than other cell types. In addition, they are not poisoned by carbon monoxide, which can even be used as fuel. This allows SOFCs to use gases made from coal.

The SOFC's high-temperature operation has disadvantages. It results in a slow startup and requires significant thermal shielding to retain heat and protect personnel, which may be acceptable for utility applications but not for transportation and small portable applications. The high operating temperatures also place stringent durability requirements on materials. The development of low-cost materials with high durability at cell operating temperatures is the key technical challenge facing this technology.

Scientists are currently exploring the potential for developing lower temperature SOFCs operating at or below 800°C that have fewer durability problems and cost less. Lower temperature SOFCs produce less electrical power, however, and stack materials that will function in this lower temperature range have not been identified.

Regenerative (Reversible) Fuel Cell

Regenerative fuel cells produce electricity from hydrogen and oxygen and generate heat and water as by-products, just like other fuel cells; however, regenerative fuel cell systems can also use electricity from solar power or some other source to divide the excess water into oxygen and hydrogen fuel, by electrolysis (Figure 12.11). This is a comparatively young fuel cell technology being developed by NASA and others.

The cells consist of an electrolyte solution, usually potassium hydroxide (KOH), and porous carbon electrodes containing platinum or some other suitable metal. The electrolyte solution is dissociated in one cell into hydrogen and oxygen during the day by the addition of electricity, which is generated by means of a solar panel. The hydrogen and oxygen would flow into a storage container. At night, the hydrogen and oxygen would flow into the fuel cell where they would be converted into water and release energy. The closed-loop system could

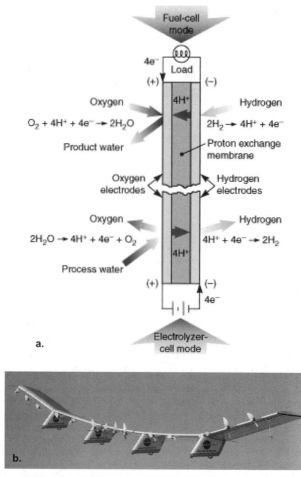

Figure 12.11 (a) Renewable regenerative fuel cell. (b) NASA's HELIOS solar-electric airplane uses a regenerative fuel cell to keep it flying through the night. Courtesy of Lawrence Livermore National Laboratory, and the U.S. Department of Energy.

have significant advantages because it could enable the operation of a fuel cell without requiring a new hydrogen infrastructure.

Each type of fuel cell operates at a different temperature, and each is best suited to a particular application. PEM fuel cells are well suited to transportation because they can provide a continuous electrical energy supply from fuel at relatively low operating temperatures and are relatively lightweight. PAFCs have been used in the initial commercialization of stationary fuel cell systems to generate electricity. Over 200 PAFC units, 200 kW each, are currently in operation around the world.

"Distributed power" is a new approach that utility companies are beginning to investigate: locating small, energy-efficient power generators closer to factories or cities. Because the fuel cells are modular and more efficient than existing systems, these small units can be placed onsite. Installation is less of a financial risk. The waste heat from these units can be used to head industrial or residential buildings. As demand for power increases, more fuel cell modules can be added to meet that demand.

■ Clean Cars for the Future

Two factors have been responsible for the development of automobiles that are not powered by a gasoline engine: the eventual exhaustion of oil supplies and the need to reduce tailpipe emissions. Legislation passed in 1998 in California required that 10% of all cars sold in the state meet zero-emission standards by 2003. Automakers sued the state and won an injunction banning its implementation. After years of automakers claiming this was an unreasonable goal, in early 2008, the California Air Resources Board issued a ruling that automakers produce at least 7500 zero-emission vehicles a year beginning in 2014. In addition to the promising of hydrogen fuel cell cars just discussed, several other options are being developed. These include electric cars and hybrid cars that run on both gasoline and electricity.

At the global warming conference in Kyoto in Japan in 1997, Toyota introduced a hybrid car, the Prius, that can achieve 66 mpg and that cuts tail pipe emissions by almost 90%. The Prius uses a small gasoline engine and a nickel–metal battery in tandem. Unlike a completely electric car, it uses batteries to run an electric motor when the car is traveling less than 15 mph and the gasoline engine kicks in at speeds above this level. The gasoline engine turns a generator that recharges the batteries. Toyota will introduce a "plug-in" hybrid in 2010 that will replace the nickel–cadmium rechargeable batteries that are used today with lithium batteries (the type that are used in laptop computers). The lithium batteries will recharge faster and take a longer charge. The "plug-in" hybrid can be recharged at night and will have the ability to travel 40 miles without turning on the gasoline engine. Recently, General Motors, and Chrysler have unveiled hybrid electric cars, which they started producing in 2008. General Motors plans to introduce the "Volt," a totally electric car in 2010. It will use electric motors for propulsion and its batteries will give it a range of 40 miles. If the Volt is driven more than 40 miles, a small onboard gas-powered generator provides the electricity needed to extend its range. Many major automobile manufacturers are developing fuel cell technology for cars of the future. Most of the systems are fueled with pure hydrogen gas, which is stored onboard as a compressed gas. As mentioned earlier, hydrogen is a noncondensible gas; its critical point ($-240°C$, $-400°F$) is an extremely low temperature. This means that ambient temperature hydrogen gas cannot be liquefied, no matter how high the pressure applied and that the hydrogen fuel tank holds a smaller amount of fuel than if it were holding a liquid.

This is a significant problem for fuel cell vehicles, which need to have a driving range of 300 to 400 miles between refueling to be competitive with gasoline vehicles.

Another approach involves storing the hydrogen in a solid form. For years, scientists have known that boron hydrides store a significant amount of energy; however, direct combustion of boron hydrides presented a difficult engineering problem and, thus, work on these compounds as a fuel source was largely abandoned in the 1960s. In 2001, the Daimler Chrysler process, called "Hydrogen on Demand," was introduced. It safely generates high-purity hydrogen from environmentally-friendly raw materials. The produced hydrogen can then be consumed in a fuel cell or hydrogen-burning engine to produce useful power.

The "hydrogen" is stored at ambient conditions in a nonflammable liquid "fuel," which is an aqueous solution of sodium borohydride ($NaBH_4$). Sodium borohydride is made from borax, a material that is found in substantial natural reserves globally. The fuel solution itself is nonflammable, nonexplosive, and easy to transport. The Hydrogen on Demand system releases the hydrogen stored in sodium borohydride solutions by passing the liquid through a chamber containing a catalyst. The hydrogen is liberated in this reaction:

$$NaBH_4 + 2\,H_2O \xrightarrow{\text{CAT}} NaBO_2 + 4\,H_2 + \text{Heat}$$

The only other reaction product, sodium metaborate, is water soluble and environmentally benign. The sodium metaborate can be either disposed of or recycled as the starting material for the generation of new sodium borohydride. The reaction occurs rapidly in the presence of catalyst; there is no need to supply external heat to access the hydrogen. The heat generated is sufficient to vaporize a fraction of the water present, and as a result, the hydrogen is supplied with co-generated moisture. Moisture in the H_2 stream is an added benefit for both fuel cells (humidifying the proton exchange membrane) and internal combustion engines (reducing autoignition, slowing the combustion flame speed).

Another approach to storing hydrogen as a solid involves the use of recycled plastics.

Powerballs are small solid balls or pellets of sodium hydride (NaH) that are coated with a waterproof plastic coating or skin. Powerballs are stored directly in water and can remain in water for months with little or no change to the coatings. As soon as a Powerball plastic collating is fractured, the sodium hydride inside can react with the water to produce hydrogen.

$$NaH + H_2O \rightarrow NaOH + H_2$$

The sodium hydride/water reaction is fast. Sodium hydride Powerballs react with water to release hydrogen on demand.

A Powerball facility (Figure 12.12) will pelletize NaH and coat it with recycled polyethylene shells to create Powerballs. Sodium hydride can be pelletized with a process similar to that used to pelletize charcoal briquettes, salt pellets, and wood pellets. Once the reactive NaH powder has been pelletized, each pellet is individually coated with a flexible polyethylene coating. The finished pellets are then dried and shipped as a safe and energy-dense form of hydrogen for use in industrial applications or, perhaps someday, for fuel cell electric vehicles.

A recycling facility could be constructed that will produce NaH from the waste sodium hydroxide (NaOH) solution left behind in Powerball tanks. This facility greatly reduces the cost for NaH used for Powerballs by producing it from waste NaOH solution.

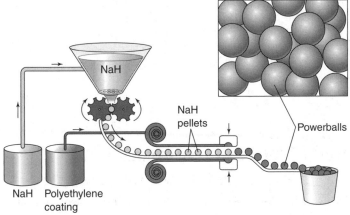

Figure 12.12 Powerballs and a schematic of a plant for producing Powerballs.

■ Energy Sources for the 21st Century

As oil supplies in the United States dwindle, the United States will become almost entirely dependent on imported oil. By the end of the 21st century, as recoverable oil and natural gas reserves worldwide are used, the United States and other industrialized nations will be forced to shift to other sources of energy.

Coal is plentiful in the United States and, in the short term, more will be used, primarily to generate electricity. Some oil is expected to be obtained from oil shale, tar sands, and coal liquefaction; however, all of these fossil fuel sources have a high environmental cost.

The best immediate option is energy conservation. Because as yet there is not any suitable substitute for liquid fuels, developing cars that are more fuel efficient to reduce consumption should be a top priority. Many believe that if another energy crisis is to be avoided, new laws must be passed mandating substantially higher gas mileages than the present requirement of 27.5 mpg.

The most promising source of renewable energy for the future is solar energy. Already solar cells show great promise, and as they become cheaper and more efficient, solar-powered electric-generating plants, automobiles, and numerous other products are likely to become common. Fuel cells and solar production of hydrogen are other promising long-term options. In suitable areas, wind power, geothermal energy, and hydroelectric power from small generating plants are expected to make an increasingly large local contribution. The future of nuclear energy is uncertain, but there is renewed interest in building new nuclear power plants now that there is worldwide concern about global warming and the carbon imprint of electricity generated by burning fossil fuels.

Without a strong commitment by the federal government to provide incentives for conservation and to fund the research and development of economically and environmentally acceptable new technologies, the transition from an economy based on fossil fuels to one based on alternative energy sources will not be easy. As the following quotation shows, the difficulties inherent in changing the status quo have been understood for centuries.

There is nothing more difficult to carry out, nor more doubtful of success, nor more dangerous to handle, than to initiate a new order of things. For the reformer has enemies in all who profit by the old order, and only lukewarm defenders in all those who would profit by the new order. This lukewarmness arises partly from fear of their adversaries, who have the law in their favor, and partly from the incredulity of mankind, who do not truly believe in anything new until they have had actual experience of it.

(Machiavelli, *The Prince*, 1517)

■ Additional Sources of Information

Gipe P. *Wind Energy Comes of Age.* New York: John Wiley & Sons, 1995.

Manwell JF, McGowan JG, Rogers AL. *Wind Energy Explained: Theory, Design and Application.* New York: John Wiley & Sons, 2002.

Thomas S, Zalbowitz M. *Fuel Cells-Green Power.* Los Alamos, New Mexico: Los Alamos National Laboratory. Accessed December 2008 at http://www.lanl.gov/orgs/mpa/mpa11/Green%20Power.pdf.

U.S. Department of Energy. Hydrogen, Fuel Cells & Infrastructure Technologies Program, 2008. Accessed December 2008 at http://www1.eere.energy.gov/hydrogenandfuelcells/.

U.S. Department of Energy. *International Energy Outlook: Hydroelectricity and Other Renewable Resources.* Accessed December 2008 at http://www.eia.doe.gov/oiaf/archive/ieo00/hydro.html.

■ Keywords

active solar heating system
alkaline fuel cells (AFCs)
bioconversion
biogas
biomass
direct methanol fuel cells (DMFCs)
doping silicon
energy conservation
energy efficiency
fuel cells

geothermal energy
heliostats
hydrogen gas
hydropower
molten carbonate fuel cells (MCFCs)
parabolic dish
passive solar heating
phosphoric acid fuel cells (PAFCs)
photovoltaic cell (solar cell)

polymer electrolyte membrane (PEM) fuel cells
Powerballs
regenerative fuel cells
solar energy
solar thermal power (STP)
solid oxide fuel cells (SOFCs)
tidal power
wind turbines

■ Questions and Problems

1. How do you differentiate between active and passive solar heating systems?
2. Describe how the three different STP designs collect solar energy. For what electrical need are these designs optimal?
3. Why are STPs used for peak electricity demand?

4. Which of the three STPs gives the highest efficiency? Why?
5. Why are parabolic STPs suited for remote locations?
6. Draw a schematic diagram of a photovoltaic cell.
   a. Which elements can be added to silicon to make a p-type semiconductor?
   b. Which elements can be added to silicon to make an n-type semiconductor?
   c. Which element may eventually replace silicon?
7. Suggest another element that could do the following:
   a. Replace Ga in a p-type semiconductor
   b. Replace As in an n-type semiconductor
8. List four forms of biomass. What is the energy yield obtained from burning biomass?
9. How is biogas produced?
10. Why is the CO_2 released from burning biomass considered recent CO_2?
11. Why is CO_2 from burning coal considered to be burning ancient carbon?
12. Write chemical equations to show how each of the following are produced from biomass:
    a. Methyl alcohol
    b. Ethyl alcohol
    c. Methane
13. List the advantages and disadvantages of using biomass as a source of energy.
14. What is gasohol?
    a. Do automobile engines need to be modified to burn methanol?
    b. Do automobile engines need to be modified to burn ethanol?
15. Will tidal power ever be a major source of energy in the United States? Explain.
16. For wind power
    a. How hard must the wind blow to make a wind turbine cost-effective?
    b. Where are most of the wind turbines in the United States?
    c. List three countries that are investing heavily in wind power.
    d. Which country leads the world in the production of wind turbines?
17. How much electricity can a single wind turbine produce? Is this enough electricity for a farm?
18. List three major disadvantages of wind power.
19. Will geothermal energy ever be a major source of energy worldwide? Explain.
20. For water power,
    a. What efficiency can be achieved by producing electricity by using hydropower? How does this compare to coal fired electric generators?
    b. Is hydropower a renewable resource?
    c. List two environmental problems that hydropower dams cause.
21. Write a balanced chemical reaction for the photodissociation of water into hydrogen gas.
22. What would be the theoretical maximum voltage generated by a hydrogen fuel cell that is operated at
    a. 80°C
    b. 200°C
    c. 800°C
23. Why does a single fuel cell only produce 0.7 V rather than the 1.23 V that is theoretically possible?

24. When a fuel cell is operating at 0.7 V and is generating approximately 60% of the maximum energy available from the cell, how can you account for the remaining 40%?
25. What would be the theoretical maximum voltage generated by a hydrogen fuel cell that is operated at 4°C?
26. Draw a schematic of a PEM fuel cell.
    a. What reaction is occurring at the cathode?
    b. What reaction is occurring at the anode?
    c. Why is the PEM often referred to as a "proton exchange membrane" fuel cell?
27. Describe the membrane that is used in the polymer electrolyte fuel cell.
    a. Why is a fluorocarbon used as the backbone of the membrane?
    b. Why are side chains attached?
    c. Why is a sulfonate attached?
28. What electrode material is used for the PEM fuel cell?
    a. List the disadvantages of using platinum.
    b. Suggest how the amount of platinum needed for the electrode could be reduced.
29. The critical temperature of hydrogen gas is −240°C. What does this mean?
    a. Compare the density of gaseous and liquid hydrogen.
    b. At what temperature would a container need to be to keep hydrogen a liquid?
    c. What is the heat of vaporization of hydrogen?
    d. Would it be practical to fuel a car with hydrogen in a liquid form?
30. What is a noncondensable gas?
31. Draw a schematic of a PAFC.
    a. What reaction is occurring at the cathode?
    b. What reaction is occurring at the anode?
    c. How is the phosphoric acid electrolyte held between the anode and cathode?
32. Compare the efficiency of a PAFC to a coal-fired electrical plant.
    a. Will PAFCs probably be used to power cars? Why or why not?
    b. Are PAFCs less expensive than PEMs?
    c. Can they be used with reformed hydrogen?
33. Draw a schematic of a DMFC.
    a. What reaction is occurring at the cathode?
    b. What reaction is occurring at the anode?
    c. Why is methanol fed into the fuel cell with steam?
34. List the advantages and disadvantages of the DMFC.
35. Draw a schematic of an AFC.
    a. What reaction is occurring at the cathode?
    b. What reaction is occurring at the anode?
    c. What is the electrolyte used in the AFC?
36. Past AFCs have operated at high temperature. Newer designs are made to operate at lower temperatures. What advantage is achieved by low-temperature operation?
37. Draw a schematic of a MCFC.
    a. What reaction is occurring at the cathode?
    b. What reaction is occurring at the anode?
    c. Describe the molten electrolyte used in this fuel cell.

38. Why can nonprecious metals be used as catalysts in the MCFC?

39. The MCFC does not require organic fuels to be reformed. Why?
   a. What is the efficiency of the MCFC?
   b. Are MCFCs poisoned by CO_2?
   c. What is the main liability of the MCFCs?

40. Draw a schematic of a SOFC.
   a. What reaction is occurring at the cathode?
   b. What reaction is occurring at the anode?
   c. Describe the SOFC electrolyte.

41. The SOFC does not require organic fuels to be reformed. Why?
   a. What is the efficiency of the SOFC?
   b. Are SOFCs poisoned by CO_2, CO, S?
   c. What is the main liability of the MCFCs?

42. Draw a schematic of a regenerative fuel cell. When it is making electricity,
   a. what reaction is occurring at the cathode?
   b. what reaction is occurring at the anode?
   c. Describe the regenerative fuel cell electrolyte.

43. List an application in which a regenerative fuel cell would have great advantages over the other types of fuel cells.

44. Draw a schematic of a regenerative fuel cell. When it is regenerating (electrolyzer)
   a. what reaction is occurring at the cathode?
   b. what reaction is occurring at the anode?

45. What is meant when an electrical utility uses "distributed power?"

46. How would "Hydrogen on Demand" work in an automobile?
   a. How is hydrogen stored as a solid or liquid?
   b. How is hydrogen gas liberated when it is needed?

47. How is sodium borohydride used in a "Hydrogen on Demand" system?
   a. How is hydrogen gas liberated from the sodium borohydride?
   b. How could the waste products of the reaction be disposed?

48. What is a "Powerball?"
   a. How is hydrogen gas liberated from a Powerball?
   b. How is the Powerball manufactured?
   c. How can the waste products from the spent Powerballs be disposed?

49. What is the current Federal gas mileage requirement for passenger cars?

50. What are the best options for energy sources for the 21st century?

CHAPTER

13

Inorganic Metals
in the Environment

THIS CHAPTER FOCUSES ON HEAVY METALS, THEIR PROPERTIES, AND HOW THEY ARE MEASURED IN ENVIRONMENTAL SAMPLES. This group is named *heavy metals* because their densities are higher than those of other common metals. The four heavy metals that are the focus of this chapter—lead (Pb), mercury (Hg), cadmium (Cd), and arsenic (As)—are those of greatest concern because of their extensive use by industry, their toxicity, and their pervasiveness in the environment. Metals are elements that are nondegradable, and because they are virtually indestructible, they accumulate in the environment.

Although heavy metals are often considered to be water pollutants, they are primarily transported through the environment by the movement of air. Chapter 6 discussed that particulates are released from anthropogenic sources such as fossil fuel combustion, or waste incineration; they are also released by plants as pollen. Heavy metals can absorb on the surface of particulates and travel great distances in the air. The particulates are then deposited on the surface of the Earth, lakes, and oceans.

■ Bioaccumulation of Heavy Metals

Because metals are elements, they cannot be broken into simpler, less toxic forms. They persist unchanged in the environment for many years and are biomagnified by aquatic organisms. Toxic metals can cause brain damage, kidney and liver disorders, and bone damage. Many are carcinogens. Toxic metals enter waterways from two main sources: industrial waste discharges and particulates in the atmosphere that settle and are carried in runoff.

The extent of bioaccumulation depends on the rate (R) at which it is ingested and the mechanism by which it is eliminated. Many routes of

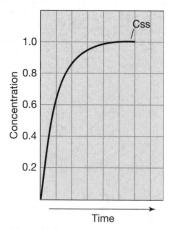

Figure 13.1 Increase in mercury concentration to a steady state concentration (Css).

elimination are discussed in Chapter 16. Usually, the rate of elimination is directly proportional to the organism's concentration (C) of the substance.

Rate of ingestion $= R$

Rate of elimination $= kC$

As an organism ingests a chemical, its concentration in the organism increases. Figure 13.1 shows that as the concentration of mercury rises, so does its rate of elimination. Eventually a steady state is established:

Rate of ingestion $=$ Rate of elimination

$$kC = R$$

Where

$$C = mg/L \text{ and } R = mg/day$$

The steady-state accumulation (C_s) is

$$C_s = R/k$$

It is convenient to describe the rate of elimination in terms of the half-life ($t_{1/2}$) of the chemical. First-order kinetics indicate that the relationship between the rate constant (k) and the $t_{1/2}$ is as follows:

$$k = 0.69/t_{1/2}$$

Substituting for k

$$C_s = (R)(t_{1/2})/0.69$$

The equation tells us that the longer the $t_{1/2}$ of the chemical, the higher its steady-state accumulation will be. Next we consider four metals—mercury, lead, cadmium, and arsenic—that are particularly dangerous because of their widespread use and toxicity.

EXAMPLE 13.1

Mercury (Hg^{2+}) has a $t_{1/2}$ of 6 days. If a person ingests 2 mg per day, calculate the steady-state concentration of mercury.

Use this equation:

$$C_s = (R)(t_{1/2})/0.69$$

$$C_s = (2 \text{ mg/day})(6 \text{ days})/0.69 = 18 \text{ mg}$$

■ Toxic Metals

Many metals—including sodium, potassium, calcium, magnesium, copper, and zinc—are essential for the normal development and well-being of humans and other animals. Other metals, even if present in an organism at very low concentrations, are toxic. The Environmental Protection Agency (EPA) has established maximum permissible levels in drinking water for the following toxic metals: antimony, arsenic (which is not a metal but is usually listed with them), barium, beryllium, cadmium, chromium, copper, lead, mercury, selenium, and thallium (Table 13.1). Municipal water supplies are monitored constantly to check that these limits are not exceeded. Atomic spectroscopy is one technique that is commonly used to monitor the concentration of dissolved metals.

Atomic Spectroscopy

In atomic spectroscopy, metal ions in solution are transformed into gaseous atoms by a flame, furnace, or **plasma** that operates between 2000°K and 11,000°K. The concentration of each metal is measured by the absorption or emission of ultraviolet or visible light from these gaseous atoms. Atomic spectroscopy techniques are highly sensitive, with detection limits in the parts per million (ppm) to parts per billion (ppb) level for environmental water samples and can easily determine the concentration of one element in an environmental sample without interference from other dissolved metals.

Flame Atomic Absorption Spectroscopy

Some **atomic absorption spectroscopy (AAS)** methods use a flame as a sample preparation chamber and a holder for the sample as the spectrophotometric measurement is made. Flame

---

Table 13.1

Maximum Permissible Levels for Metals in Drinking Water

Metal	Maximum Level (ppb)
Antimony	6
Arsenic	10
Barium	2000
Beryllium	4
Cadmium	5
Chromium (total)	100
Copper	1300
Lead	15
Mercury (inorganic)	2
Selenium	50
Thallium	2

Source: www.epa.gov.

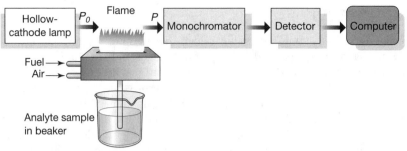

Figure 13.2 A flame atomic absorption spectrometer. Adapted from D.C. Harris. *Quantitative Chemical Analysis, Sixth edition*. New York: W.H. Freeman, 2003.

AAS is currently the most widely used of all of the atomic spectroscopy methods. Figure 13.2 shows that the AAS instrument is relatively simple. Figure 13.3 is a schematic of an atomic absorption spectrometer with a premix burner. The fuel and oxidant for the flame are mixed with the aqueous sample in the premix chamber of the burner. The rapid flow of the reactant gases, along with the spinning baffles in the chamber, acts to reduce the droplet size of the sample to a fine aerosol. Excess liquid that is no longer in the gas stream collects on the bottom of the premix chamber and flows out to waste. The fine aerosol spray is nebulized (sucked) into the burner where the water evaporates and the compounds present in the sample are decomposed into gaseous atoms.

The most common fuel is acetylene, and the most common oxidizer is air, which produces a flame with a temperature of 2400°K to 2700°K.

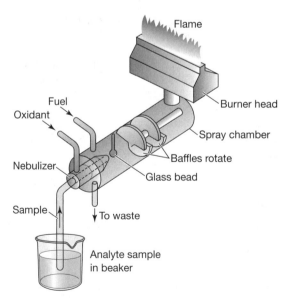

Figure 13.3 A premix nebulizer burner for a flame atomic absorption spectrometer. Adapted from D.C. Harris. *Quantitative Chemical Anaylsis, Sixth edition*. New York: W.H. Freeman, 2003.

The flame is thermally hot and can transfer energy to the sample ions. It also contains free electrons that are formed during the combustion of the fuel. When the electrons in the flame are captured by the metal ion, a single metal atom is present in the gas phase. EPA Method 220.1, which is used to determine the concentration of copper in drinking, used the flame AAS technique. A water sample containing the Cu^{2+} ion would undergo the following reaction in the flame:

$$Cu^{2+}\ (l)\ +\ 2\,e^-\ \to\ Cu\,(g)$$

Because the absorbance measurement that is going to be made involves the interaction of a beam of light with the outermost electrons in the Cu atom, the addition of electrons to reduce the Cu^{2+} ion into a neutral gaseous atom is an important step in preparing the sample for analysis.

The beam of light that is shot into the flame by the atomic absorption spectrometer comes from a hollow-cathode lamp. The cathode of the hollow-cathode lamp is made from the same element that is being determined; in this case, it would be made of lead. In the lamp, lead is vaporized and excited, and when this happens, the lamp emits a beam with a wavelength that is exactly what is needed to measure the absorbance of the Cu atoms in the flame. If the sample to be measured contains many different metal ions, each element that is to be determined by atomic absorption spectrometry has to be done sequentially. The process for each element is as follows: The **hollow cathode lamp** for the first element of interest is installed in the instrument. The monochromator of the instrument is tuned for the appropriate wavelength (λ) for that element. The optimum fuel and oxidant are fed to the nebulizer burner, and the flame is lit. Water is aspirated into the nebulizer to establish a background. Calibration samples that contain different, known concentrations of the analyte of interest are aspirated sequentially to establish a plot of absorbance (ν).

Each sample of interest is then sequentially measured, being careful to aspirate water between each sample. Because each element needs its own hollow-cathode lamp, these steps must be repeated for each element in the sample.

Metal atoms in the flame absorb some of the light from the hollow-cathode lamp.

$$M\ +\ h\nu\ \to\ M^*$$

Remember the following:

$$Transmittance\ =\ T\ =\ P/P_0$$
$$Absorbance\ =\ A\ =\ -\log T$$

P_0 is the amount of light that reaches the detector when distilled water is being analyzed, and P is measured when the sample is being analyzed. The amount of light reaching the detector is proportional to the concentration of the element in the sample over a limited concentration range. A calibration curve is constructed plotting absorbance versus concentration. The concentration of unknown samples is interpolated from the calibration curve.

The limit of detection for flame AAS is different for each element, but on average, it tends to be approximately 0.5 mg/L (ppm). This technique is not sensitive enough to certify that a certain drinking water sample contains less than 15 ppb of lead or 10 ppb of arsenic. The flame AAS also requires a relatively large sample (a 2 mL minimum); it aspirates 1 mL/min.

Because of these limitations, a graphite furnace is often used in place of the flame. The graphite furnace technique is discussed later in the chapter.

Lead

Lead has many useful properties. It has a low melting point, does not corrode, and is malleable and dense. The Latin name for lead is *plumbum,* which explains its chemical symbol (Pb) and why people who repair water pipes are called plumbers. Its major use in the United States today is for the manufacture of lead–acid storage batteries. It is used as a component of some types of solder and is present in glazes used on some ceramic ware. Until restricted by law in 1978, lead compounds were used to color paint. Tetraethyl lead was added to gasoline as an octane booster. Although lead has been banned from gasoline since 1987, lead-containing emissions from automobiles remain a problem in third-world cities where leaded gasoline is still used. Lead in auto emissions adheres to dust particles in the air. The particles settle onto streets, and the lead enters water supplies in runoff.

Of all of the toxic metals, lead is the most widely distributed in the environment and the one that the average person is the most likely to encounter. At one time, lead poisoning was common among workers involved in the production of lead or the manufacture of lead products, but now precautions are taken to limit workers' exposure. Today, those at the greatest risk are members of the general public, particularly young children, who live in older buildings containing lead-based paint. Others in danger live in an area close to a battery-recycling plant, a lead smelter, or a factory that is likely to release lead-containing particulates.

Although lead is a toxic metal, it does not pose much of a threat to adults. The average adult can excrete 2 mg of lead per day, and most of us do not generally take in this much from food, water, and air. If intake exceeds excretion, however, the excess lead is rapidly transported to bone marrow and then stored in the bones. Some lead remains in the liver and kidneys, but 90% eventually ends up in the bones. Lead disrupts hemoglobin metabolism, causing anemia, and like mercury, it inhibits enzymes containing sulfhydryl groups, which causes damage to the central nervous system. The concentration of lead in blood is measured to detect lead poisoning. Concentrations as low as 10 µg/dL are considered to be dangerous.

Lead is particularly toxic to young children, who are at greater risk than adults because they cannot excrete as much lead; consequently, they retain a higher percentage of ingested lead than adults. Also, their growing bones do not absorb lead as rapidly as full-grown bones; the lead remains in the bloodstream longer and has more opportunity to damage developing organs. Exposure to lead can stunt a child's intellectual, behavioral, and physical development. Studies have shown that infants exposed to lead have IQ scores that are 5% lower by age 7 than the scores of unexposed children; the children are six times more likely to have reading disabilities and seven times more likely to drop out of school.

Today, old lead paint causes most childhood lead poisoning. Although the use of lead-based paints was banned in 1978, children continue to be poisoned because many older buildings contain lead-based paint under layers of newer paint. If the painted surfaces are sanded to remove the paint, lead-containing dust contaminates the air in the building. In buildings where paint is peeling from the walls, lead poisoning is common. Even paint on the outside walls of a home can release dust that can be blown through the windows into its living space. Undernourished, often hungry, children frequently develop an appetite for strange things, a condition that is known as *pica*. Children with this syndrome often eat lead paint because it tastes sweet. Farmers were warned that if they painted their fences with lead paint, the cows would eat it off the wood.

Some older paints, from the early 1900s, contained as much as 25% lead by weight. Paints produced in the 1940s contained 50,000 ppm of lead. Thus, if a child consumed 100 mg of this paint (the weight of an aspirin tablet), he or she would ingest 5000 µg of lead, of which almost 4000 µg would be absorbed.

The **Lead-Based Paint Poisoning Prevention Act of 1971**, as amended by the Housing and Community Development Act of 1987, established 1.0 mg/cm² of lead in paint on a surface, as the federal threshold requiring abatement of lead-based paint in public and Native American housing developments nationwide. For paint samples, lead levels expressed in mg/cm² and lead levels expressed in percentage lead by weight were roughly equivalent. Thus, a level of 1.0 mg/cm² is equivalent to a lead concentration of 1% by weight. The **Residential Lead-Based Paint Hazard Reduction Act of 1992**, which is known as **Title X**, mandated the evaluation and reduction of lead-based paint hazards in the nation's existing housing. Title X also established 0.5% (0.5 mg/cm²) as the new federal threshold. Portable X-ray fluorescence instruments are used to measure the concentration of lead in paint that is intact and still on a surface (a wall, ceiling, or floor).

X-Ray Fluorescence

X-ray fluorescence (XRF) spectrometry is a spectrometric method that is based on the detection of X-ray radiation that is emitted from the sample being analyzed. XRF was discussed in Chapter 6 as a method for analyzing the metallic composition of airborne particulates. XRF is a two-step process that begins with a focused X-ray beam striking the sample. The incident X-ray strikes an inner shell electron in the sample atoms, which causes the electron to be ejected like a pool ball being struck by the cue ball. The lowest electron shell is the "K" shell. The second lowest is the "L" shell and the next the "M" shell. The vacancy caused by the loss of the emitted electron is filled by an outer-shell electron that drops to the lower level (see Figure 6.25). Because the outer-shell electron is at higher energy, when it drops to the lower level, it loses excess energy by releasing a photon of electromagnetic radiation. The fluorescent photon has an energy that is equal to the difference between the two electron energy levels. The photon energies are designated as K, L, or M X-rays, depending on the energy level filled; for example, a K shell vacancy filled by an L level electron results in the emission of a Kα X-ray (Figure 6.25). Because the difference in energy between the two electron levels is always the same, an element in a sample can be identified by measuring the energy of the emitted photon (or photons if there is more than one electron emitted). The intensity of the emitted photons is also directly proportional to the concentration of the element emitting the photon in the sample. The XRF instrument measures the photon energy to identify which element is present and the intensity of that photon to measure the amount of the element in the sample. Because this technique measures energy differences from inner shell electrons, it is insensitive to how the element (being measured) is bonded. The bonding shell electrons are not involved in the XRF process. XRF will not detect every element; the elemental range is limited to elements larger than beryllium, and the detection of low atomic number ($Z < 11$, Na) elements is difficult.

Figure 6.26 shows a schematic diagram of the XRF instrument. An X-ray tube produces X-rays that are directed at the surface of the sample. The incident X-rays cause the sample to release photons. The emitted photons released from the sample are observed at 90° to

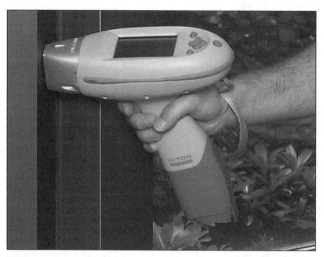

Figure 13.4 A portable XRF measuring a painted surface. Courtesy of NITONÇ LLC.

the incident X-ray beam. An energy-dispersive XRF detector collects all of the photons simultaneously. Each photon strikes a silicon wafer that has been treated with lithium and generates an electrical pulse that is proportional to the energy of the photon. The pulse height is proportional to the energy of the photon. The concentration of the element is determined by counting the number of pulses.

The XRF spectrometer is very sensitive and can determine transition metals at the ppm level. It is not as sensitive for lighter elements (Z < 19), as they have higher detection limits ranging from 10s to 100s of ppm. XRF is considered to be a bulk analysis technique; the X-ray source does not focus well into a narrow beam, and thus, it bombards the entire surface of the sample with X-rays. Figure 13.4 shows an XRF being used to measure the concentration of lead in paint on a wall.

Title X directs the U.S. EPA, the U.S. Department of Housing and Urban Development, and other agencies to develop programs that would ultimately reduce lead hazards in housing. Because children are more at risk when lead paint is friable (able to be inhaled or ingested), a measurement of the concentration of lead in dust is also needed. Title X and Title IV programs have resulted in increased dust sampling in residences across the country. One of the preferred methods for sampling residential dust for lead uses baby or hand wipes. The baby wipes are used to collect dust in the home in composite samples. The term *composite* is used when two or more physical samples are combined for laboratory analysis. In the case of wipe sampling, two or more wipes collected from common components (such as, floors or window sills) in a dwelling are combined in the field and then analyzed as a single sample. Flame AAS is not sensitive enough to measure lead in wipe samples. **Graphite furnace atomic absorption spectrometry (GFAAS)** is usually used to make this measurement.

Graphite Furnace Atomic Absorption Spectrometry

The flame atomizer is removed from the AA spectrometer and is replaced with an electrically headed graphite furnace. Light from the lead hollow cathode lamp produces a 283.3-nm absorption line that travels through a window at each end of the graphite furnace (Figure 13.5).

The wipe samples are carefully packaged, labeled, stored, and returned to the laboratory for analysis. The wipes are placed in a mixture of nitric acid and hydrogen peroxide that is then heated and diluted to a constant volume. The nitric acid/hydrogen peroxide solution digests (dissolves) the wipe and any dust on the wipe. The resulting solution is ready for analysis by GFAAS. Under normal circumstances, the wipe will not contain much lead, and flame atomic absorption spectrometry is not sensitive enough to determine the concentration of lead.

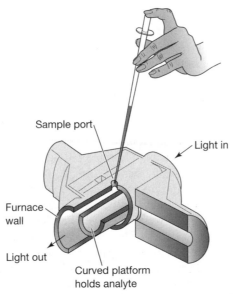

Sample port

Light in

Furnace
wall

Light out

Curved platform
holds analyte

Figure 13.5 The graphite furnace for AA. Adapted from D.C. Harris. *Quantitative Chemical Analysis, Sixth edition.* New York: W.H. Freeman, 2003.

A graphite furnace atomizer must be used in place of the flame to increase the sensitivity of the method.

When using flame AA, the residence time of the metal atoms in the optical path is only a fraction of a second as the flame gases rise. The graphite furnace, on the other hand, holds the atomized sample in the optical path for several seconds, therefore allowing higher sensitivity. Using GFAAS, as little as 5 µg of lead can be detected per wipe. A calibration curve for GFAAS is constructed by analyzing a minimum of three calibration standards and a blank. A calibration curve is prepared to relate absorbance to concentration. Calibration and quantitative lead measurement of sample digests should be performed in sequential order.

EPA Method 239.2, which is used to measure the concentration of dissolved lead in drinking water, uses GFAAS to make the measurement. The limit of detection for this method is 1 µg/L (1 ppb) and only a 20-µL sample is needed.

Traces of lead can be found in drinking water. In older buildings that have lead pipes, small amounts of lead are dissolved from the pipes as the water passes through them, particularly if the water is acidic. Lead also may be dissolved from lead solder on copper pipes. As a precaution, hot water from the faucet should not be used for cooking, and cold water should be allowed to run for several minutes first thing in the morning. The longer that stagnant water sits in the pipes that contain lead, the greater its concentration. Children should be taught to turn on the tap and count to 20 before drawing a glass of water.

In the past, plumbing was a source of unacceptably high lead levels in drinking water. Today, although plastic piping has replaced metal piping in many buildings and is used in new construction, some older buildings still receive water through lead pipes or copper pipes sealed with lead solder. The current limit for lead in drinking water is 15 ppb (see Table 13.1).

In 2004, a scandal broke in Washington, DC, when the Washington Area Sanitary Authority announced that of 6000 DC houses that were tested, 2287 had water-lead levels exceeding 50 ppb. An additional 157 homes were measured at 300 ppb and one house had an astonishing concentration of 48,000 ppb. Typically, elevated lead levels in houses that are only slightly higher than the EPA limit stem from problems inside the house, not from the water company. The culprit used to be lead-based solder used in older houses to connect pipe fittings. But that solder (called *fifty-fifty* due to its composition being half lead and half tin), was banned in 1987 and replaced with a lead-free type called tin-antimony solder. With lead levels as high as have been found in Washington, DC, suspicion focused on service lines, the pipes that bring the water from the water main in the street to the house. DC government officials have estimated that of the city's 107,000 service lines, 23,000 are made of a lead alloy, and that replacement of these lines by the city would take 15 years.

In 2000, the Washington Area Sanitary Authority changed the disinfectant that was used to treat drinking water from chlorine to chloramines. The chemistry of water disinfection was described in Chapter 8. Immediate studies focused on whether the chloroamine made the water more corrosive and more likely to dissolve the inner layer of the service pipes. Addition of zinc orthophosphate to passivate the inner surface of the pipes has been proposed as an interim solution until the lead service lines can be removed. Washington, DC may not be the only city with this problem. Chicago and other major cities are known to have more lead pipes in their municipal distribution systems than does Washington, DC.

Mercury

Mercury is the only common metal that is a liquid at room temperature. It is a component of many rocks and is released continually, but very slowly, into natural waters by normal chemical weathering processes. Small quantities of mercury also vaporize into the atmosphere from mercury-containing rocks in the Earth's crust. Concentrations of mercury in water from natural sources generally amount to a few ppb.

Mercury and its compounds are used in the production of chemicals, paints, plastics, pharmaceuticals, steel, and electrical equipment. They are used also as algicides in paper manufacture and as fungicides to treat seeds. Mercury and its compounds can thus reach natural bodies of water from manufacturing or farming sources as well as from sites where mercury-rich zinc and silver ores are mined. The major source of mercury emissions into the atmosphere is the burning of coal, which always contains some mercury, by electric power plants. Another source is waste incineration. In both cases, mercury-containing fly ash is released from smokestacks and can enter bodies of water as it falls back to earth in rainfall.

In bodies of water, mercury and mercury compounds tend to settle at the bottom and adhere to sediments. Elemental mercury and its inorganic compounds are not readily absorbed by aquatic creatures; however, anaerobic bacteria in bottom sediments can convert these forms of mercury to the organic methylmercury ion (CH_3Hg^+), which is very soluble and readily absorbed. Unless swallowed in a considerable quantity, mercury and most of its inorganic salts cause little harm because they are rapidly excreted from the body. Methylmercury, however, remains in the body for months and can damage the nervous system, kidneys, and liver and can cause birth defects. Methylmercury bioaccumulates as it moves up the food chain; for this reason, it can be dangerous to eat large quantities of large ocean fish such as tuna and swordfish.

Inhalation of mercury vapor or absorption of mercury salts through the skin causes serious neurologic problems. In the 19th century, workers making felt hats from animal pelts suffered brain damage as a result of absorbing mercury while they washed skins in solutions of mercury (II) nitrate. This is the origin of the expression "mad as a hatter" and the explanation for the name and behavior of the Mad Hatter in Lewis Carroll's *Alice in Wonderland.*

A case of environmental poisoning with mercury occurred in the late 1950s in Japan when a chemical plant discharged mercury-containing wastes into Minimata Bay. Aquatic organisms became contaminated with mercury, and a toxic organic mercury compound was formed and passed up the food chain to small fish and then to larger fish. Through **biomagnification**, the mercury became more concentrated at each level, and more than 100 people who ate fish taken from the bay were poisoned. Forty-four people died, and many children were born with birth defects.

A worse tragedy occurred in Iraq in 1972. Over 6000 people became sick, and nearly 500 died after eating bread made from wheat that had been treated with a methylmercury fungicide. The imported grain was intended for planting, but the warning label was written in a language that the Iraqis did not understand.

In the United States, the main concern is the mercury content of fish, which is so high in fish caught in the Great Lakes that people are advised not to eat them. Because the conversion of mercury to methylmercury is accelerated in acidic water, the acidification of lakes can increase concentrations of this toxic form of mercury in fish.

Cold-Vapor Atomic Absorption Determination of Mercury

Because elemental mercury is a volatile metal, it does not need to be heated in a flame or furnace during spectroscopic analysis. Mercury can be analyzed by a "**cold-vapor**" **atomic absorption** technique. A diagram of the apparatus that can be used to determine the concentration of mercury is shown in Figure 13.6. Ionic mercury in an aqueous sample is reduced by reaction with Sn^{2+} to elemental mercury, which is then transferred to a gas-tight cell in a pretreatment step.

$$Hg^{2+} + Sn^{2+} \rightarrow Hg + Sn^{4+}$$

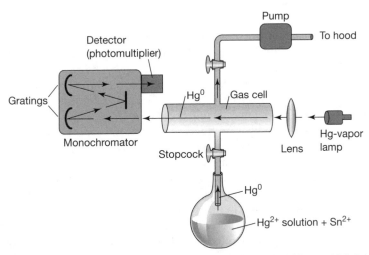

Figure 13.6 "Cold-vapor" atomic absorption spectrometer. Adapted from K.A. Rubison and J.F. Rubison. *Contemporary Instrumental Analysis.* Upper Saddle River, NJ: Prentice Hall, 2000.

The amount of light reaching the detector is proportional to the concentration of the mercury in the sample over a limited concentration range. A calibration curve is constructed plotting absorbance versus concentration. The concentration of unknown samples is interpolated from the calibration curve.

EPA Method 245.1, which is used to determine the concentration of mercury in water, calls for the use of the "cold-vapor" AAS technique. The limit of detection of this method is 0.5 µg/L (ppb). Because of its low boiling point (356°C), mercury is the only metallic element that is volatile enough to be atomized at room temperature.

Cadmium

Cadmium is used in electroplating and in the manufacture of paints, plastics, nickel–cadmium batteries, and control rods used in the nuclear power industry. These products are potential sources of cadmium that contaminate water, but the main cause for concern is cadmium in zinc products. Most zinc ores contain small amounts of cadmium; as a result, some cadmium is usually present as an impurity in zinc metal and zinc compounds produced from the ores. Cadmium is chemically very similar to zinc (both are in group 12 of the periodic table), and it is thought that the mechanism of cadmium poisoning may involve the substitution of cadmium for zinc in certain enzymes. When cadmium enters the body, it accumulates in the liver. Cadmium is a carcinogen; long-term effects include increased blood pressure and liver, kidney, and lung disease.

Zinc and its compounds have many commercial uses. Large quantities of zinc are used to galvanize iron, a process in which iron is coated with zinc to protect it from corrosion. Because zinc is an essential element for plants, it is included in most inorganic fertilizers. Leafy vegetables and tobacco leaves absorb zinc—along with any cadmium associated with it—from water in the soil. Smokers run a much greater risk of cadmium poisoning than do nonsmokers.

In the 1950s, numerous cases of cadmium poisoning occurred when effluents from a zinc mine were discharged into the Zintsu River in northern Japan. Water contaminated with the discharge was used to irrigate rice fields. People who ate the rice developed an extremely painful skeletal disorder known as itai-itai byo (or "ouch-ouch" disease). Cd^{2+} can replace Ca^{2+} in the bone. Their bones became brittle from loss of calcium and broke very easily. Many also suffered from abdominal pain, diarrhea, vomiting, liver damage, and kidney failure.

Inductively Coupled Plasma Optical Emission Spectroscopy

Cadmium and zinc can be individually analyzed in water samples by AAS. EPA Method 213.2 calls for GFAAS for the analysis of cadmium in water, and EPA Method 289.2 uses it for dissolved zinc in water. The **inductively coupled plasma optical emission spectroscopy (ICP-OES)** technique, however, is a technique that will simultaneously measure with detection limits that are a thousand times lower than flame AAS.

A plasma is a mixture of gases that conducts electricity. It is conductive because it contains significant concentrations of cations and electrons. The ICP technique uses an argon plasma; argon gas is injected, and argon ions and electrons conduct the electricity. Figure 13.7 is a schematic of a typical inductively coupled plasma torch. Argon ions, once formed in the plasma, are capable of absorbing sufficient power from the induction coil that is sending electromagnetic radiation into the torch in the form of a 2-kW radiofrequency signal at 27 MHz. The argon ions are accelerated by the oscillating radiofrequency and form a closed annular "torch" that reached temperatures as high as 10,000°K and has the shape of a donut.

Samples are carried into the torch by argon flowing through the central quartz tube. Once in the torch, the solvent is stripped from the metal ions. Electrons are captured by the ions, and the extreme heat of the torch thermally excites the metal atom into an excited state. As the excited metal atom leaves the torch, it relaxes to ground state and releases a photon of light, a process that is called optical emission. The cadmium ion in solution undergoes the following processes in the argon plasma:

$$Cd^+ + e^- \rightarrow Cd$$

$$Cd + heat \rightarrow Cd^*$$

$$Cd^* \rightarrow Cd + h\nu$$

Because the major gas in the plasma is the inert gas argon, the Cd^* has a long $t_{1/2}$ because there are not other gases present to react with it. If this analysis was attempted in an atomic absorption flame, there would be reactive oxygen radicals and ions that would react with the Cd and Cd^* to reduce the emission signal (and raise the limit of detection).

The most commonly used device for sample injection is a **nebulizer** (Figure 13.8). In the nebulizer, the sample solution is sprayed into a stream of argon, which produces small droplets that are carried by the argon stream into the plasma.

A typical argon plasma has a very intense, bright white appearance. If the emission signal is measured 10 to 30 mm above the core of the plasma, the plasma is optically transparent. This means that there is less interference that might distort the signal and give an erroneous reading.

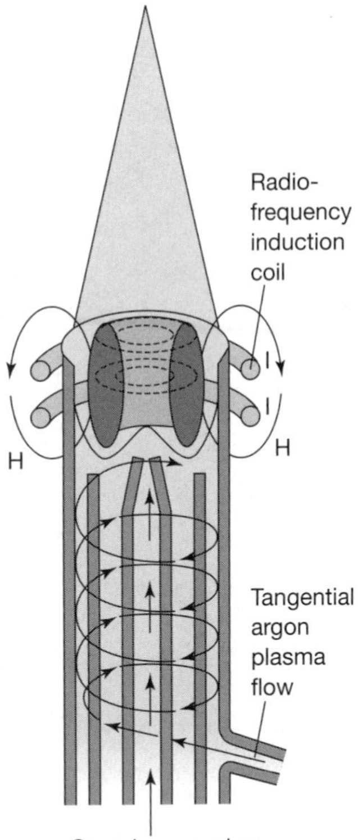

One type of ICP-OES spectrometer is a simultaneous multichannel instrument (Figure 13.9). In this type of instrument, as many as 60 photo detectors are located along the curved focal plane of a concave grating monochromator. The circumference of the spectrophotometer is called a *Rowland Circle;* the curvature corresponds to the focal curve of the concave grating. The grating monochromator separates the emission of light from each element present in the sample and disperses each individual wavelength (λ) to the appropriate photo detector. The emission from each metal ion in solution can be determined simultaneously. Each photo detector has to be placed in exactly the precise location on the Rowland Circle to measure the specific emission from the analyte of interest. These simultaneous instruments are expensive (more than $200,000). As each is made by the manufacturer, the exact metals to be determined must be known in advance because the photo detectors must be placed exactly for the element of interest.

Emission measurements tend to be more sensitive than absorbance measurements. When an absorbance measurement is made, the difference in intensity of the light from the source is being measured. Because the light source is bright, it is difficult to measure by how much its intensity was reduced. Emission measurements, on the other hand, do not have to contend with the light source. In the case of ICP-OES, if there is no Cd present in the sample, there will be no emission at the wavelength (λ) for Cd (226.5 nm). It is much easier to measure a small emission against

Figure 13.7 ICP torch. Adapted from V.A. Vassell, *Science* 202 (1978): 185.

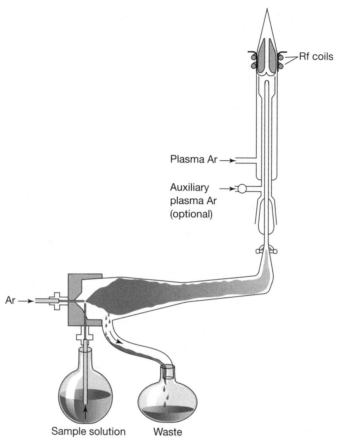

Figure 13.8 A typical nebulizer for sample injection into a plasma source. Adapted from D.A. Skoog, F.J. Holler, and T.A. Nieman. *Instrumental Analysis, Fifth edition*. Philadelphia: Saunders College, Harcourt Brace College, 1998.

a blank background than it is to measure the difference in intensity in absorbance. The ICP-OES methods have detection limits in the ppb range for many metals.

The EPA has developed a method that uses ICP-OES to screen water samples for dissolved metals. This is EPA Method 200.7, which will determine whether a water sample contains any (and at what concentration) of a list of more than 20 metal analytes. Simply by tuning the ICP-OES from the wavelength of detection for one metal to the next, each metal can be determined sequentially. Table 13.2 lists the ICP-OES emission wavelength and detection limits for the dissolved metals that can be determined by EPA Method 200.7.

Arsenic

Arsenic is a by-product in the manufacture of certain chemicals and of some mining operations. Many subsoils naturally contain arsenic compounds, and arsenic can seep into

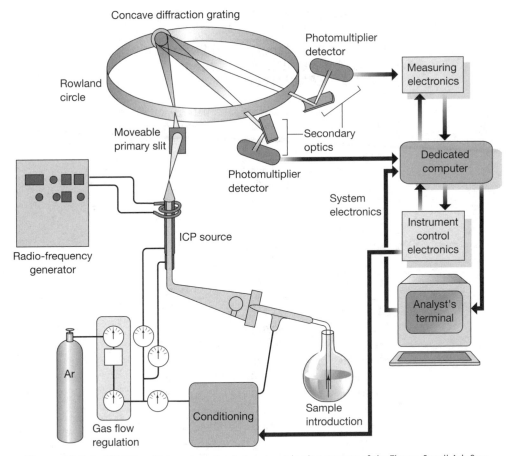

Figure 13.9 An ICP-OES multichannel spectrophotometer. Adaption courtesy of the Thermo Jarrell Ash Corp.

groundwater from this source. In parts of India and Bangladesh, where many wells for irrigation were dug during the last 25 years, there have been epidemics of acute arsenic poisoning because of severe contamination of groundwater with naturally occurring arsenic. Most cases of arsenic poisoning occur slowly over many years and are difficult to diagnose. Arsenic, a known carcinogen, greatly increases the risk of bladder cancer. As a result of a 1999 National Academy of Science study, the EPA lowered the limit to 10 ppb in 2001.

In the past, arsenic compounds were used extensively as pesticides. Because of their toxicity, inorganic arsenic salts containing arsenate ions (AsO_4^{3-}) or arsenite ions (AsO_3^{3-}) were used to destroy rodents, insects, and fungi. In recent years, there has been a decline in the use of these arsenic compounds in today's pesticides. The less toxic organic compound arsphenamine was the first drug used successfully to treat syphilis.

Table 13.2

Emission Wavelengths and Detection Limits for Metal Ions in Aqueous Solution, EPA Method 200.7

Element	Wavelength, nm	Detection limit, µg/L (ppb)
Aluminum	308.215	45
Arsenic	193.696	53
Antimony	206.833	32
Barium	455.403	2
Beryllium	313.042	0.3
Boron	249.773	5
Cadmium	226.502	4
Calcium	917.933	10
Chromium	267.716	7
Cobalt	228.616	7
Copper	324.754	6
Iron	259.940	7
Lead	220.353	42
Magnesium	279.079	30
Manganese	257.610	2
Molybdenum	202.030	8
Nickel	231.604	15
Selenium	196.026	75
Silica (SiO)	288.158	58
Silver	328.068	7
Sodium	588.995	29
Thallium	190.864	40
Vanadium	292.402	8
Zinc	213.856	2

Like heavy metals, arsenic compounds inactivate enzymes by reacting with sulfhydryl groups in enzyme systems.

arsenite ion   enzyme active site   inactivated enzyme

During World War I (WWI), the Allied Army was caught off guard when the German army first used "mustard gas." The word *mustard* refers to the yellow-green color of this gas, which was chlorine (Cl_2), the first chemical agent used in modern warfare. Chlorine is a corrosive poison that forms hydrochloric acid in the wet mucous membranes of the nose and lung. Extensive research by the U.S. War Department to find a "more effective" chemical warfare agent led to the production of an arsenic-containing gas, **Lewisite**, named for W. Lee Lewis, who discovered it.

Lewisite

The War Department of the United States established the American University Experimental Station (AES) on the American University campus in 1917 in response to the use of mustard gas in Europe. The AES was established by the government to develop new chemical warfare agents and an effective synthetic method for producing Lewisite. The threat of this new and more deadly mustard agent was partially responsible for Germany accepting the conditions of an Armistice in 1918.

WWI saw the development of trench warfare (Figure 13.10). Because airplanes had not been developed to drop bombs as yet, the weapon of choice was the mortar (Figure 13.11). Much of the work at AES involved the development of a mortar-based chemical agent delivery system. When WWI ended, AES was disbanded and the American University Campus reopened.

Figure 13.10 WWI Trench Warfare. Courtesy of American University.

Figure 13.11 WWI mortars that may contain mustard gas.

In the late 1990s, over 70 years later, a real estate developer digging a hole for a basement near the American University campus found a buried cache of mortar shells (Figure 13.12). Subsequent investigations showed that the U.S. War Department buried mortars and other munitions when they left the campus in 1918.

Because some of these shells contained Lewisite, the surrounding soil was contaminated with arsenic from leaking munitions. The American University and the surrounding neighborhood were declared a formerly used defense site. A search for buried munitions began in earnest in 1998.

Figure 13.12 WWI mortars removed from burial pit.

Figure 13.13 Geophysical investigation using ground piercing radar. Courtesy of U.S. Army Corps of Engineers.

In an attempt to locate buried WWI munitions, a geophysical investigation (Figure 13.13) using radar that can detect metallic objects below ground was begun. The geophysical investigations were correlated with 1917 era aerial maps to identify areas where munitions were used or stored and where they might still remain. Because the buried objects might be fused munitions, vapor containment structures (Figure 13.14) were constructed over the areas being excavated in case of an unintended detonation. To date, the Army Corps has removed over 70 motor shells that were buried. The Army Corp of Engineers has removed more than 600 dump truck loads of arsenic contaminated soil from the university's athletic fields. The university and its surrounding neighborhood, unwilling victims of the War Department's irresponsible disposal of these deadly weapons, have unknowingly lived with arsenic-contaminated soil for over 85 years.

AA Hydride Method for Arsenic

Although arsenic is a low-boiling (613°C) element, it is not volatile enough at room temperature to be analyzed by the cold-vapor technique that is used for mercury. It is, however, too volatile to be analyzed by flame or GFAAS. It vaporizes so quickly in a flame or plasma that the AA spectrometer gives a poor As detection limit. Arsenic is determined by a method in which the arsenic is converted to a hydride (AsH_3) that is volatile at room temperature (boiling point −62°C); the hydride gas is passed into a specialized AA spectrometer cell where it is decomposed into metal atoms that are then measured by AA spectrometry.

Table 13.3 shows that a number of metals form metal hydrides that are gases at ambient temperature. Volatile metal hydrides are generated by mixing a solution of sodium borohydride ($NaBH_4$) with a solution of the dissolved metal in acid. Arsenous acid and selenous acid, the As (III) and Se (IV) oxidation states of arsenic and selenium, respectively, are instantaneously converted by $NaHBH_4$ to a volatile hydride. The hydrides are continuously purged by argon into a heated quartz cell, which decomposes the metal hydride into gaseous metal

Figure 13.14 A vapor containment structure built over an excavation pit. Courtesy of U.S. Army Corps of Engineers.

atoms and hydrogen. This quartz cell sits in the optical path of an atomic absorption spectrometer (where the flame nebulizer is located in Figure 13.2). A calibration curve is constructed of absorbance versus concentration of the metal. The concentration of arsenic in a water sample is interpolated from the calibration curve.

Because the metal hydride boiling points are so low, the reaction with $NaBH_4$ also nicely separates the metal of interest from the matrix. EPA Method 206.3 uses the AAS hydride generation techniques to determine arsenic in water. It has a detection limit of 2 µg/L (ppb). EPA Method 270.3 is used to determine selenium in drinking water using the hydride technique.

Table 13.3			

Boiling Points of Selected Elements and Their Hydrides

Element	Boiling Point, C	Hydride Formula	Boiling Point of Hydride, C
Arsenic	613	AsH_3	−62
Antimony	1380	SbH_3	−18
Selenium	682	H_2Se	−41
Tellurium	990	H_2Te	−2

Additional Sources of Information

Christian GD. *Analytical Chemistry*, 6th ed. New York: John Wiley & Sons, 2003.

Clesceri LS, Greenberg AE, Eaton AD, editors. *Standard Methods for the Examination of Water and Wastewater*, 21st ed. Washington, DC: American Public Health Association, 2005.

Harris DC. *Quantitative Chemical Analysis*, 7th ed. New York: W.H. Freeman and Company, 2006.

Rubison KA, Rubison JF. *Contemporary Instrumental Analysis*. Upper Saddle River, NJ: Prentice Hall, 2000.

Skoog DA, Holler FJ, Crouch SR. *Instrumental Analysis*, 6th ed. Philadelphia, PA: Saunders College, Harcourt Brace College, 2007.

Keywords

arsenic
atomic absorption spectroscopy (AAS)
biomagnification
cadmium
cold-vapor atomic absorption determination of mercury
graphite furnace atomic absorption spectrometry (GFAAS)
hollow-cathode lamp
inductively coupled plasma optical emission spectroscopy (ICP-OES)

lead
Lead-based Paint Poisoning Prevention Act of 1971
Lewisite
mercury
nebulizer
plasma
Residential Lead-based Paint Hazard Reduction Act of 1992
Title X
X-ray fluorescence (XRF)

Questions and Problems

1. Why are the metals that are discussed in this chapter called *heavy metals*?
2. What is its steady-state accumulation of methylmercury in a person who eats 500 g of tuna daily that contains 0.2 ppm of methylmercury (the $t_{1/2}$ is 70 days)?
3. What is its steady-state accumulation of mercury in a person who eats 500 g of tuna daily that contains 0.2-ppm mercury?
4. What is the EPA maximum permissible level of the following metals in drinking water?
    a. Lead
    b. Arsenic
    c. Cadmium
    d. Mercury
5. Draw a schematic diagram that shows all of the components of an atomic absorption spectrometer (AAS).
    a. How is an aqueous sample introduced into the AAS?
    b. What happens to the water in the sample?
    c. The metal ion analyte has a positive charge. How does it become a neutral atom?
6. What is the most common fuel/oxidizer mixture used for flame AA? How hot is it?

7. If a sample gives the following result:
   a. 85 T %, what is the absorbance?
   b. 0.11 A, what is the transmittance?
   c. What is the maximum absorbance that the AAS will report?
8. What is the purpose of the premix chamber of the nebulizer burner that is used for AA analysis?
9. List the processes that take place from the time an aqueous sample of Pb^{2+} is aspirated into the premix chamber to the time the absorbance measurement is made in AAS.
10. Describe the source lamp that is used for AAS.
    a. From what element would the cathode be made?
    b. Why is this design used?
    c. Why not use a deuterium lamp that is usually used in an ultraviolet/visible (UV/VIS) spectrometer?
11. You are asked to analyze for mercury at the 0.5-ppm level in cans of tuna that have been imported. Outline a procedure, including sample preparation, AAS instrumental conditions, vaporization technique, and method of calibration.
12. To determine the sensitivity of an AAS method, a chemist determines the concentration of analyte that absorbs 1% of the light from the lamp. A water sample containing 1.5 µg/mL of lead gave an absorbance of 0.055. Estimate the sensitivity of this AAS method for lead.
13. Cadmium exhibits a sensitivity of 0.05 ppm on the AAS instrument in your laboratory. What would be the expected absorption for a sample containing 0.80 ppm of Cd? What chemists mean by sensitivity is explained in problem 12.
14. A series Cd standard solutions was prepared to determine the atomic absorbance of each. A separate set of mercury standard solutions was prepared and measured. The absorbance of each standard solution is given in this table.

Cd (mg/L)	A_{226}	Hg (mg/L)	A_{283}
1.00	0.086	1.00	0.142
2.00	0.177	2.00	0.292
3.00	0.259	3.00	0.438
4.00	0.350	4.00	0.579

   a. Is the AAS method more sensitive for Cd or Hg?
   b. Can the analysis for the two metals be done simultaneously by AAS?
   c. Will the absorbance from Cd in the sample interfere with the absorbance measurement for Hg?
   d. A waste water sample gives an A = 0.218 for Cd and an A = 0.269 for Hg. Calculate the concentration of Cd and Hg in the sample and express the results in ppm.
15. Lead has been widely used during the past.
    a. What is the major industrial use for lead today?
    b. List the physical properties that make it so useful.
    c. Does lead corrode?
16. Lead in drinking water is a public health issue.
    a. What is the current limit for lead in drinking water?

b. How does lead enter the municipal water system?

c. Does the acidity of the water effect the concentration of lead?

d. What is "first draw" water?

17. How much lead was in older paints form the 1940s?

a. Why are children at greater risk than adults for lead poisoning?

b. What physical effect does lead poisoning have on children?

c. What is the mechanism by which lead affects metabolism?

18. When was lead-based paint banned?

a. What was the new standard that was set for lead in paint?

b. If a liquid paint contains 2% lead, how much lead will be in the dried paint on the wall in mg/cm^3?

c. What is Title X?

19. What is an XRF used to measure?

a. Why is XRF used to measure lead paint on a wall or ceiling rater than an AAS?

b. Title X established a federal threshold for lead in paint on a wall. What is it?

c. If the dried paint on the wall has a lead concentration of 3.5 mg/cm^3, what percentage of lead was in the paint?

20. Draw a schematic diagram of the XRF spectrometer.

a. What is used to generate the incident beam?

b. Why are the released X-rays observed at 90° to the incident beam?

c. Describe the detector.

d. Can the XRF be used to determine whether one small spot on a wall has lead present?

21. Describe the process that produces an X-ray for XRF analysis.

a. Why does XRF not detect light elements?

b. Would $PbCO_3$ give a different XRF signal for lead than would PbO?

c. Can XRF determine elements at the ppb level?

22. Why are dust samples analyzed for lead content by GFAAS and not by XRF?

23. Why does the GFAAS increase the sensitivity of the AA?

a. What volume of sample is used for GFAAS?

b. You need to determine whether a drinking water sample exceeds the current limit. Can you use flame AAS or do you need GFAAS? Why?

c. You need to determine the amount of lead in a child's blood. Can you use flame AAS or do you need GFAAS? Why?

24. A drinking water sample was analyzed for lead. When 50 μL of the sample was injected into the GFAAS, it gave a reading of 0.080 absorbance. The drinking water sample was spiked with a solution containing a known amount of lead. The concentration of the spiking solution was 5.0 μg/L, and 25 μL was added to 25 μL of the drinking water. When the spiked sample was measured by GFAAS, it gave an absorbance of 0.120 absorbance. What was the concentration of lead in the drinking water. Did this sample exceed EPA limits?

25. EPA Method 239.2 is used to measure lead in water samples. Go the EPA Web site and find answers to the following questions:

a. What wavelength is used?

b. The calibration curve is made over what concentration range?

c. How big of a sample is used?

26. EPA Method 239.2 is used to measure lead in water samples. Go the EPA Web site and find answers to the following questions:
    a. What wavelength gives the greatest sensitivity?
    b. How can interference from sulfate be minimized?
    c. Can containers contaminate the sample?
27. Which is more toxic: elemental mercury and its inorganic salts or organic mercury compounds?
    a. How is methylmercury formed in the environment?
    b. Do inorganic mercury salts bioaccumulate?
    c. In an aquatic setting, where can mercury compounds be found?
28. What is the major source of mercury emissions into the atmosphere?
    a. Does the inhalation of mercury cause health problems?
    b. Does methylmercury bioaccumulate?
    c. List five industrial uses of mercury.
29. Describe the atomic absorption technique that is used to measure mercury.
30. Mercury is measured by the "cold-vapor" AAS technique.
    a. Why not use an acetylene/air flame to atomize mercury in AAS?
    b. How is the Hg^{2+} reduced to Hg before analysis?
    c. Is the cold-vapor technique used for the measurement of other metals?
31. EPA Method 245.1 is used to measure mercury in water samples. Go to the EPA Web site and find answers to the following questions:
    a. What wavelength is used?
    b. The calibration curve is made over what mass range?
    c. How big of a sample is used?
32. EPA Method 245.1 is used to measure mercury in water samples. Go the EPA Web site and find answers to the following questions:
    a. How is a possible interference from sulfide removed?
    b. Is special treatment needed for seawater or other saline water?
    c. Can certain organics interfere with the method? Explain.
33. What is the main source of cadmium pollution in the environment?
    a. List four industrial uses of cadmium.
    b. What is galvanized iron?
    c. Are smokers more at risk for cadmium poisoning?
34. List three advantages that ICP-OES has over flame AAS.
    a. Does ICP-OES use a hollow-cathode lamp?
    b. What is a Rowland Circle?
    c. Could you accurately measure mercury concentration in an aqueous sample with ICP-OES?
35. Draw a diagram of the argon torch used in ICP-OES.
    a. What temperature is reached in the argon torch?
    b. What is the shape of the argon plasma?
    c. Why is the optical measurement made 10 to 30 mm above the core of the plasma?
36. Write a series of chemical equations that show what happens to Cd^{2+} when it enters the plasma torch.
    a. Is the ICP-OES an absorbance or emission technique? Explain.

b. Why are emission techniques more sensitive than absorbance techniques?

c. Why is argon used as the plasma gas?

37. Cadmium can be determined by ICP-OES by observing the emission from the torch at 225.5 nm. A calibration curve was constructed by measuring the emission from solutions of known concentration. The standard cadmium solutions (in ppb) gave the following results:

[Cd]	Detector Response
20	12
45	26
90	52
120	71
150	88

The ground water sample was measured and gave a response of 58. What is the concentration of Cd in the ground water sample?

38. Atomic emission in ICP-OES is the result of excited state atoms losing energy. At $2000°K$, only 9.9×10^{-6} of all sodium ions in a sample are in the excited state. Why is ICP-OES such a sensitive technique if only such a small fraction of atoms are excited?

39. A series of Cd and Hg standard solutions was prepared to determine the ICP-OES response for each dissolved metal. The response of each standard solution is given in this table.

Cd (µg/mL)	E_{226}	Hg (µg/mL)	E_{283}
1.00	0.086	1.00	0.142
2.00	0.177	2.00	0.292
3.00	0.259	3.00	0.438
4.00	0.350	4.00	0.579

a. Is the ICP-OES AAS method more sensitive than AAS (see Problem 12)?

b. Can the analysis for the two metals be done simultaneously by ICP-OES?

c. Will the absorbance from Cd in the sample interfere with the absorbance measurement for Hg?

d. A waste water sample gives an $A = 0.218$ for Cd and an $A = 0.269$ for Hg. Calculate the concentration of Cd and Hg in the sample and express the results in ppb.

40. Explain how a sample is introduced into the ICP-OES.

a. How would a drinking water sample be prepared for analysis?

b. How would a soil sample be prepared for analysis of Cd?

c. What is a nebulizer?

41. EPA Method 200.7 is used to measure dissolved metals in water samples. Go to the EPA Web site and find answers to the following questions:

a. How are the metals measured sequentially?

b. How are the calibration curves made for more than one metal?

c. How big of a sample is used?

42. EPA Method 200.7 is used to measure dissolved metals in water samples. Go to the EPA Web site and find answers to the following questions:

a. What is a calibration blank?

b. What is a reagent blank?

c. What is an interference check sample?

43. What is the EPA limit for arsenic in drinking water?
   a. Were arsenic compounds ever consumer products? For what use?
   b. Name an antidote for arsenic poisoning. How does it work?
   c. What is the mechanism by which arsenic poisons a person?
44. Can arsenic be determined by cold-vapor AAS?
   a. What is the boiling point of As? Of Hg?
   b. Can As be determined by flame AAS?
   c. Can As be determined by ICP-AES?
45. Write a chemical equation that shows how arsenic hydride is formed before analysis by AAS.
46. Describe the hydride-AAS technique that is used to measure arsenic in environmental samples.
47. EPA Method 206.3 is used to measure arsenic in water samples. Go to the EPA Web site and find answers to the following questions:
   a. What wavelength is used?
   b. The calibration curve is made over what concentration range?
   c. How big of a sample is used?
48. List four elements that can be measured by hydride-AAS.
   a. List the boiling point of each.
   b. Can they be determined by flame AAS?
   c. Can they be determined by ICP-OES?
49. How is the arsenic in an environmental sample measured by absorbance in the AAS if it is a hydride? Does the AAS method not measure atomic As?
50. An area of environmental interest may be contaminated with lead and/or arsenic. Devise an AAS method for the analysis of these two metals.

A mining quarry. ©AbleStock.

CHAPTER

14

Organic Chemicals
in the Environment

ENDOCRINE DISRUPTORS ARE SYNTHETIC CHEMICALS THAT BLOCK, MIMIC, OR OTHERWISE INTERFERE WITH NATURALLY PRODUCED HORMONES, THE BODY'S CHEMICAL MESSENGERS THAT CONTROL HOW AN ORGANISM DEVELOPS AND FUNCTIONS. Wildlife and humans are exposed daily to these pervasive chemicals that have already caused numerous adverse effects in wildlife and are most likely affecting humans as well.

Hormones play a crucial role in the proper development of the growing fetus, which is vulnerable to even the most trace concentrations of introduced substances. Substances that have no effect in an adult can become poisonous in the developing embryo. Chemicals are passed from mother to offspring via the womb and breast milk in mammals and via the egg in reptiles, amphibians, fish, and chickens, leading to transgenerational effects.

The World Wildlife Fund reports that the effects of endocrine disruptors on animals are varied and range from alligators born with abnormally small penises and birds with crossed beaks to the sudden disappearance of entire populations. Wildlife researchers over the last few years have unearthed a variety of endocrine disruptor-related effects: interrupted sexual development, thyroid system disorders, inability to breed, reduced immune response, and abnormal mating and parenting behavior. Species such as terns, gulls, harbor seals, bald eagles, beluga whales, lake trout, panthers, alligators, turtles, and others have suffered more than one of these effects.

Some endocrine-disrupting chemicals are persistent in the environment and bioaccumulate; they accumulate in the fatty tissue of organisms and increase in concentration as they move through the food web. Because of their persistence and mobility, they accumulate in and harm species far from their original source.

Thousands of different organic chemicals are synthesized each year for use as insecticides, herbicides, detergents, and insulating materials, and for many other purposes. Some of them are not adequately tested for **toxicity** before being put on the market. Many of these chemicals persist in the environment for long periods of time, and if they enter waterways, they can cause serious health and environmental problems. There is growing concern that these **persistent organic pollutants (POPs)** may be acting as hormone disrupters. POPs also are suspected of causing neurologic disorders, suppressing the immune system, and increasing the risk of cancer.

Because they can be transported by wind and water, most POPs generated in one country can affect people and wildlife far from where they are used. They are very stable molecules and persist for long periods in the environment and can accumulate and pass from one species to the next through the food chain. To address this concern, a conference was held in Stockholm, Sweden, in May 2001. A treaty, known as the Stockholm Convention, was signed by more than 90 countries that promised to reduce or eliminate the production, use, and release of 12 key POPs, which are known as "the dirty dozen."

The major impetus for the Stockholm Convention was the finding of POP contamination in Arctic regions, which are thousands of miles from any place where the POPs were manufactured or used. Some POPs evaporate from water or land surfaces into the air. Then POPs return to Earth in snow, rain, or mist. Tracing the movement of POPs in the environment is complex because they can attach to particulate matter. Satellites passing from Asia across the Pacific Ocean to North America have monitored clouds of moving dust.

Most of the 12 key POPs (listed in Table 14.1) are no longer produced in the United States. Although most developed nations have taken strong action to control the dirty dozen, a great number of developing nations have only

Table 14.1

The 12 Key POPs—the Dirty Dozen

POP	Use
Aldrin	crop insecticide (corn, cotton)
Chlordane	crop insecticide (vegetables, citrus, cotton, potatoes)
DDT	crop insecticide (cotton)
Dieldrin	crop insecticide (cotton, corn)
Endrin	crop insecticide (cotton, grains)
Heptachlor	insecticide (termites and soil insects)
Hexachlorobenzene	fungicide for seed treatment
Mirex	insecticide (termites, fire ants)
Toxaphene	insecticide (livestock and crops)
PCBs	industrial chemical (heat exchange fluid for electrical transformers, paint and plastic additive)
Dioxins	unintentionally produced during combustion
Furans	unintentionally produced during combustion

www.epa.gov/international/toxics/pop.htm.

recently begun to restrict their production, use, and release. Of the 12 chemicals, 10 were intentionally produced by industry, and 9 were produced as insecticides or fungicides. Only 2 of the 12 chemicals, dioxins and **furans**, are unintentionally produced in combustion processes.

The chemical formulas of **dichlorodiphenyltrichloroethane (DDT)**, 2,3,7, 8-tetrachlorodibenzo-*p*-dioxin (TCDD), and related dioxins and representative **polychlorinated biphenyls (PCBs)** are shown in Figure 14.1.

■ Polychlorinated Hydrocarbons

All of the "dirty dozen" are certain organochlorine compounds, called **polychlorinated hydrocarbons.** These compounds pose a threat to aquatic life because they are stable and do not readily break into simpler, less toxic forms. They persist in the environment for long periods of time and, like toxic metals, bioaccumulate through the food chains. Polychlorinated hydrocarbons are insoluble in water but soluble in fats. They become concentrated in the fatty tissues of fish and of birds and humans who eat the fish. We consider three polychlorinated hydrocarbons that are of particular concern as environmental contaminants: DDT, dioxin, and PCBs.

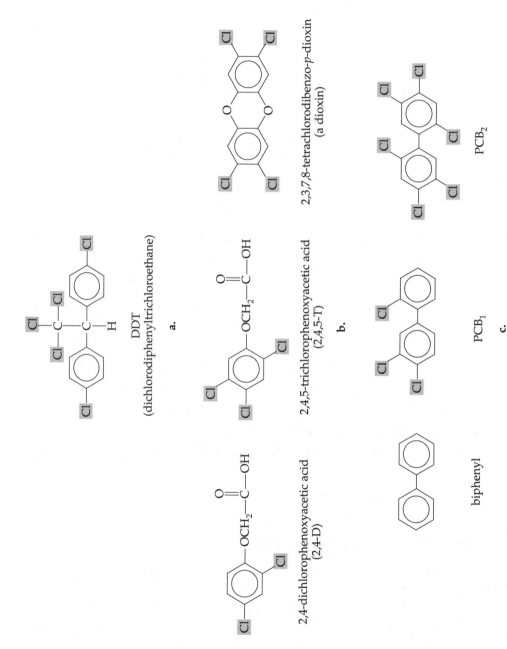

Figure 14.1 Chemical structures of (a) DDT; (b) TCDD and other dioxins; (c) biphenyl and two PCSs that can be derived from it.

DDT

Ever since humans first appeared on the Earth, they have had to contend with pests. The biblical account of the liberation of the Hebrews from bondage in Egypt in the 13th century BC includes descriptions of plagues of lice, flies, and locusts in the Nile Valley (Exodus 7:20–11:10). The Black Death, which killed millions of people in the Middle Ages, was transmitted to humans by fleas on rats carrying bubonic plague. The failure of the potato crop in Ireland in 1845, which caused widespread starvation and mass emigration to the United States, was caused by the potato blight. Malaria, spread by the anopheles mosquito, still incapacitates or kills millions of people annually in the less developed countries, and swarms of locusts continue to devastate crops in many areas of the world. Plants on which we depend for food are under attack from insects, fungi, bacteria, viruses, and other microorganisms and must compete with weeds for nutrients. Mice, rats, rabbits, and other animals also take a share of the crop.

Before 1940, only a few pesticides were available. Among them were several naturally occurring insect poisons extracted from plants, including **pyrethrins**, compounds obtained from the pyrethrum flower (a member of the chrysanthemum family). Other early insecticides were nicotine sulfate obtained from tobacco, rotenone from the tropical derris plant, and garlic oil. A number of inorganic chemicals, primarily compounds of arsenic and lead, were used also. They were less desirable and are rarely used today because they persist in the environment and are toxic to humans and other animals besides insects.

The massive use of pesticides began at the end of World War II with the introduction of DDT and escalated as other synthetic organic pesticides were developed. Today, in the United States, approximately 500,000 metric tons of pesticides (insecticides and herbicides) are applied annually to control pests.

When DDT was introduced in the late 1940s, it appeared to be an ideal insecticide. It was cheap to produce and apparently nontoxic to humans and other mammals. Because it did not break down easily, it continued to kill insects for a long period after application. Later, however, it was discovered that water contaminated with DDT from aerial spraying and runoff from treated land was having a devastating effect on fish-eating birds. By the 1950s and 1960s, populations of bald eagles, peregrine falcons, and brown pelicans had shrunk to disturbingly low levels. Also, insects became increasingly resistant to the pesticide, and increasingly larger amounts had to be applied to achieve the desired results.

The first person to draw public attention to the dangers of pesticides was biologist **Rachel Carson.** In her classic best seller, *Silent Spring,* published in 1962, she warned that indiscriminate reliance on synthetic chemicals to control insects could lead to a spring without songbirds and, eventually, to the disruption of all life. She was vigorously attacked by the chemical and agricultural industries but was supported by other scientists and the general public. Rachel Carson is credited with launching the environmental movement of the 1970s, which led directly to the creation of the **Environmental Protection Agency (EPA).**

Research has shown that DDT interferes with calcium metabolism in birds, and as a result, eggshells (which are composed primarily of calcium compounds) become thin and break when parent birds attempt to incubate the eggs. In 1973, the use of DDT was banned in the United States, and since that time, bald eagles and other fish-eating birds have made a dramatic recovery.

Dioxin

Dioxin is a family of chlorinated hydrocarbons that has come to be associated with one particular compound, 2,3,7,8-tetrachlorodibenzo-*p*-dioxin, usually abbreviated as TCDD. Like DDT, TCDD persists in the environment for a very long time and bioaccumulates up the food chain.

TCDD is not produced intentionally. Its major source is the burning of chlorine-containing medical and municipal wastes. With the introduction of efficient high-temperature waste incinerators, which convert dioxins to harmless products, emissions from this source have decreased substantially. In the past, TCDD also reached waterways in the effluent from paper mills, which used chlorine to bleach paper pulp. Now most mills have switched to a bleaching agent that does not produce TCDD. TCDD also is formed as a by-product of the manufacture of trichlorophenol, a chemical that is used in the manufacture of a variety of herbicides, including 2,4-D (Silvex). Silvex is widely used on agricultural and suburban lawns to control broad-leaved weeds.

During the Vietnam War, 2,4-D was mixed in equal quantities with the related herbicide, 2,4,5-T, to make the defoliant **Agent Orange**. Present in the latter as an impurity was TCDD. Because TCDD is extraordinarily toxic to guinea pigs and is a known carcinogen for many animals, there was concern that humans might be affected also. As a result of these concerns, the EPA banned the use of 2,4,5-T in 1985.

Agent Orange's effects on humans have been debated for many years. In 1993, a comprehensive study linked Agent Orange to three types of cancer and two skin conditions but failed to support Vietnam veterans' claims that the chemical caused birth defects, infertility, and other disorders. However, more recently, the EPA concluded that exposure to dioxin, in addition to increasing the risk of cancer, may also disrupt some reproductive mechanisms and suppress the immune system.

PCBs

PCBs are structurally similar to DDT and, like DDT and dioxin, are fat soluble and bioaccumulate at the upper levels of food webs. PCBs present in water in a negligible concentration can become over a million times more concentrated in fish. In 1977, production of PCBs in the United States was halted, and disposal of PCB-containing products is now strictly regulated. As a result, there has been a dramatic decrease in PCB contamination of fish; however, because PCBs are stable, they persist in the environment for many years, and fish in many lakes still contain significant levels of contaminants. PCBs continue to be produced and used in Russia and many developing countries.

PCBs are fire resistant and stable at high temperature and have a high electrical resistance. They were widely used as insulating materials in transformers, electrical capacitors, and condensers, and are still present in older equipment. PCBs also were used extensively as platicizers in the plastics industry. Plastics tend to be brittle; the addition of PCBs makes them more flexible and resistant to cracking.

PCBs were spread widely in the environment and entered surface waters in industrial discharges and in particulates from incinerators. When plastic wastes and other PCB-containing materials are burned, PCB vapors condense on airborne particles that then fall directly onto water or reach water in runoff from the land. PCBs are now found in the body fat of animals living in the furthest corners of the Earth, including polar bears in the Arctic and albatrosses on remote Pacific Islands.

PCBs cause eggshell thinning and neurologic damage in birds and impair reproduction of aquatic species. In humans, they cause chloracne (a serious form of acne) and liver damage, and most important, they can be transferred from mother to fetus through the placenta and from mother to infant through breast milk. PCBs cause stillbirths and retard growth, and—like other POPs—have been linked to reproductive disorders, birth defects, and cancer.

In response to growing concerns about POPs, an international conference to plan ways to phase out all production of POPs was held in Montreal in 1998. A major problem was the need to find a cheap pesticide to replace DDT that is so valuable for controlling insect-borne diseases in developing countries. There is growing concern that many of the POPs, which are now widespread in the environment, may be interfering with the endocrine system in both animals and humans and disrupting reproduction and fetal development. These pollutants include PCBs, dioxins, DDT, methoxychlor and other pesticides, and phthalates, which are components in many plastics. They have been shown to cause infertility, reduced sperm counts, cancer, and neurologic disorders.

Observations on wildlife exposed to POPs and studies on laboratory animals have shown that even at extremely low levels many of these chemicals mimic natural estrogens and cause feminizing in males and accentuation of female characteristics in females. For example, male fish exposed to estrogen mimic produce ovaries and eggs instead of sperm. There is also growing evidence that POPs are responsible for some of the increasing numbers of abnormalities found in wildlife, such as alligators with underdeveloped reproductive organs.

Because hormones are common to many different species, it would be expected that chemicals that disrupt biochemical processes in animals also would disrupt them in humans. It is now thought that the dramatic drop in sperm counts found in large groups of men in the United States and Europe over the last 50 years (Figure 14.2) and the increase in the number of cases of undescended testicles may be due to these same chemicals.

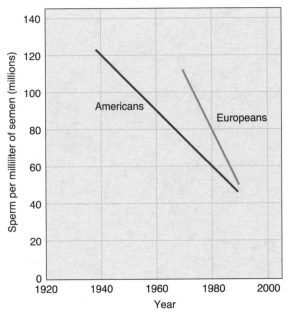

Figure 14.2 The average sperm count of American and European men dropped between 1938 and 1990. From *Vital Signs*, 1999, p. 148. Used by permission of World Watch Institute, Washington, DC, www.worldwatch.org.

Although the chemical structures of estrogen mimics are unlike those of natural estrogens (Figure 14.3), research has shown that they are able to bind to estrogen receptor sites. Once bound to these sites, they initiate the action of the natural hormone. Other hormone disruptors block receptors and depress normal hormone action. It is known that some combinations of hormone-disrupting chemicals act synergistically; that is, the effect of exposure to certain combinations of these chemicals far exceeds the sum of the effects of exposure to each chemical singly.

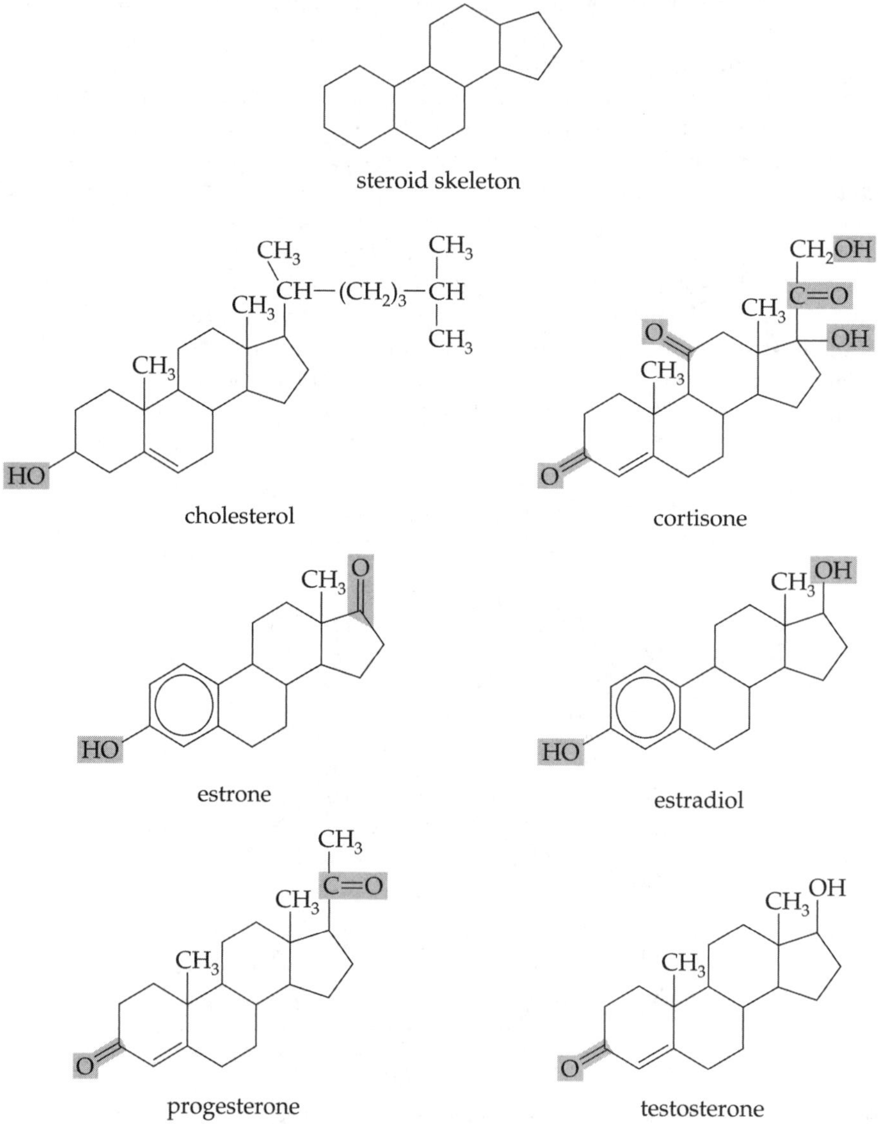

steroid skeleton

cholesterol

cortisone

estrone

estradiol

progesterone

testosterone

Figure 14.3 Steroids all share the same basic four-ring structure.

The widely used DDT substitute methoxychlor has been shown to block the male hormone system, affecting mating behavior, sperm counts, and the structure of the prostate. Three contaminants commonly found in Wisconsin drinking water (the pesticides aldicarb and atrazine and nitrate residues) appear to work interactively and unpredictably to produce endocrine, immune system, and behavioral changes in animals. The new findings show that some endocrine-disrupting effects, previously thought to be estrogenic, are in fact **antiandrogenic** effects (the ability to block the male hormone system), opening an entirely new area of research and suggesting new, previously unknown hazards from chemicals.

Approximately 80,000 synthetic chemicals are on the market today, and most of them have never been tested for toxicity. Some have been screened to determine whether they are carcinogens, but only a minute fraction has been tested for their effects on the endocrine system. Now, by law, the EPA is being required to test 15,000 chemicals for their possible role as endocrine disruptors.

■ Persistent, Bioaccumulative, and Toxic Pollutants

In 1998, the EPA launched the **Persistent, Bioaccumulative, and Toxic (PBT) Pollutants Program**, which has a wider scope than the **Stockholm Convention** and has the same goals of reducing the use and release of PBT pollutants while making sure that they are disposed of properly. The PBT program is also focusing on a list of 12 priority pollutants, which are listed in Table 14.2. Some of these pollutants, such as aldrin, dieldrin, PCB, and dioxins, are listed

Table 14.2	
EPAs Priority Level-1 PBTs	
PBT Compound	**Use**
Aldrin/Dieldrin	crop insecticide (corn, cotton)
Alkyl-lead	octane booster in leaded gasoline
Benzo(a)pyrene	unintentionally produced during combustion
Toxaphene	insecticide (livestock and crops)
Chlordane	crop insecticide (vegetables, citrus, cotton, potatoes)
DDT	crop insecticide (cotton)
Dioxins/Furans	unintentionally produced during combustion
Hexachlorobenzene	fungicide for seed treatment
Mercury and mercury compounds	incineration of medical and municipal waste
Mirex	insecticide (termites, fire ants)
Octachlorostyrene	produced from carbon electrodes used in electrolytic process for producing chlorine
PCBs	industrial chemical (heat exchange fluid for electrical transformers, paint and plastic additive)

www.epa.gov/pbt.

in the Stockholm Convention "dirty dozen," but the PBT list also contains inorganic elements, such as mercury or organometallic compounds that contain an inorganic atom.

Goals of the EPA's PBT program are as follows:

1. To prevent the introduction of new PBTs into the marketplace
2. To reduce the risk to human health and the environment from exposure to priority PBTs already in the environment
3. To halt the transfer of PBTs by air, water, and land
4. To assess PBT's long-term effect on the environment

To prevent the introduction of new PBTs, the EPA exercised its authority under the Toxic Substances Control Act (TSCA) to strengthen the process by which it screens new chemicals that are intended for industrial markets. It also established a similar policy for pesticides under the Federal Insecticide, Fungicide and Rodenticide Act (FIFRA). Under TSCA, chemical manufacturers are required by law to submit notifications (**premanufacture notifications [PMNs]**) to the EPA for new chemicals that they are planning to manufacture domestically or import on an industrial scale. The EPA uses a sophisticated computer program that evaluates a chemical's characteristics to ascertain whether it might pose an unreasonable risk to human health or the environment. Based on this evaluation, the EPA can stop the production of these chemicals until the manufacturer can prove that they will not pose an unreasonable risk if released into the environment.

In 2007, more than 56,000 chemicals were screened online using the PBT Profiler and after screening, the EPA received 1071 PMNs. The EPA will identify which PMNs might pose a PBT risk and after EPA issues its findings, the manufacturer may conduct further testing that might refute PBT concerns.

The PBT computer program that the EPA used has been modified and released to chemical manufacturers so that they can test their compounds for PBT properties. The program is known as the "PBT Profiler," and companies can access it at http://www.pbtprofiler.net. The PBT Profiler uses established screening models to estimate PBT characteristics based on chemical structure and physical and chemical properties. Later in this chapter we will learn how to use the Web-based PBT Profiler to determine the PBT characteristics of organic chemicals.

The query begins with a chemical's name, structure, or CAS number. **CAS registry numbers** are unique numerical identifiers for chemical compounds. They are also referred to as **CAS numbers (CAS RNs)**. The Chemical Abstract Service, a division of the American Chemical Society, assigns these identifiers to every chemical that has been described in the literature. Approximately 41 million compounds have received a CAS number so far, with approximately 4000 new ones being added each day. The intention is to make database searches more convenient, as chemicals often have many names. Almost all molecule databases today allow searching by CAS number. To find the CAS number of a compound given its name, formula, or structure, the following free resource can be used: the National Institute of Standards and Technology (NIST) Chemistry WebBook at http://webbook.nist.gov/chemistry/ or the National Cancer Institute (NCI) Database Browser at http://cactvs.cit.nih.gov/.

With this information, the PBT program estimates the chemical's melting point, boiling point, and vapor pressure at 25°C. A high melting point tends to indicate low water solubility, and a low melting point indicates that increased absorption is possible through the skin, gastrointestinal tract, or lungs. Chemicals with vapor pressure of greater than 10^{-4} mm Hg exist mostly in the vapor phase and often have higher potential inhalation exposures than chemicals with low vapor pressure. The program also uses the octanol/water partition coefficient as an estimation tool; chemicals that are fat soluble (low solubility in water) are likely to bioaccumulate.

■ Octanol/Water Partition Coefficient

In the study of the environmental fate of organic compounds, the **octanol/water partition coefficient** (K_{ow}) has become a key parameter. It has been shown to correlate with water solubility, soil/sediment **sorption coefficient**, and bioconcentration. Of these three properties that can be estimated from the K_{ow}, water solubility is the most important because it affects both the fate and transport of chemicals. Highly soluble compounds are quickly distributed by the hydrologic cycle, have low sorption coefficients for soils and sediments, and tend to be more easily degraded by microorganisms. In addition, chemical reactions, such as hydrolysis and photolysis, tend to occur more readily if a compound is soluble.

The solvent 1-octanol is an amphiphilic solvent and has both hydrophobic and hydrophilic natures. Figure 14.4 shows that the amphiphilic nature of octanol gives it a solvating capability that is similar to that of humic acid or other naturally occurring colloids that have the ability to associate with both polar and nonpolar compounds. The K_{ow} provides a convenient experimental test that will predict whether an organic compound will dissolve in water or be absorbed by organic material contacting the natural water.

As organic molecules increase in size, they decrease in polarity and water solubility; they are becoming more hydrophobic, which can be measured by the octanol/water partition coefficient. K_{ow} is defined as the ratio of the molar concentrations of a chemical in n-octanol and water, in dilute solution.

$$K_{ow} = C_{octanol}/C_{water}$$

The K_{ow} is determined by measuring the concentration of a particular compound in the water and the octanol phases after a period of mixing. The K_{ow} is constant for a given compound at a certain temperature. Because the K_{ow} is a ratio of two molar concentrations, it has no units. K_{ow} values can be in the millions for important environmental contaminants (PCBs, chlorinated pesticides, dioxins, and furans); they are usually reported as the base 10 logarithm, log K_{ow}. The K_{ow} expresses a tendency of an organic contaminant to move from the water phase to the immiscible 1-octanol phase. Octanol, a long-chain alcohol, mimics fat (lipids), and the K_{ow} is an indication of the tendency of an organic contaminate to bioaccumulate.

The K_{ow} has two characteristics that make it especially useful in environmental assessments. First, it varies in predictable ways within classes of organic compounds. For example, Figures 14.5 and 14.6 show that if K_{ow} is known for one member of a class of compounds, it can be used to estimate a value for other members of the same family. In these figures, the K_{ow} can be correlated to the number of chlorine atoms in a compound or the number of aromatic rings in polynuclear aromatic compounds (polyaromatic hydrocarbons [PAHs]). The log K_{ow} of some nonpolar organic molecules is listed in Table 14.3.

The second characteristic is that the K_{ow} can be correlated with the sorption of organic compounds on soil. By using the K_{ow}, it is possible to estimate the potential sorption of organic contaminants based on the structure of the compounds and the organic carbon content of the soil.

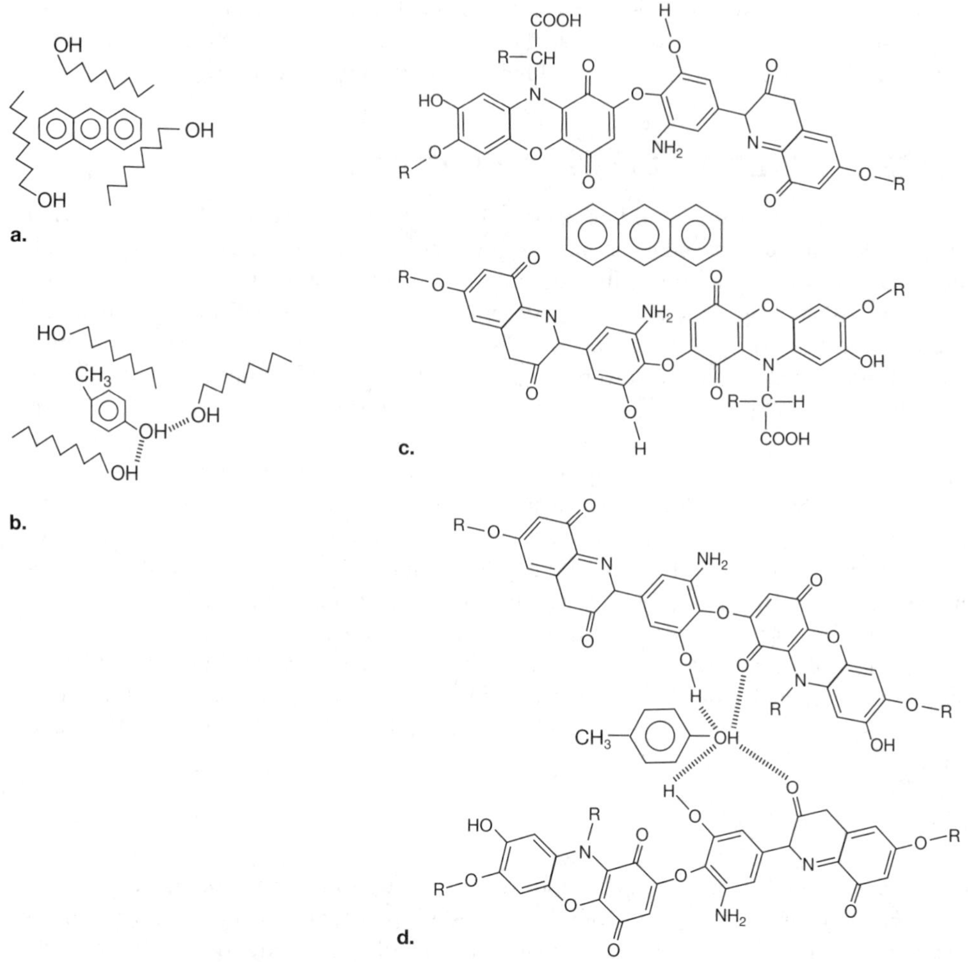

Figure 14.4 The solvent *n*-octanol is used to mimic the behavior of humic substances. Octanol solubilizes nonpolar (a) (naphthalene) and polar (b) (phenol) solutes in different ways. Humic materials (c and d) interact with the same solutes. Hydrophobic portions of the octanol and humic material associate with hydrophobic organic solutes (naphthalene). Hydrogen bonding dominates their interaction with polar groups of the hydrophilic solutes (phenol).

Sorption on Soils

Generally speaking, nonpolar organic molecules are sorbed by soils as a function of their **hydrophobicity** (K_{ow}) and the organic content of the soil. The sorption coefficient (K_p) is defined as follows:

$$K_p = \frac{\text{concentration solid phase}}{\text{concentration solution}} = \frac{mg/kg}{mg/L} = \frac{L}{kg}$$

Experimental studies using flow-through techniques produce K_p values that are sensitive to the flow rate and suggest that increased sorption occurs with longer exposure times. The sorption is also dependent on the amount of organic carbon in the soil aquifer.

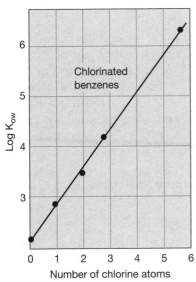

Figure 14.5 Relationship of number of chlorine atoms in a compound and log K_{ow}.

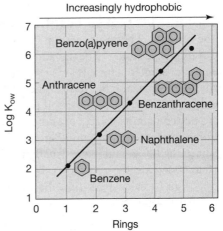

Figure 14.6 Relationship of number of aromatic rings in a compound and log K_{ow}. Adapted from M. D. Piwoni and J. W. Keeley, *Basic Concepts of Contaminant Sorption at Hazardous Waste Sites.* EPA Ground Water Issue 6, EPA/540/4-90/053, October 1990.

Table 14.3	
Log K_{ow} of Some Nonpolar Organic Molecules	
Compound	**Log K_{ow}**
aldrin	6.5
DDT	6.2
hexachlorobenzene	5.6
chlordane	5.5
heptachlor	5.3
toxaphene	4.8
fluoranthene	4.7
Endrin	4.6
phenanthrene	4.5
Dieldrin	4.3
1,2,4-trichlorobenzene	4.2
dibutyl phthalate	4.1
dibenzyl	4.0
1,4-dichlorobenzene	3.6
ortho-dichlorobenzene	3.7
chlorobenzene	2.8
toluene	2.7
benzene	2.1
phenol	1.5

Carbon-Normalized Sorption Coefficient

The sorption of nonpolar organics also depends on the amount of soil organic carbon, which was produced from the degradation of naturally occurring organic matter (leaves, detritus). Soils and aquifer materials are very heterogeneous, and the organic carbon content can vary considerably in both the vertical and horizontal dimensions. Fortunately, the variability tends to be greatest in the vertical soil profile. Most of the studies of the movement of nonpolar organic contaminants are concerned with the contaminant movement in ground water away from the source (dump, underground storage tank). Although the soil organic carbon content in the horizontal plane usually varies by a factor of 10 or less, it can vary by a factor of 10 to 100 vertically.

K_p can be calculated from the **carbon-normalized sorption coefficient (K_{oc})**, which is defined as follows:

$$K_{oc} = \frac{\text{sorption coefficient, } K_p}{\text{fraction organic carbon, } f_{oc}}$$

To determine the soil organic carbon content at a site, samples are taken back to the laboratory, where they are first acidified to dissolve and remove the inorganic carbon (carbonates). Weighed samples of the soil are then burned in an oxygen atmosphere, and the CO_2 that is produced is quantitatively measured by infrared spectroscopy. If, for instance, the soil organic content is 0.2% for a specific sample, then the f_{oc} is 0.002.

The octanol/water partition and soil/water partition experiments are very similar, with soil replacing octanol in the test. As a result, there have been several empirical relationships developed for estimating sorption using K_{ow} and the organic carbon content of the soil. One of these is commonly used for estimating the sorption of common environmental contaminants in an aquifer with low carbon content.

$$\log K_{oc} = 0.69 \log K_{ow} + 0.22$$

Usually empirical relationships such as this require the selection of reference organic compounds in which structure is similar to those of interest. Figure 14.7 shows that there is a linear relationship between $\log K_{ow}$ and $\log K_{oc}$ for a series of chlorinated benzenes. Even when a structurally dissimilar compound, such as anthracene, which contains no chlorines and is a PAH, is compared with the chlorinated benzenes, its K_{oc} can be estimated with reasonable accuracy.

Experimentally Measuring Absorption

To obtain the best information on which to base an estimate of sorption, tests should be made with the contaminates of concern using soil and aquifer material from the site of interest. The goal is to experimentally measure the partition coefficient (K_p) for use in the prediction of contaminant movement.

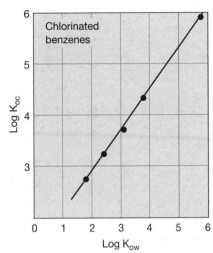

Figure 14.7 Partitioning on soil organic carbon as a function of the octanol/water coefficient.

The standard approach to determine K_p is to generate a sorption isotherm, a graphical representation of the amount of material sorbed at a variety of solute concentrations. The Freundlich isotherm is the most common relationship used:

$$S = K_p C$$

Where

S is the mass of organic contaminant per mass of sorbent in the units mg/kg.
C is the concentration of the contaminant in mg/L.

EXAMPLE   14.1

An industrial landfill is contaminated with 1,4-dichlorobenzene. Preliminary tests indicate that most of the contamination is below the water table. The contaminant concentration in ground water averages 1 mg/L. The measured soil organic carbon is 0.2%, and pore water occupies 50% of the aquifer volume. Estimate the K_{oc} and the K_p and S.

Solution
From Table 14.3, we find that the K_{ow} for 1,4-dichlorobenzene is 3.6.

Using this equation

$$\log K_{oc} = 0.69 \log K_{ow} + 0.22$$

Therefore:

$$\log K_{oc} = 0.69(3.6) + 0.22$$
$$\log K_{oc} = 2.70$$
$$K_{oc} = 506$$

Since

$$K_p = K_{oc}\left(f_{oc}\right)$$

$$K_p = 506\left(0.002\right) = 1.01$$

Thus, the 1,4-dichlorobenzene is equally distributed between the solid and liquid phases.

$$S = K_p C$$

$$S = 1.01\left(1.0\right) = 1.01 \, mg/kg$$

Thus, we would expect the solid to have a 1.0-mg/kg concentration.

Henry's Law

If the contaminant is a gas, one of the physical properties that the PBT Profiler uses is its solubility in water. The solubility of a gas in water is described by Henry's law, which states that at constant temperature the solubility of a gas in a liquid is proportional to the partial pressure of the gas that is in contact with the liquid. The **Henry's law constant (K_H)** is really the air/water partition coefficient. The EPA expresses K_H in two ways; the first is a nondimensional value that relates the chemical concentration in the atmosphere (gas phase) to its concentration in the water phase. The K_H also can be determined by dividing the vapor pressure of a chemical (in atmospheres) by its solubility in water (in mol/m^3) to give K_H in atm-m^3/mol.

The PBT Profiler uses K_H to predict the fate of a chemical once it is released into surface water. A large K_H (of more than 10^{-3}) indicates that the chemical is likely to evaporate from the water and partition into air. A low K_H (10^{-5} to 10^{-7}) indicates that the chemical is not likely to volatize and will remain in surface water. Table 14.4 lists the K_Hs for several chemical contaminants of interest. K_H provides an indication of the partition between air and water at equilibrium and is also used to calculate the rate of evaporation from water.

Table 14.4

Henry Law Constants (K_H)

CAS Number	Chemical	K_H (atm-m^3/mole)
75092	Dichloromethane	3.0×10^{-3}
50000	Formaldehyde	6.1×10^{-5}
67641	Acetone	4.0×10^{-5}
67561	Methanol	4.4×10^{-6}
60571	Dieldrin	5.4×10^{-7}

■ Using the EPA's PBT Profiler

Companies can access the **PBT Profiler** at http://www.pbtprofiler.net. The PBT Profiler uses established screening models to estimate PBT characteristics based on chemical structure and physical and chemical properties. The PBT profile is intended to answer three questions for manufacturers:

1. Once released, will a chemical go into the air, water, soil, or sediment?
2. How long will this chemical stay in the media (air, water, soil, or sediment)?
3. Will the chemical present a hazard?

Persistence

The EPA has established criteria that categorize a chemical as not persistent, persistent, or highly persistent. These criteria are specific for each medium and reflect the half-life of the chemical in days.

The PBT Profiler calculates an atmospheric half-life by determining the importance of a chemical's reaction with two of the most prevalent atmospheric oxidants: hydroxyl radicals and ozone. The half-life is calculated directly from gas-phase hydroxyl radical and ozone reaction rate constants. These rate constants are obtained from a database of measured values or, if no experimental values are available, they are estimated.

The half-life for degradation of a chemical in water, soil, and sediment is determined by the PBT Profiler by using the ultimate biodegradation expert survey module of the BIOWIN estimation program. This estimation program provides an indication of a chemical's environmental biodegradation rate in relative terms such as hours, hours to days, days, days to weeks, and so on; the terms represent the approximate amount of time needed for degradation to be "complete." The mean value within the estimated time range returned by the ultimate biodegradation survey model is converted to a half-life using a set of conversion factors. This program assumes that six half-lives constitute "complete" degradation of a chemical substance (assuming first-order kinetics).

The PBT Profiler assesses the persistence of the chemical of interest from data stored in its database and issues a report. When viewing the report, a green font indicates that the chemical is not persistent. An orange font indicates that the chemical is persistent, and a red font indicates that the chemical is highly persistent. The EPA persistence criteria are listed in Table 14.5.

Table 14.5

EPA's Persistence Criteria for PBT Chemicals

Media	Half-Life of Chemical		
	Not Persistent	**Persistent**	**Highly Persistent**
Water	< 2 months	≥ 2 months	> 6 months
Soil	< 2 months	≥ 2 months	> 6 months
Air	≤ 2 days		> 2 days
Sediment	< 2 months	≥ 2 months	> 6 months

Bioconcentration

The PBT Profiler determines a chemical's potential to bioaccumulate directly from an estimated **bioconcentration factor (BCF)**, which is estimated using SRC's BCFWIN estimation program.[1] The estimated BCFs are compared with those contained in the EPA Criteria. The BCFWIN program yields a prediction of BCF based on a chemical's octanol/water partition coefficient and one or more chemical structure-based correction factors, if applicable. The model does not explicitly address a variety of factors that may influence bioaccumulation under field conditions, such as the possible metabolism of the chemical in exposed organisms, which could lead to actual bioaccumulation being lower than predicted.

The PBT Profiler assesses the bioconcentration of the chemical of interest from data stored in its database and issues a report. When viewing the report, a green font indicates that the chemical is not expected to bioaccumulate (BCF, 1000); an orange font indicates that the chemical will bioaccumulate (BCF of 1000 or more), and a red font indicates that the chemical is highly bioaccumulative (BCF of 5000 or more).

Toxicity

The PBT Profiler considers only the potential chronic toxicity of a substance to fish and estimates it using the Ecological Structure Activity Relationships (ECOSAR) program, which predicts the toxicity of chemicals to aquatic organisms such as fish, invertebrates, and algae by using structure activity relationships (SARs). ECOSAR uses SARs to predict the aquatic toxicity of chemicals based on their structural similarity to chemicals for which aquatic toxicity data are available. SARs express the correlations between a compound's physical/chemical properties and its aquatic toxicity. SARs measured for one compound can be used to predict the toxicity of similar compounds belonging to the same chemical class. The SARs contained within the ECOSAR are based on test data, and many of the SAR predictions have been evaluated. More information on ECOSAR, as well as the program itself, is available from the EPA.

For the PBT Profiler, the fish chronic (ChV) estimates from ECOSAR are used to predict toxicity because there is the potential for long-term exposure to persistent chemicals. If the ChV cannot be estimated using the equations available in ECOSAR, the PBT Profiler will return "not estimated." The PBT Profiler identifies chemicals that exceed the log octanol/water partition coefficient cutoffs used by ECOSAR. If the cutoffs are exceeded for a specific chemical, the PBT Profiler will return a ChV value of "not estimated" and will not run the ECOSAR estimation. Because of the algorithms that ECOSAR uses, this program estimates a water solubility separately using the octanol/water partition coefficient.

PBT chemicals are those that persist in the environment. They generally occur in low concentrations and can be transported throughout the biosphere. They can be bioconcentrated in aquatic organisms and transported up food chains to humans, birds, and wild mammals. Exposure to PBT chemicals is generally through the diet and is the result of chronic exposure. Chronic exposures may lead to chronic toxicity and not acute toxicity. That is, chronically toxic chemicals affect processes other than survival. It would be possible, therefore, for organisms to survive the acute effects of such chemicals (that is, not die) yet still undergo adverse effects from long-term exposure (for example, chronic exposure may adversely affect growth or reproduction). In order to best estimate the effects of PBT chemicals on the environment, the fish chronic toxicity is used in the PBT Profiler. The ECOSAR program also estimates a variety of other aquatic toxicity endpoints depending on the structure of the chemical. These include the acute toxicity of a chemical to fish (both fresh and saltwater),

water fleas (daphnids), and green algae. For some chemical classes, endpoints for other organisms may be estimated (such as earthworms).

The PBT Profiler assesses the toxicity of the chemical of interest from data stored in its database and issues a report. When viewing the report, a green font indicates that the chemical has a low toxicity concern (fish ChV of more than 10 mg/L or no effects when water is saturated with the chemical). An orange font indicates a moderate toxicity concern (a ChV of less than 10 mg/L), and a red font indicates a high toxicity concern (ChV of less than 0.1 mg/L).

To demonstrate how the PBT profiler works, a common industrial chemical o-xylene will be entered into the PBT profiler.

EXAMPLE 14.2

Determine the PBT characteristics of o-xylene. Using the PBT Profiler, indicate whether o-xylene is expected to be persistent. Determine the half-life of o-xylene in air, water, soil, and sediment. Determine whether it will bioaccumulate and whether it should be considered toxic.

Solution

1. The PBT Profiler works more efficiently if the CAS number for the chemical of interest is entered (rather than structure or chemical name). To look up the CAS number, go to NIST's Chemistry WebBook at http://webbook.nist.gov/chemistry/ or the NCI Database Browser at http://cactvs.cit.nih.gov/. The CAS number for o-xylene is 95476.
2. Enter the CAS number into PBT Profiler at www.pbtprofiler.net.
3. The following report is generated:

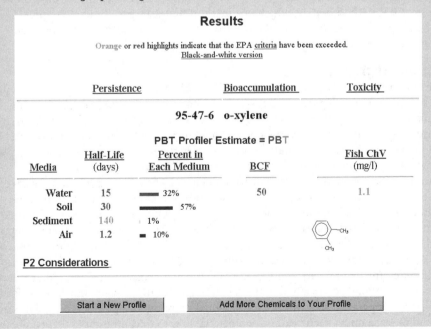

USING THE EPA's PBT PROFILER **421**

4. Persistence results: The PBT Profiler has estimated the half-life of o-xylene in the various media and has indicated that it is expected to be found predominately in soil and that its half-life does not exceed the EPA criteria. Its half-life in sediment does exceed EPA criteria (persistent), but because only 1% partitions to sediment, this is not considered to be a major determinant; therefore, o-xylene is estimated not to be persistent in the environment.
5. Bioaccumulation: The estimated BCF of 50 for o-xylene does not exceed the EPA criteria. Therefore, The PBT Profiler estimates that o-xylene is not expected to bioaccumulate in the food chain.
6. Toxicity: the estimated fish ChV for o-xylene, 1.1 mg/L, exceeds the EPA criteria. The PBT Profiler indicates that o-xylene is expected to be toxic to fish.

■ Analysis of POPs in Environmental Samples

The EPA searches many contaminated sites in the United States for the PBTs that are listed in Table 14.2. The EPA publishes a comprehensive list of the test methods that should be used for specific environmental analysis. The list of analytes and methods can be found on the Web site of the National Environmental Index (http://www.nemi.gov). The test methods are written for specific types of samples. Some EPA methods are written for the analysis of solid waste and others for drinking water, etc. A search of the EPA Web site will give a specific method for a specific analyte and application. The concentration of organochlorine pesticides and PCBs, which are also on the POP "dirty dozen" list, is determined in environmental samples by EPA Method 8081A, which is a method that uses gas chromatography (GC) to separate the specific organochlorine analyte from other chemicals in the sample before it is quantitatively measured by an **electron capture detector (ECD)**.

Gas Chromatography

Gas chromatography (GC) is a dynamic method of separation and detection of volatile organic compounds. GC separates the gaseous components of a mixture by partitioning them between the inert gas mobile phase and a stationary phase. Chromatography functions on the same principle as extraction, only one phase is held in place while the other moves past it. If one component of the gaseous mixture is more strongly absorbed to the stationary phase, then it will move through the system more slowly than components that are not.

Figure 14.8 is a schematic of the major components of a GC. The major components of the GC are the carrier gas, the injection port, the **separation column**, the detector, and the data-acquisition system. The column is held inside an oven that can be raised to 350°C (680°F), and the injector and detector each have separate heaters. The liquid environmental sample is injected by a syringe through a septum (rubber seal) into a heated injection port, in which the sample is rapidly vaporized. The gaseous sample is swept out of the injector into the separation column by the helium carrier gas that is flowing at a constant flow rate. After the components of the sample are separated in the separation column, they are individually detected as they pass through the detector. The individual components produce Gaussian-shaped peaks, and the area under each peak is proportional to the concentration of that component.

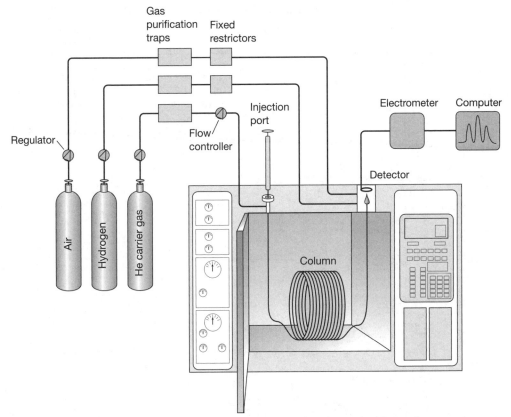

Figure 14.8 Typical gas chromatograph. Courtesy of Hewlett-Packard Corporation.

The separation column is a long (30 m), capillary tube (0.25 mm, internal diameter) made of fused silica (SiO_2). Most methods call for a column that is between 30 to 60 m long. The inner wall of the capillary is coated with a 1.0-μm thick film of stationary liquid phase. The choice of liquid stationary phase is based on the "like dissolves like" rule. Nonpolar stationary phases are best for nonpolar molecules such as the organochlorine pesticides. Silicone stationary phases are commonly used because they are thermally stable at the elevated temperatures in the GC oven. EPA Method 8081A calls for two separate columns, one with a methyl silicone stationary phase and the other with a 35% diphenyl dimethyl polysiloxane; the structure can be seen in Figure 14.9.

A GC separation can be completed by keeping the separation column a constant temperature (isothermal separation) or by increasing the temperature in a predetermined way. A good general rule that organic compounds will elute from the GC in order of increasing boiling points. A temperature-programmed GC separation is one in which the injection is made while the column is at relatively low temperature (50°C, 122°F), and then the temperature of the oven is increased with time (dT/dt). The temperature of the column is raised during the separation to increase the vapor pressure of the components still in the column and to bring them out of the column more quickly. Figure 14.10 shows that if the temperature is

Common stationary phases in capillary gas chromatography

Structure	Polarity		Temperature ran (°C)
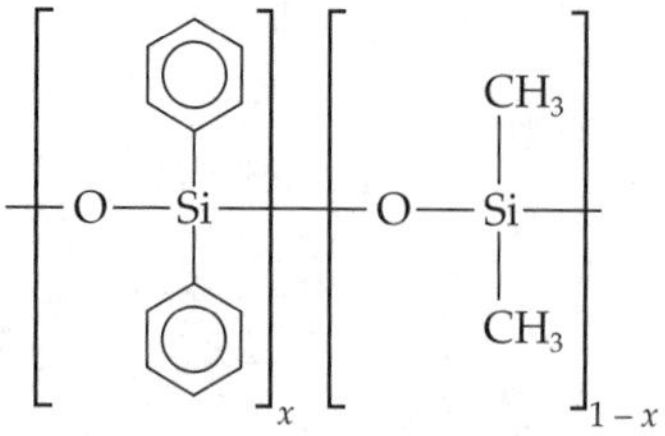	$x = 0$	Nonpolar	–60°–320°
	$x = 0.05$	Nonpolar	–60°–320°
	$x = 0.35$	Intermediate polarity	0°–300°
	$x = 0.65$	Intermediate polarity	50°–370°

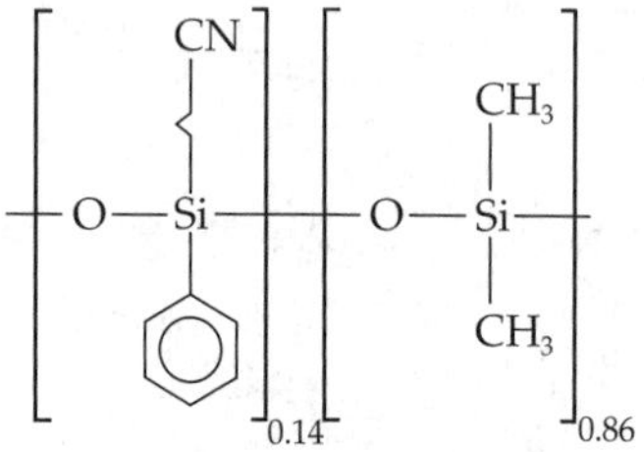

(Diphenyl)$_x$(dimethyl)$_{1-x}$
polysiloxane

(Cyanopropylphenyl)$_{0.14}$
(dimethyl)$_{0.86}$ polysiloxane

Intermediate polarity   –20°–280°

H_2CH_2——O⟩$_n$ Strongly polar   40°–250°

(Carbowax
(polyethylene glycol))

Strongly polar   0°–275°

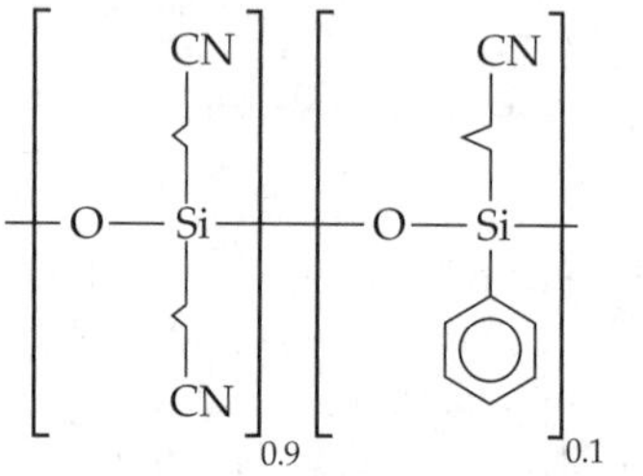

(Biscyanopropyl)$_{0.9}$
(Cyanopropylphenyl)$_{0.1}$ polysiloxane

Figure 14.9 Common stationary phases used for GC of organochlorine pesticides. Adapted from D.C. Harris. *Quantitative Chemical Analysis, Sixth edition.* New York: W.H. Freeman, 2003.

raised from 50° to 250°C (122° to 482°F) at a rate of 8°C per minute, the late eluting peaks in the isothermal injection are eluted more quickly and the peaks are more symmetrical.

The GC is operated with the carrier gas helium flowing at a constant flow rate (F). Each component requires a specific volume of gas, the retention volume (V_r), to push it through the GC. Thus, the time needed to push a specific component through the GC is the retention time (t_r).

$$V_r = t_r F$$

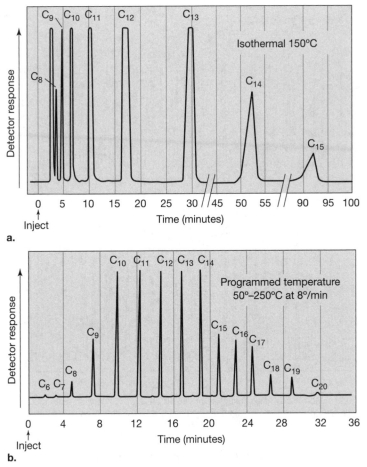

Figure 14.10 Comparison of (a) isothermal and (b) temperature programmed GC separation. Adapted from H.M. McNair and E.J. Bonelli. *Basic Gas Chromatography*. Palo Alto, CA: Varian Instrument Division, 1968.

The t_r is directly proportional to the V_r and is more convenient to describe the elution of components by t_r than by other retention volumes. Once the flow rate and temperature program have been set for the separation, the standard is injected into the GC, and the retention times of all of the components are measured. Each component in an unknown sample is identified by its unique retention time. The EPA Method 8081A requires that the sample stream be split after the injector and the sample simultaneously be separated on two different columns. Figure 14.11 shows that the separation of the same organochlorine pesticide sample by two GC columns containing different stationary phases gives different retention times for the same compound. By comparing the results from the two separations, this technique assures than there is not some small contaminant lying under one of the peaks that is giving a higher concentration than is really there. The retention times and identity of the compounds are listed in Table 14.5.

The ECD is particularly sensitive to halogen-containing molecules but relatively insensitive to hydrocarbons, ketones, and alcohols. Figure 14.12 shows that the ECD is constructed

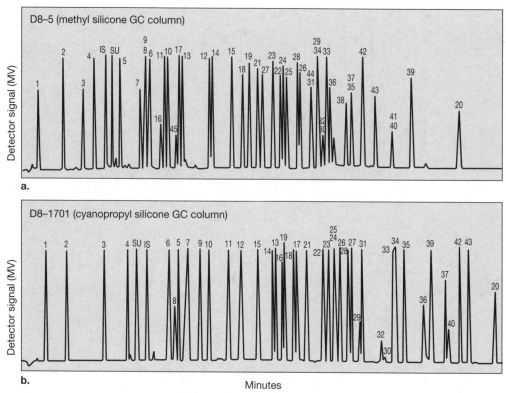

Figure 14.11 Organochlorine pesticides analyzed on two different GC columns: (a) methyl silicone (b) 35% phenyl silicone. Adapted from EPA Test Method 8081A, Revision 1, December 1996, p. 43.

of a cavity with two electrodes and a radiation source. A ^{63}Ni source releases β radiation, which collides with a reagent gas (5% methane in Ar) that is flowing through the detector, producing a plasma containing electrons and positive ions. The electrons are attracted to the anode producing a small steady current. When sample molecules with a high electron affinity enter the detector, they capture some of the electrons, and the rate of electron collection at the anode decreases. The detector senses the drop in the current and records the change in signal as the separated compound passes through the detector. The response generated by the detector is proportional to the amount of compound.

It is important to calibrate the GC-ECD with the compound(s) of interest. The response generated by a fixed amount of a compound is a function of the number of electronegative atoms (chlorines) in that molecule. For example, the response for polychlorinated organic pesticides can be as much as a million times higher than that for a hydrocarbon. Once the GC-ECD is calibrated with the appropriate standard, its response is linear with a variation less than 3%. Organochlorine pesticides can easily be measured at the low parts per billion and mid-parts per trillion level using this method.

Because the detector does not give an equal response to every compound, to calculate the amount of each component in a sample, each constituent must be calibrated relative to the response of the detector. The calibration factor (response factor) must be determined for each

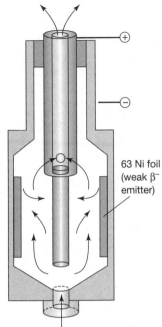

Figure 14.12 The ECD detector. Adapted from K.A. Rubison and J.F. Rubison. *Contemporary Instrumental Analysis.* Upper Saddle River, NJ: Prentice Hall, 2000.

63 Ni foil (weak β⁻ emitter)

Effluent in, N_2 or Ar + 10% CH_4

component in which the concentration is to be determined. The calibration factor is defined as follows:

$$CF = \frac{\text{peak area of the compound in the standard}}{\text{mass of the compound injected (in nanograms)}}$$

It is experimentally measured for every organochlorine compound that is going to be measured by injecting a standard that contains a known amount of material.

To determine the concentration of each organochlorine pesticide, the GC is first calibrated with a standard that contains known quantities of the pesticides to be measured. The retention time and area under each peak are recorded. If an environmental water sample is to be measured, then the sample is extracted with methylene chloride (CH_2Cl_2). If a solid environmental sample (soil) is to be measured, it is extracted with a 1/1 hexane-acetone mixture. The extracts are then injected into the GC. The concentration of each organochlorine pesticide is determined from the following equation:

$$\text{concentration}\,(\mu g/L) = \frac{(Ax)(Vt)(D)}{(CF)(Vi)(Vs)}$$

Where

Ax = area of the peak for one of the organochlorine pesticides.
Vt = total volume of the concentrated extract (μL).
D = dilution factor (if the sample was diluted before analysis).
CF = calibration factor for the organochlorine pesticide.
Vi = volume of the extract injected (μL).
Vs = volume of the sample extracted in mL.

Although EPA Method 8180A calls for a GC that is configured with an ECD detector, other methods use a GC with the flame ionization detector, which is much more sensitive for the analysis of hydrocarbons (gasoline and diesel oil spills). The GC is a very efficient way of separating mixtures and can be attached to a number of different detectors. EPA Method 8257A, for instance, describes how to analyze for PCBs and PAHs in soil and solid waste by using a GC that is attached to a mass spectrometer (MS).

Analysis of PCBs and PAHs by GC-Mass Spectrometry

The analysis of some environmental pollutants, such as polychlorinated biphenyls, PCBs, and PAHs, is complicated by the fact that the material being measured is not just one chemical. PCBs, for instance, were commercial products that were prepared as mixtures of chlorinated

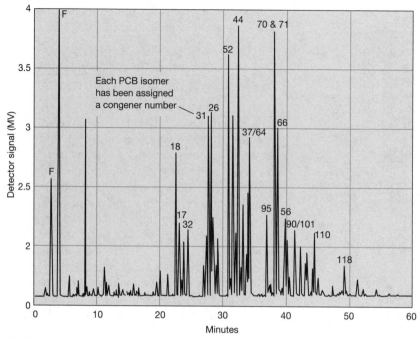

Figure 14.13 A GC separation of PCBs. Adapted from K. Ballschmiter and M. Zell, *J. Anal. Chem.* 302 (1980): 20–31.

biphenyl molecules that have certain physical properties. Each mixture was prepared to have properties for a specific industrial use. GC analysis of commercial PCBs gives complex chromatograms similar to the one shown in Figure 14.13 because they contain more than 50 different chlorinated biphenyl molecules. Because of this complex pattern, there is a greater probability that some other compound may be present in the environmental sample that would elute from the GC at the same time as one of the PCB peak (co-elution). The co-elution of two peaks would cause the GC to measure the sum of the signal for each compound and to report a value for the PCB concentration that is too high. This is what is called a *method error*; the analytical method does not accurately measure the amount of analyte present in the sample. Changing the GC detector to one that would measure the analyte only would improve the method and eliminate the error. Using a MS as the detector is one way to modify the method.

Mass Spectrometry

Mass spectrometry (MS) is a technique that is used to measure the mass of atoms or molecules. To obtain a mass spectrum, gaseous sample atoms are introduced into the source of the MS where they are ionized. Figure 14.14 shows that the positive ions are accelerated into the analyzer of the MS by ion acceleration plates that have a very large negative charge (20,000 V) applied to it. The ions are then separated in the analyzer by a magnetic field that separates them according to the mass-to-charge ratio (m/z). The entire MS has to be kept at a high vacuum (10^{-6} Pa) so that ions do not encounter collisions with background gas molecules. If the

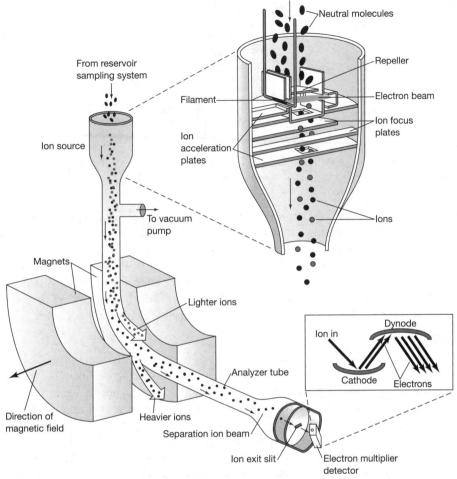

Figure 14.14 Magnetic sector mass spectrometer. Adapted from F.W. McLafferty. *Interpretation of Mass Spectra.* New York: Benjamin, 1966.

ions have a +1 charge, then the m/z is equivalent to the mass of the ion. The magnet of the MS can be used to scan from low to high mass as a sample is introduced. Figure 14.15 is a typical mass spectrum, which is a display of detector response versus m/z. This figure shows a mass spectrum for the pesticide Methomyl, which has a formula of $C_5H_{10}N_2O_2S$ and shows a molecular ion at m/z of 162.

The mass spectrum contains the molecular ion ($M^{\cdot+}$), which is formed by impact with an energetic electron:

$$M + e^- \rightarrow M^{\cdot+} + 2\,e^-$$

The mass of the molecular ion is the same as the analyte and is useful in verifying the identity of the sample. The mass spectrum also contains fragment ions, which are formed in the source

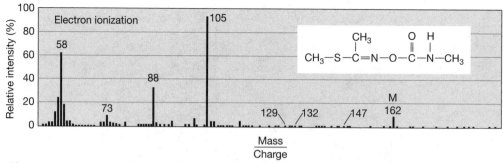

Figure 14.15 Separation of ions by mass in a mass spectrometer. Adapted from K.A. Rubison and J.F. Rubison. *Contemporary Instrumental Analysis.* Upper Saddle River, NJ: Prentice Hall, 2000.

of the MS. The combination of these ions forms a pattern of peaks that is unique and provides a "fingerprint" for the molecule. The computer attached to the MS has a database of spectra of known compounds that can be searched to match the spectra of an unknown sample and to provide a conclusive identification of the unknown. Figure 14.16 shows the molecular ions in the mass spectra of biphenyl and chlorinated analogs (PCB congeners). Biphenyl (C_6H_{10}) has one molecular ion at 154. The chlorinated biphenyls, on the other hand, have complicated mass spectra because chlorine has a ^{37}Cl and a ^{35}Cl isotope with a naturally occurring 1:3 ratio.

GC-MS

The combination of GC and MS provides an instrument (GC-MS) that combines the best features of the two techniques and eliminates each technique's major weakness. GC can separate volatile compounds with great efficiency but cannot unambiguously identify them. MS provides detailed structural information about most compounds so that they can be identified, but it must be given pure samples. The combination of the two techniques allows the GC to separate and vaporize the components of the sample and to introduce them sequentially into the MS where the unique mass spectra of each component will be measured and each component can be identified. Capillary GC columns have low enough flow rates that the GC can be directly coupled to the MS (Figure 14.17).

GC-MS can be used to identify one component in a complex environmental sample. The "fingerprint" MS spectra obtained from the GC-MS can be matched with spectra stored in the NIST/EPA/NIH Mass Spectral Database, which contains the mass spectra of more than 190,000 compounds. The computer algorithm matches the GC-MS spectra from the sample to the five best spectral matches in the database and presents the analyst with a "fit index" that assesses how close the match is between the unknown and the known.

GC-MS is also used to measure the concentration of one or more analytes in a complex environmental mixture. Quantitation can be based on peak areas from the mass chromatogram or from selected ion monitoring. When the selected ion monitoring technique is used, the MS does not scan over a range of masses; instead, the MS scans one mass of interest and then jumps to another. Because the MS can spend more time monitoring the selected ions, the sensitivity of the MS is 100 to 1000 times greater. For example, Table 14.6 lists the PAH compounds that are measured by EPA Method 8275A and the mass of the ion that is measured to quantify each. When using this technique, the MS jumps from one quantitation

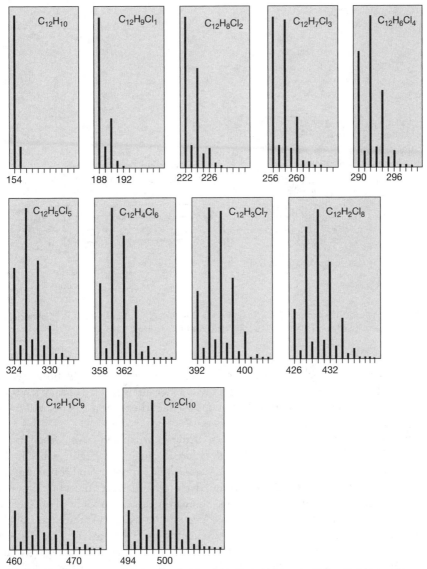

Figure 14.16 The mass spectra of biphenyl and various chlorinated biphenyls. Notice that biphenyl has one molecular ion at 154. The chlorinated biphenyls have complicated mass spectra because chlorine has a 37CL and a 35CL isotope with a 1:3 ratio.

ion to another as the analysis proceeds. The detection of picograms (10^{-12} g) of analyte is possible using this technique.

To convert the peak area of one measured mass to concentration of analyte, whether using selected ion monitoring or mass chromatograms, the instrument must be calibrated. The two main techniques that are used are external or internal calibration. When using external calibration, the area of the mass chromatogram, for one or more compounds, is

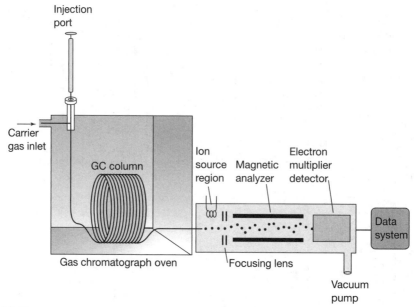

Figure 14.17 Schematic of a typical GC-MS. Adapted from D.A. Skoog, F.J. Holler, and T.A. Nieman. *Instrumental Analysis, Fifth edition*. Philadelphia: Saunders College, Harcourt Brace College, 1998.

determined by injecting a known amount of analyte into the GC-MS. The unknown sample is then injected in a separate experiment, and the MS measures the area of each mass selected and computes the concentration of each analyte based on the response generated by the calibration standard.

Oftentimes, the best analysis is achieved by using internal standards, which are known amounts of standard that is added to the sample before the analysis is begun, which in this case the internal standard added to the soil sample. After sample extraction and concentration, only the ratio of the response between the analyte and the internal standard must be measured. This ratio multiplied by the amount of the internal standard gives the amount of analyte injected into the GC-MS. This number is converted to concentration using the dilution factors.

The best internal standards are very similar chemically to the analyte. Any losses of analyte during the sample preparation and analysis are then duplicated by the losses of internal standards. Homologues of the analyte can be used as internal standards, but the very best standards are isotopically labeled versions of the analyte.

Analysis of Dioxins and Furans by GC-MS

EPA Method 613 is used to measure the amount of 2,3,7,8 tetrachlorodibenzo-*p*-dioxin (2,3,7,8-TCDD) in municipal and industrial wastewater. The "native" 2,3,7,8-TCDD, which has the formula $C_{12}H_4O_2Cl_4$, has a molecular ion of 320. This method calls for the addition of one of two possible isotopically labeled internal standards; the 2,3,7,8-TCDD standards are synthesized using isotopes that will make the standard heavier than the native 2,3,7, 8-TCDD. In this case, either all of the carbons are ^{13}C, giving a 2,3,7,8-TCDD that weighs 332, or the chlorines are ^{37}Cl, giving a 2,3,7,8-TCDD that weighs 328. Because the native

Table 14.6

Selected Ions for the Quantitation of PAHs by EPA Method 8275A

Compound	Quantitation Ion
1,2,4-Trichlorobenzene[2]	180
Naphthalene	128
Acenaphthylene	152
Acenaphthene	153
Dibenzofuran	168
Fluorene	166
4-Bromophenyl phenyl ether	248
Hexachlorobenzene[2]	284
Phenanthrene	178
Anthracene	178
Fluoranthene	202
Pyrene	202
Benzo[a]anthracene	228
Chrysene	228
Benzo[b]fluoranthene	252
Benzo[k]fluoranthene	252
Benzo[a]pyrene	252
Indeno(1,2,3-cd)pyrene	276
Dibenz[a,h]anthracene	278
Benzo[g,h,i]perylene	276

and isotopically labeled 2,3,7,8-TCDD are chemically equivalent, any losses of analyte during sample preparation and analysis will be the same for the analyte and standard. By monitoring the 320/332 or 320/328 ratio and multiplying it by the amount of the internal standard, the amount of analyte in the environmental water sample is detected. Detection limits of 2 parts per trillion of 2,3,7,8-TCDD in water are attained using this method.

■ References

1. Boethling RS, Howard PH, Meylan WM, Stiteler W, Beauman J, Tirado N. Group contribution method for predicting probability and rate of aerobic biodegradation. *Environmental Science and Technology* 1994;28:459–465.

2. Meylan WM, Howard PH, Boethling RS, Aronson D, Printup H, Gouchie S. Improved method for estimating bioconcentration factor (BCF) from octanol-water partition coefficient. *Environmental Toxicology and Chemistry* 1999;18:664–672.

■ Additional Sources of Information

Clesceri LS, Greenberg AE, Eaton AE, editors. *Standard Methods for the Examination of Water and Wastewater,* 21th ed. Washington, DC: American Public Health Association, 2005.

Environmental Protection Agency (EPA). Homepage. Accessed December 2008 at http://www.epa.gov.

Harris DC. *Quantitative Chemical Analysis,* 7th ed. New York: W. H. Freeman and Company, 2006.

Hemond HF, Fechner-Levy EJ. *Chemical Fate and Transport in the Environment,* 2nd ed. San Diego, CA: Academic Press, 1999.

Rubison KA, Rubison JF. *Contemporary Instrumental Analysis.* Upper Saddle River, NJ: Prentice Hall, 2000.

Skoog DA, Holler FJ, Crouch SR. *Instrumental Analysis,* 6th ed. Philadelphia, PA: Saunders College Publishing, Harcourt Brace College Publishers, 2007.

■ Keywords

Agent Orange

antiandrogenic

bioconcentration factor (BCF)

carbon-normalized sorption coefficient (K_{oc})

CAS registry numbers (CAS RNs)

dichlorodiphenyl-trichloroethane (DDT)

dioxin

electron capture detector (ECD)

Environmental Protection Agency (EPA)

furans

gas chromatography (GC)

Henry's law constant (K_H)

hormones

hydrophobicity (K_{ow})

mass spectrometry (MS)

octanol/water partition coefficient (K_{ow})

PBT Profiler

persistent organic pollutants (POPs)

Persistent, Bioaccumulative, and Toxic (PBT) Pollutants Program

polychlorinated biphenyls (PCBs)

polychlorinated hydrocarbons

premanufacture notifications (PMNs)

pyrethrins

Rachel Carson

separation column

sorption coefficient

Stockholm Convention

toxicity

■ Questions and Problems

1. What function do hormones carry out?
   a. Give two examples of organs affected by hormones.
   b. Name the male sex hormone.
   c. Name the female sex hormone.
2. What are endocrine disruptors?
   a. Do they accumulate in the food chain?
   b. What are transgenerational effects?
   c. Are they a concern to only wildlife?

3. Are the chemical structures of estrogen mimics similar to the structure of estrogen?
   a. Do all hormone-disrupting chemicals act alone?
   b. List the effects of estrogen mimics on males.
   c. Do estrogen mimics have any effect on women?
4. What are POPs?
   a. How are they transported?
   b. How do POPs manage to travel to the Arctic regions?
   c. How did the Stockholm Convention propose dealing with POPs?
5. The Stockholm Convention focused on the dirty dozen chemicals.
   a. Are the dirty dozen chemicals still manufactured in the United States?
   b. Name the four types of chemicals that make up the dirty dozen.
   c. Of the 12 chemicals, 10 were intentionally produced. What is the source of the other two?
6. Polychlorinated hydrocarbons
   a. Do they bioaccumulate?
   b. Are they soluble in water?
   c. Do they decompose?
7. Describe the process by which polychlorinated hydrocarbons are bioaccumulated in the food chain.
8. List four naturally occurring insecticides.
9. What is the major source of 2,3,7,8-TCDD?
   a. Does it bioaccumulate?
   b. Is it soluble in water?
   c. Does it decompose while in the environment?
10. What is Agent Orange?
    a. It is a mixture of what?
    b. Synthesize 2,4,5-T from trichlorophenol.
    c. How could 2,3,7,8-TCDD be formed as a by-product in the synthesis of trichlorophenol?
11. List four industrial uses for PCBs.
    a. Draw the structure of biphenyl.
    b. Draw the structure of several PCB molecules.
    c. Do PCBs burn?
12. POPs have been shown to mimic natural estrogen.
    a. Draw the structure of estrogen.
    b. Is the structure of the 2,3,7,8-TCDD similar to that of estrogen?
    c. Can POPs bind to estrogen binding sites?
13. POPs may interfere with the endocrine system.
    a. List four observations made in animal species that support this hypothesis.
    b. List two observations made in humans that support this hypothesis.
    c. What is the function of the endocrine system?
14. How does the EPA's PBT pollutants list compare with the Stockholm Convention's dirty dozen list?
15. What are EPA's premanufacture notifications?
    a. From what law does EPA get the authority to screen new chemicals?
    b. From what law does EPA get authority to screen new pesticides?
    c. What is the EPA's PBT Profiler?

16. The EPA's PBT Profiler uses a chemical's physical properties to estimate many of its physical properties.
    a. If the melting point of a compound is high, what does this suggest about its solubility in water?
    b. If a compound has a high vapor pressure, will its potential inhalation exposure be greater than that of a compound with a low vapor pressure?
17. The environmental fate of organic compounds can be estimated using the octanol/water partition coefficient.
    a. How is the octanol/water partition coefficient experimentally determined?
    b. What are the units of the K_{ow}?
    c. Why is 1-octanol used in this test?
18. If an organic compound has a high solubility in water, then
    a. Will it rapidly absorb on soil?
    b. Is it likely to be rapidly degraded by microorganisms?
    c. Is it likely to undergo hydrolysis and photolysis reactions rapidly?
19. Arrange the log K_{ow} of DDT, aldrin, chlordane and dieldrin, and 1,4 dichlorobenzene in decreasing order (see Table 14.3). Find correlations between the structures of these molecules and the K_{ow} values. Which of these molecules would be least likely to bioaccumulate?
20. Arrange the log K_{ow} of benzene, ortho-dichlorobenzene, 1,2,4 trichlorobenzene, and DDT in decreasing order. Find correlations between the structures of these molecules and the K_{ow} values. Which of these molecules would be least likely to bioaccumulate?
21. A Midwest farm is contaminated with Aldrin. Preliminary tests indicate that most of the contamination is below the water table. The contaminant concentration in ground water averages 1 mg/L. The measured soil organic carbon is 0.2%, and pore water occupies 50% of the aquifer volume. Estimate the K_{oc}, K_p, and S.
22. State Henry's law.
    a. What is a Henry's law constant (K_H)?
    b. Give two different ways to calculate the K_H.
    c. What does a large K_H indicate about the fate of a chemical?
23. What is an antiandrogenic effect?
24. State the three categories that the PBT Profiler uses to describe the persistence of a chemical.
    a. How is the atmospheric half-life determined?
    b. How is the half-life for degradation in water, soil, and sediment determined?
    c. How many half-lives must pass before a chemical is considered to be degraded?
25. How does the PBT Profiler determine potential chronic toxicity of a chemical?
    a. What is ECOSAR?
    b. What is a SAR?
    c. Why are aquatic organisms used for comparison?
26. Which of the following will be more likely to vaporize from water and partition into the air above the water: formaldehyde or Dieldrin?
27. Determine the PBT characteristics of hexachlorocyclohexane. Using the PBT Profiler, indicate whether it is expected to be persistent. Determine its half-life in air, water, soil, and sediment. Determine whether it will bioaccumulate and if it should be considered toxic.
28. Determine the PBT characteristics of benzidine. Using the PBT Profiler, indicate whether it is expected to be persistent. Determine its half-life in air, water, soil, and sediment. Determine whether it will bioaccumulate and whether it should be considered toxic.

29. Determine the PBT characteristics of isosafrole.
    a. Using the PBT Profiler, indicate whether it is expected to be persistent.
    b. What is its half-life in air, water, soil, and sediment?
    c. Will it bioaccumulate?
    d. Should it be considered toxic?
30. Determine the PBT characteristics of 3-methylphenol.
    a. Using the PBT Profiler, indicate whether it is expected to be persistent.
    b. What is its half-life in air, water, soil, and sediment?
    c. Will it bioaccumulate?
    d. Should it be considered toxic?
31. Why is the separation column kept in a separate oven?
    a. What is an isothermal separation?
    b. Why is the temperature of the oven sometimes programmed?
32. Draw a schematic of a gas chromatograph that could be used for analyzing the POP dirty dozen.
    a. How is the sample introduced into the GC?
    b. Why are silicone stationary phases used?
    c. Why is helium used as the mobile phase?
33. For GC columns, describe the following:
    a. Typical lengths
    b. Typical diameters
    c. The structure of the stationary phase used in EPA Method 8180
34. Draw a schematic of the ECD.
    a. Describe how it works.
    b. Why is it extremely sensitive for chlorinated compounds?
    c. Could it be used to detect gasoline in soil? Why or why not?
    d. What GC detector could be used to detect gasoline in soil?
35. When 1.00 mmol of chlorobenzene and 1.50 mmol of 1,2,4 trichlorobenzene were separated by GC, they gave relative peak areas of 915 and 1485 units, respectively. When 0.50 μmol of chlorobenzene was added to an environmental sample containing 1,2,4 trichlorobenzene, the relative GC peak areas were 838 for chlorobenzene and 814 for 1,2,4 trichlorobenzene. How much 1,2,4 trichlorobenzene did the environmental sample contain?
36. Why does EPA Method 8081 require the GC separation to be confirmed by a second GC column? Why does this method not use a GC-MS with a MS detector?
37. A standard solution containing 5.5×10^{-8} M tetrachlorobenzene and 2.0×10^{-7} M chlorotoluene (an internal standard) in methylene chloride gave peak areas of 398 and 768, respectively. A 1-L sample of river water was extracted with 50 mL of methylene chloride, and after separation of the water layer, methylene chloride was reduced to 2.0 mL; 2.0×10^{-7} moles of the internal standard chlorotoluene were added the methylene chloride extract. A GC separation of the extract gave peak areas of 625 and 530 for chlorotoluene and tetrachlorobenzene, respectively. What was the concentration of tetrachlorobenzene in the river water in parts per billion?
38. Go to the Web site for the National Environmental Methods Index (http://www.nemi.gov) and find the following:
    a. Two EPA methods that are used to determine chlorinated pesticides in water by using a gas chromatograph and ECD

b. An EPA method for determining organics in water by GC with a MS detector

c. An EPA method for determining the concentration of benzene in water

39. Define what is meant by the term "retention volume" as it applies to gas chromatography.

  a. What is retention time?

  b. Why is the flow rate kept constant?

  c. How do you determine the identity of specific GC peaks?

40. Go to the Web site for the National Environmental Methods Index (http://www.nemi.gov) and find the following:

  a. Two EPA methods that are used to determine Heptachlor in water by using a gas chromatograph and ECD

  b. An EPA method for determining Toxaphene in water by GC with an MS detector

  c. An EPA method for determining the concentration of Mirex in water

41. Draw a schematic of an MS.

  a. Describe how ions are made in the source of the MS.

  b. Does the MS operate at atmospheric pressure?

  c. How are the ions separated by mass?

42. What is a molecular ion?

  a. What is a fragment ion?

  b. What is the molecular ion of benzo(a)pyrene?

  c. What is the molecular ion of phenol?

  d. What is the molecular ion of dibenzyl?

43. The compound chlorobenzene has been found in contaminated sediments. What would be the expected m/z and intensity of its molecular ion(s) (suggestion: look up natural abundance of chlorine isotopes)?

44. Draw a schematic diagram of a GC-MS.

  a. How is the sample introduced into the GC-MS?

  b. Does the helium carrier gas from the GC interfere with the MS measurement of the mass of sample component?

  c. How is a component of a sample quantified by GC-MS?

45. Mercury undergoes reaction in sediments to form dimethylmercury. What would be the expected m/z and intensity of its molecular ion(s) (suggestion: look up natural abundance of mercury isotopes)?

46. What is the difference between a total ion chromatogram and a selected ion chromatogram?

47. EPA Method 613 calls for isotopically labeled 2,3,7,8-TCDD.

  a. Determine the molecular ion for 2,3,7,8-TCDD that is isotopically labeled with all carbons being ^{13}C.

  b. Determine the molecular ion for 2,3,7,8-TCDD that is isotopically labeled with all chlorines being ^{37}Cl.

48. Go to the Web site for the National Environmental Methods Index (http://www.nemi.gov) and find the following:

  a. Two EPA methods that are used to determine the concentration of dioxin

  b. An EPA method for determining furans

  c. What are the detection limits for the methods in part a?

49. Methyl t-butyl ether has been added to gasoline to improve air quality (see Chapter 10). Leaking gasoline storage tanks have released it into the environment, and it has found

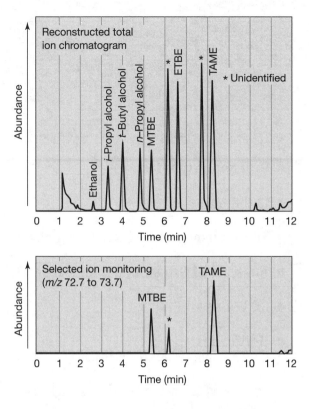

its way into groundwater. Methyl t-butyl ether can be measured at the parts per billion level by means of a GC-MS technique. Methyl t-butyl ether is an industrial chemical and contains impurities. Two of those impurities are ethyl t-butyl ether (ETBE) and t-amyl methyl ether (TAME).

a. What is the m/z expected for ETBE and TAME?

b. The total ion chromatogram of the sample is shown below. Why is m/z 73 being measured?

The major ions (m/z) for the three compounds are listed below:

MTBE	ETBE	TAME
73	87	87
57	59	73
	57	71
		55

c. Why are there not any peaks with m/z above 87?

d. Why does ETBE not give an m/z = 73 peak?

50. Tetrachlorodibenzodioxin (several isomers) and hydroxytetrachlorodibenzofuran (several isomers) all have a molecular formula of $C_{12}H_4O_2Cl_4$. Devise a method that would allow you to analyze for all simultaneously in environmental sediments.

51. Give two examples of environmentally important chlorinated furans.

CHAPTER

15

Insecticides, Herbicides, and Insect Control

- **Additional Sources of Information**

- **Keywords**

- **Questions and Problems**

THE THREE MAIN CLASSES OF SYNTHETIC ORGANIC INSECTICIDES IN USE TODAY ARE (1) CHLORINATED HYDROCARBONS, (2) ORGANOPHOS-PHATES, AND (3) CARBAMATES. Chlorinated hydrocarbons kill a wide range of insect types—including many that are beneficial—and are termed **broad-spectrum insecticides**. In Chapter 14, we saw how many organochlorine pesticides have been banned and have been labeled persistent organic pollutants. Figure 15.1 shows the structures of dichlorodiphenyltrichloroethane (DDT), chlordane, aldrin, and dieldrin, four organochlorine pesticides that are on the "dirty dozen" list. Most organophosphates and carbamates are narrow-spectrum insecticides, which means that they are toxic to only a few types of insects. They also are considered to be nonpersistent because they break down rapidly once they have been released into the environment. This chapter discusses less persistent insecticides, the analytic methods that the Environmental Protection Agency (EPA) uses to assess their presence in the environment, and alternate methods of insect control.

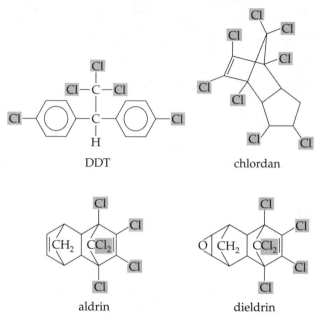

DDT

chlordan

aldrin

dieldrin

Figure 15.1 Examples of persistent organochlorine pesticides that have been banned. Adapted from M.D. Joesten, et al. *World of Chemistry*. Philadelphia: Saunders College Publishing, 1991.

■ Organophosphates

Organophosphate insecticides all contain a pentavalent phosphorous atom. There are three subclasses of organophosphates that are categorized by structural features of the molecules. Figure 15.2 shows the structures and examples of the three classes. The phosphonates (type A) are phosphonate esters and contain no sulfur. The organophosphate insecticide dichlorvos, which is used for insect control in food storage areas, green houses, and barns as well as on livestock, is an example of a type A organophosphate. Dichlorvos is fairly toxic to mammals; it has an LD_{50} of 25 mg/kg. In the phosphorothioates (type B), a sulfur replaces the doubly bonded oxygen to phosphorus. **Parathion**, a type B organophosphate that has an LD_{50} of only 3 mg/kg, is more than 20 times as toxic to rats as is DDT. Parathion is available for application by only licensed pesticide applicators. **Diazinon**, which is used to control cockroaches, silverfish, ants, and fleas in residential, nonfood buildings is also a type B organophosphate. Diazinon is very highly toxic to birds (an LD_{50} of 1.4 mg/kg for mallards). Citing excessive risk to children, in 2000 the U.S. EPA banned consumer use of diazinon, the nation's #2 selling home and garden insecticide. The phosphorodithioates (type C) have two sulfurs bonded to phosphorus. **Malathion**, which is the active ingredient in household fly spray, is a type C organophosphate. Malathion is not as toxic (an LD_{50} of 880 mg/kg) to mammals and has been sprayed in California and Florida to control the Mediterranean fruit fly.

Organophosphates are cheap to produce and are very effective against many different insects. Organophosphate pesticides are **neurotoxins**, nerve poisons that inactivate cholinesterase, the

Type A 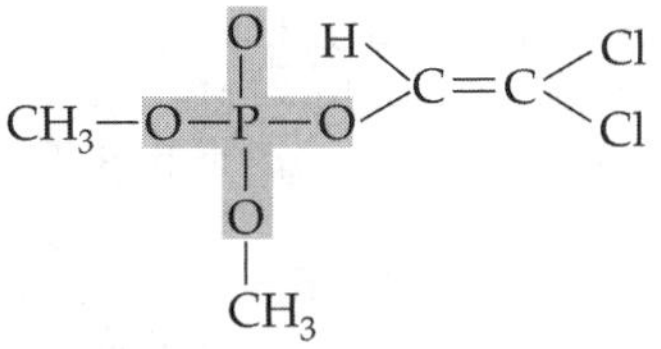   dichlorvos

Type B 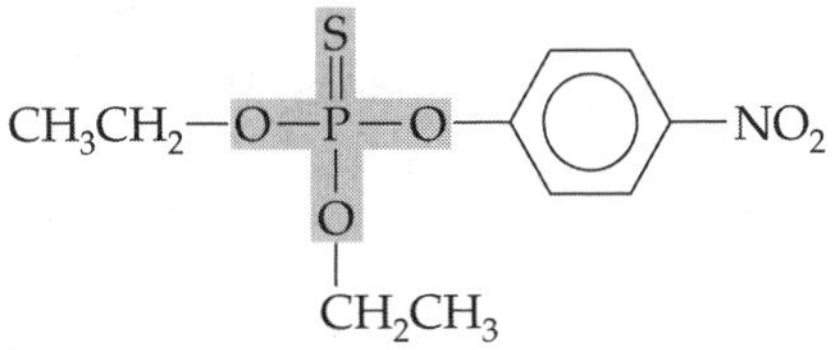   parathion 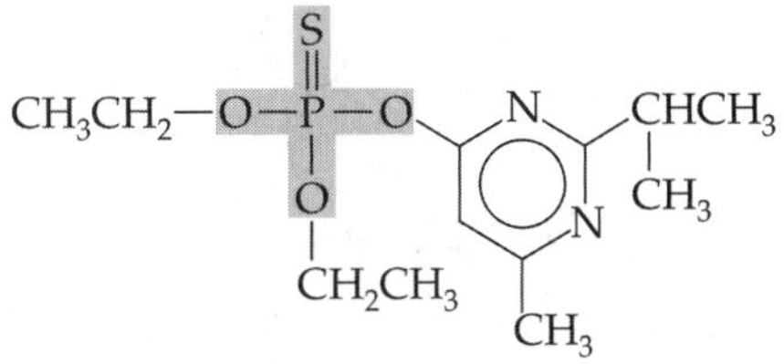   diazinon

Type C 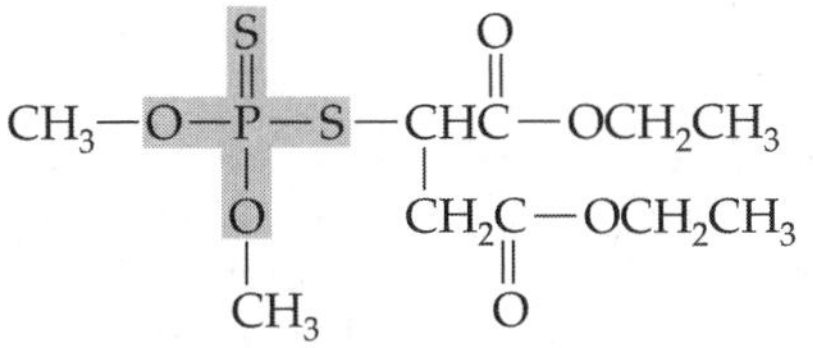   malathion

Figure 15.2 Types of organophosphate insecticides.

enzyme that plays a vital role in the transmission of nerve impulses between nerve fibers. Ingestion of organophosphates by humans can result in irregular heartbeat, convulsions, and even death. Because of this, farmers must wear respirators and protective clothing when applying them. Another disadvantage with organophosphates is that because they break down rapidly in the environment they must be applied frequently to be effective.

During World War II, German chemists developed the organophosphate nerve gases Tabun and Sarin. Tabun has a fruity odor, whereas Sarin, which is four times as toxic as Tabun, is odorless and thus more difficult to detect. After the war, the U.S. army began manufacturing both of these poisons (which they named agent GA and agent GB) and other related poisons.

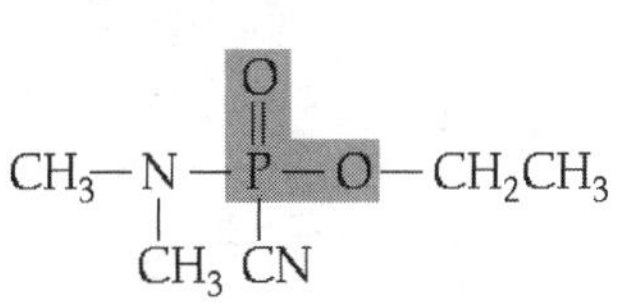

Tabun (GA) 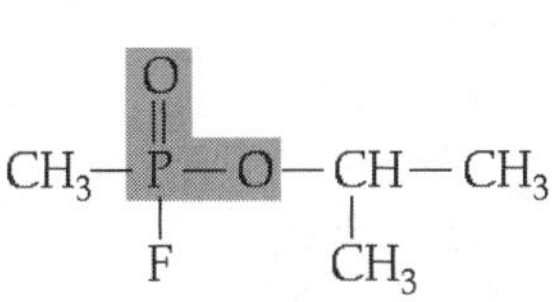   Sarin (GB)

Tabun and Sarin are also cholinesterase inhibitors. Both contain an alkyl group bonded to a PO_2 group. If troops were exposed to the gases, the gases would be absorbed through the skin as well as through the lungs. Victims would lose muscle control and die in a few minutes from suffocation.

In 1995, the Aum Shinrikyo, a Japanese religious cult obsessed with the apocalypse, released Sarin into the Tokyo subway system. The attack came at the peak of the Monday morning rush hour in one of the busiest commuter systems in the world. Witnesses said that subway entrances resembled battlefields as injured commuters lay gasping on the ground with blood gushing from their noses or mouths. The attack killed 12 people and sent more than 5000 others to hospitals. As a result of this attack and the September 11th terrorist attacks, chemical detection systems have been installed in the subway systems in Washington and New York.

■ Carbamates

Carbamates, which are derivatives of carbamic acid (Figure 15.3), include the following functional group in their structures:

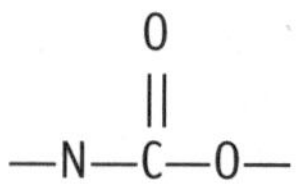

$$
\begin{array}{c}
O \\
\parallel \\
-N-C-O-
\end{array}
$$

Examples are **carbaryl** (Sevin) and **aldicarb** (Temik) (Figure 15.3).

Chlorinated hydrocarbons and most organophosphates kill a wide range of insect types—including many that are beneficial—and are termed broad-spectrum insecticides. Most carbamates are **narrow-spectrum insecticides**, which means that they are toxic to only a few types of insects. Unfortunately, one of the insects that they kill is the honeybee.

Carbamates, like organophosphates, inactivate cholinesterase and are rapidly broken down in animal tissues and in the environment, and most are much less toxic to humans and other mammals than are organophosphates.

■ The Transmission of Nerve Impulses

Nerve impulses travel along nerve fibers by electrical impulses. To pass from the end of one nerve fiber to receptors on the next nerve fiber, the impulse must cross a small gap, called the **synapse** (Figure 15.4). When an electrical impulse reaches the end of a nerve fiber, chemi-

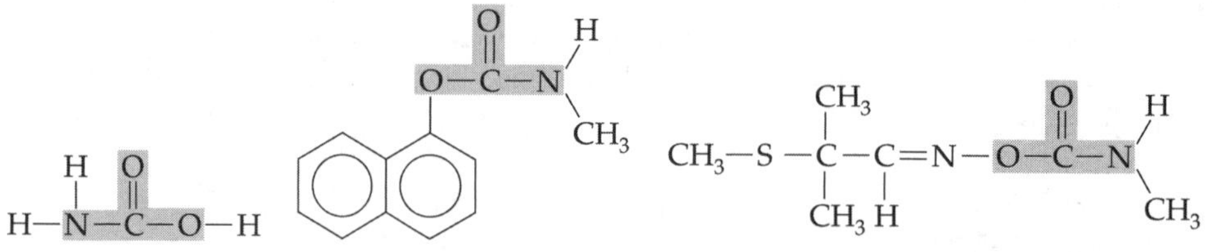

carbamic acid

carbaryl
(Sevin)

aldicarb
(Temik)

Figure 15.3 Carbamate insecticides such as carbaryl and aldicarb are derived from carbamic acid.

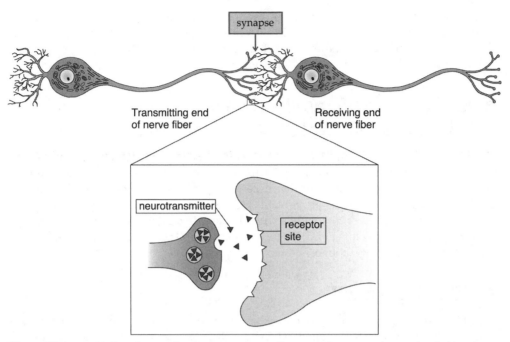

Figure 15.4 Transmission of a nerve impulse from one nerve fiber to another occurs when an electrical impulse stimulates the release of neurotransmitter molecules from the transmitter end of one nerve fiber. The neurotransmitter molecules cross the synapse and fit into receptor sites on the receiving end of another nerve fiber.

cals called **neurotransmitters** are released that allow the impulse to cross the gap and travel to receptor cells on the receiving nerve fiber. Each neurotransmitter must fit into a specific receptor to bring about the transfer of the message. Once the nerve impulse has been received, the neurotransmitter is destroyed; the synapse is cleared and is ready to receive the next electric signal.

An important neurotransmitter is acetylcholine. Once it has mediated the passage of an impulse across the synapse, it is broken down to acetic acid and choline. The reaction is catalyzed by the enzyme cholinesterase. Other enzymes convert acetic acid and choline back to acetylcholine, which can then transmit another impulse across the synapse.

$$\underset{\text{acetylcholine}}{CH_3\overset{\overset{\displaystyle O}{\|}}{C}OCH_2CH_2\overset{\overset{\displaystyle CH_3}{|}}{\underset{\underset{\displaystyle CH_3}{|}}{N^+}}-CH_3} + H_2O \xrightarrow{\text{cholinesterase}} \underset{\text{acetic acid}}{CH_3\overset{\overset{\displaystyle O}{\|}}{C}-OH} + \underset{\text{choline}}{HOCH_2CH_2\overset{\overset{\displaystyle CH_3}{|}}{\underset{\underset{\displaystyle CH_3}{|}}{N^+}}-CH_3}$$

Different neurotoxins disrupt the acetylcholine cycle at different points. They either block the receptor sites, block the synthesis acetylcholine, or inhibit the enzyme cholinesterase.

Organophosphate and carbamate insecticides are **anticholinesterase poisons** that prevent the breakdown of acetylcholine by inactivating the enzyme cholinesterase. These poisons bond to the enzyme and thus prevent the breakdown of acetylcholine. As the acetylcholine builds, nerve impulses are transmitted in quick succession, and nerves, muscles, and other organs are overstimulated. The heart begins to beat erratically, causing convulsions and death.

■ Analysis of Organophosphate Insecticides in Water

EPA Method 614 describes an analytical method for the determination of organophosphate insecticides in municipal and industrial wastewater. This analysis uses a **gas chromatography (GC)** method that is similar to that used for organochlorine pesticides. Because the organophosphate insecticides do not contain chlorine atoms, the electron capture detector is relatively insensitive to these compounds. EPA Method 614 describes the use of a GC that has a **flame photometric detector (FPD)**. This is a very specific test because the FPD detector only detects organosulfur or organophosphorous compounds.

■ GC Using an FPD

The FPD is used for pesticide analysis in water and food. The detector is sensitive to compounds containing sulfur or phosphorus. Figure 15.5 shows that the separated compounds that are separated in the capillary GC column are mixed with H_2 and O_2 as they flow into the FPD. In the FPD, they are burned in the hydrogen flame where the sulfur and phosphorus compounds give off light in a process called *chemiluminescence*. An optical filter is inserted between the flame and the detector to make the detector specific for ether sulfur or phosphorus. A filter that passes 526 nm is used to make the detector sensitive to only organophosphorus compounds (which is called for in EPA Method 614), and one passing 394 nm is used to make it sensitive to only organosulfur compounds. The amount of light reaching the detector is proportional to the amount of compound in the flame.

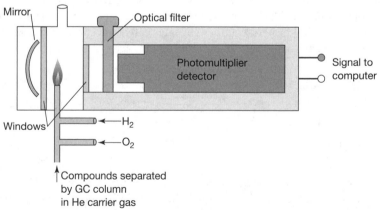

Figure 15.5 The FPD. Adapted from K.A. Rubison and J.F. Rubison. *Contemporary Instrumental Analysis*. Upper Saddle River, NJ: Prentice Hall, 2000.

■ Herbicides

In addition to animal pests, farmers must contend with weeds, which compete with crops for nutrients and water. The traditional method for controlling weeds is **tillage**. Typically, the soil is plowed, which turns over the top 10 to 15 inches of soil and buries and smothers the weeds. The soil is then harrowed to break large clods before the crop is planted. During the growing season, row crops must be cultivated to destroy weeds that grow between the rows.

With the introduction of selective **herbicides** in the 1950s, the need for many of these steps was reduced, and today, **no-till agriculture** is practiced extensively in most of the developed countries. Herbicides are applied to a field to kill weeds, and then, in a single operation, a planting machine cuts a furrow, adds fertilizer and seeds, and covers the seeds. The weeds that the herbicide killed remain as mulch and protect the soil from erosion and water loss. The amount of energy and manual labor saved by using herbicides in no-till farming is enormous. No-till farming also reduces the loss of topsoil caused by erosion. No-till farming methods are good both for the farmer and the aquatic ecosystem.

The first chemicals used for weed control were nonselective and included solutions of sodium chlorate ($NaClO_3$), sulfuric acid, copper salts, and various other inorganic chemicals. These herbicides were of limited value in agriculture because they could kill both weeds and crop plants with which they came in contact. The massive use of herbicides began in the 1950s with the introduction of the selective organic herbicides **2,4-D** and **2,4,5-T** (Figure 15.6a) and other related chlorophenoxy compounds. Today, approximately 50% of all pesticides used in the world are herbicides.

Selective herbicides kill only a particular group of plants, such as grasses or broad-leaved plants. Most selective herbicides are **systemic herbicides**; that is, they are absorbed by leaves or roots and then are transferred throughout the plant. They act like growth hormones, causing a plant to grow so fast that it dies. 2,4-D is very effective against emerging broad-leaved weeds; 2,4,5-T is used to control woody plants. A mixture of these two herbicides, known as Agent Orange, was used extensively as a defoliant during the Vietnam War. Because of health problems associated with its use, the EPA banned it in 1985.

A popular systemic herbicide used to control perennial grasses is **glyphosate**, which is sold under the trade name Roundup® (Figure 15.6b). It is a phosphate derivative of the amino acid glycine. Because of its ability to form an ion, it is attracted by ion exchange sites in the soil where it sticks, even though it is very water soluble. Glyphosate does not bioaccumulate. It is metabolized by soil microorganisms and has a half-life of approximately 60 days in the environment.

It works by inhibiting the enzymes necessary for the synthesis of the essential amino acids tyrosine and phenylalanine. Stopping this essential biosynthetic pathway kills the plant. Animals are unaffected by glyphosate because their biosynthetic pathways are different from those of plants. All plants are affected by glyphosate, and it cannot be used to kill weeds selectively. Farmers use it to clear all vegetation from a field and to prepare fields for no-till agriculture.

Another important herbicide is **atrazine** (Figure 15.6c), which is used extensively in the production of no-till corn crops. Atrazine acts by interfering with photosynthesis but has no effect on corn and a number of other crop plants because these plants have the ability to change the herbicide into a harmless form. Weeds, however, do not have this ability and

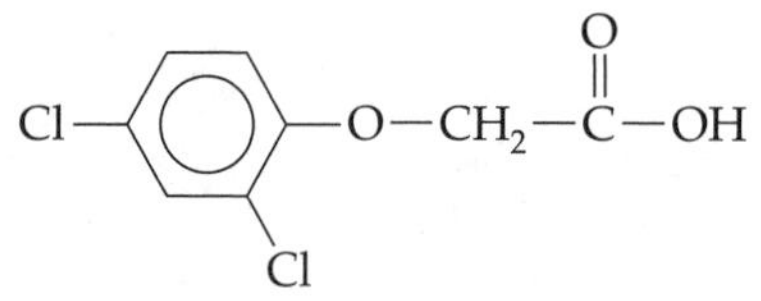

2,4-dichlorophenoxyacetic acid (2,4-D) 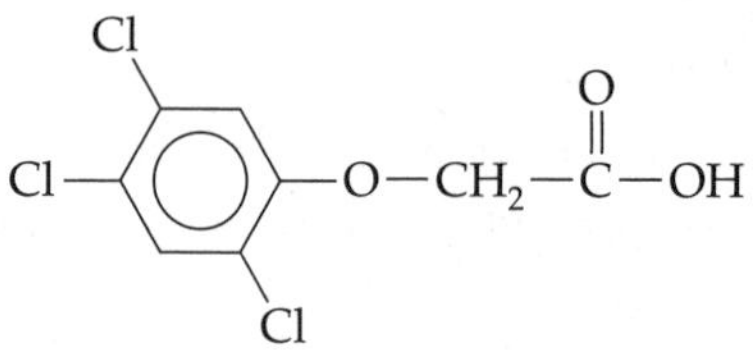   2,4,5-trichlorophenoxyacetic acid
(2,4,5-T)

a.

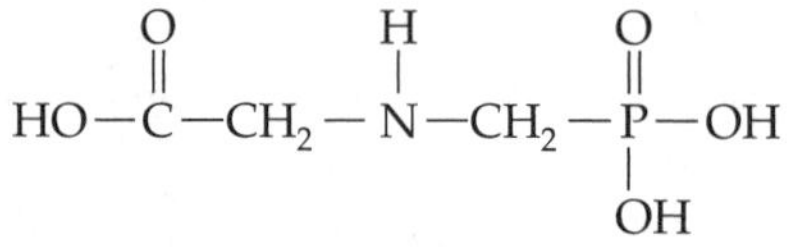

glyphosate   atrazine

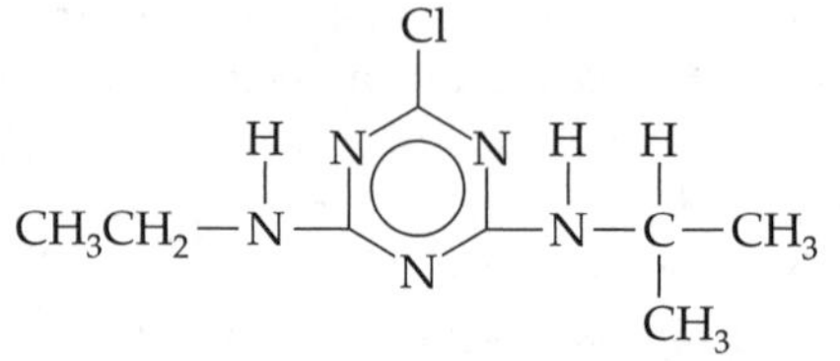

b.   c.

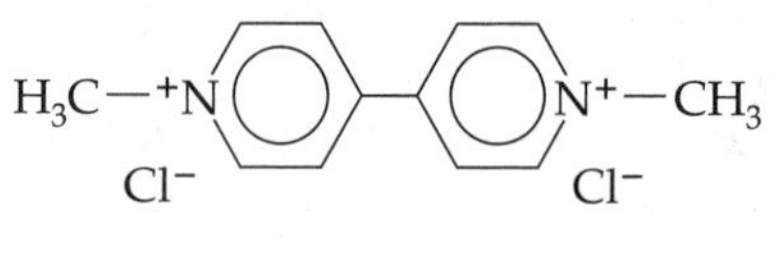

paraquat dichloride

d.

Figure 15.6 Some common herbicides.

quickly die. Atrazine does not bioaccumulate and is moderately soluble in water. It does, however, affect amphibians. Airborne atrazine from California is implicated in damage to frog populations in the Rocky Mountain states.

Paraquat, 1,1'-dimethyl-4,4'-bipyridilium dichloride salt (Figure 15.6d), and several related compounds, such as diquat and difenzoquat, also are widely used as herbicides. These bipyridilium compounds contain two pyridine rings per molecule. These herbicides are toxic to most plants, but because they break down rapidly in the soil, they can be used to kill weeds that emerge before the crop plant. Paraquat is toxic to humans and must be used with care. Because of its widespread use, the contamination of food or drinking water by paraquat has been reported.

Analysis of 2,4-D and 2,4,5-T herbicides (EPA Method 5.15.1) is accomplished by extraction of the herbicide from water followed by separation and quantitation using a GC with electron capture detector. This method is similar to that described in Chapter 14 for the analysis of chlorinated pesticides. Paraquat, on the other hand, is a very polar molecule that cannot be measured by a GC technique. The temperatures needed to vaporize paraquat into the gas phase in the GC are so high that decomposition of the molecule occurs. Because of this, a **high-pressure liquid chromatography (HPLC)** technique is used to measure the concentration of paraquat in drinking water.

High-Pressure Liquid Chromatography

HPLC is a technique that is used for compounds that are not easily made into gases for separation by GC. Most biological molecules have too high of a boiling point or decompose before boiling and therefore cannot be separated by GC. HPLC uses a high-pressure pump to force solvent through separation columns that contain very fine particles that give high-efficiency separations. EPA Method 549.1 uses a HPLC method to determine the biprydidilium herbicides paraquat and diquat.

In HPLC, each module can be purchased separately. Figure 15.7 shows that the components are: a solvent reservoir, solvent pump, sample injector, separation column, detector, and data system. It is very similar to an ion chromatograph (described in Chapter 9). A separation that uses a single solvent of constant composition is called an *isocratic elution*. Separation efficiency can sometimes be improved by using gradient elution; two solvent systems are used to change the ratio of solvents continuously. Unlike GC, where the mobile phase was a gas, in liquid chromatography, a liquid moves across the stationary phase.

HPLC separation columns are constructed from stainless steel tubing that can withstand pressures up to 5000 psi. The columns are filled with stationary phase particles with typical diameters from 5 to 30 μm of silica, alumina, glass, polystyrene-divinylbenzene resin, or ion-exchange resin depending on the type of the HPLC separation. For EPA Method 549.1, the separation column is filled with polystyrene-divinylbenzene resin beads, which are 5 μm in diameter.

Many different types of detectors can be used in HPLC: a differential refractive index, ultraviolet (UV) absorbance, fluorescence, evaporative light scattering, and electrochemical and mass spectrometry. EPA Method 549.1 calls for a detector that can measure UV absorbance. The wavelength of UV light (λ) is adjusted to 308 nm for diquat and 257 nm for paraquat. The UV absorbance at 308 nm is directly proportional to the concentration of paraquat in the sample. The minimum detection limit for this method is less than 1 part per billion of paraquat.

The liquid mobile phase is prepared by mixing a buffered aqueous solution with 1-hexanesulfonic acid. Once injected into the HPLC, the bipyridilium herbicides, which

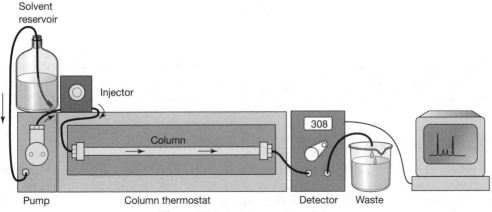

Figure 15.7 Components of an HPLC.

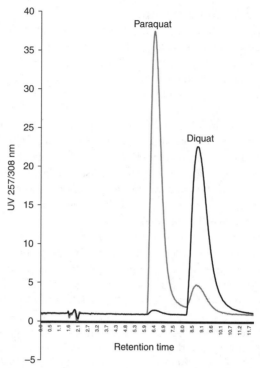

Figure 15.8 HPLC separation of paraquat and diquat. Courtesy of Analytical Sales and Services, Inc., Pompton Plains, NJ.

are quaternary ammonium cations, form ion pairs with the conjugate base of the hexane sulfonic acid.

The hexane tail of the sulfonic acid is attracted to the nonpolar stationary phase and the ion-pair partitions between the liquid and stationary phase. Figure 15.8 shows an HPLC chromatogram of the separation and a UV spectra for paraquat and diquat.

■ Problems with Synthetic Pesticides

Synthetic pesticides have enabled the control of insect-borne diseases and have been responsible for large increases in crop yields, but their use also has created many problems. The major problems are that the pests develop resistance and the pesticides pose threats to the environment and to human health.

The Pesticide Treadmill

Because of genetic variations, some insects in a population have more **resistance to pesticides** than do others. When an area is sprayed with a pesticide for the first time, most insects are killed, but the resistant ones survive to breed the next generation. With repeated spraying of the same pesticide, each generation contains a higher percentage of resistant insects. Control can be maintained only if increasingly larger doses of pesticide are applied. Eventually, a new, more potent pesticide may be required. Another problem is that after a pest has been brought under control, a **resurgence** often occurs, with the pest returning in even greater numbers.

If beneficial insects are destroyed together with the target pests, the natural ecosystem is upset, and other problems result. Honeybees that normally pollinate crops are killed, and minor pests that previously were controlled by natural predators multiply, causing serious **secondary pest outbreaks**. Farmers find themselves on a **pesticide treadmill** (Figure 15.9), continually spending more money for larger quantities and more varieties of pesticides, which become increasingly ineffective.

Health Problems

The people at the greatest risk from pesticides are farmworkers and workers in pesticide manufacturing plants. Numerous pesticide-related illnesses and deaths have occurred, particularly in less developed countries, among farmworkers who have failed to take adequate precautions when applying pesticides or who have entered sprayed areas before the pesticide had broken down to a harmless form. Organophosphates are especially dangerous in this respect.

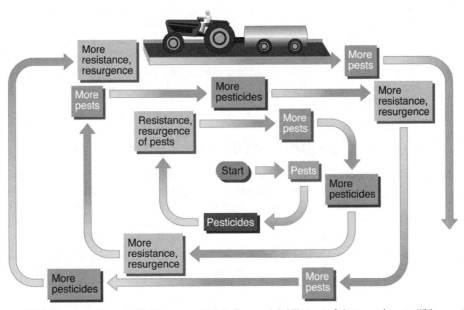

Figure 15.9 The pesticide treadmill. When a pesticide is first used, it kills most of the target insects. With repeated use, the number of insects resistant to the pesticide increases, and a more potent pesticide must be applied. Once an insect population is controlled, resurgence often occurs, and more and more pesticide of ever increasing potency must be applied in an unending cycle. Adapted from B.J. Nebel. *Environmental Science: The Way the World Works, Third edition*. Englewood Cliffs, NJ: Prentice Hall, 1990.

The worst pesticide accident on record occurred in 1984 when highly toxic **methyl iso-cyanate** (CH_3NCO), a gas used in the manufacture of carbaryl (Sevin) and other pesticides, leaked from a Union Carbide plant in Bhopal, India. More than 3000 people were killed; 14,000 suffered severe injuries, including blindness, brain damage, sterility, and liver and kidney damage. Many thousands more suffered less severe injuries.

The possibility of water contamination by persistent pesticides that leach through soil into groundwater is a continuing concern for the general public. A disturbing example was the **ethylene dibromide** (**EDB**) ($BrCH_2CH_2Br$) contamination of well water that occurred in several areas where the pesticide had been used as a soil fumigant to control root nematodes. EDB was also used to protect stored grain, and unacceptable levels were detected in flour and other foods. Although it had been shown in 1970 that EDB caused cancer in laboratory animals, opposition from users and manufacturers delayed an EPA ban on its use until 1984.

There is great concern today that pesticide residues in foods may cause adverse health effects, and the EPA is continuously reassessing the legal limits it sets for pesticides in food. Consumer groups have called for a ban on most uses of methyl parathion on fruits such as peaches, which in processed form are eaten frequently by infants and young children. Children are of particular concern because, weight for weight, they usually eat more fruits and vegetables than do adults and because they tend do be more sensitive to neurotoxins and carcinogens than adults.

Biochemist Bruce Ames, who developed a screening test for carcinogens (discussed in Chapter 16), believes that the risk from pesticide residues is exaggerated. He has pointed out that because many of the foods we eat, including potatoes, tomatoes, peanut butter, mushrooms, celery, wine, and beer, contain toxic chemicals, including carcinogens, our exposure to natural toxins far exceeds any exposure from pesticide residues.

In addition to possible exposure via food and water, people are often exposed to pesticide mists that drift beyond the target areas. Other problems are caused by the enormous quantities of herbicides and other pesticides that are applied annually to private lawns, golf courses, and other nonagricultural land. In the United States, legislation requires that the EPA registers and approves all pesticides.

■ Alternative Methods of Insect Control

Each year pests destroy one-third of all of the food grown in the world. In the United States alone, crop losses caused by insects amount to over $4 billion annually. The total world population of insects is estimated to be approximately a billion-billion (10^{18}), and 0.01% (10^{14}) of them are agricultural pests. Obviously, without some form of pest control, there would be a drastic drop in food production, and our ability to feed the world would be jeopardized. Fortunately, some alternatives to the massive use of synthetic pesticides are now available. In this section, we study biological and chemical methods that control insects in ways that do not threaten either the environment or human health.

Chemical Communicating Substances

Pheromones
Pheromones are chemical substances that are released by insects and other animals as a means of communication. They are used to mark trails and territory and to attract mates. A number

of species-specific pheromones secreted by female insects to attract males have now been extracted from insects or synthesized in the laboratory and used to control insect pests. The pheromones, which are effective in picogram (1×10^{-12} g) amounts, can be used to lure male pest insects into baited traps or can be sprayed over an infested area. Males in a sprayed area smell a female in all directions and are unable to follow a pheromone trail to locate a female, thus fail to mate. Approximately 30 insect pheromones are now available for insect control.

Pheromones represent an environmentally attractive approach to pest control. They are species specific and thus affect only the target pest and are biodegradable, nontoxic, and effective at very low concentrations; however, they are costly and difficult to produce. Many thousands of insects must be collected to obtain sufficient material for analysis, and identification and synthesis of the desired pheromone present many problems.

Sex Pheromones

The gypsy moth (*Lymnantria dispar*) is a species that is not native to North America and was brought to the United States by individuals that wanted to start a silk industry. When gypsy moths escaped, they multiplied rapidly, and because they can feed on over 500 species of vegetation, they ravaged forests of the northeastern United States. The sex attractant of the female gypsy moth was identified as (Z)-7,8-epoxy-2-methyloctadecene and has been given the trade name disparlure. The U.S. Forest service baits traps with disparlure and places them in forests to study the movement of the gypsy moth infestation. Knowing the movement of the gypsy moth allows them to plan a strategy to control the infestation.

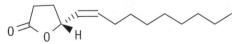

Disparlure, gypsy moth male attractant
(Z)-7,8-epoxy-2-methyloctadecene

The sex pheromone of the female Japanese beetle (*Poplilla japonica*) is R-(Z)-5-(1-decenyl)dihydro-2(3H)-furanone, and the synthetic version is sold under the trade name Japonilure. Japanese beetle traps, which use Japonilure to attract the male Japanese beetle, are familiar objects in many American front-yard gardens (Figure 15.10). Putting the trap in your front yard might actually cause more damage to your garden because the attracted beetles often feed on the way to the trap. The trap is effective; you should give it to your neighbor as a gift. That way the Japanese beetles are attracted out of your yard and into your neighbor's yard (nice gift).

Japanese beetle (*Poplilla japonica*)
Japonilure, male sex attractant

Figure 15.10 A Japanese beetle trap. Courtesy of the Tanglefoot Company.

Notice that the structures of the sex attractants of the gypsy moth and Japanese beetle are complex organic molecules with specific geometry and stereochemistry. Because the pheromone molecule must fit into an active site in the insect's antennae, a stereoisomer similar in structure but with different geometry will not fit the receptor. Because of this, the synthesis of these pheromones is a complex, multistep synthetic process, and the cost of the synthetic pheromone is high. Luckily, only a minute amount is needed for each trap.

The 3M™ Corporation produces 3M™ Sprayable Pheromone® for insect mating disruption. Figure 15.11 shows that the pheromone is microencapsulated within microscopic polymeric capsules. When sprayed into the environment, the active ingredients are released slowly over time. In areas treated with 3M Sprayable Pheromone, the air is permeated with pheromone so that the males cannot follow the female's plume, so mating is disrupted. By suppressing mating, fewer egg masses and fewer eggs per mass are laid, resulting in fewer larvae to cause crop damage. 3M and other companies sell sprayable pheromones for the control of many insects that infest fruit trees: the grape berry moth, black-headed fireworm, sparganothis fruitworm, peachtree borers, and the oriental fruit moth. The sprayable time-released pheromones are very effective in controlling pests in commercial orchards.

Sex attractants target adult insects to disrupt mating (Figure 15.12). If mating does occur, pheromones do not control juvenile forms, such as larvae and caterpillars, which often do the most harm. Commercial pesticides, on the other hand, target the larvae of the insect.

Aggregation and Trail Pheromones

Social insects, such as ants and bees, send chemical messages to attract other members of the species for defense or to lead the way to a new food supply. Termites (*Zootermopsis nevaensis*) use hexanoic acid to mark their trails. Each member taps the tip of his or her abdomen on the ground to reinforce the trail scent.

(*Zootermopsis nevaensis*)

Pest control contractors could use a trail of hexanoic acid to lead termites away from your home and into a trap.

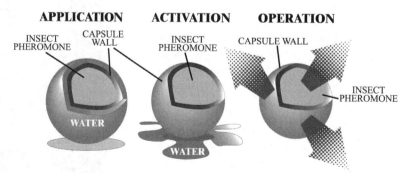

APPLICATION **ACTIVATION** **OPERATION**

Figure 15.11 The encapsulated pheromone is slowly released from the chamber in the center of the capsule through the capsule wall. Water does not permeate through the capsule wall to hydrolyze the pheromone. Material developed by 3M, May 1998. This material is reproduced by courtesy of 3M. No further reproduction is permitted without 3M's prior written consent.

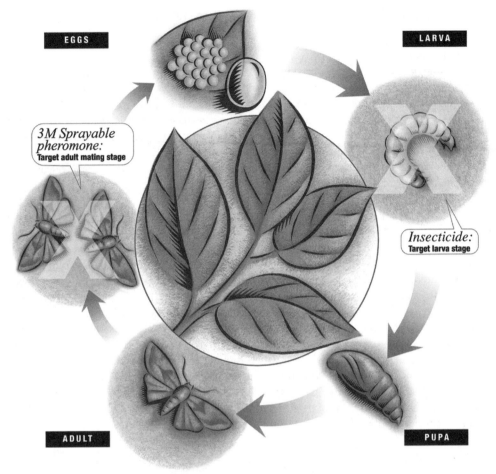

Figure 15.12 Pheromones target adult insects whereas commercial pesticides target insect larvae. Material developed by 3M, May 1998. This material is reproduced by courtesy of 3M. No further reproduction is permitted without 3M's prior written consent.

Some species of termites release an alarm pheromone when they sense danger. The alarm pheromone signals trouble, and they begin to flee. The alarm pheromone of the termite species *Drepanotermes rubiceps* and *Nasutitermes exitiosus* is shown here.

Drepanotermes rubiceps   *Nasutitermes exitiosus*

An application of the alarm substance to the foundation of your home would frighten termites away.

Juvenile Hormones

Another biological approach to insect control is the use of **juvenile hormones (JHs)**. Hormones are chemicals that are produced in special plant and animal cells. They are transported in the bloodstream to specific target tissues where they control and mediate many physiologic activities. JHs secreted by immature insects regulate the early stages of development. As the insect matures, production of JH ceases, and the insect develops into an adult. If immature insects are sprayed with JH, they remain at an immature stage and are unable to mate.

JH is a vital component of the insect endocrine system. JHs are produced and released by the corpus allatum, a gland at the base of the insect's brain, secreted into the hemolymph, and transported by binding proteins; they enter cells by diffusion across the cell membrane, and there the products of the corpus allatum interact in some way with genome, probably via nuclear receptors of the steroid family.

In 1972, the author of this book worked with a group of scientists that isolated and identified the JH of the house fly, *Mustica bullata*. More than 500 pounds of house flies were raised, ground up, and extracted. The extract was subjected to sequential chromographic separations that yielded only 200 ng of JH.

At least five naturally occurring structurally similar compounds with JH activity have been found in different insects: JH-0, JH-I, JH-II, JH-III, and 4-methyl JH-I. JHs are a unique group of sesquiterpenoid compounds that have been identified definitively in only insects. They are structurally unique among animals and constitute a family of derivatives of farnesoic acid. They appear to have many endocrine functions, including insect growth, development, and metamorphosis. In many adult insect species, JH regulates reproductive maturation and affects sexual behavior. JH is involved in other activities as well, targeting many tissues, including brain, fat body, flight muscle, and accessory glands.

JH-0

JH-I

JH-II

JH-III

4-methyl JH-I

The epoxide functional groups in all naturally occurring JHs are easily hydrolyzed to a diol, which reduces the effectiveness of the JH. Companies have synthesized JH mimics; molecules that are structurally similar to JH are stable molecules and function like JH when applied to insect larvae. The first commercially available JH mimic is called methoprene (Figure 15.13), which the EPA approved in 1975. The labile epoxide functional group in JH was replaced by a hydrolytically stable methoxy in methoprene, which makes methoprene more stable and cheaper to produce than a natural JH. In various formulations, methoprene and related compounds have been used to control mosquitoes, fleas, cockroaches, and a number of other insects that are pests in the adult stage. JHs are not as useful against insects that are agricultural pests in their immature stage. When used against caterpillars, for example, the immediate effect is to cause the caterpillars to grow larger and eat more of the crop for a longer period of time.

Allomones

Allomones are chemical communicating substances that act between species, and the chemical communicating substance provides a benefit to the species releasing it. Skunks, for instance, produce foul-smelling mercaptans that they spray with their urine to discourage predators from attacking them. Plants and trees can also protect themselves from insects with toxic chemicals that are synthesized in response to stress. The walnut tree, for instance, synthesizes the allomone juglone as a natural insecticide.

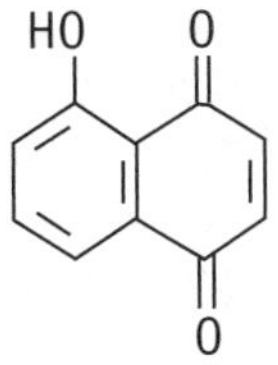

juglone
walnut leaves (*Juglans* sp.)

Other **insect growth regulators** that disrupt development at various stages are also available, and new ones are constantly being developed. Chitin synthesis inhibitors, which prevent the growth of chitin, the substance that forms an insect's shell, are a promising recent development. Without a protective shell, the insect dehydrates and dies.

■ Biological Controls

In **biological control** methods, pests are kept in check by natural enemies, including predators, parasites, and disease-causing bacteria or viruses. To be effective, predator insects must be carefully chosen. Praying mantises and ladybugs that are sold commercially to home gardeners are

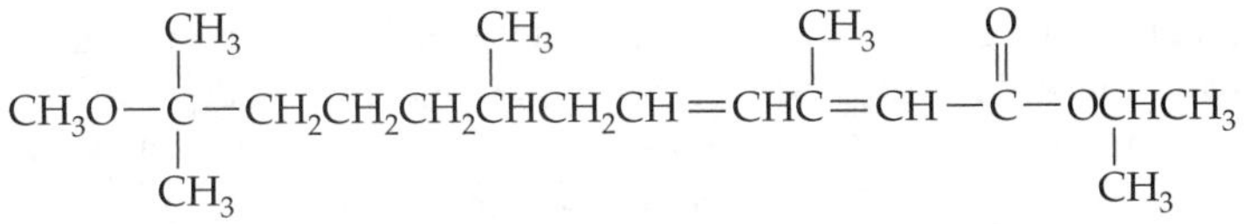

methoprene

Figure 15.13 Methoprene, the first commercially available JH mimic.

generally of little value in controlling pests. Praying mantises are rapidly reduced in number because they devour each other and are as likely to eat useful insects as they are to eat predator insects. Ladybugs, if not at the feeding stage when released, simply fly away.

The naturally occurring soil bacterium *Bacillus thuringiensis* is toxic to many leaf-eating caterpillars. Marketed as a powder under the name DiPel®, it has been used by home gardeners for many years. It is also used as a spray to control gypsy moths. In 1972, a virus was approved for use against the *Heliothus zea* species of bollworms, the moth larvae that attack the tips of corn kernels (corn earworms) and burrow into cotton bolls (cotton bollworms).

Although a great deal of research is needed to identify suitable biological agents and production is difficult and costly, biological controls have many advantages. In general, they destroy only the target species and are nontoxic to other species. Once established, they are self-perpetuating, and pests do not become resistant to their natural enemies.

■ Sterilization

Insect sterilization is another option for pest control. In this technique, large numbers of the pest insects are bred and then sterilized by exposure to radiation. The irradiated insects are released in the target area where they mate with normal insects but produce no offspring. This method is only effective if the number of sterile insects released is sufficient to overwhelm the natural population of nonsterile insects. The procedure is expensive and of limited applicability, but it has been used with great success against the screwworm fly, a deadly pest that lays its eggs in the open wounds of cattle. The developing larvae feed on the wound, and resulting infections can kill the animal. Sterilization also has been used successfully to control certain tropical fruit flies.

The various alternative methods of insect control techniques just described are not as immediately effective or as easy to apply as the commonly used synthetic organic insecticides, but they have many advantages for the environment.

■ Future Farming: Less Reliance on Agricultural Chemicals

Pressure from environmental groups and consumers and fear that supermarkets may reject products that contain pesticide residues have encouraged farmers to adopt farming practices that rely less heavily on the massive use of synthetic fertilizers and pesticides. In California, where approximately 50% of the vegetables and 49% of the fruits and 90% of nuts produced in the United States are grown, the number of farmers using no synthetic agrichemicals has greatly increased during the last 10 years. To control pests, these farmers rely on **integrated pest management (IPM)**, a technique that was first introduced in the 1960s.

In IPM, several pest control methods are integrated into a carefully timed program. Success depends on knowledge of the metabolism of the crop plant and of the life cycles of the pests to which the crop is susceptible and on an understanding of the interactions of the pests with their whole environment. IPM makes use of sex attractants, male sterilization, natural pesticides such as **pyrethrins** and nicotine, and the release of natural predators.

pyrethrin
(*Chrysanthemum*)

nicotine
(*Nicotiana tobaccum*)

Because many pests feed on only one particular crop, crop rotation and diversity are introduced to prevent insects, fungi, and microbes from building up in the soil and in crop residues from year to year. Planting and harvesting times are adjusted to maximize the number of insects that die of starvation between successive crops.

The aim of IPM is not to eliminate all pests but to keep them at a level low enough to prevent economic loss. A program does not exclude the use of synthetic pesticides but reserves them for selective use at the lowest possible dose and only where absolutely necessary. IPM programs, which usually must be individualized to suit soil and climate conditions for each crop, are expensive to set up and more labor-intensive than conventional methods but on a long-term basis are less expensive.

In 1989, the IPM approach received support from the Board of Agriculture of the National Research Council, which recommended greater emphasis on biologically based methods of farming, including IPM, the use of manure, and crop rotations to return fertility to the soil, and mulch and other nonchemical means of controlling weeds. In 1996, IPM was endorsed by the National Research Council, which issued a report recommending a reduction in the use of synthetic pesticides on farmland.

Consumers could play an important part in reducing pesticide use if they would tolerate weeds in their lawns (or switch to a ground cover other than grass) and if they would accept fruits and vegetables with some blemishes. Currently, many pesticides are applied, not to increase yield, but solely to achieve the perfect appearance that many consumers demand.

■ New Varieties of Crop Plants

World grain production increased dramatically during the last 50 years (Figure 15.14). In addition to the use of synthetic fertilizers and pesticides, the third major factor responsible for the increase has been the development of new varieties of crop plants. In the 1960s, new high-yielding grains developed by agricultural scientists through selective cross-breeding were introduced into India, South America, and other developing countries, enabling them to triple yields per acre. Equally important has been the development of crop plants resistant to attack by fungi and insect pests.

Breeding plants with a new trait is a slow process that may take at least 10 years of painstaking work. Newer genetic engineering techniques can reduce the time required to as little as one year. In 1994, after Food and Drug Administration approval, the first genetically engineered tomato, called the Flavr Savr, was sold commercially. It was engineered to postpone the normal softening and decay process, thus making it possible for the farmer to wait until the

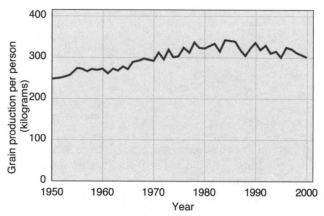

Figure 15.14 World grain production per person from 1961 to 2004.

fruit ripened on the vine before picking it. Commercial sales of the Flavr Savr tomatoes flagged and after a few years it was removed from the marketplace because it was not profitable.

■ Can We Feed Tomorrow's World?

The demand for food is expected to double its present level by the middle of the 21st century. Meeting this demand will be very difficult because the technologies that enabled farmers to increase grain production so dramatically during the last 50 years—fertilizers, new crop varieties, pesticides, and irrigation—already have been fully exploited in many countries. Plowing more land would seem to be an obvious option, but because of soil degradation and conversion of land to other uses, very little new cropland is available. Because water tables are falling on all continents, and many rivers are running dry before reaching the sea; irrigating arid land to grow more crops is no longer possible in most countries.

Additional use of fertilizer can be expected to increase yields in some developing nations, but the developed nations are already using the maximum amount of fertilizer that existing crop varieties can use effectively.

Genetic engineering is not expected to increase yields much more because already in today's crop plants the share of the plant's photosynthate (the product of photosynthesis) going to the seed is between 50% and 55%. Plant physiologists do not believe this share can be increased above approximately 60%. One way scientists hope to increase grain production is by developing crop plants that are resistant to drought and will grow in salty soils.

Since 1996, scientists have been developing a wide variety of crops, including corn, potatoes, rice, and cotton, that are genetically engineered to produce *B. thuringiensis*. These plants produce high levels of the insect-killing toxin throughout the growing season. Many people oppose planting these crops because they fear that insects will develop resistance to *B. thuringiensis,* which is one of very few effective natural pesticides.

As gains in grain productivity continue to slow, the world demand for food continues to expand. Unless there is a reduction in the rate of population growth, food shortages in the 21st century appear to be inevitable.

■ Additional Sources of Information

David Pimentel, editor. *Techniques for Reducing Pesticide Use: Economic and Environmental Benefits.* New York: John Wiley & Sons, 1997.

Environmental Protection Agency (EPA). Homepage. Accessed December 2008 from http://www.epa.gov.

Fong WG, Moye HA, Seiber JN, Toth JP. *Pesticide Residues in Foods: Methods, Techniques, and Regulations.* New York: John Wiley & Sons, 1999.

Lee PW, Aizawa H, Barefoot AC, Murphy JJ. *Handbook of Residue Analytical Methods for Agrochemicals.* New York: John Wiley & Sons, 2003.

Whitford F. *The Complete Book of Pesticide Management: Science, Regulation, Stewardship, and Communication.* New York: John Wiley & Sons, 2002.

■ Keywords

2,4-D
2,4,5-T
Aldicarb
allomones
anticholinesterase poisons
atrazine
biological control
broad-spectrum insecticides
carbamates
carbaryl
chlorinated hydrocarbons
diazinon
ethylene dibromide (EDB)
flame photometric detector (FPD)
gas chromatography (GC)
glyphosphate
herbicides
high-pressure liquid chromatography (HPLC)
insect growth regulators
insect sterilization
integrated pest management (IPM)

juvenile hormones (JHs)
malathion
methyl isocyanate
narrow-spectrum insecticides
neurotoxin
neurotransmitters
no-till agriculture
organophosphate
paraquat
parathion
pesticide treadmill
pheromones
primary nutrient
pyrethrins
resistance to pesticides
resurgence of insects
secondary pest outbreaks
synapse
systemic herbicides
tillage

■ Questions and Problems

1. There are three major classes of synthetic organic insecticides. In which of these does parathion belong?
2. What is meant by the term *persistent pesticide*? Give an example.
3. Are any organophosphates or carbamate pesticides on the persistent organic pollutant "dirty dozen" list?

4. Describe the basic structural feature that all organophosphate insecticides possess. What characterizes a
   a. Type A organophosphate insecticide
   b. Type B organophosphate insecticide
   c. Type C organophosphate insecticide
5. List the following organophosphate insecticides in the order of their toxicity: malathion, diazinon, parathion, or dichlorvos?
6. Are carbamates broad-spectrum insecticides? Draw the carbamate functional group.
7. How do the structures of the insecticide diazinon and the nerve gas Tabun differ?
8. Are carbamate insecticides persistent? Explain.
9. Define *neurotoxin*. Give three examples of three naturally occurring chemicals that are neurotoxins.
10. What is the function of acetylcholine?
11. Show how organophosphate insecticides inactivate the enzyme cholinesterase.
12. Draw a schematic of a gas chromatograph that is configured to carry out the analysis of organophosphate insecticides by EPA Method 614.
13. Describe EPA Method 614. Start with the collection of the water sample and end with the analytic result (concentration of organophosphate insecticides in the sample).
14. Why does EPA Method 614 use a FPD for the GC. Why not use an electron capture detector as is used for chlorinated insecticides?
15. Describe how the FPD operates. How is it made sensitive for only organophosphorus compounds?
16. With regard to herbicides,
    a. What were the first chemicals used for weed control?
    b. What is a systemic herbicide?
    c. What is the composition of Agent Orange? For what purpose and when was it used?
17. What is meant by no-till farming?
18. What is tillage?
19. Draw the structure of the herbicide glyphosphate.
    a. What is its brand name?
    b. Does it selectively kill weeds?
    c. Why does it stick to soil?
    d. What is its half-life in the environment?
20. Why does glyphosphate kill plants and not mammals?
21. Draw the structure of the herbicide atrazine.
    a. To what class of organic molecules does it belong?
    b. Does it kill corn or other crop plants?
    c. Does it bioaccumulate?
22. Draw the structure of the herbicide paraquat.
    a. To what class of organic molecules does it belong?
    b. How is it used to kill weeds?
    c. Is it toxic to humans?
23. How would you analyze a ground water sample for the herbicide 2,4,5-T? Describe your method in detail.
24. Draw a schematic diagram of an isocratic HPLC system. Label all parts.

25. How does a gradient HPLC differ from an isocratic HPLC? Describe how the gradient is made.

26. In Chapter 9, ion chromatography was used to separate ions. Did the ion chromatography described there use a gradient technique?

27. When separating paraquat, why is hexanesulfonic acid added to the mobile phase?

28. List four different stationary phases that are used for HPLC separations. What stationary phase does EPA Method 549.1 use?

29. How would you analyze a ground water sample for the herbicide paraquat? Describe your method in detail.

30. List six different detectors that can be used with a HPLC. What detector does EPA Method 549.1 call for?

31. Describe in detail what is meant by the *pesticide treadmill*?

32. Is EDB a persistent pesticide?
  a. Has the EPA banned it?
  b. For what was it used?

33. Why is there so much concern for children's exposure to pesticide residues?

34. What is a pheromone?
  a. How is a pheromone used to control insects?
  b. How much pheromone do you need?
  c. Are all pheromones sex attractants?
  d. Do sex pheromones control insect larvae?

35. List five advantages of using sex pheromones to control insects.

36. Are trail-marking substances used by ant pheromones? How could such a substance be used as a method of controlling ant infestations?

37. Are alarm substances used by termite pheromones? How could such a substance be used as a method of controlling termite infestations?

38. What is an insect JH? What function does it play in the development of an insect? Its structure contains what functional groups?

39. Describe how sex attractants are used to control insect populations.

40. What is a JH mimic?
  a. Why synthesize a mimic?
  b. Have JH mimics been effective against insects?
  c. What happens if JH mimics are applied to an immature insect?

41. Are JHs effective in controlling the population of agricultural pests?

42. What is an allomone? Give two examples.

43. What is an insect growth regulator, IGR?

44. The powder DiPel is a biological control agent. Describe how it is used to control leaf-eating caterpillars.

45. Describe how sterilization can be used as a technique to reduce an insect population.

46. How do alternative methods of insect control differ from those used by insect exterminators?

47. How does the objective of IPM differ from that of applying synthetic pesticides?

48. Two natural insecticides used in IPM are nicotine and pyrethrin. What is the source of these chemicals?

49. When did world grain production reach its maximum?

50. Why not irrigate arid land to increase the amount of grain produced in the world?

51. How can genetic engineering be used to increase world grain production?

52. What is *B. thuringinesis*? How can it be used in IPM?

CHAPTER

16

Toxicology

HUMANS HAVE BEEN FAMILIAR WITH THE HARMFUL EFFECTS OF ANI-MAL VENOMS AND POISONOUS PLANTS FOR THOUSANDS OF YEARS. This knowledge was used in hunting and in warfare. The earliest known document to give information about poisons is the Ebers papyrus (circa 1500 BC), which describes early medical practice in Egypt. This document includes nearly 1000 recipes, many containing recognizable poisons. For example, it mentions hemlock, the poison used in the execution of the Greek philosopher Socrates (470–399 BC); aconite, an arrow poison that the Chinese used; opium, used both as a poison and an antidote; and plants containing belladonna alkaloids, which can cause cardiac arrest. During the Roman Empire and continuing through the Middle Ages and into the Renaissance, poisoning was a common practice. The Borgia family in Italy removed many of their enemies in this way.

Not until the beginning of the 19th century was any systematic attempt made to identify the agents responsible for the toxicity of venoms and poisonous plants. One of the first to identify the chemical makeup of a poison was the French physiologist François Magendie. In 1809, while studying arrow poisons, he isolated the alkaloid strychnine from the plants used to make the poison and showed that it was responsible for the convulsions that victims suffered before they died.

Today, we are concerned with the risk of exposure to toxic substances that may be present in water supplies, the atmosphere, our homes, and the workplace. Particular concern exists about the possible harmful effects of pesticide residues in foods. In this chapter, we examine the sources and effects of exposure to toxic substances. We focus on what makes certain substances toxic and how the risk associated with each is measured.

■ What Is Meant by Toxic?

A substance is said to be toxic if it causes harm to a biological system. Only substances that cause harm when present at a very low concentration (parts per million [ppm] or less) are generally described as toxic. Many substances, including common table salt, are harmful at a high concentration but are not labeled toxic. Toxic substances act in many different ways: Some upset metabolism and others the nervous system; still others cause genetic changes, cancers, or birth defects.

Although the terms poison and toxin have different meanings, they are often used interchangeably. **Poison** is a general term for any toxic substance, whether synthetic or of natural origin. The term **toxin** refers to poisons of biological origin. Toxins can be present in spoiled foods. Aflatoxins, for example, are an extremely potent group of toxins that is present in molds that can grow on peanuts and grains while they are in storage. Insect stings and snakebites can also introduce toxins into humans and other animals.

Aspirin, the most widely used drug in the world, is an invaluable medicine but is toxic in large doses. For many years before the introduction of childproof containers, aspirin caused more cases of accidental poisoning in young children than any other substance. Because the attractively shaped tablets looked like candy, unsupervised children often ingested a toxic quantity of them.

■ Types and Routes of Exposure

The determination of the toxicity of a specific substance requires knowledge of the conditions of exposure. The time and frequency of exposure, route by which the exposure occurs, physical state of the substance (gas, liquid, or solid) as well as the dose of the substance are all important factors.

Exposure to a specific chemical may be **acute**, that is, an exposure in which the dose is delivered in a single event and absorption is rapid. Although this type of exposure is usually short, an acute exposure may cause severe health problems. An acute exposure can occur in a very short period of time, for example, the time during which an injection is made by a syringe. Other acute exposures may be prolonged for hours, such as exposure of factory workers to airborne chemicals during a working day.

On the other hand, an exposure is **chronic** if the individual is exposed to the specific chemical at some frequency (usually daily or weekly) over a period of time. Severity of the exposure depends on the toxicity of the specific chemical, the amount in which the individual was exposed, and the frequency of the exposure. Generally, we can expect that the effects of a particular amount of a substance will be reduced if the total amount is divided into smaller portions and then given over a longer period. Thus, a single dose of 500 mg of a toxic substance might seriously affect a person's health or even cause death. If the 500 mg is divided in half and administered twice in one day, the effect will usually be less severe; if the 500 mg is divided into ten 50-mg amounts with each given weekly, the individual might not observe any deleterious effect.

■ Routes of Administration

Percutaneous

The most common exposure of humans to chemicals of all types is by exposure through contact of the chemical with the skin (**percutaneous**). In order to pass into the skin, the chemical must traverse the epidermal cells. Generally, it is thought that gases penetrate the epidermal tissues quite freely; liquids go through less freely, and solids, which are insoluble in water, are probably incapable of penetrating the sign to any significant degree. It appears that nonpolar liquids pass through the skin more readily than polar molecules. Absorption of chemicals is faster through skin that is abraded or inflamed; for this reason, chemicals that are not normally considered hazardous may be dangerous to persons suffering from active inflammatory dermatoses.

Different species have different rates of percutaneous absorption. The toxicity of DDT is approximately equal in an insect and mammal when the insecticide is injected into each, but it is much less toxic to the mammal when applied to only the skin. This happens because DDT is poorly absorbed through the skin of a mammal but passes readily through the chitinous exoskeleton of the insect, and the insect has a greater body surface area in relationship to weight than do mammals.

Inhalation

Exposure to chemicals in the atmosphere occurs through unavoidable **inhalation**; however, the chemical must be either a gas or solid of sufficiently small particle size so that it is not removed in the airway to the lungs. The potential hazards associated with exposure to chemicals via the respiratory tract are particularly dangerous in some industrial working environments and in urban areas with high-density populations.

The U.S. Occupation Safety and Health Act of 1970 spelled out the need for developing standards that would define acceptable conditions for inhalation exposure of individuals to chemicals in the working environment. Since that time, the National Institute for Occupational Safety and Health (NIOSH) has developed and published these standards. NIOSH sets standards for the **Occupational Safety and Health Administration (OSHA)**, which oversees workplace safety. Each standard is formulated for a single chemical compound and specifies the maximal allowable concentration in the workplace air measured as a time-weighed average (TWA) concentration for exposure of humans during an 8-hour work-day and a 40-hour workweek to which workers will be exposed day after day without adverse effect. OSHA also sets a permissible exposure level (PEL) which sets a legal exposure limit for TWA exposures in any 8-hour work shift during a 40-hour workweek. In some cases, the standard also includes a ceiling level of maximal concentration allowable for the compound in air. The **short-term exposure limit (STEL)** is the highest concentration to which workers can be exposed for short periods of time. No more than four excursions per day are permitted with this limit, with 60 minutes between exposure periods at the STEL and provided that the daily threshold limit value (TLV) is also not exceeded. STELs are recommended when only toxic effects for a specific chemical have been reported for high short-term exposures in either animals or humans. TLVs range from 0.0002 ppm for osmium tetroxide, an extremely poisonous compound, to 5000 ppm for carbon dioxide. Table 16.1 lists the TLVs for some commonly encountered environmental chemicals.

Table 16.1

TLVs for Some Commonly Encountered Environmental Chemicals

Compound	TLVs (ppm)	mg/M^3
Butyl alcohol	100	
Carbon dioxide	10,000	18,000
Carbon monoxide	35	40
Chlorine	0.5	1.5
Chlorine dioxide	0.1	0.3
Chloroform	2	9.8
Formaldehyde	20	30
Gasoline	300	900
Hexane	50	180
Methyl alcohol	200	260
Ozone	0.1	0.2
Sulfur dioxide	2	5
Trichloroethylene	50	270

Source: http://www.cdc.gov/niosh/npg/nengapdx.html#f

Oral

The gastrointestinal tract is the third most common means by which a chemical enters the body. Many environmental toxicants enter the food chain and are absorbed from the gastrointestinal tract, which can be viewed as a tube going through the body (Figure 16.1). Although it is within the body, its contents can be considered to be separated from the body. Toxicants within the gastrointestinal tract do not produce injury until they pass through the wall of the gastrointestinal tract and are absorbed into the bloodstream.

Absorption of toxicants can take place along the length of the gastrointestinal tract from the mouth to the rectum. Some chemicals are absorbed in the stomach; because the stomach is acidic (a pH of approximately 2), the **Henderson-Hasselbalch equation** can be used to determine what fraction of the toxicant is in the nonionized, lipid-soluble form available for absorption. The intestine is basic (a pH of approximately 6) and has an enormous surface area for absorbing chemicals.

The molar ratio of ionized to nonionized molecules of a weak organic acid or base in solution depends on the ionization constant. The pH at which a weak organic acid or base is 50% ionized is called the pK_a or pK_b. Like pH, pK_a and pK_b are both defined as the negative logarithm of the ionization constant. Because $pKa = 14 - pK_b$, if either ionization constant is known (Appendix B), the other can be calculated.

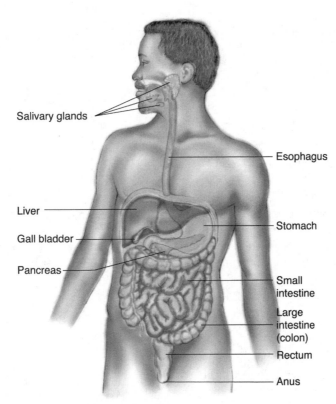

Figure 16.1 The gastrointestinal tract.

EXAMPLE 16.1

Benzoic acid is often used as a preservative in processed foods. When it is ingested, will it be absorbed in the stomach or in the intestine?

To determine the amount of benzoic acid that is ionized use the Henderson-Hasselbalch equation, which can be expressed as follows:

$$\text{For acids}: pK_a - pH = \log\frac{[\text{nonionized}]}{[\text{ionized}]}$$

$$\text{For bases}: pK_a - pH = \log\frac{[\text{ionized}]}{[\text{nonionized}]}$$

From Appendix B, we see that the K_a of benzoic acid is 6.28×10^{-5}. Its pK_a is 4.2.

To determine the amount of each form in the stomach, substitute the pH of the stomach (a pH of approximately 2) and the pK_a for benzoic acid ($pK_a = 42$).

Substituting into the Henderson-Hasselbalch equation

$$pK_a - pH = \log\frac{[\text{nonionized}]}{[\text{ionized}]}$$

$$4.2 - 2 = \log\frac{[\text{nonionized}]}{[\text{ionized}]} = 2.2$$

or taking the antilog

$$10^2 = \frac{[\text{nonionized}]}{[\text{ionized}]} = 100$$

One hundred times more unionized benzoic acid exists in the stomach; 99% of the benzoic acid is in the lipid-soluble, nonionized form.

To determine the amount of each form in the intestine, substitute the pH of the intestine (a pH of approximately 6) and the pK_a for benzoic acid ($pK_a = 4.2$).

Substituting into the Henderson-Hasselbalch equation:

$$pK_a - pH = \log\frac{[\text{nonionized}]}{[\text{ionized}]}$$

$$4 - 6 = \log\frac{[\text{nonionized}]}{[\text{ionized}]} = -2$$

or taking the antilog

$$10^{-2} = \frac{[\text{nonionized}]}{[\text{ionized}]} = \frac{1}{100}$$

In the intestine, only 1% of the benzoic acid is in the nonionized, lipid-soluble form. We would expect that benzoic acid would primarily be absorbed in the stomach.

Parenteral

An injection of chemicals into the body by means of a syringe with a hollow needle is a common procedure that is used in the administration of drugs. Toxicologists use intravenous injection because the absorption step (through the skin, lung, stomach, or intestine) is eliminated and injection is the most rapid means of achieving a high concentration of a chemical within the body. Some drug addicts also inject illegal drugs, such as heroin, because the introduction of a high concentration of heroin rapidly gives a euphoric "rush" (chemically induced high).

■ Dose and Response

Toxicology is the study of the effects of chemical compounds on living creatures. The most fundamental concept in toxicology states that a relationship exists between the dose of an agent and the response that is produced in a biological system. The amount of chemical (per unit of body mass) with which the living creature is exposed is the **dose**. The **response** is the effect that is manifested in the living creature as a result of the exposure. In order to study dose–response relationships, several assumptions must be made. The first is that the observed response is due to the chemical and not caused by some other factor. The second assumption is that the response is related to the dose and that (a) the chemical interacts with receptor site in the living creature, (b) the response is related to the concentration of the chemical at the receptor site, and (c) the concentration of the chemical at the receptor site is proportional to the dose administered. The third assumption is that there is a quantifiable way to measure and express the results.

To generate toxicologic data for an uncharacterized chemical, a study is usually begun that uses rats or mice as the living organisms. The uncharacterized chemical is administered to the rats or mice by either an oral or intraperitoneal injection. An intraperitoneal injection is a common practice because the administered chemical is rapidly absorbed because of the rich blood supply to the peritoneal cavity and its large surface area. A range of dose is administered over a fixed period of time. The number of animals dying after administration is recorded. When the results (the percentage of animals killed for the dose given) are plotted, a sigmoid curve similar to Figure 16.2 is obtained. The slope of the dose–response curve also provides useful information. It reports the range of doses between the no-effect level and a

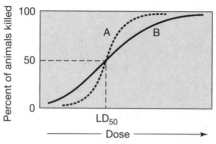

Figure 16.2 The dose-response for two different chemicals. Chemical A has a narrow margin of safety. Chemical B has a wider margin of safety.

lethal dose, a concept known as the margin of safety. If the dose–response curve is especially steep, the margin of error is small. In Figure 16.2, chemicals A and B both have the same LD_{50} (LD stands for lethal dose), but chemical B has a wider margin of error than substance A. Chemicals with a narrow margin of error are more dangerous. In the hospital, anesthesiologists are required to administer anesthetics, which are drugs with a narrow margin of error. The statistical analysis of the data allows the calculation of the dose expected to be lethal in 50% of the animals in a group.

Testing for Toxicity: The LD_{50} Test

Since 1920, the standard method for determining the toxicity of a chemical has been the **LD_{50} test.** In this procedure, different doses of the chemical being tested are administered to large groups of laboratory animals. The result is expressed as the LD_{50} value, which is the dose that kills 50% of the animals in a group. The LD_{50} values of some common chemicals are listed in Table 16.2. Comparing the LD_{50} values allows us to assess the relative toxicity of the chemicals.

The smaller the LD_{50} value, the greater is the toxicity of the substance. For example, the LD_{50} for aflatoxin-B fed to rats is 0.009 mg/kg, whereas that for caffeine is 130 mg/kg. This means that aflatoxin-B is approximately 10,000 times as toxic to rats as caffeine.

The LD_{50} values in Table 16.2 are a measure of the toxicity of the listed chemicals to rats or mice. Other animal species may vary in their response to these substances. For example, guinea pigs are 10,000 times as sensitive to dioxin (TCDD) as are dogs. Because of these species differences, LD_{50} values for rats, or other animals, cannot be extrapolated to determine the amount of a chemical that would be lethal to 50% of a human population. Nevertheless, the values for rats are useful in predicting the probable toxicity in humans. A substance that is extremely toxic to animals is likely to be toxic to humans, too.

The method used to introduce a chemical into an animal's body affects the LD_{50} test and must be taken into account. For example, in mice, the LD_{50} for procaine, a local anesthetic, is 800 mg/kg when the drug is injected under the animal's skin, 500 mg/kg when given by mouth, and only 45 mg/kg when injected directly into a vein.

The chemical also may be administered by inhalation of air containing different concentration levels of the chemicals for a specific period of time. The term calculated from this experiment is the LC_{50}, the atmospheric concentration expected to be lethal to 50% of a group of animals of that species exposed for the specified time period. The LD_{50} or LC_{50} for a chemical provides an initial index of comparable toxicity.

Table 16.2

LD$_{50}$ Values of Selected Chemicals

Chemical	LD$_{50}$ (mg/kg)*
Sugar	29,700
Ethyl alcohol	14,000
Vinegar	3310
Sodium chloride	3000
Malathion (insecticide)	1200
Aspirin	1000
Caffeine	130
DDT (insecticide)	100
Arsenic	48
Strychnine	2
Nicotine	1
Aflatoxin-B	0.009
Dioxin (TCDD)	0.001
Botulinum toxin	0.00001

*For rats or mice.

■ Excretion of Chemicals from the Body

Detoxification of Chemicals by the Liver

Because the liver has the ability to detoxify many chemicals, humans can tolerate moderate amounts of some poisons. The liver can detoxify a chemical by oxidizing or reducing it or by coupling it to a chemical such as a sugar or an amino acid, which are naturally present in the liver.

Oxidation–Reduction Reactions

Ethyl alcohol (ethanol [CH_3CH_2OH]) is detoxified by an oxidation reaction. After an alcoholic beverage is consumed, ethyl alcohol is absorbed into the bloodstream from the intestines and transported to the liver, where enzymes catalyze the oxidation of the alcohol, first into acetaldehyde, which is further oxidized into acetic acid, and then into carbon dioxide and water.

Chronic alcohol consumption causes liver enzymes to build up. The same enzymes that oxidize alcohol also oxidize the male sex hormone testosterone. Alcoholic impotence, a well-known symptom of alcoholism, is a direct result of the oxidation of testosterone by the high concentration of liver enzymes.

The end product of liver oxidation is not always less toxic than the chemical being oxidized. For instance, methyl alcohol (methanol [CH_3OH]) is oxidized to the more toxic chemical formaldehyde (HCHO). Methyl alcohol poisoning is known to cause blindness, respiratory failure, and death. It is not the methyl alcohol itself that causes the problems, but its oxidation product, formaldehyde. Ethyl alcohol is often administered intravenously as an antidote for methyl alcohol poisoning and for ethylene glycol (CH_2OHCH_2OH; antifreeze) poisoning. Because ethyl alcohol competes with these other alcohols in the oxidation reactions in the liver, it can be used to slow their conversion to more harmful oxidation products, thus giving the body a chance to excrete them before damage is done.

Coupling Reactions

The cytochrome P-450 enzymes in the liver play an important role in detoxifying poisons. These enzymes act on fat-soluble substances to make them water soluble and thus more easily excreted. For example, benzene (C_6H_6), a component of gasoline, is a toxic, aromatic compound that is insoluble in water. If benzene is ingested, it tends to be deposited in fatty tissues in the liver. The cytochrome P-450 enzymes oxidize it to phenol, which is more soluble in water. The phenol then couples to glucuronic acid, a sugar that is naturally present in the liver, to form phenyl glucuronide, which is even more water soluble than phenol, and is readily excreted in the urine.

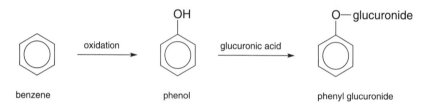

■ Teratogens, Mutagens, and Carcinogens

The toxic chemicals that we have discussed so far act quickly and cause harm almost immediately. Harmful effects of other toxic substances are often not apparent until as much as 10 years or more after exposure to them. Toxic substances of this kind include **carcinogens** (chemicals that cause cancer), **teratogens** (chemicals that cause birth defects by damaging embryonic cells), and **mutagens** (chemicals that are capable of altering genes and chromosomes sufficiently to cause inheritable abnormalities in offspring). Some chemicals are both mutagens and carcinogens, whereas others are only carcinogens.

Teratogens

From the third to the eighth week of pregnancy, the embryo is particularly sensitive to the harmful effects of teratogens. During this critical period, the different parts of the body are differentiated, and the limbs, eyes, ears, and the internal organs are developed. If the embryo is exposed to teratogens at this time, abnormalities in development can result. Examples of teratogens are listed in Table 16.3.

Probably the most notorious teratogen is thalidomide, a drug that was prescribed extensively as a tranquilizer and sleeping pill in Europe and Japan in 1960 and 1961. On the basis of animal studies, thalidomide was considered so safe at the time that it was sold without a prescription in what was then West Germany. In 1961, it became shockingly apparent that the drug could cause deformities. If it was taken between days 34 and 50 of pregnancy, children were born with no arms or legs or with abnormally short limbs and other deformities. Approximately 10,000 children were affected.

Table 16.3	
Teratogenic and Mutagenic Substances	
Teratogens	**Mutagens**
Arsenic (As)	Aflatoxin
Cadmium (Cd)	Benzo(a)pyrene
Cobalt (Co)	Lysergic acid diethylamide
Mercury (Hg)	Nitrous acid (HNO_2)
Diethylstilbestrol	Ozone (O_3)
Polychlorinated	Tris(2,3-dibromopropyl)
biphenyls	phosphate
Retinoic acid	
Thalidomide	

thalidomide

Thalidomide was never approved for distribution in the United States, thanks to the action of Frances Kelsey, an investigator with the Food and Drug Administration. Dr. Kelsey was not convinced that the animal testing done by the company that was promoting the drug was complete and, despite pressure from the company, refused to issue approval. As a result, only 20 "thalidomide babies" were born in the United States.

Another drug that has been shown to be a teratogen is Accutane (retinoic acid), which is used to treat acne. In 1988, the Food and Drug Administration concluded that Accutane could cause heart defects, facial abnormalities, and mental disability if a developing embryo was exposed to the drug for just a few days. Accutane may have caused birth defects in approximately 1000 children born to women who took the drug between 1982 and 1986.

One of the most dangerous and most commonly used teratogens is alcohol. Consumption of even small quantities of alcohol can lead to the development of fetal alcohol syndrome. Infants with this syndrome are abnormally small, often mentally disabled, and have facial deformities. Cigarette smoking during pregnancy also affects the fetus. Smoking raises a woman's blood levels of carbon monoxide, nicotine, and benzo(a)pyrene (a known carcinogen), and these chemicals can then pass into the developing infant's blood. Because any chemical that can pass across the placenta from mother to child is a potential teratogen, pregnant women should not drink alcohol or smoke and should take only medications that their doctors prescribed.

Mutagens

Mutagens are chemicals that alter the sequence of bases in the nucleic acids that make DNA, the material that carries genetic traits from one generation to the next. Changes in DNA also may be brought about by exposure to radiation, and some occur spontaneously. Changes in DNA are called **mutations**. Examples of mutagens are listed in Table 16.3.

Carcinogens

Chemical carcinogens are compounds that cause cells in an organism to reproduce and grow abnormally to produce a **malignant tumor.** Cells in a malignant tumor may divide more rapidly or more slowly than do normal cells. The growth of the tumor is usually uncontrolled, and tumor cells migrate through the body to other tissues, which they destroy. Tumors are characterized as **benign tumors** if they grow slowly and do not spread to other tissues. Benign tumors, unlike malignant tumors, often regress spontaneously.

Chemicals That Cause Cancer

Chemicals that are known to cause cancer vary widely in their structures, and scientists still do not know which chemical structures are likely to make a compound carcinogenic. In 2007, approximately 50 agents have been shown conclusively to cause cancer in humans (Table 16.4). These agents have been designated by the World Health Organization's International Agency for Research on Cancer (IARC) as Group 1, "the agent is definitely carcinogenic to humans." Another 166 had been listed as Group 2A "the agent is probably carcinogenic to humans." As with mutagens, there is concern that any animal carcinogens may also be human carcinogens.

As early as 1775, soot was suspected to be the cause of the high incidence of testicular cancer in chimney sweeps in England. Later it was shown that 3,4-benzo(a)pyrene, a component of soot and coal tar, is a carcinogen.

3,4-benzo(a)pyrene

Multiringed compounds such as 3,4-benzo(a)pyrene are called polycyclic aromatic hydrocarbons. Polycyclic aromatic hydrocarbons (PAHs) have been studied extensively, and many of them have been shown to be carcinogenic. (There appears to be some correlation between the size and shape of the PAH molecule and carcinogenicity.) Often formed when organic materials are burned, they are present in automobile exhaust, cigarette smoke, charcoal-grilled meat, and stack gases from industrial incinerators.

Other compounds that were studied because they were suspected of causing cancer in humans were the aromatic amines. In 1895, a German physician noted a number of cases of bladder cancer among workers in a dye-making factory. In 1937, it was shown that 2-naphthylamine, one of the amines to which the workers were exposed, caused bladder cancer in dogs. Since then, related compounds, including benzidine, which is an intermediate in the manufacture of magenta-colored dyes, have been shown to cause bladder cancers in animals.

Other dyes that have received attention are the azo dyes, several of which have been found to be carcinogens. The structures of these compounds include two benzene rings joined by

Table 16.4

Substances Known to be Human Carcinogens

Aflatoxins
Alcoholic Beverage Consumption
4-Aminobiphenyl
Analgesic Mixtures Containing Phenacetin
  (See Phenacetin and Analgesic Mixtures
  Containing Phenacetin)
Arsenic Compounds, Inorganic
Asbestos
Azathioprine
Benzene
Benzidine
Beryllium and Beryllium Compounds
1,3-Butadiene
1,4-Butanediol Dimethanesulfonate
  (Myleran®)
Cadmium and Cadmium Compounds
Chlorambucil
1-(2-Chloroethyl)-3-(4-methylcyclohexyl)-
  1-nitrosourea (MeCCNU)
bis(Chloromethyl) Ether and Technical-Grade
  Chloromethyl Methyl Ether
Chromium Hexavalent Compounds
Coal Tar Pitches (See Coal Tars and Coal
  Tar Pitches)
Coal Tars (See Coal Tars and Coal Tar Pitches)
Coke Oven Emissions
Cyclophosphamide
Cyclosporin A
Diethylstilbestrol
Erionite
Estrogens, Steroidal
Ethylene Oxide
Hepatitis B Virus
Hepatitis C Virus

Human Papillomas Viruses: Some
  Genital-Mucosal Types
Melphalan
Methoxsalen with Ultraviolet A Therapy (PUVA)
Mineral Oils (Untreated and Mildly Treated)
Mustard Gas
2-Naphthylamine
Neutrons (See Ionizing Radiation)
Nickel Compounds
Radon
Silica, Crystalline (Respirable Size)
Smokeless Tobacco (See Tobacco-Related
  Exposures)
Solar Radiation (See Ultraviolet Radiation
  Related Exposures)
Soots
Strong Inorganic Acid Mists Containing
  Sulfuric Acid
Sunlamps or Sunbeds, Exposure to (See
  Ultraviolet Radiation-Related Exposures)
Tamoxifen
2,3,7,8-Tetrachlorodibenzo-*p*-dioxin
  (TCDD); "Dioxin"
Thiotepa
Thorium Dioxide (See Ionizing Radiation)
Tobacco Smoking (See Tobacco-Related
  Exposures)
Vinyl Chloride
Ultraviolet Radiation, Broad Spectrum UV
  Radiation (See Ultraviolet Radiation-
  Related Exposures)
Wood Dust
X-Radiation and Gamma Radiation
  (See Ionizing Radiation)

Source: National Institute of Environmental Health Sciences "Eleventh Annual Report on Carcinogens, 2007."
Bold entries indicate new or changed listing in *The Report on Carcinogens, Eleventh Edition.*

a −N = N− group. Azo dyes have been used for many years to color foods. Recently, a number of them, including FD&C Yellow No. 3, were banned because they can react with stomach acid to produce known carcinogens.

Not all carcinogens are aromatic compounds. The nonaromatic organic compound vinyl chloride ($CH_2 = CHCl$), which is used in huge quantities to make the polymer polyvinyl chloride, is a well-known carcinogen. Its carcinogenicity was confirmed after it was suspected of causing a rare form of liver cancer in workers who were exposed to large quantities of it in the atmosphere in their workplace.

Inhalation of very fine asbestos fibers causes lung cancer and various respiratory diseases (Chapter 17). Certain metals and metal compounds are also carcinogenic. Metals that can cause cancer in animals include beryllium, chromium, lead, nickel, and titanium (Chapter 13).

Carcinogens are present in the natural environment in many plants and microorganisms. Aflatoxin-B, which causes liver cancer in rats, is present in molds that grow on grains. Food items that contain carcinogens include certain mushrooms, pepper, mustard, celery, and citrus oils. It is generally accepted that the concentration of carcinogens in these foods is too low to be of concern.

The Development of Cancer

Metabolic poisons and neurotoxins produce effects that are almost immediate on entering the body, but carcinogens behave differently. Many years may elapse before a person exposed to a carcinogen shows signs of cancer. For example, poly(vinyl chloride) was first manufactured in the 1940s, but it was not until approximately 25 years later that liver cancer began to appear in workers who had been involved in its production. Similarly, workers exposed to asbestos and benzidine did not develop cancers until 20 years after exposure.

Some chemicals cause cancer at the point where they make contact with the body; for example, inhaled carcinogens in cigarette smoke and inhaled asbestos fibers cause lung cancers. Other chemicals cause cancer in an organ that is distant from their entry point into the body. The liver, which receives most of the toxic chemicals that enter the body, and the colon, where solid wastes collect, are particularly susceptible to carcinogens.

How cancers develop is still not understood. The pathways leading to the growth of a tumor are complex, and there is evidence that different carcinogens produce cancers in different ways. It is generally agreed that carcinogenesis is a two-step process. In the first step, which is called **initiation**, a reaction occurs between the carcinogen and the DNA in a cell, and as a result, an abnormal cell is formed.

Initiation occurs most often by a reaction of the DNA with genotoxic carcinogens, substances that bind irreversibly with DNA. They can be electrophilic or metabolically activated to form electrophilic species, such as the preservative sodium nitrite, which was discussed earlier. Substances that require metabolic activation are called **procarcinogens**. Benzo(a)pyrene and vinyl chloride are organic procarcinogens that must be metabolically activated. Carcinogens that do not require biochemical activation are classified as **primary** or **direct-acting carcinogens**. Dimethylsulfate and bis(chloromethyl) ether are primary carcinogens that do not require bioactivation.

The abnormal cell has the potential to grow into a malignant tumor. It may start to divide and grow into a tumor immediately after it is formed, but more often, it continues its normal functions and does not proliferate until a second stage, called *promotion,* occurs months or years later. Promotion is thought to occur in response to exposure to a toxic agent (the promoter). The abnormal cell becomes a cancerous cell and develops into a tumor.

The World Health Organization has estimated that lifestyle and environmental factors are responsible for the development of up to 75% to 80% of all cancers. Approximately 5% are thought to be caused by inherited genetic traits. Cigarette smoking is blamed for 30% of cancers, diet for 25% to 30%, occupational exposure for 10% to 15%, and environmental pollutants for 5% to 10%. Cancer strikes more than a million Americans each year (1,437,180 in 2008), and one-third of those now living will eventually develop some type of cancer. As the life expectancy of the average person continues to rise above 70 years, the number of deaths caused by cancer will increase. Fortunately, in the United States, the five-year survival rate between 1996 and 2003 is 66%. Many people could reduce the risk of developing cancer by making lifestyle changes.

■ Nucleic Acids

The complex compounds called **nucleic acids** are found in all living cells except mammalian red blood cells. There are two kinds of nucleic acids: **deoxyribonucleic acid (DNA)** and **ribonucleic acid (RNA)**. DNA is found primarily in the **nucleus** of cells, whereas RNA is found primarily in the **cytoplasm**, the part of the cell surrounding the nucleus.

Functions of Nucleic Acids

DNA and RNA are responsible for the storage and transmission of genetic information in all living organisms. They hold the key to how genetic information is transferred from one cell to another and how genetic traits are transmitted, via sperm and eggs, from parents to offspring. The major function of DNA, and one in which RNA is involved, is the control and direction of protein synthesis in body cells. Chemical information stored in the DNA of genes specifies the exact nature of the protein to be made and thus dictates the character of the organism.

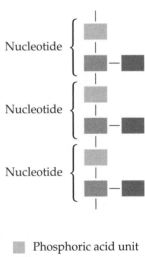

The Primary Structure of Nucleic Acids

In much the same way that polysaccharides are composed of monosaccharides and proteins are composed of amino acids, nucleic acids are composed of long chains of repeating units called **nucleotides**. DNA molecules are the largest of the naturally occurring organic molecules.

Each nucleotide in a chain is made up of three components: (1) a **sugar**, (2) a **nitrogen-containing heterocyclic base**, and (3) a **phosphoric acid** unit (Figure 16.3). The sugar is a pentose, either **ribose** or **deoxyribose** (Figure 16.4). The only difference between these two sugars is at carbon 2, where ribose has a hydrogen atom and an –OH group and deoxyribose has

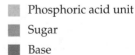

Phosphoric acid unit

Sugar

Base

Figure 16.3 Nucleic acids consist of long chains of nucleotides. Each nucleotide in a chain is composed of a sugar, a base, and a phosphoric acid unit.

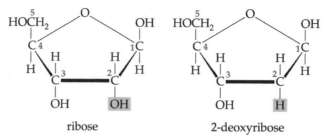

Figure 16.4 The pentoses ribose and deoxyribose differ at carbon 2, where ribose has an H atom and an OH group and deoxyribose has two H atoms. Ribose is present in RNA; deoxyribose is present in DNA.

two hydrogen atoms. As the names indicate, the sugar in DNA is deoxyribose, whereas that in RNA is ribose.

The five different bases found in nucleic acids are shown in Figure 16.5. Two—**adenine** (A) and **guanine** (G)—are double-ring bases and are classified as **purines**. The other three—**cytosine** (C), **thymine** (T), and **uracil** (U)—are single-ring bases and belong to the class of compounds called **pyrimidines**. Purines and pyrimidines are bases because their nitrogen

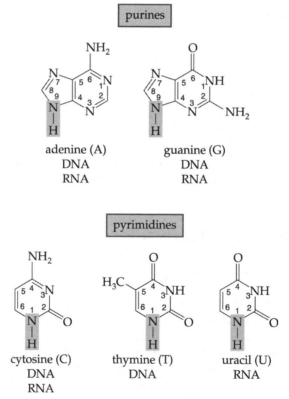

Figure 16.5 Of the five bases found in nucleic acids, adenine and guanine are purines, while cytosine, thymine, and uracil are pyrimidines.

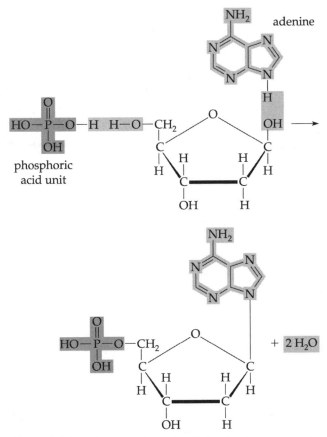

Figure 16.6 The formation of a nucleotide from its three components.

atoms can accept protons. Adenine, guanine, and cytosine are found in both DNA and RNA. Thymine is found in DNA, whereas uracil is found in RNA.

The formation of a representative nucleotide from its three components is shown in Figure 16.6. The pentose forms an ester bond with a phosphoric acid unit through the –OH on carbon 5 and forms another bond with the base through the –OH group on carbon 1. Water is eliminated as the bonds are formed. When successive nucleotides bond to form long-chain nucleic acids, the phosphoric acid units form a second ester bond through the –OH group on carbon 3 of the pentose unit of another nucleotide (Figure 16.7). The alternating sugar–phosphoric acid units form the backbone of the nucleic acids.

The Double Helix

The sequence of the four bases in a nucleic acid, like the sequence of amino acids in proteins, is its primary structure. Like proteins, nucleic acids also have a secondary structure. A DNA molecule has two polynucleotide chains wound around one another to form a **double helix**, a structure that can be compared with a spiral staircase. The phosphate–sugar backbone represents the handrails, and the pairs of bases linked by hydrogen bonds represent the steps

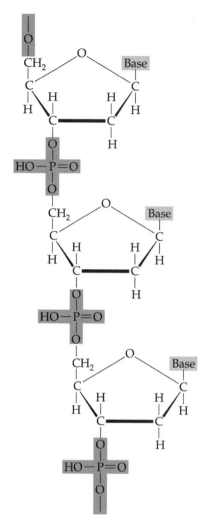

Figure 16.7 A three-nucleotide segment of a DNA strand, showing the alternating phosphate-sugar units that form the backbone of the molecule.

(Figure 16.8a). If one chain of the helix has a G base, it must always be bonded to C base on the other chain, and an A base must always be bonded to T base. This arrangement gives "steps" of almost equal lengths and explains why amounts of C and G and amounts of A and T in DNA molecules are always equal (Figure 16.8b). Although hydrogen bonds are relatively weak, there are so many of them in a DNA molecule that under normal physiologic conditions they hold the two chains together. The C-G (and G-C) base pairs are held together by three hydrogen bonds, the A-T (and T-A) base pairs by two (Figures 16.8a and 16.8b). Figure 16.8c shows a space-filling model of DNA.

DNA, Genes, and Chromosomes

The DNA in the nuclei of cells is coiled around proteins called histone molecules to form structures known as **chromosomes** (Figure 16.9). The number of chromosomes varies with species. Humans have 46; each parent contributes 23.

Long before the structure of DNA was understood, **genes** were defined as sections of chromosomes that determined inherited characteristics such as blue eyes and dark hair in humans. We now define genes as the segments of DNA molecules that control the production of all of the different proteins in an organism. Proteins, in turn, control all of the chemical reactions of life that occur in an organism, including the development of blue eyes and dark hair. Humans have approximately 80,000 genes.

DNA molecules vary in the number and sequence of the base pairs that they contain. The precise sequence of base pairs in the DNA molecule is the key to the genetic information that is passed from one generation to the next; it is this sequence that directs and controls protein synthesis in all living cells. Each organism begins life as a single cell. The unique DNA in the nucleus of that cell determines whether the cell as it multiplies develops into a human, a bird, a rose, or a bacterium. The DNA carries all of the information needed for making and maintaining the different parts of the organism whether those parts are hearts, legs, wings, or petals.

Cell Replication

The double-helix structure of DNA that Crick and Watson proposed in 1953 explains very simply and elegantly how cells in the body divide to form exact copies. The two intertwined chains of the double helix are complementary. An A on one always pairs with a T on the other and similarly a C pair with a G. Just before a cell divides, the double strand begins to unwind

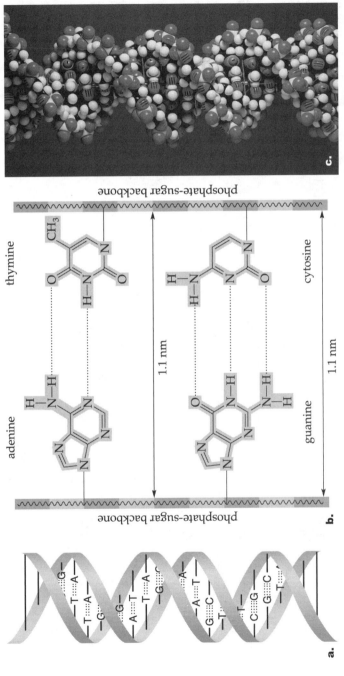

Figure 16.8 (a) Schematic representation of the double helix model of DNA. Hydrogen bonding between adenine (A) and thymine (T) and between cytosine (C) and guanine (G) is represented as dotted lines. (b) The steps in the spiral staircase structure of the DNA molecule are formed by two base pairs: Adenine (a purine) pairs with thymine (a pyrimidine) through two hydrogen bonds; guanine (a purine) pairs with cytosine (a pyrimidine) through three hydrogen bonds. These pairings give steps of equal width (1.1 nm) separating the two strands. (c) A space-filling model of DNA in which carbon atoms are blue, hydrogen atoms are white, oxygen atoms are red, nitrogen atoms are dark blue, and phosphorus atoms are yellow.

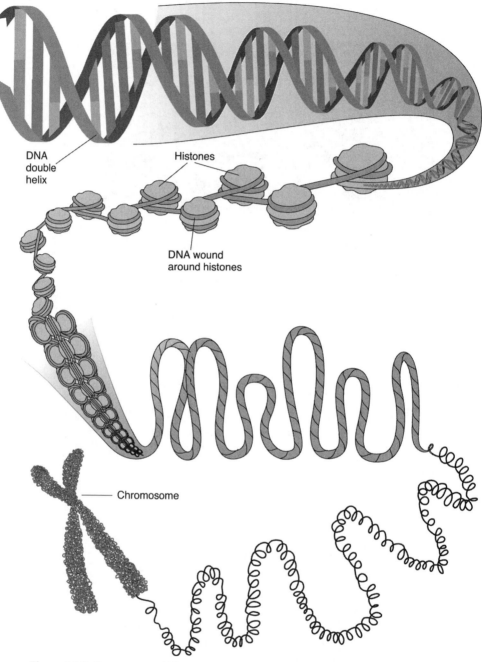

DNA
double
helix

Histones

DNA wound
around histones

Chromosome

Figure 16.9 Chromosomes, which are found in the cell nuclei, are tightly coiled strands of DNA.

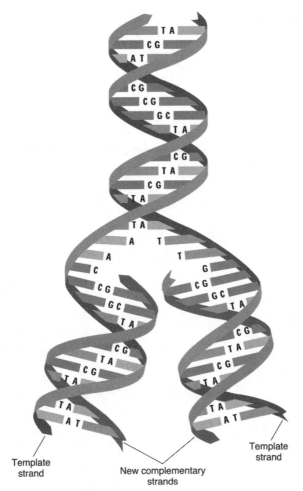

Template strand

New complementary strands

Template strand

Figure 16.10 In DNA replication, the double helix gradually unwinds, and each strand acts as a template. Complementary nucleotides are attached to the single strands, forming two new DNA molecules identical to the original one.

(Figure 16.10). Each unwinding strand serves as a pattern, or template, for the formation of a new complementary strand. Nucleotides, which are always present in the cell fluid surrounding DNA, are attracted to the exposed bases and become hydrogen bonded to them: A to T, T to A, C to G, and G to C. In this way, two identical DNA molecules are formed, one for each of the two daughter cells.

■ Protein Synthesis

Protein synthesis is carried out in a series of complex steps involving RNA. As we noted earlier, the sugar in RNA is ribose, and the bases are uracil (U), adenine (A), guanine (G), and cytosine (C) (see Figure 16.5). The primary structure of RNA is similar to that of DNA

(Figure 16.3): Ribose–phosphoric acid units form the backbone, and each ribose unit is bonded to one of the four bases. RNA molecules are much smaller than DNA molecules and contain from 75 to a few thousand base pairs. RNA molecules exist primarily as single strands rather than in a double-helix form.

Protein synthesis proceeds in two main steps:

$$DNA \xrightarrow[\text{transcription}]{} mRNA \xrightarrow[\text{translation}]{\text{tRNA}} protein$$

In the first step, **transcription**, a single strand of RNA is synthesized inside the cell nucleus as follows. A segment of the DNA double helix separates into single strands, and the exposed bases on one strand act as the template for the synthesis of a molecule of RNA. The base sequence in the RNA, which is known as *messenger RNA (mRNA)*, complements the base sequence on the DNA strand (except that wherever there is an adenine in the DNA, the RNA transcribes a uracil instead of thymine) (Figure 16.11).

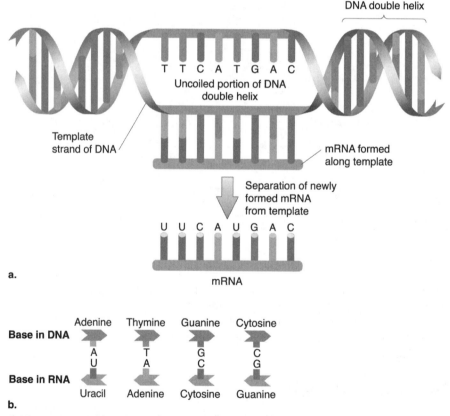

Figure 16.11 (a) In transcription, a segment of the DNA double helix unwinds, and one strand acts as the template for synthesis of messenger RNA (mRNA). Note that where there is an adenine (A) in the DNA, the mRNA transcribes uracil (U). (b) The base pairings between DNA and RNA.

EXAMPLE 16.2

Show the sequence of bases in the mRNA that would be synthesized from the lower strand of DNA with the following sequence of bases:

- T - A - C - G - G - T - T - C - A - C -

- A - T - G - C - C - A - A - G - T - G - template strand

Solution

The bases pair as follows: A-U, T-A, G-C, and C-G; therefore, the mRNA would have the following sequence of bases.

- U - A - C - G - G - U - U - C - A - C -

Notice how the sequence in the top strand of DNA corresponds to the sequence in the mRNA.

Show the sequence of bases in the DNA segment that acted as the template for the synthesis of the following section of mRNA:

- C - G - G - U- A - U - C - U - A -

Answer

DNA: - G - C - C - A- T - A - G - A - T -

The next step is the **translation** of the code that has been copied from the DNA strand into the synthesis of a new protein. To direct this synthesis, the mRNA leaves the nucleus and takes its chemical message to the cytoplasm of the cell where it binds with cellular structures called **ribosomes**. Taking part in translation are molecules of another kind of RNA called *transfer RNA* (*tRNA*), which are responsible for delivering amino acids one by one to the mRNA. Each tRNA molecule carries a three-base sequence called an **anticodon**, which determines which specific amino acid it will deliver. Three-base sequences along the mRNA strand, called **codons**, determine the order in which tRNA molecules bring the amino acids to the mRNA.

Guided by the first codon on the mRNA strand, a tRNA molecule with an anticodon that is complementary to this codon transports a specific amino acid to the mRNA codon. For example, if the bases are lined as shown in Figure 16.12a, where the first codon is A-U-G, a tRNA with a U-A-C anticodon will transport the amino acid methionine (met) to the mRNA, where the tRNA anticodon will pair up through hydrogen bonding with the complementary codon. Similarly, a second tRNA molecule will bring the amino acid phenylalanine (phe) to the next mRNA codon, U-U-U. A peptide bond then forms between

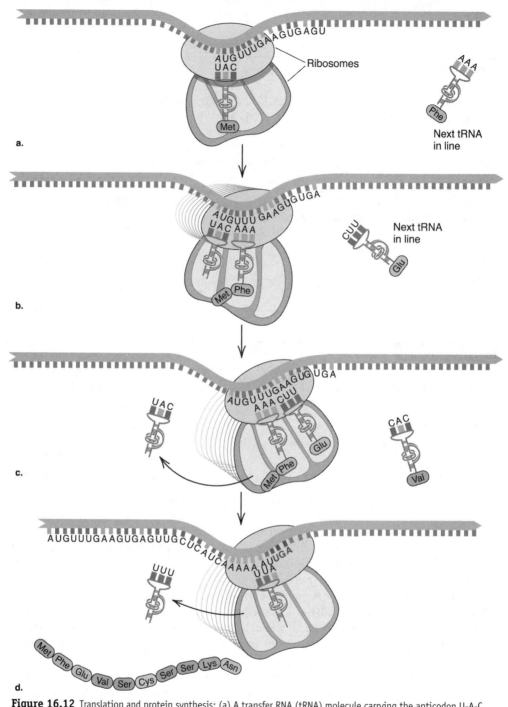

Figure 16.12 Translation and protein synthesis: (a) A transfer RNA (tRNA) molecule carrying the anticodon U-A-C delivers the amino acid methionine (met) to the mRNA strand, where the anticodon bonds with the codon A-U-G. Another tRNA molecule carrying the anticodon A-A-A brings the amino acid phenyl alanine (phe) to the next mRNA codon, U-U-U. (b) A peptide bond forms between Met and Phe, and Met separates from the tRNA. A third amino acid, glutamic acid (Glu), is brought to the next mRNA codon, G-A-A. (c) The tRNA molecule carrying the anticodon is released from the mRNA. (d) When synthesis of the protein is complete, the protein molecule separates from the mRNA.

the two amino acids, and the methionine separates from the tRNA (Figure 16.12b). When a third amino acid, glutamic acid (glu), has been added, the first tRNA molecule is released from the mRNA (Figure 16.12c). The actual protein synthesis occurs in the ribosomes that move along the mRNA one codon at a time as the amino acid chain grows. In this way, the mRNA is read codon by codon and the protein is built one amino acid at a time in the correct sequence. When the synthesis is completed, the protein separates from the mRNA (Figure 16.12d).

■ The Genetic Code

The specific amino acid that each of the three-base sequences in mRNA codes for has been established. This information (Table 16.5) is the **genetic code**. Sixty-four possible three-base codons can be formed from the four bases in mRNA (4^3 or $4 \times 4 \times 4 = 64$). Because all proteins are made from only 20 amino acids, there is some redundancy in the system; that is, several amino acids are designated by more than one codon. Some codons act as "stop"

Table 16.5

The Genetic Code

First Base	Second Base								Third Base
	U		**C**		**A**		**G**		
U	U-U-U	Phe	U-C-U	Ser	U-A-U	Tyr	U-G-U	Cys	U
	U-U-C	Phe	U-C-C	Ser	U-A-C	Tyr	U-G-C	Cys	C
	U-U-A	Leu	U-C-A	Ser	U-A-A	Stop	U-G-A	Stop	A
	U-U-G	Leu	U-C-G	Ser	U-A-G	Stop	U-G-G	Trp	G
C	C-U-U	Leu	C-C-U	Pro	C-A-U	His	C-G-U	Arg	U
	C-U-C	Leu	C-C-C	Pro	C-A-C	His	C-G-C	Arg	C
	C-U-A	Leu	C-C-A	Pro	C-A-A	Gln	C-G-A	Arg	A
	C-U-G	Leu	C-C-G	Pro	C-A-G	Gln	C-G-G	Arg	G
A	A-U-U	Ile	A-C-U	Thr	A-A-U	Asn	A-G-U	Ser	U
	A-U-C	Ile	A-C-C	Thr	A-A-C	Asn	A-G-C	Ser	C
	A-U-A	Ile	A-C-A	Thr	A-A-A	Lys	A-G-A	Arg	A
	A-U-G	Met	A-C-G	Thr	A-A-G	Lys	A-G-G	Arg	G
G	G-U-U	Val	G-C-U	Ala	G-A-U	Asp	G-G-U	Gly	U
	G-U-C	Val	G-C-C	Ala	G-A-C	Asp	G-G-C	Gly	C
	G-U-A	Val	G-C-A	Ala	G-A-A	Glu	G-G-A	Gly	A
	G-U-G	Val	G-C-G	Ala	G-A-G	Glu	G-G-G	Gly	G

signals to terminate protein synthesis. The codon AUG not only encodes for methionine, it is also a "start" signal for protein synthesis. Most of the approximately 100,000 genes in each cell of our bodies encode for a specific protein.

If a chemical substance or some other agent causes a mutation, a protein with an incorrect sequence of amino acids may be formed. If the mutation occurs in a body (somatic) cell, very little damage may result, or there may be uncontrolled cell growth: a cancer. If the mutation occurs in a germ cell (egg or sperm), the alteration in the amino acid sequence is passed on to a person's offspring and may result in a child being born with a hereditary disease such as sickle cell anemia, hemophilia, or cystic fibrosis. The abnormal genes responsible for cystic fibrosis and several other hereditary diseases have been identified, but how they cause the symptoms associated with the diseases is still not understood.

Many chemicals that alter chromosomes and produce mutations in plants, viruses, bacteria, insects, mice, and other animals have been identified. Although there is no conclusive proof that any chemical has caused a mutation in human germ cells, there is concern that chemicals that cause mutations in microorganisms and animals may also cause them in humans.

Mechanisms of DNA Damage

The damage caused in DNA by ionizing radiation, nonionizing radiation, and chemicals is different. The damage can involve the breaking of the DNA backbone or the cross-linking of the two complementary strands or an alteration to the chemical structure of the DNA bases. Figure 16.13 shows the various ways that DNA is damaged.

Ionizing Radiation

Ionizing radiation, such as X-rays and γ rays, can break the DNA strand. Ionizing radiation, when it hits a DNA strand, imparts enough energy to break the sugar–phosphate backbone. Depending on its energy, it can break only one strand or both.

Ultraviolet Light

Ultraviolet light (nonionizing radiation) has sufficient energy to induce reactions of the DNA bases. These reactions can involve the formation free radicals and often produce pyrimidine dimes the photoproducts.

Chemicals

Chemicals can damage DNA by a reaction with the DNA bases to form a new compound (*adduct*), by inserting themselves between the two strands of the helix (*intercalation*), or by a reaction with both strands to cause cross-linking.

One chemical that is a focus of concern is sodium nitrite ($NaNO_2$), which is used to retard spoilage and give a pink color to processed meats such as hot dogs, bologna, bacon, and smoked ham. When these foods are eaten, sodium nitrite is converted to nitrous acid (HNO_2) by stomach acid.

$$NaNO_2 \; + \; HCl \; \longrightarrow \; HNO_2 \; + \; NaCl$$
sodium   nitrous
nitrite   acid

An examination of Figure 16.5 shows that three of the bases that code DNA—adenine, guanine, and cytosine—all contain an amino group ($-NH_2$). When nitrous acid acts on these bases, it replaces the amino with a hydroxyl group ($-OH$). When this occurs, the base that

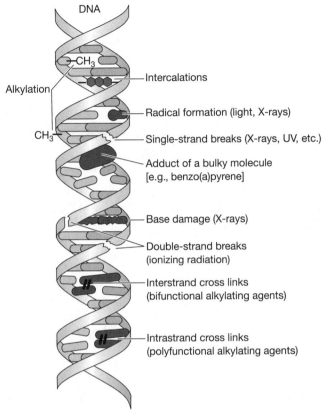

Figure 16.13 DNA damage by physical and chemical agents.

underwent a reaction can no longer function normally. Miscoding is more likely to occur at this base's position during DNA replication or protein synthesis.

Nitrous acid has been shown in laboratory studies to cause mutation in bacteria, and as a result of this finding, the Food and Drug Administration has recommended limiting the amount of nitrite allowed in foods and is considering a complete ban on its use.

Alkylation of DNA

Aliphatic and aromatic secondary amines easily form nitroamines (n-nitrosamines) in the presence of nitrous acid or nitrogen oxides. Nitrosamines have been found widely in the environment and have raised considerable concern because of their unexpected toxicity. In the mid-1950s N,N-dimethylnitrosamine (DMN), which was being used as a rocket fuel, was found to be very carcinogenic. DMN and other nitrosamines were subsequently found in food, tobacco products, and bacon that had been cured with nitrites.

Alkylnitrosamines are powerful carcinogens. They are metabolically activated by oxidation and generate a carbocation that can alkylate DNA. The mechanism for the formation of the methyl carbocation from DMN is as follows:

$$N=O \quad\quad\quad NADPH \quad\quad N=O$$
$$CH_3-N-CH_3 \ + \ O_2 \ \longrightarrow \ CH_3-N-CH_2OH$$

$$CH_3-N=N-OH \quad\longleftarrow \quad CH_3-N-N=O \ (H)$$

$$CH_3^+ \ + \ N_2 \ + \ OH^- \quad\quad\quad + \ HCHO$$

The CH_3^+ that is produced attacks the nitrogenous bases in DNA nucleotides. One of the most common mechanisms is an attack of the carbocation on one of the nitrogens in the base guanine forming predominately 7-methylguanine.

It can also attack the nitrogen in the three position, forming 3-methylguanine.

In a separate attack, the CH_3^+ can also methylate the oxygen in guanine forming O^6-methylguanine.

These alkylation reactions block sites that would normally hydrogen bond with a complementary nucleotide on the other strand of the helix. The result of the alkylation is miscoding; it occurs when the cell divides. Each DNA strand trying to form a new DNA molecule that is identical to the original is unable to hydrogen bond (in the way it normally does) at the alkyl-ated site. Guanine would normally code for a cytosine, but instead, another base is inserted; the result is a mutation at that site.

■ Ames Test

In the mid-1970s, the Environmental Protection Agency (EPA) began a risk analysis to determine the cancer risks associated with toxic chemicals and pesticides. The EPA has a four-step procedure for risk analysis: (1) hazard assessment, (2) dose–response assessment, (3) exposure assessment, and (4) risk characterization. Because the information available to the EPA is often incomplete, the analysis is often imprecise. By constantly reassessing the risks as new information becomes available, the EPA can review and rewrite regulations to protect public health.

The process of hazard assessment involves examining evidence and determining whether a potential hazard has caused harm. Often historical data are available and have been used to calculate risks. For example, it has been determined that 3.6 of every 1000 people who smoke a pack of cigarettes a day will die from smoking-related diseases each year. Although historical data describe the exposure of certain individuals to chemicals, it is very difficult to correlate that information with cancer deaths. Because the onset of cancer usually occurs long after exposure, many other factors affecting the person's health for the last 20 years may be involved.

In cases in which the cause-and-effect relationship is obvious, epidemiologic studies are used. To establish a link between a particular chemical and a certain cancer, an epidemiologic study compares a large number of people who were exposed to the chemical and their cancer rates with a large number of people who were not exposed and their cancer rates. The statistical correlation of this type of information has been used by the National Cancer Institute to label particular chemicals as *known carcinogens*.

If no historical data exist, a second way to assess a cancer risk is to test the chemical of interest on animals. For example, before a new artificial sweetener can be marketed, it must be tested on several thousand mice or rats. Rodents have much shorter lifetimes (2 to 4 years) than humans, and thus, even though they may develop cancers late in their lives, the test takes only three to five years. If a significant number of rodents develop tumors after being fed the sweetener, the test indicates that the sweetener may be a carcinogen. The three main criticisms of these studies are (1) the large numbers of animals that are needed, (2) that rodents and humans may respond differently to a particular chemical, and (3) that the doses fed to the animals are often very high. Proponents of animal testing point that all chemicals shown to be human carcinogens are also carcinogenic to animals.

A third way to assess the risk of cancer is to conduct a bacterial screening test. The **Ames test** uses *Salmonella typhimurium* bacteria to test for mutations. Although the test is simple and inexpensive to perform, it is the least accurate of the three methods for assessing cancer risk. The Ames test assumes that the first step in the development of cancer is a mutation and, therefore, a chemical that causes a mutation is likely to be a carcinogen as well.

The bacteria that are used in the Ames test are modified biologically so that they are unable to synthesize the amino acid histidine. In the absence of histidine, the bacteria cannot grow. In the test, the modified bacteria are placed in a medium that contains all of the ingredients that they need to grow, except histidine. As can be seen in Figure 16.14, the suspected chemical carcinogen is then added to the medium. If the test chemical is not a mutagen, no bacterial growth will occur. If it is a mutagen, however, a mutation will occur, and the bacteria will revert to a form that can synthesize histidine and grow. The growth of bacteria in the test dish identifies the chemical as a mutagen.

Because of the strong correlation between chemicals that cause cancer and those that cause mutations (approximately 90% appear to do both), the Ames test is useful for screening for potential chemical carcinogens. If a chemical is identified as a mutagen in the Ames test, it is then tested on animals to confirm its carcinogenicity.

The process of dose–response assessment, the second step in analyzing risk, begins with establishing a correlation between exposure of animals to a chemical and a harmful effect. Next, the relationship between the dose (concentration of chemical) and the effect is studied. Both the frequency of incidence (the number of times the chemical produces an adverse effect) and the severity of the effect are noted. Once this information is available, statisticians use it to predict the effect of the chemical on human health.

If the chemical being studied is shown to be harmful, an **exposure assessment** is undertaken. Information on groups of people who have already been exposed, where and how they came to be exposed, the dose to which they were exposed, and the duration of exposure must be collected. Only when this process is completed can the risks be characterized.

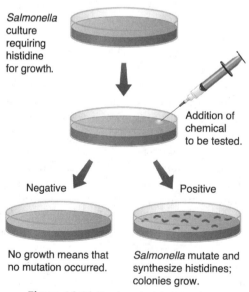

Salmonella culture requiring histidine for growth.

Addition of chemical to be tested.

Negative

Positive

No growth means that no mutation occurred.

Salmonella mutate and synthesize histidines; colonies grow.

Figure 16.14 The Ames test for mutations.

The information gathered in the first three steps of the risk analysis is compiled, and the risk posed by the chemical in question is calculated. This step is called *risk characterization,* and the risk is expressed as a probability. The 1990 Clean Air Act regulates chemicals that pose a cancer risk greater than 1 in 1 million people who are exposed to high doses. The Food and Drug Administration uses the same standard for regulating chemicals in food and drugs. Because it is difficult to express this information, risks are often presented as a reduction in life expectancy or the increased probability of dying (Table 16.6).

Genetic Tests for Cancer

Genetic tests for susceptibility to diseases such as cancer are becoming commercially available at clinics around the country. At the same time, much debate surrounds whether genetic testing should be used outside of research studies and families with a known history of the disease. Meanwhile, public knowledge and interest in genetic evaluation grow as news media and Internet resources provide an increasing amount of information on the scientific breakthroughs in this area. As a result, many patients are eager to discuss with their physicians the prospect of genetic DNA testing for cancer and other diseases.

New evidence indicates that specific groups characterized by predisposing genetic traits or ethnicity may have a heightened risk from exposures to certain carcinogens. Recent epidemiologic studies of environmental carcinogens have shown that people with certain genetic traits may be more likely to develop cancers when exposed to PAHs and aromatic amines.

Table 16.6	
A Ranking of Hazards According to the Degree of Risk	
Hazard	**Annual Risk***
Cigarette smoking (1 pk/day)	3.6 per 1000
All cancers	2.8 per 1000
Motor vehicle accident	2.4 per 10,000
Police killed in action	2.2 per 10,000
Air pollution, eastern United States	2.0 per 10,000
Home accidents	1.1 per 10,000
Alcohol, light drinker	2 per 100,000
Radiation, sea level	2 per 100,000
Electrocution	5.3 per 1,000,000
Drinking water containing the EPA limit of chloroform	6 per 10,000,000

*Probability of dying.

Although screening for genetic traits such as breast or prostrate cancer is becoming more common, a test for susceptibility to carcinogens is not available; however, knowledge of both acquired genetic and environmental susceptibility will be instrumental in developing health and regulatory policies that will protect the more susceptible groups from risks of environmental carcinogens.

■ Additional Sources of Information

Crosby DG. *Environmental Toxicology and Chemistry.* New York: Oxford University Press, 1998.

Hodgson E, Smart RC. *Introduction to Biochemical Toxicology.* 3rd ed. Hoboken, NJ: John Wiley & Sons, 2001.

Klaassen CD, Watkins J. *Casarett & Doull's Essentials of Toxicology.* New York: McGraw-Hill, 2003.

Manahan SE. *Toxicological Chemistry and Biochemistry,* 3rd ed. Boca Raton, FL: Lewis, 2002.

Williams PL, James RC, Roberts SM, editors. *Principles of Toxicology: Environmental and Industrial Applications,* 2nd ed. Hoboken, NJ: John Wiley & Sons, 2000.

■ Keywords

acute exposure
adenine
Ames test
anticodon
benign tumor
carcinogens
chromosomes
chronic exposure
codons
coupling reactions
cytoplasm
cytosine
deoxyribonucleic acid (DNA)
deoxyribose
detoxification
direct-acting carcinogens
dose
double helix
exposure assessment
genes
genetic code
guanine
guanosine

Henderson-Hasselbalch equation
inhalation
initiation of cancer
LD_{50} test
malignant tumor
mutagens
National Institute for Occupational Safety and Health (NIOSH)
nitrogen-containing heterocyclic base
nucleic acids
nucleotides
nucleus (cells)
oral
Occupational Safety and Health Administration (OSHA)
oxidation-reduction reactions
parenteral
percutaneous
phosphoric acid
poison
primary carcinogens
procarcinogens
purines

pyrimidines
response
ribonucleic acid (RNA)
ribose
ribosomes
Salmonella typhimurium
short-term exposure limit (STEL)
sugar

teratogens
thymine
time-weighed average
toxin
transcription
translation
uracil

■ Questions and Problems

1. Define the following:
   a. Toxin
   b. Poison
   c. Acute exposure
   d. Chronic exposure
2. Why is percutaneous application of DDT more toxic to insects than mammals?
3. Define the following:
   a. Time-weighed average
   b. STEL
   c. Occupation Safety and Health Act of 1970
4. The weak base trimethylamine is often found in spoiling fish. When it is ingested, will it be absorbed in the stomach or in the intestine?
5. Citric acid is often used as a preservative in processed foods. When it is ingested, will it be absorbed in the stomach or in the intestine?
6. Name the three main ways a chemical can enter the body.
7. Define the following:
   a. Dose
   b. Response
   c. LD_{50}
   d. Margin of error
8. A commercial pain relief medication contains 500 mg of acetaminophen per tablet. Assume that the LD_{50} of 338 mg/kg for mice applies to humans as well. How many tablets, taken all at once, would produce a 50% chance of a lethal dose of acetaminophen in a 154-pound (70 kg) person?
9. A commercial anti-inflammatory preparation contains 350 mg of ibuprofen per tablet. Assume that the LD_{50} of 1050 mg/kg for rats applies to humans as well. How many tablets, taken at once, would produce a 50% chance of a lethal dose in a 44-pound (20 kg) child?
10. The herbicide Paraquat was once sprayed on marijuana plants by the U.S. government to kill the plants so that they could not be sold. Paraquat has an LD_{50} of 100 mg/kg for rats. Another herbicide, Silvex, has an LD_{50} of 650 mg/kg in rats. Is Paraquat more toxic to humans that Silvex?
11. Dioxin was a toxic by-product formed in the production of Agent Orange during the Vietnam War. Dioxin has an LD_{50} of 0.022 mg/kg in rats. How much dioxin would be lethal to a 154-pound (70-kg) person?

12. The liver removes toxins from our bodies. How is ethyl alcohol removed from the body?

13. Why is ethyl alcohol used as an antidote for methyl alcohol poisoning?

14. What is the P-450 system? How does it participate in detoxifying water-insoluble poisons?

15. Describe the difference between a teratogen and a mutagen.

16. Describe the difference between a mutagen and a carcinogen.

17. What are the three components of every nucleotide? Draw a simple diagram to show how the components are bonded.

18. Name two types of nucleic acids. How do the sugar molecules in these two types differ from each other?

19. Name the base that forms a pair with each of the following bases:
    a. Uracil
    b. Guanine
    c. Adenine
    d. Cytosine

20. The base sequence along one strand of DNA is ACTGT. What would be the sequence on the complementary strand of DNA?

21. Why does cytosine bond only with guanine?

22. Describe two ways in which RNA differs from DNA.

23. Which nucleotides have double rings?

24. Where in a living cell is DNA found? How many chromosomes do humans have?

25. Describe how replication occurs.

26. Describe the two main steps in protein synthesis.

27. Define the following:
    a. Codon
    b. Anticodon
    c. Ribosomes
    d. tRNA

28. Write a reaction that shows how sodium nitrite is converted to nitrous acid by stomach acid.

29. Which three nucleotide bases react with nitrous acid?
    a. Write equations showing the reactions.
    b. Draw the products of the three reactions.
    c. How might these reactions cause a mutation?

30. Write a mechanism for the formation of the methyl carbocation alkylating agent from DMN.

31. Write a mechanism for the attack of the methyl carbocation alkylating agent on guanine. Why does attack happen primarily at the 7-position? Why does it also happen at the 3-position?

32. What are the differences between benign and malignant tumors?

33. What are the sources of PAHs?

34. Give two examples of aromatic compounds that are carcinogens.

35. If natural foods contain many natural carcinogens, why do many young people not die from cancer?

36. What is an initiator, and what is a promoter of cancer growth?

37. Are all carcinogens also mutagens?

38. The city of New Orleans takes its drinking water form the Mississippi River. It has been shown that the water in the Mississippi contains many carcinogens at low levels (parts per billion). Why is the cancer rate in New Orleans not much higher than for the rest of the country?

39. Describe the EPA's four-step procedure for risk analysis.

40. Why are tests that assess the cancer risk of a chemical performed on rodents?

41. Describe how to perform an Ames test.

42. With regard to the Ames test
   a. How are the bacteria biologically modified?
   b. Does the Ames test directly test for a carcinogen?
   c. Which nutrient is absent from the bacterial growth medium?

43. How is an exposure assessment undertaken?

44. Devise an epidemiologic study that would show whether there was a potential cancer-causing effect caused by chlorination of drinking water in the United States.

CHAPTER

17

Asbestos

ASBESTOS IS A GENERIC NAME THAT HAS BEEN GIVEN TO SIX TYPES OF NATURALLY OCCURRING SILICATE MINERALS THAT HAVE BEEN USED IN COMMERCIAL PRODUCTS. The first reported use of asbestos can be traced to Finland in 2500 BC, where asbestos from a local deposit was used to reinforce clay utensils and pottery. Since then, asbestos has been used for protection from fire and as an insulator in high-temperature industrial applications.

Asbestos has been used in hundreds of products that collectively are frequently referred to as **asbestos-containing material (ACM)**. Asbestos became widely used because it is plentiful, readily available, and low in cost. Because of its unique properties (fire resistance, high tensile strength, poor heat and electric conduction, and resistance to chemical attack), asbestos was well suited for industrial applications.

The invention in 1900 of the Hatschek machine permitted the continuous fabrication of sheets from an asbestos–cement composite. Asbestos was used in the manufacture of siding and roofing shingles, wallboard, joint compounds, cement pipes, insulation for pipes, floor tiles, ceiling tiles, high-temperature gaskets, blown-in insulation, boiler insulation, and woven products such as asbestos gloves and blankets. The U.S. Navy used asbestos extensively during World War II to insulate boiler pipes and to protect welders that were repairing damaged ships from hot metal. Some time later, the development of asbestos brakes, clutches, and gaskets for the rapidly growing automobile industry dramatically increased the demand for asbestos.

During the first three-quarters of the 20th century, the worldwide production of asbestos increased dramatically, increasing at a rate of 3% to 4% per year, reaching a maximum in 1977 (Figure 17.1). During the 1960s and 1970s, the worldwide concern about the health problems associated with exposure to asbestos fibers led to a large reduction in their use. In the 1980s and 1990s, consumption declined 5% annually. In 1989, the Environmental Protection Agency (EPA) issued a ban on most asbestos-containing products. In 1991, a federal court overturned the ban on asbestos because there was insufficient proof that asbestos gaskets and packing were as dangerous as other asbestos-containing products. As a result of the court's decision, the following asbestos-containing products remain banned: flooring felt, rollboard, and corrugated, commercial, or specialty paper. In addition, the regulation continues to ban the use of asbestos in products that have not historically contained asbestos.

Asbestos is a general term that describes six different asbestos types, all of which belong to the class of minerals called silicates.

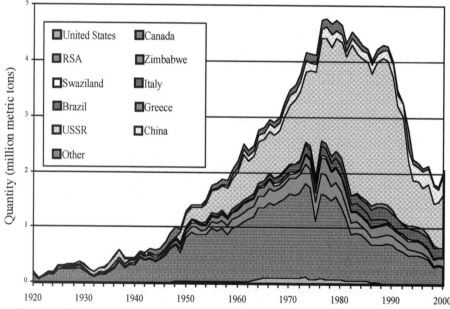

Figure 17.1 Worldwide production of asbestos 1920–2000. Courtesy of the U.S. Geological Survey.

■ Silicates

The basic unit of all **silicates** is the SiO_4 tetrahedron, in which a silicon atom at the center is bonded to four oxygen atoms located at the corners (Figure 17.2). Because the radius of the silicon atom is only approximately one-third that of the oxygen atom, the silicon atom fits comfortably between the four oxygen atoms. Because each oxygen atom can bond to two silicon atoms, adjacent SiO_4 units can join by sharing oxygen atoms. Linking allows long chains, sheets, or complex three-dimensional networks to be formed.

Three-Dimensional Networks and Sheets

The mineral **quartz** (Figure 17.3a) is a silicate that is composed entirely of SiO_4 units that are joined to form a huge three-dimensional array (Figure 17.3b). In **micas**, the SiO_4 units combine to form sheet-like arrays in which each tetrahedron is joined to three others (Figure 17.4). Because of the planar arrangement of the SiO_4 tetrahedra, micas are easily cleaved into thin sheets (Figure 17.4). Common mica is muscovite, a pearly white mineral that was used in medieval Europe to make windowpanes before glass became readily available. Other micas include talc, a soft mineral used to make talcum powder, and kaolinite, a clay mineral that is used to make china tableware.

Long-Chain Silicates: Asbestos

Asbestos is a general term for six natural silicates formed from double chains of SiO_4 tetrahedra (Figure 17.5). The family of asbestos minerals can be subdivided into serpentine and amphibole fibers (Figure 17.6). The **serpentine** group contains a single

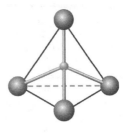

● Silicon atom

● Oxygen atom

b.

Figure 17.2 (a) Tetrahedron. (b) The tetrahedral arrangement of the SiO_4 unit in silicates.

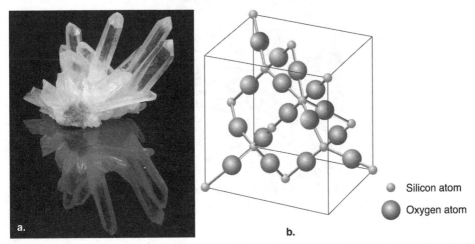

● Silicon atom

● Oxygen atom

Figure 17.3 (a) A specimen of quartz. (b) In quartz, silicon and oxygen atoms join to form three-dimensional networks.

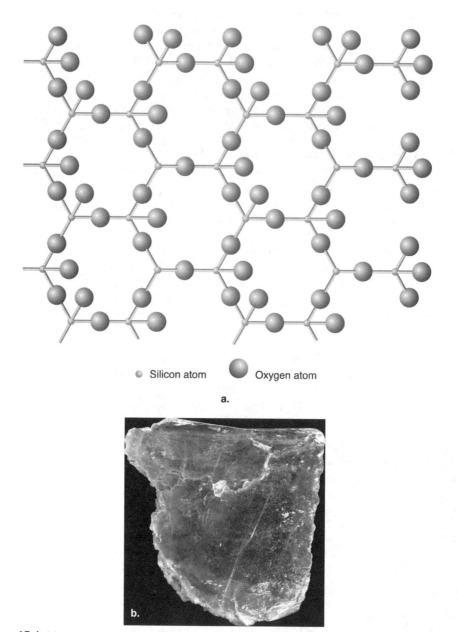

Silicon atom   Oxygen atom

a.

b.

Figure 17.4 (a) In micas, silicon and oxygen atoms join to form sheets. (b) A specimen of mica showing how it can be separated into thin sheets.

Silicon atom   Oxygen atom

b.

Figure 17.5 (a) Asbestos fibers. (b) In asbestos, silicon and oxygen atoms join to form long chains. Two parallel rows of linked SiO_4 tetrahedrals join through oxygen atoms to form double bonds. Used with permission of the California Department of Conservation, California Geological Survey.

member, chrysotile. As can be seen in Figure 17.6, five varieties of amphiboles exist: **anthophyllite** asbestos, grunerite asbestos (amosite), riebeckite asbestos (crocidolite), tremolite asbestos, and actinolite asbestos.

Amphiboles

The framework for all **amphiboles** is a double chain that is composed of two rows of SiO_4 tetrahedra aligned side by side (Figure 17.7). Metal atoms are attached to these tetrahedra. This double-ribbon arrangement creates a plane of negative charges that is neutralized by the metal cations. Amphiboles are distinguished from one another by the amount of sodium, calcium, magnesium, and iron that they contain. The chemical composition of amphiboles

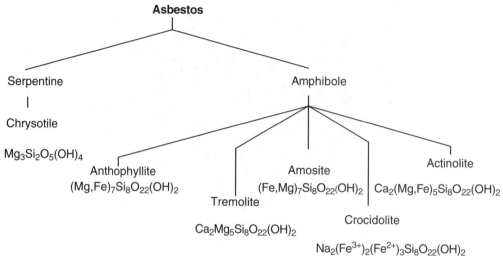

Figure 17.6 The classification of types of asbestos.

reflects the environment in which they were formed. The average chemical composition of amphibole minerals may be represented as follows:

$$A_{0-1}, B_2C_5T_8O_{22}(OH)_2$$

Where

A = Na, K
B = Na, Ca, Mg, Fe^{2+}, Mn, Li
C = Al, Fe^{2+}, Fe^{3+}, Ti, Mg, Mn, Cr
T = Si, Al

The amphiboles can be viewed as a series of minerals in which one cation is progressively replaced by another at a given site. For example, if the Mg in tremolite is partly replaced by Fe^{2+} in the C position, the replacement yields **actinolite**. The crystal structure can host a wide variety of metal cations without introducing strain into the lattice. Only three of the amphiboles are discussed farther because crocidolite and amosite were the only two amphiboles that were mined for industrial use, and **tremolite**, although having no industrial application, is found as a contaminant in chrysotile or talc.

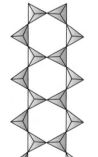

$(Si, Al)_4O_{11}$

Figure 17.7 The silicate framework of amphiboles. From U.S. Geological Survey, Open-File Report 02-149, R. L. Virta, *Geology, Mineralogy, Mining, and Uses,* Online Version 1.0. Accessed 5/3/04.

Serpentines: Chrysotile

In **chrysotile**, the only serpentine, the element magnesium is coordinated with the oxygen atoms in the silicate tetrahedra. The silicate tetrahedra is arranged to form a sheet silicate

(Figure 17.8). Chrysotile fibers are found as veins in rocks. Figure 17.9 shows that the chrysotile fibers have grown at right angles to the walls of the cracks. Most of the chrysotile fibers are extracted from deposits in which the fiber length is most often approximately 1 cm, although some deposits yield fibers that are several centimeters long.

Chrysotile is a hydrated magnesium silicate, and its stoichiometric chemical composition can be given as Mg_3Si_2-$Si_5(OH)_4$. The geothermal processes that formed chrysotile usually also co-deposit other nonfibrous minerals such as magnetite, brucite, calcite, dolomite, and talc. Chrysotile fibers can be extremely thin, having an average diameter of approximately 25 nm. Industrial chrysotile fibers are aggregates of these unit fibers that usually have a diameter from 0.1 to 100 µm; their lengths range from a fraction of a millimeter to several centimeters, although most chrysotile fibers used are shorter than 1 cm.

Chrysotile has an octahedral brucite layer with the formula $(Mg_6O_4(OH)_8)^{4+}$ intercalated between each silica tetrahedra sheet (Figure 17.10). The silicate and brucite layers share oxygen atoms, which would normally be separated by

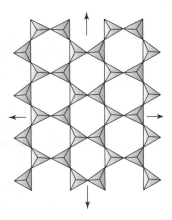

$(Si, Al)_4O_{11}$

Figure 17.8 The silicate framework of serpentines. From U.S. Geological Survey, Open-File Report 02-149, R. L. Virta, *Asbestos: Geology, Mineralogy, Mining, and Uses,* Online Version 1.0. Accessed 5/3/04.

distances of 0.305 nm in the silicate layer and 0.342 nm in the brucite layer. The mismatch of the O—O distances induces a curvature into the sheets, which propagates along a preferred axis, leading to the formation of the tubular structure found in chrysotile. The substitution of other cations in the brucite layer is limited by the structural strain that would result from replacement of magnesium with metal ions having different radii.

Chrysotile, the most common type of asbestos in the United States, forms as curly fibers. Currently, chrysotile is the only type of asbestos mined on a large scale. In 1999, one firm in California accounted for all U.S. chrysotile production. Most of the chrysotile sold in the

Figure 17.9 Host rock with a vein of the serpentine mineral chrysotile.

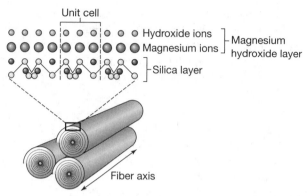

Figure 17.10 Surface structure of chrysotile fibers. From U.S. Geological Survey, Open-File Report 02-149, R. L. Virta, *Asbestos: Geology, Mineralogy, Mining, and Uses,* Online Version 1.0. Accessed 5/3/04.

United States was mined in Quebec, Canada. Small amounts of tremolite asbestos are mined in India, and commercial production of crocidolite (blue asbestos) and amosite ended in the mid-1990s in South Africa.

■ Physical Properties of Asbestos Minerals

Asbestos was widely used in industrial applications because of its resistance to heat and chemicals and its low cost compared with man-made materials. It was used as a spray-on fire-proofing material, in cement building materials (roofing, pipes), in friction materials (brakes and clutches), in gaskets, as well as asphalt coatings and sealants. As a result, an estimated 20% of buildings, including hospitals, schools, and other public and private structures, contain ACM.

The industrial applications for chrysotile take advantage of its useful properties: resistance to heat and corrosion, high tensile strength, low electrical conductivity, and a high coefficient of friction.

Thermal Properties of Asbestos

Asbestos fibers can be viewed as hydrated silicates. They are stable and resistant to heat, and depending on fiber type, they can withstand repeated exposure to very high temperatures. Chrysotile, for instance, is stable up to 550°C, at which point dehydroxylation of the brucite layer begins. This process continues as the temperature increases and is complete at approximately 1200°C. At that temperature, the chrysotile has been converted to magnesium silicate that at this point is recrystallized to form the mineral forsterite (a magnesium silicate). Depending on the impurities (tremolite) present in the chrysotile, the temperature of this conversion may be even higher.

Another reason that chrysotile is such a useful material is its high tensile strength. As chrysotile is heated to 500°C, its tensile strength actually increases slightly. As the temperature is raised above 500°C, however, the tensile strength drops sharply.

The behavior of amphibole fibers under continuous heating is similar to that observed with chrysotile. Because the amphiboles have a lower water (hydroxyl) content, their dehydroxylation reaction begins at a lower temperature (usually between 400°C to 600°C, depending on the amphibole type). The products generated from the thermal decomposition of amphiboles are the minerals magnetite, hematite, and silica. Unlike chrysotile, the tensile strength of amphiboles begins decreasing at approximately 200°C.

Chemical Resistance of Asbestos

Because it is resistant to chemical attack, asbestos has found wide use in the manufacture, storage, and transport of chemicals by many industries. Because of its brucite surface, which is a magnesium hydroxide mineral, chrysotile is a basic mineral and is very resistant to attack by strong alkali; on the other hand, it reacts rapidly with strong acids.

Crocidolite is the strongest of the asbestos fibers. It has high tensile strength and is acid-resistant. Amosite is highly resistant to heat and very flexible. It is, however, susceptible to acids and alkali. Also, it has less tensile strength than either chrysotile or crocidolite and is more difficult to spin into yarn or cloth.

■ Asbestos Diseases

Because asbestos fibers are naturally occurring and extremely aerodynamic, virtually everyone is exposed to asbestos. To cause disease, asbestos fibers must be inhaled over an extended period of time. Asbestos fibers then accumulate in the lungs and pleura, despite the natural clearing mechanisms by which the lung removes foreign particles. As exposure increases, the risk of disease also increases; therefore, measures to minimize exposure and consequently minimize accumulation of fibers will reduce the risk of adverse health effects. Unfortunately, small amounts of asbestos can cause cancer.

Asbestos is most dangerous when it becomes airborne. As long as ACMs are not damaged or disturbed, the asbestos fibers do not become airborne. During an asbestos building survey, inspectors assess the condition of ACMs. These conditions do deteriorate over time. The biggest concern is friable asbestos, which the EPA defines as ACMs that can be "crumbled, pulverized, or reduced to powder by hand pressure when the material is dry"; friable also can indicate flaking of paint or deterioration of material. Many asbestos contractors recommend encapsulating asbestos in a plastic coating so that it is more difficult to crumble or pulverize.

As asbestos fibers accumulate in the lungs, several types of diseases may occur. Asbestosis is a scarring of the lung tissue that impairs the elasticity of the lung and hampers its ability to exchange gases, leading to inadequate oxygen intake to the blood. Asbestosis restricts breathing, leading to decreased lung volume and increased resistance in the airways. It is a slowly progressive disease, with a latency period of 15 to 30 years. Another disease attributed to asbestos exposure is mesothelioma, which is a cancer of the pleural lining. It is a disease in which the only known cause is asbestos exposure. By the time it is diagnosed, it is almost always fatal. Similar to other asbestos-related diseases, mesothelioma has a longer latency period of 10 to 20 years. Lung cancer is a malignant tumor of the bronchi that grows through surrounding tissue, invading and often obstructing air passages. The time between exposure to asbestos and the occurrence of lung cancer is 20 to 30 years. It should be noted that there is a synergistic effect between smoking and asbestos exposure that creates an extreme susceptibility to lung cancer. Cigarette smokers who work in asbestos-related fields are at a

50 to 90 times greater risk of developing lung cancer than nonsmokers because the two substances together multiply the risk and increase the likelihood of disease.

Respiratory System

Each adult male human inhales approximately 3000 mL of air per breath, adding up to between 10 to 20 m^3 of air a day. The lung is a specialized organ that takes oxygen out of ambient air and transfers it to the blood while carbon dioxide is released from the blood and is returned to the air that is exhaled. Figure 17.11 shows that the upper respiratory tract begins at the nose and mouth.

To protect the lung, mechanisms exist to expel harmful components from the respiratory system. Nasal hairs act as filters for large particles. The upper passages of the respiratory system—the bronchi—are lined with cells that contain small hairs called cilia. When a particle is inhaled, the cilia help transport that particle on a blanket of mucous upward toward the larynx, where it can be expelled by coughing. The higher in the respiratory tract that particles are deposited, the greater the chance that they will be removed by these lung clearance mechanisms. Particles most likely to be retained in the lung are those that penetrate deeply into the alveoli of the lung before being deposited.

When particles are inhaled deeply into the lung, systems of lymphatic vessels (found throughout the lung) can remove a variety of materials from the small air spaces by transporting them through ducts into the lymph nodes. There is some evidence in animals that long fibers are retained in the lungs for longer periods than short fibers. This relationship may be associated with the inability of the body's macrophages to engulf and remove or destroy fibers that are significantly larger than themselves. Analyses of autopsied human lungs often show that higher numbers of short (<5 μm) fibers are present deep in the lung. Another mechanism that protects the lung is a process by which a foreign particle is sequestered by the lung coating it in protein. The coated fiber, which also contains a considerable amount of iron, is known as a *ferruginous body*. People who are employed in exceptionally dusty workplaces may have thousands of ferruginous bodies in their lungs. Some controversy exists about the ability of short asbestos fibers to cause disease.

Fiber Drift

Airborne particles (and asbestos fibers) are deposited in the respiratory system by the action of four different mechanisms:

1. Particles are removed from the inhaled air by settlement under gravity.
2. When a change in the direction of airflow occurs, the particle's inertia causes it to continue on its original path until it hits the wall of the lung.
3. Particles suspended in the air are under bombardment by gas molecules, which constantly displace them and may cause their deposition on surfaces in the respiratory system.
4. For a particle to remain suspended in the air, its center must be separated from all surfaces by a distance greater than the particle radius. Deposition occurs when the distance from the surface is equal to the particle radius. The rate of deposition increases as the particle moves into smaller and smaller airways and the ratio of the particle radius to tube diameter increases.

Particles that are found in significant numbers in the alveoli of the lung are usually smaller than 10 μm, but 50 to 200 μm long asbestos fibers have been found deep in the lung. Because

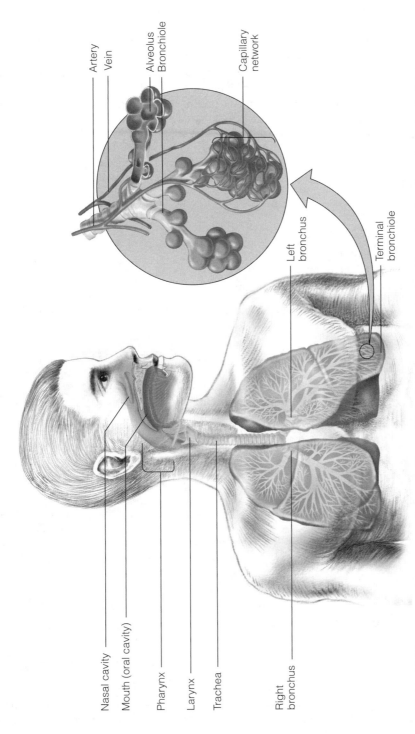

Nasal cavity

Mouth (oral cavity)

Pharynx

Larynx

Trachea

Right bronchus

Left bronchus

Terminal bronchiole

Artery

Vein

Alveolus

Bronchiole

Capillary network

Figure 17.11 Human respiratory system. The average diameter of the air flow systems are: trachea, 20 mm; bronchus, 8 mm; terminal and respiratory bronchioles, 0.5 mm and alveolar duct, 0.22 mm.

asbestos fibers are subjected to the same deposition mechanisms as other particles—namely sedimentation and inertial precipitation—it is surprising that such long fibers are transported deep in the lung. The rate at which sedimentation and inertial precipitation mechanisms clear the air of particles is proportional to the free-falling speed of the particle or fiber. Because the magnitude of the particles falling is proportional to its aerodynamic diameter, it is important to understand how to estimate a fiber's aerodynamic diameter.

Studies of the deposition of fibers in air have produced an expression that relates the aerodynamic equivalent diameter (De) to the actual diameter of the fiber. That expression is as follows:

$$De = 66D\left[\frac{B}{2 + 4B}\right]^{2.2}$$

Where

De is the aerodynamic equivalent diameter.
D is the actual diameter.
B is the aspect ratio (length/diameter) of the fiber.

This equation has an important bearing on the respirability of fibers and on their ability to drift in the air.

EXAMPLE 17.1

What is the aerodynamic equivalent diameter of an asbestos fiber that has a diameter of 1.5 µm and is 6 µm in length?

Step 1. Substitute values into the equation:

$$De = 66D\left[\frac{B}{2 + 4B}\right]^{2.2}$$

$$De = 66(1.5)\left[\frac{4}{2 + 4(4)}\right]^{2.2}$$

Step 2. Compute De

$$De = 3.5 \text{ µm}$$

The falling speed of an asbestos fiber is predominately determined by its diameter and shows little sensitivity to length, especially when the aspect ratio is high. If a fiber has a sufficiently small diameter, the falling speed can be low enough for the fiber to drift and escape deposition by settlement and inertial precipitation mechanisms. If particles with diameters of 10 microns are the largest that can reach the pulmonary airspaces, then asbestos fibers with diameters of 3.5 μm, regardless of their length, can also.

Regulation of Asbestos

Because so much of asbestos exposure occurred in the workplace, the regulation of airborne asbestos is now the responsibility of the **Occupational Safety and Health Administration (OSHA)**. The current OSHA standard for asbestos fibers in the air is referred to as the **permissible exposure limit (PEL)**. The current OSHA PEL is 0.1 fiber/mL (or 0.1 fiber/cm^3) measured over an eight-hour period. This PEL is a **time-weighted average (TWA)** that is intended to measure the exposure of a worker during a normal eight-hour workday. Two different concentration indices have been used to express the amount of airborne asbestos. Both express the number of fibers per unit volume of air. Initially, the unit millions of particles per cubic foot (mp/ft^3) was used. Today, the unit fibers per cubic centimeter (ft/cm^3) is used. The PEL was lowered over the years as the risks of asbestos exposure became better known. The history of the asbestos regulatory standards is listed in Table 17.1. PEL were slowly lowered not because of scientific reasons, but because of opposition from asbestos-producing companies.

OSHA was established as a federal agency in 1971. With the establishment of OSHA, regulations were enacted requiring avoidance of asbestos concentrations above those recommended by various voluntary associations such as the **American Conference of Government Industrial Hygienists**, which had issued proposed standards in the hope of persuading industry to lower exposure limits to protect workers from exposure. The American Conference of Government Industrial Hygienists recognized as early as 1959 that workers could get

Table 17.1	
History of PEL for Asbestos	
Year	**TWA**
1948–1968	5 mp/f^3 (5-million particles per cubic feet)*
1968–1969	12 f/cm^3
1970–1978	5 f/cm^3
1978–1991	2 f/cm^3 chrysotile
	0.5 f/cm^3 amosite and tremolite
	0.2 f/cm^3 crocidolite
1991–1997	0.2 f/cm^3 all types of asbestos
1997–present	0.1 f/cm^3

*Equivalent to 12 f/cm^3.

mesothelioma from exposures even at the lowest level recommended. The initial use of existing standards was intended to get OSHA off to a running start. Because of the health risks, in 1974, the EPA banned the use of asbestos for insulation and fireproofing, and many communities began removing asbestos from public buildings, particularly from schools. Since 1980, EPA has listed asbestos as a substance known to be a carcinogen.

The process of asbestos removal is not only extremely costly but also likely to release fibers into the air and expose removal workers to unusually high fiber levels. The EPA now recommends that, unless it is crumbling, asbestos be left in place, encapsulated in a plastic coating.

In 1986, the Asbestos Hazard Emergency Response Act (AHERA) was signed into law. It requires public and private nonprofit primary and secondary schools to inspect their buildings for asbestos-containing building materials. The EPA has published regulations that require schools subject to AHERA to perform an original inspection and periodic reinspections every three years for ACM. Schools that fail to meet the AHERA requirements are subject to civil enforcement action by the EPA or designated state environmental agency. Furthermore, the EPA has established a "clean" standard for schools to be 0.01 fiber/cm^3.

■ Analytical Methods that Identify Asbestos Fibers

All of the analytic methods that have been previously described in this book are techniques that determine the concentration of the chemical of interest (*analyte*) in the environment by measuring one of its fundamental characteristics (mass, ultraviolet absorbance, X-ray emission, etc.). In the case of asbestos, however, the airborne asbestos concentration is determined by the measurement of a macroscopic property, the size, and shape of the fiber. Microscopes are tools to both observe and identify asbestos fibers as well as the quantity present. The analytical technique must be able to differentiate fibers. Many industrial applications, for instance, use glass wool or fiberglass, both of which have fibrous structures. Air sampling in an industrial workplace also will capture other fibers such as hair, wool, and cotton. The analytical techniques must be able to determine accurately whether asbestos fibers are present and if so, precisely how many.

Methods to Determine Bulk Asbestos

OSHA not only issues standards for workplace safety but also analytic methods to determine potentially hazardous material. OSHA Method 191 is a prescribed method that uses **polarized light microscopy** to identify whether a material contains bulk asbestos.

Light Microscopy

The **light microscope** has been used to determine the type of mineral present in geologic samples for over 100 years. Minerals that contain atoms that are arranged in random order are classified as amorphous materials. Those minerals in which atoms are arranged in a distinct order are classified as crystalline materials. By observing how light interacts with thin sections of the minerals, classification of minerals can be accomplished.

Isotropic materials, which include gases, liquids, and cubic crystals, have the same optical properties when observed from all directions. They have only one refractive index and show no difference to light passing through them from different directions. The six types of asbestos are crystalline and are **anisotropic.** When an anisotropic material is observed, the arrangement of atoms in the substance appears different, as the direction of observation is changed. Photons of light passing through asbestos fibers from different directions will encounter different electrical neighborhoods, which will affect the path and time of travel of the light beam. Different types of asbestos all are silicates and have a similar array of atoms, but the arrangement is not the same in all directions. The techniques described here rely on the fact that light traveling through asbestos fibers in different directions will behave differently but predictably.

The sample to be analyzed must be dry and thin enough to be placed on a microscope slide under a coverslip. If the material is too thick, a portion can be shaved off with a scalpel.

The analysis consists of three parts: to determine if asbestos is present, identify the type of asbestos, and quantify how much of the sample is asbestos.

Polarized Light Microscopy

Besides providing information on the shape, color, and size of different minerals, **polarized light microscopy** can distinguish between isotropic and anisotropic materials. The technique used to identify asbestos exploits the anisotropic properties of asbestos fibers.

Plane-Polarized Light

According to the wave theory of light, light waves oscillate (vibrate) at right angles to the direction in which the light is traveling through space. The oscillations occur in all planes that are perpendicular to the path traveled by the light. Figure 17.12a schematically shows oscillations present in a flashlight beam, as viewed by an observer looking directly into the beam. If the beam of light is passed through a sheet of Polaroid material, the light that is transmitted through the Polaroid sheet oscillates in one plane only (Figure 17.12b). Such light is called **plane-polarized light.** If a second Polaroid lens is inserted before your eye, that lens must be aligned in a parallel lane so that plane-polarized light is passed. If the second Polaroid lens is rotated, no plane-polarized light is transmitted through the second lens.

Polarized Light Microscope

Figure 17.13 shows that there are two polarizing filters in a polarizing microscope: the polarizer lens, which is fixed in place, and the analyzer lens. The sample stage, which is placed between the two filters, can be easily rotated.

Besides information about the gross fiber morphology and color of the sample, plane-polarized light determines whether the sample exhibits *pleochroism,* which is the property of a substance to show different absorption colors when exposed to polarized light coming from different directions. The observed colors change with the orientation of the crystal and can be seen with only plane-polarized light.

Asbestos fibers are anisotropic and they exhibit pleochroism. Chrysotile asbestos fibers appear crinkled, like damaged hair, under plane-polarized light, whereas crocidolite and amosite asbestos are straight. Figure 17.14 shows that if the microscope stage is rotated, the colors transmitted by the asbestos fibers begin to change colors (blue, yellow, and red). If the sample contained glass fibers rather than asbestos, there

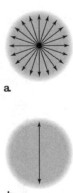

a

b.

Figure 17.12 (a) Light passing through the first Polaroid lens is polarized in one plane and it easily passes through the second Polaroid lens that is set in the same direction. (b) If the second Polarizing lens is rotated, no light is transmitted to the eye.

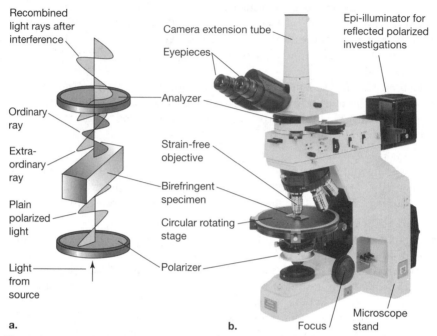

Recombined light rays after interference

Camera extension tube

Eyepieces

Epi-illuminator for reflected polarized investigations

Analyzer

Ordinary ray

Extra-ordinary ray

Strain-free objective

Birefringent specimen

Plain polarized light

Circular rotating stage

Light from source

Polarizer

a.

b.

Focus

Microscope stand

Figure 17.13 A polarized light microscope. Courtesy of Nikon Instruments Inc., Melville, New York.

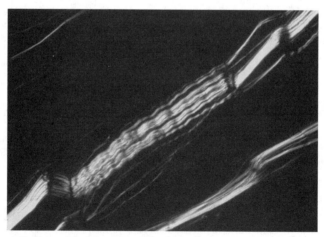

Figure 17.14 Chrysotile fibers as observed in a polarized-light microscope. As the sample stage is rotated, the color of the fiber changes.

would be no change in color as the microscope stage was rotated. Because glass fibers are isotropic, they are unaffected by rotation under polarized light.

This method is capable of estimating the amount of asbestos in the sample. In order to do so, a microscopist with considerable experience estimates the percentage of asbestos by using observation and estimation methods.

Methods to Determine the Amount of Airborne Asbestos

Analyzing for airborne asbestos is technically difficult. Asbestos fibers can be extremely small and may number several million for an average size room where ACM is present. It is not practical to pass all of the air in a room through a filter, and thus, only a small fraction of the total asbestos fibers present is trapped and counted. Significant errors can be introduced when the results of a few isolated samples are extrapolated to describe the contents of the entire room. Furthermore, fibers of asbestos may closely resemble those of hair, cloth, glass, paper, and other nonasbestos materials. As a result, identifying and counting asbestos fibers require sophisticated instruments and highly trained technicians.

The research arm of OSHA, the **National Institutes of Occupational Safety and Health (NIOSH),** has developed two protocols for determining airborne asbestos. OSHA Method 160 (formerly known as NIOSH Method 7400) is a method that uses phase contrast microscopy to determine the concentration of airborne asbestos.

OSHA Method 160 calls for the collecting of airborne fibers on a filter. A pie-shaped section of the filter is then analyzed by dissolving the filter and placing it in a polarized light microscope. The microscopist counts only fibers that are longer than 5 μm with a 3:1 aspect ratio (three times longer than their width).

Air Sampling

To determine the concentration of asbestos fibers in air, sampling is conducted by drawing air through a filter at constant flow rate. Rotary vane pumps (Figure 17.15) are those in which the flow rate can be adjusted from 3 to 20 L/min. The sampling time and flow rate are chosen to give a fiber density of 100 to 1300 fibers/mm on the filter. The most significant problem when sampling for asbestos is overloading the filter with nonasbestos dust. While sampling, the filter can be checked using a small flashlight; if there is a visible dust layer on the filter, it should be removed and replaced with a new one. After the sample is collected, the filter is closed and sealed for transport back to the laboratory.

Blank samples are used to determine whether any contamination has occurred during sample handling. Clean filters are taken to the sampling site and are handled in the same way as the sample filter with one exception: No air is drawn through the filter. The blank filter is opened in the place where the sample is being drawn. Blank filters are held open for 30 seconds and then closed and sealed for transport back to the laboratory.

The filters can be set out to measure the background level of airborne asbestos in an area. The filter also can be attached to a worker so that their breathing zone is sampled. A battery-operated pump is attached to the employee's belt and a hose is run over the shoulder to a filter that is attached to the collar; the employee performs normal tasks while the amount of asbestos

Figure 17.15 A rotary vane air pump with a filter attached.

fibers in the breathing zone is measured. The samples are prepared for analysis by carefully opening the filters and mounting the collected material by a specific protocol to produce an optically smooth background.

PCM

Phase contrast microscopy (PCM), first described in 1934 by Dutch physicist Frits Zernike, is a contrast-enhancing optical technique that can be used to produce high-contrast images of transparent specimens. In effect, the phase contrast technique employs an optical mechanism to translate minute variations in phase into corresponding changes in amplitude, which can be visualized as differences in image contrast. The eyepiece of the phase contrast microscope is fitted with a special device (**Walton-Beckett graticule**) that allows the microscopist to measure the length and width of the objects in the microscopic field of view. The prepared sample is placed on the mechanical stage of the microscope, and the microscopist focuses the image. The PCM makes the asbestos fibers appear black against the background (Figure 17.16). The microscopist counts the fibers that have a diameter of greater than 0.2 μm using a technique described in Example 17.2.

EXAMPLE 17.2

An example of an image that would be observed through the microscope is seen in Figure 17.17. OSHA has a very definite counting method. An asbestos fiber is defined as one that has a 3:1 or greater aspect ratio and is longer than 5 μm. That is, the fiber is at least three times longer than wide and longer than 5 μm. The fibers in this figure are numbered. Fibers 1 to 6 are all single fibers that are contained in the circle and do not cross the boundary; they are all counted as one

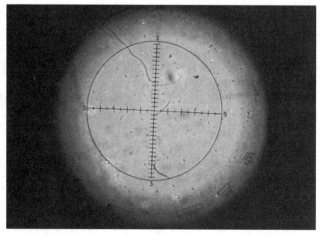

Figure 17.16 A typical image from a filter that is used to measure airborne asbestos.

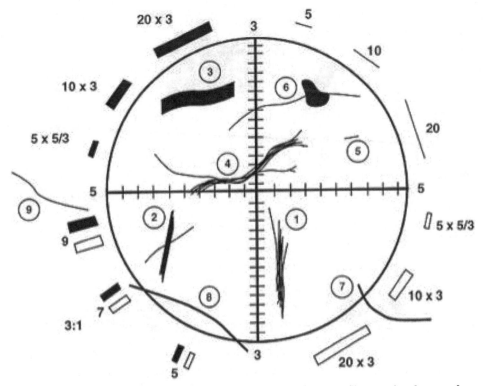

Figure 17.17 A hypothetical example from OSHA that is used to teach asbestos fiber counting. Courtesy of Jim Murray/Criterion Laboratories, Inc.

fiber each. Fiber 7 crosses the circle once, and because it lies partially outside the circle, it is counted as one half of a fiber. Fiber 8 is too short to be counted. Fiber 9 is actually two crossing fibers and is counted as two fibers. Fiber 10 is outside of the circle and is counted as zero fiber. Fiber 11 crosses the circle twice and is counted as zero fiber. Fiber 12, although split, crosses the circle and is counted as one-half a fiber. Thus, this example would have nine total fibers counted.

One additional OSHA counting rule is this: If there is a question of whether a fiber is asbestos or not, remember that "when in doubt, count."

To calculate the airborne concentration, the following equation is used:

$$\frac{\left[\left(\frac{FB}{FL}\right) - \left(\frac{BFB}{BFL}\right)\right] \times 49}{ft} = AC$$

Where

FB = the total number of fibers greater than 5 μm.

FL = the total number of fields counted on the filter.

BFB = the total number of fibers greater than 5 μm counted in the blank.

BFL = the total number of fields counted in the blank.

f = the flow rate of collection pump (cc/min).

t = the sample collection time (min).

The PCM technique has four main advantages:

1. It is specific for fibers. Nonfibrous particles are not counted.
2. PCM equipment is inexpensive.
3. The analysis can be performed quickly and on site for rapid determination of the concentration of airborne asbestos fibers.
4. The PCM technique has been in use for more than 20 years, and the results can be correlated with prior studies.

The PCM technique has three major disadvantages:

1. It does not positively identify the fiber as asbestos. No test is used to verify that the fibers counted are asbestos; fibers other than asbestos may be included in the count.
2. The PCM technique cannot measure fibers smaller than 0.2 μm in diameter. The smallest asbestos fibers may be as small as 0.02 μm. These fibers would be missed by PCM.
3. Asbestos fibers that are released from friction products (brakes and clutches) or from gasket maintenance may be shorter than 5 μm. These fibers would be missed by PCM.

To overcome the shortcomings of the PCM technique, NIOSH has issued Method 7402 that uses transmission electron microscopy to identify and count the asbestos fibers on the filter.

TEM

Transmission electron microscopy (TEM) is the preferred analytical method for asbestos samples that are collected outdoors because of its ability to detect small fibers (0.0002 μm in diameter) and to distinguish asbestos fibers from nonasbestos fibers. TEM is the method recommended by the California Office of Environmental Health Hazard Assessment (OEHHA).

A TEM operates on the same basic principle as the light microscope but uses electrons rather than light. Electrons, which are produced in the electron source, travel through a specimen in a way that is similar to a beam of light passing light through a sample in a light microscope. Figure 17.18 shows that instead of glass lenses directing light wavelengths through a specimen, the electron microscope's electromagnetic lenses direct electrons through a specimen. Because the wavelength of electrons is much smaller than the wavelength of light, the resolution achieved by the TEM is many times greater than that of the light microscope. Thus, the TEM can reveal the finest details of structure. Figure 17.19 is a TEM image of chrysotile asbestos from an asbestos mine in Quebec that is magnified 10,000 times; magnifications of 1000 to 500,000 times are routine. Because the resolution of the TEM is so high, the beam can be focused on one fiber in a complex jumble of fibers collected in a dirty industrial setting.

Some of the electrons striking the surface are immediately reflected back toward the electron source; these are called *backscattered electrons*. An electron detector can be placed above the sample to scan the surface and produce an image from the backscattered electrons. Some

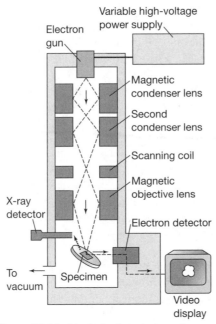

Figure 17.18 Schematic of an electron microscope.

of the electrons striking the surface penetrate the sample and can be measured by an electron detector that is placed beneath the sample. In addition, some of the electrons striking the surface cause X-rays to be emitted. By measuring the energy of the emitted X-ray, the elemental composition of the surface can be determined by a technique called **energy-dispersive X-ray spectroscopy (EDX)**.

Figure 17.19 A SEM image of chrysotile asbestos from a mine in Quebec, Canada.

EDX

Electrons bombarding the atoms on the surface of the sample are able to strike an inner-shell electron in the sample atoms, which causes that electron to be ejected like a pool ball being struck by the cue ball. The lowest electron shell is the "K" shell, the second lowest the "L" shell and the next the "M" shell. The vacancy caused by the loss of the emitted electron is filled by an outer shell electron that drops to the lower level (see Figure 6.25). Because the outer shell electron is at higher energy when it drops to the lower level, it loses excess energy by releasing a photon of electromagnetic radiation. This X-ray photon has an energy that is equal to the difference between the two electron energy levels. The photon energies are designated as K, L, or M X-rays, depending on the energy level filled; for example, a K shell vacancy filled by an L level electron results in the emission of a $K\alpha$ X-ray. Because the difference in energy between the two electron levels is always the same, an element in a sample can be identified by measuring the energy of the emitted photon (or photons if there is more than one electron emitted). The intensity of the emitted photons is also directly proportional to the concentration of the element emitting the photon in the sample.

Figure 17.20 shows an EDX detector that measures the energy of the X-ray photons; it can be attached to a scanning electron microscope (SEM). The spectrum obtained is used to identify which element is present and the intensity of that photon to measure the amount of the

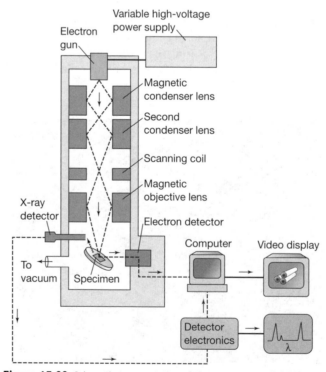

Figure 17.20 Schematic representation of the components of the SEM-EDX.

element in the sample. Because this technique measures energy differences from inner shell electrons, it is insensitive to how the element (being measured) is bonded. The bonding shell electrons are not involved in the process. EDX will not detect every element; the elemental range is limited to elements larger than beryllium, and the detection of a low atomic number (Z < 11, Na) elements is difficult.

Figure 17.21 is a SEM image of an amosite fiber that was collected on a membrane filter and an EDX spectrum of the X-rays ejected from the surface of the fiber. The x-axis of the spectrum displays the energy of the X-ray in Kev; the y-axis displays the intensity of the X-ray signal. The X-rays from silicon atoms in the amosite fiber give the peak with highest intensity, iron, and magnesium next. Oxygen and hydrogen, which are two of the principal elements in amosite, are not seen because their atomic number is less than 11. SEM-EDX is able to identify this fiber as amosite, and not chrysotile, based on the EDX spectrum obtained.

Many different fibers look similar under the light microscope (Figure 17.22), which makes it difficult to identify them positively by physical characteristics alone. On the other hand, the EDX, when coupled with TEM, will definitively identify fibers as being asbestos.

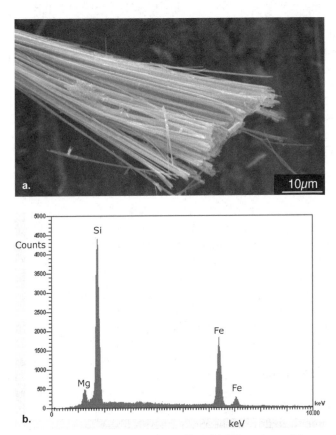

Figure 17.21 (a) Amosite fibers collected on a membrane filter, (b) EDX spectrum of the amosite fiber.

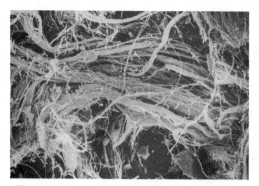 

Figure 17.22 SEM images of chrysotile fibers on the left are 50 times smaller than the glass fibers that are on the right.

Comparing PCM and TEM Measurements

One major limitation of PCM analysis, especially in outdoor environments, is that the analyst cannot distinguish asbestos from nonasbestos fibers, such as cellulose, talc, or gypsum. Also, PCM cannot detect fibers that have a diameter of approximately 0.2 microns or less, which could substantially underestimate the asbestos fiber concentrations.

These limitations make PCM impractical for the analysis of ambient asbestos samples. TEM is the preferred analytic method for outdoor asbestos samples because of its ability to detect small fibers (less than 0.2 microns in diameter) and to distinguish between asbestos fibers and nonasbestos fibers. TEM measurements cannot be directly related to risk factors, however, because the studies on which the risk assessment was based used the less expensive PCM analysis. The TEM measurements are usually converted to PCM-equivalent units, using the following equation:

$$1\,\text{PCM fiber} \;=\; 320\,\text{TEM structures}$$

After the events of September 11, 2001, the EPA, the State of New York, and NIOSH have all performed extensive studies on the dust remaining at the World Trade Center. NIOSH collected a total of 804 air samples. No asbestos fibers were detected in over 57.9% (446) of these air samples, as determined by PCM. The remaining 358 air samples indicated TWA concentrations of asbestos ranging up to 0.89 fibers/cc as determined by PCM. Twenty-five of World Trade Center samples exceeded the OSHA PEL of 0.10 fibers/cc.

The particle size of the dust released in that tragedy is extremely small, and TEM was also used to characterize its composition; 114 of these air samples were analyzed by TEM. No asbestos fibers were detected in over 49% (56) of these air samples as determined by TEM. The remaining 58 air samples indicated TWA concentrations ranging up to 0.024 asbestos fibers/cc as determined by TEM. These TEM results indicate that the vast majority of the fibers reported from the PCM analyses are nonasbestos fibers. Further analyses are being performed to determine the nature of these nonasbestos fibers.

■ Asbestos Litigation

By 1935, asbestosis was widely recognized as a health threat that affected a large fraction of those who had regularly worked with the material. It was known that the disease would not become obvious for the first few years of exposure. By the time the disease became apparent, ending exposure would not halt the progress of the disease caused by the fibers already trapped in the lung. During the 1940s and 1950s, further reports of the carcinogenicity of asbestos appeared in medical articles.

Companies producing and using asbestos in industrial products contracted scientists to perform experiments that would show that exposure to asbestos fibers was harmless. When the results of their studies showed just the opposite—that asbestos fibers caused asbestosis and cancer—the research results were withheld from the public. Leaders in the asbestos industry realized that their near-term financial interest was best served by having asbestos hazards receive a minimum of publicity.

The groundbreaking work on asbestos exposure-induced disease among U.S. workers was conducted by Dr. Irving Selikoff at the Mt. Sinai School of Medicine in New York during the late 1960s and 1970s. Litigators still use his work and that of his colleague William Nicholson as a standard reference on occupational exposure to asbestos and to form projections of occupational disease due to asbestos exposure. A result of their pioneering research was the realization by government agencies of the dangers posed by asbestos exposure and the necessity to regulate exposure in the workplace.

The asbestos industry's failure to respond adequately to the evolution of medical knowledge about the dangers of asbestos exposure has been well documented. It was this failure decades ago that set the stage for the current litigation. The long chain of adverse medical and legal consequences has reinforced the principle that the employer, the producer of chemical materials, always has had the responsibility to not endanger the workers, community, or society in general. This responsibility when ignored brings about an irreversible legacy of human and economic disaster in subsequent decades.

■ Additional Sources of Information

Castleman BL. *Asbestos: Medical and Legal Aspects,* 4th ed. New York: Aspen Publishers, 1996.

Lippmann M. *Environmental Toxicants: Human Exposures and Their Health Effects,* 2nd ed. New York: John Wiley & Sons, 1999.

National Institute of Occupational Safety and Health (NIOSH), Centers for Disease Control and Prevention. Homepage. Accessed December 2008 at http://www.cdc.gov/niosh.
NIOSH Method 7400—Asbestos and Other Fibers by PCM
NIOSH Method 7402—Asbestos by TEM
NIOSH Method 9000—Chrysotile by XRD
NIOSH Method 9002—Asbestos (Bulk) by PLM

Occupational Safety and Health Administration (OSHA), U.S. Department of Labor. Homepage. Accessed December 2008 at http://www.OSHA.gov.

Skinner HCW, Ross M, Frondel C. *Asbestos and Other Fibrous Materials.* Oxford, UK: Oxford University Press, 1988.

■ Keywords

actinolite
American Conference of Government
   Industrial Hygienists
amosite
amphibole
anisotropic materials
anthophyllite
asbestos
asbestos-containing material (ACM)
asbestosis
chrysotile
crocidolite
energy-dispersive X-ray (EDX)
friable asbestos
isotropic materials
light microscopy
lung cancer
mesothelioma

mica
National Institute of Occupational Safety
   and Health (NIOSH)
Occupational Safety and Health
   Administration (OSHA)
permissible exposure limit (PEL)
phase contrast microscopy (PCM)
plane-polarized light
polarized light microscopy
quartz
serpentine
silicates
synergistic effect
time-weighted average (TWA)
transmission electron microscope (TEM)
tremolite
Walton-Beckett graticule

■ Questions and Problems

1. What is the basic chemical unit that is present in all silicates? How are the atoms in this unit arranged? Illustrate your answer with a diagram.
2. The basic units in the following silicates are arranged as three-dimensional arrays, as two-dimensional sheets, or as chains. For each mineral, indicate which arrangement is present:
   a. Talc
   b. Chrysotile
   c. Tremolite
   d. Quartz
3. Name the members of each family:
   a. Serpentine
   b. Amphibole
4. Give the general formula for the amphibole family.
   a. What elements distinguish tremolite from actinolite?
   b. Show how the substitution of metal ions converts amosite to actinolite.
5. For the following silicates,
   a. Draw the SiO_4 framework for amphiboles.
   b. Draw the SiO_4 framework for serpentines.
   c. Draw the SiO_4 framework for talc.

6. Describe how chrysotile asbestos is found in rock structure. Give its physical characteristics.

7. Explain why chrysotile is a curved sheet. Why does chrysotile not have metal ions other than magnesium substituted into its structure?

8. a. What is the fiber type of the most common asbestos found in the United States?
   b. Where was most of the asbestos used in the United States mined?
   c. Where was crocidolite (blue asbestos) mined?
   d. Was anthophyllite ever mined for industrial use?

9. List four useful properties of asbestos that made it so widely used in industrial applications.

10. Chrysotile has been used as a high-temperature insulation:
   a. At what temperature does it begin to lose water?
   b. How does its tensile strength change as it is heated to 500°C?
   c. What happens to chrysotile that is heated above 900°C?

11. There is great debate over the hazards posed by asbestos.
   a. List three properties that make asbestos a useful material.
   b. Give three examples of the industrial uses of asbestos.
   c. Should we be more concerned about friable asbestos than asbestos that is encapsulated in a product (such as an asbestos floor tile)?

12. Describe the following asbestos diseases:
   a. Asbestosis
   b. Mesothelioma
   c. Lung cancer

13. Describe the synergistic effect that smoking of cigarettes has on an asbestos worker.

14. The human lung is a specialized organ. Name the various parts of the lung that air passes through from the time it enters the body to the time that it is absorbed in the blood.

15. What volume of air is inhaled per breath by an adult male? How much air does an adult male inhale per day in cubic meters and in liters?

16. List three mechanisms the body uses to clear particles from the lung.

17. List the four mechanisms that act to deposit airborne particles in the lung.

18. What is the diameter of most particles found deep in the lung? What is the length of some of the asbestos fibers found in the lung? Why is there a difference between particles and fibers?

19. What is the aerodynamic equivalent diameter of an asbestos fiber that has the following dimensions:
   a. 1.0 μm diameter, 6 μm long
   b. 2.0 μm diameter, 10 μm long
   c. 2.0 μm diameter, 30 μm long

20. An asbestos fiber that has a diameter of 2.0 μm is 40 μm long. What is the aerodynamic equivalent diameter of this fiber? Do you think it will be inhaled deep in the lung? Justify your answer.

21. An asbestos fiber that has a diameter of 3.0 μm is 50 μm long. What is the aerodynamic equivalent diameter of this fiber? Do you think it will be inhaled deep in the lung? Justify your answer.

22. When was OSHA established? Why has the regulation of airborne asbestos become OSHA's responsibility?

23. Define the following terms that are used to describe airborne asbestos exposure:
   a. PEL
   b. TWA
   c. ft/cm^3

24. A factory is tested for airborne asbestos. Filters are positioned throughout the facility. A flow of 4 L/min of air passes through the filters for a period of 15 minutes. The number of asbestos fibers present on the filters is determined by NIOSH Method 7400. The results of the counting are listed here. Do the results for any of the filters exceed the OSHA PEL for asbestos?
   a. Filter 1: 105 fibers
   b. Filter 2: 565 fibers
   c. Filter 3: 898 fibers

25. How does an isotropic material differ from an anisotropic material?

26. Draw a schematic of a polarized light microscope.

27. Define pleochroism.

28. Describe OSHA Method 191 for identifying asbestos in a bulk sample.

29. How is a workplace sampled for airborne asbestos fibers?

30. Describe NIOSH Method 7400 for the determination of airborne asbestos.

31. List the advantages and disadvantages of PCM for the determination of airborne asbestos fibers.

32. An industrial shop has been tested for asbestos using the NIOSH Method 7400. One of the fields from the filter can be seen in Figure 17.15. Count the number of fibers in this field. Assume that nine additional fields give an identical number of fibers. Blank samples show no fibers collected. If the pump collecting the sample had a flow rate of 5 L/min and the sample was collected for 60 minutes, did the air in the shop exceed OSHA standards?

33. A gas station that repairs brakes has been tested for asbestos using the NIOSH Method 740. One of the fields from the filter can be seen in Figure 17.15. Count the number of fibers in this field. Assume that four additional fields give an identical number of fibers. Blank samples show no fibers collected. The pump collecting the sample was attached to the belt of the auto mechanic and the filter sampled his breathing space. The flow rate was 5 L/min, and the sample was collected for 30 minutes while he did the brake repair. Does the air in the shop exceed OSHA standards? If the same sample was counted using NIOSH Method 7402, would you expect the number of asbestos fibers reported to be greater or less? Explain.

34. When would you use NIOSH Method 7402 rather than Method 7400?

35. Draw a schematic of a TEM.
   a. How is the electron beam focused?
   b. Why does the TEM have better spatial resolution than a light microscope?
   c. How is the image of the surface obtained?
   d. Why is an X-ray emitted during the electron bombardment?

36. Describe the following for X-rays:
   a. K, L, M shells
   b. EDX spectrum
   c. Why is the position of the Mg peak in the EDX spectrum of chrysotile in the same location as in the spectrum of amosite?
37. How is EDX used in conjunction with TEM? Draw a schematic, and explain the results expected.
38. Refer to Figure 17.21. Why is this fiber identified as amosite? Draw the EDX spectrum that you would expect to see for chrysotile.
39. Why is EDX insensitive to an element with an atomic number of less than 11?
40. The following TEM-EDX spectrum was obtained from a fiber on a filter. Do these results identify the fiber as crocidolite or chrysotile? Justify your answer.

a.

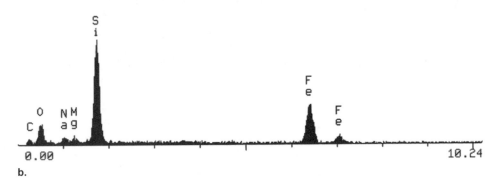

b.

41. The following TEM-EDX spectrum was obtained from a fiber on a filter. Do these results identify the fiber as crocidolite or chrysotile? Justify your answer.

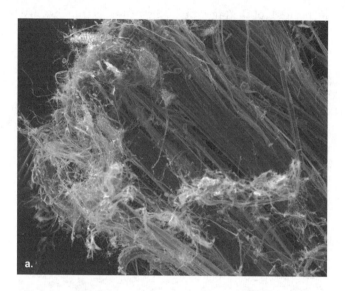

a.

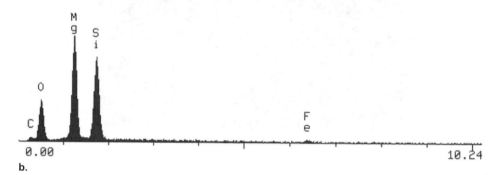

b.

42. Why does the spectrum in Figure 17.21b give two peaks in the EDX spectrum for Fe?

Blacklake asbestos mine. ©Robert St-Coeur/ShutterStock, Inc.

CHAPTER

18

The Disposal of Dangerous Wastes

- **Perpetual Storage of Hazardous Waste**
  Secured Landfill
  Deep-Well Injection
  Surface Impoundment

- **EPA Methods for Testing Solid Waste**

- **The Unsolved Problem**

- **Superfund: Cleaning Hazardous Waste Dumpsites**

- **Superfund Analytical Methods**

- **Radioactive Waste**
  Sources of Radioactive Waste
  Classification of Nuclear Waste

- **The Legacies from the Past**

- **Regulation of Radioactive Waste Disposal**
  Low-Level Radioactive Wastes
  High-Level Radioactive Wastes

- **Technologies for Radioactive Waste Disposal**

- **The Post-Cold War Challenge**

- **Additional Sources of Information**

- **Keywords**

- **Questions and Problems**

INDUSTRIAL SOCIETIES GENERATE ENORMOUS QUANTITIES OF WASTES, AND THE UNITED STATES PRODUCES MORE WASTE PER PERSON THAN ANY OTHER COUNTRY IN THE WORLD. We discard products that are worn out and items that are no longer wanted; we throw out food, paper, empty containers, and yard waste. Industrial processes produce chemical wastes, many of which contain toxic substances. The nation's nuclear weapons programs and the nuclear power industry generate wastes that are radioactive. Safe disposal of these wastes is one of the most serious environmental problems facing the United States today.

Until legislation was enacted in the 1960s and 1970s to regulate waste disposal and protect the air and water from contamination, wastes of all kinds were discarded with little concern for human health or the environment. **Hazardous wastes** were incinerated, contaminating the atmosphere with toxic emissions. They were discarded in sewers, rivers, and streams, leaking landfills, abandoned buildings, and mines, or were simply dumped on vacant lots and along roadways. Many people were exposed to polluted air and contaminated water. Today, thousands of sites in the United States remain severely contaminated with dangerous chemicals.

For purposes of management and disposal, wastes are divided into three categories: (1) solid, (2) hazardous, and (3) radioactive. The Environmental Protection Agency (EPA) has ruled that radioactive and non-radioactive wastes can be mixed for purposes of management and disposal, creating another category called (4) mixed waste. This chapter defines the different kinds of wastes and discusses the advantages and disadvantages of the disposal methods that have been recommended for each category. We examine the legislation regulating the management and disposal of wastes being generated today and to clean wastes produced in the past.

■ Careless Waste Disposal in the Past

Numerous examples of irresponsible and often criminal methods of waste disposal can be cited. For example, in 1969, oil and other flammable, greasy wastes floating on the surface of the Cuyahoga River, which flows through Cleveland, Ohio, actually caught on fire and destroyed seven bridges. In Toone and Teague, Tennessee, groundwater cannot be used for drinking because it is contaminated with pesticide wastes, which leaked from 350,000 drums that were dumped in a landfill years ago. Between 1952 and 1972, approximately 130 million liters (34 million gallons) of chemical wastes, including wastes from dichlorodiphenyl-trichloroethane (DDT) manufacture, were dumped at the Stringfellow Acid Pits, a 22-acre site near Riverside, California. Contaminated leachate from this site continues to migrate and threaten groundwater supplies. Many of the 17,000 waste drums located at a site in Kentucky known as the "Valley of the Drums" have leaked, contaminating soil and nearby water. At the huge 560-square-mile Hanford weapons complex in Washington, there are over 1000 separate sites where radioactive and hazardous wastes have been stored or dumped. Contamination at the complex is so extensive that complete cleanup may never be possible.

The most notorious case of hazardous waste dumping in the United States occurred at Love Canal in Niagara Falls, New York, where in the winter of 1978, toxic chemical wastes buried years before began oozing from the ground, forcing the evacuation of many of the town's residents.

■ Defining Solid Waste

According to the EPA, **solid waste** includes garbage, refuse, sludge from waste treatment plants, wastes from air pollution control facilities, and any other discarded material not excluded by regulation. Although termed "solid," these wastes may actually be solid, liquid, and semisolid, or may contain gaseous material. Agricultural, mining, industrial, and commercial operations and community activities all produce solid wastes that need disposal (Figure 18.1). Most agricultural wastes, which consist primarily of manure and crop residues, are plowed back into the land. Mining wastes are produced when the processed ores are left at the mine site. Nonhazardous industrial wastes, municipal refuse generated by community activities, and sludge from sewage treatment plants that is not composted into fertilizer are usually disposed of in landfills or are incinerated. Increasingly, some industrial and municipal waste is being recycled.

■ Disposal of Municipal Solid Waste

Municipal solid waste (MSW)—more commonly referred to as trash or garbage—is collected locally from homes, institutions, and commercial establishments. It includes food wastes, paper, glass, metal cans, plastic containers, yard clippings, and various other discarded household items. The composition varies depending on whether it is commercial or residential waste and whether it comes from an affluent or poor neighborhood. Typically, municipal trash collected in the United States can be divided as shown in Figure 18.2. In 2006, the United States produced 251 million tons of municipal

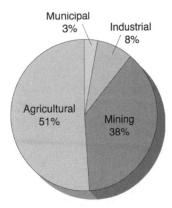

Sources of solid waste (percentages of total)

Figure 18.1 In the United States, agriculture is the major source of solid waste (51% of the total). Mining accounts for 38%, and industry for 8%. Municipal waste, which is produced by commercial operations, homes, schools, and other community activities, accounts for 3%.

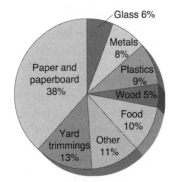

Figure 18.2 The composition of typical MSW by weight. Percentages vary, depending on the affluence of the neighborhood and whether it is primarily residential or commercial. At certain times of the year, yard waste (leaves, branches, grass clippings) account for more than 25% of the total waste. From EPA report, *Character of MSW in United States, 1997 Update.*

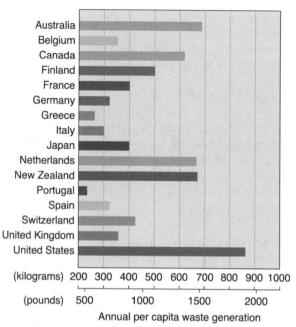

Figure 18.3 The United States produces more waste per capita than any other industrialized nation.

solid waste. MSW also includes discarded items such as furniture, household appliances, automobiles, and tires. Americans generate approximately 1600 pounds of municipal waste per person per year, far more than most other industrialized nations generate (Figure 18.3).

MSW and the Law

Until the 1960s, most MSW was trucked to open dumps, where it was discarded and often burned. The dumps became breeding grounds for rats, flies, and other pests, and the burning trash polluted the air with smoke and unpleasant odors. Since the passage of the **Solid Waste Disposal Act (SWDA)** in 1965, management and disposal of MSW have been subject to increasingly strict controls. Particularly important was the enactment of the **Resource Conservation and Recovery Act (RCRA,** pronounced "rickra") in 1976. The section of the law dealing with solid waste required that all existing open dumps in the United States be closed or upgraded; it banned the creation of new open dumps and set tough standards for landfills. In an effort to reduce the volume of waste generated, the law stressed the need for reuse and regeneration. Federal procurement agencies were directed to make use of recycled materials wherever possible, and the Department of Energy (DOE) was instructed to conduct research to develop ways of converting solid wastes into energy. Since 1983, the EPA has required that new landfills meet even stricter standards. Today, approximately 55% of all municipal waste is discarded in regulated landfills. The remainder is recycled, incinerated, or composted. As more knowledge is gained about wastes, EPA from time to time changes its regulations for their management and disposal and publishes these changes in the *Federal Register*. For instance, in mid-2003, new criteria were established for the classification of solid waste disposal facilities and practices.

Landfills

The **sanitary landfills** in use before 1983 consisted of a trench excavated in the ground in which trash was dumped and then compacted by bulldozers. Each day's load of refuse was covered with a layer of soil. Covering the waste minimizes vermin's access and the problem of odors. When completely filled, such landfills are not suitable for building because the ground continues to settle for many years, but they can be planted with grass and made into attractive recreation areas.

In sanitary landfills, there was always the danger that any metal salts and other chemicals in the liquid that leaches from rotting garbage might percolate through the underlying soil and contaminate nearby groundwater. Although MSW should not contain hazardous materials, many householders often increase the chance of groundwater contamination by discarding unused pesticides, paint, cleaning solvents, and other dangerous wastes together with their nonhazardous refuse (Table 18.1).

The EPA regulations introduced in 1983 require that all landfills meet these standards: They must be located well above the water table, and the bottom of the fill must be covered with a layer of impermeable clay or a plastic liner or both to contain the fluid (*leachate*) seeping through the trash. They must be fitted with a system for collecting this leachate and treating it, and groundwater in the vicinity must be monitored regularly. The methane produced by the anaerobic decomposition of the organic waste must be vented or preferably collected and used as fuel for generating electricity. When filled, the site must be covered with a layer of clay and then topsoil (Figure 18.4).

Table 18.1

Hazardous Materials Often Found in Household Trash

Products	Hazardous Materials
Pesticides	
Insect sprays	Insecticides, organic solvents
Other pesticides	Insecticides, herbicides
Paint	
Oil-based	Organic solvents
Latex	Organic polymers
Automotive products	
Gasoline	Organic solvents
Motor oil	Organic compounds, metals
Antifreeze	Organic solvents, metals
Batteries	Lead, sulfuric acid
Miscellaneous	
Mercury batteries	Mercury
Nail-polish remover	Organic solvents
Moth balls	*p*-Dichlorobenzene

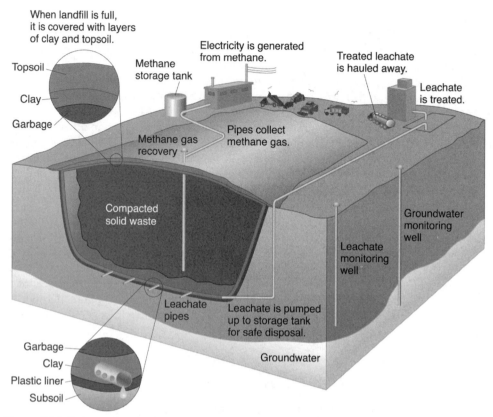

When landfill is full, it is covered with layers of clay and topsoil.

Topsoil

Clay

Garbage

Methane storage tank

Electricity is generated from methane.

Treated leachate is hauled away.

Leachate is treated.

Methane gas recovery

Pipes collect methane gas.

Compacted solid waste

Groundwater monitoring well

Leachate monitoring well

Leachate pipes

Leachate is pumped up to storage tank for safe disposal.

Groundwater

Garbage

Clay

Plastic liner

Subsoil

Groundwater

Figure 18.4 Modern sanitary landfills are designed so that any leakage is contained and cannot percolate into groundwater. Adapted from G.T. Miller, Jr. *Sustaining the Earth, Third edition*. Belmont, CA: Wadsworth, 1991.

Municipal waste disposal has become a major problem for many big cities in the United States. Not only will numerous existing landfills be filled to capacity within the next few years, but because of high costs, many municipalities have chosen to close rather than upgrade their landfills to meet the new standards. Some authorities are turning to **incineration** and recycling as alternatives. Others are trucking their MSW across state borders to state-of-the-art megafills that since the beginning of the 1990s have opened in many rural areas. Although communities receiving the waste have benefited economically, many citizens resent having other's trash in their back yard.

Incineration

Incineration of MSW has a number of advantages. The volume of the waste is reduced by approximately 85%. The high temperature kills disease-causing organisms. There is no risk of groundwater pollution, and the heat produced can be used as a source of energy. At some facilities, ferrous metals are removed from the trash before combustion, and other valuable metals, including aluminum, are recovered from the noncombustible ash.

The ash remaining after combustion must be tested, and if it contains no toxic metals or other hazardous materials, it is disposed of in a landfill. If the ash fails the test, it must be managed as hazardous waste, which is a much more expensive procedure (which will be discussed later). Metals that are most likely to be in the ash are lead and cadmium.

The main disadvantage of incinerating municipal waste is its potential for polluting the atmosphere with toxic chemicals, particularly metals. When discarded items such as cans, jar lids, and batteries are incinerated, many toxic metals, including beryllium, cadmium, chromium, lead, and mercury, are released and carried up the incinerator stack in fly ash. In the past, municipal incinerators were a significant source of air pollution, but today, in compliance with strict air quality standards, they are equipped with efficient electrostatic precipitators and other control devices to prevent the escape of fly ash.

Modern incinerators are now called **combustion facilities** or **waste-to-energy** plants if they generate electricity, burn cleanly, and consume up to 3000 tons of trash a day. They produce enough electricity to power 70,000 homes. Although combustion is never likely to contribute more than 1% to the energy needs in the United States, it can be a useful local source of energy.

Nobody wants a combustion facility in his or her neighborhood. Because these facilities often operate intermittently, there is concern that even if yearly average emissions are within allowed limits, emissions at peak periods may be excessive. Currently, approximately 14% of the over 200 million tons of MSW produced annually in the United States is incinerated.

Recycling and Resource Recovery

Increasingly larger amounts of waste materials are now being recycled. Many communities require businesses and homeowners to collect discarded items such as newspapers, office paper, aluminum cans, glass containers, and certain plastic items in separate containers so that they can be collected and recycled. Some municipalities have found it more economical to build high-technology **resource recovery plants** that sort the trash after it has been collected. Today, approximately 32.5% of MSW is recycled.

Everyone is in favor of recycling, but to be successful, recycling must be cost-effective. Dealers and manufacturers will not be willing to accept items for recycling if they cannot count on steady markets for the recycled products.

Source Reduction

Disposing of the enormous quantity of MSW that Americans produce each year is becoming increasingly difficult and costly. Obviously, the most sensible way to address this problem is for everyone to produce less trash. This would lower disposal costs, save valuable resources, and reduce adverse effects on the environment.

There are many ways to reduce waste. For example, businesses could use e-mail in place of paper for interoffice memos, both sides of paper when printing and copying, and durable mugs instead of Styrofoam cups for coffee. Homeowners would help if they let grass clippings remain on their lawns as fertilizer, used cloth napkins instead of paper, avoided wasting food and using disposal products like plastic plates and flatware, and brought their own shopping bags to the grocery store. Unwanted items such as old clothes and furniture could be donated for reuse. Already, several states, including Pennsylvania and New York, have found an effective way to reduce their trash loads: They have introduced user fees based on the volume of MSW discarded.

The Problems of Hazardous Waste

If the public is to be protected from the dangers associated with hazardous waste, two problems must be solved: (1) how to manage and dispose properly of waste generated now and in the future and (2) how to clean up wastes produced in the past. The major pieces of legislation that address these two issues are RCRA (amended in 1984) and the **Comprehensive Environmental Response, Compensation, and Liability Act (CERCLA)** of 1980, better known as the **Superfund** program.

RCRA: Regulation of Hazardous Waste

Before passage of RCRA, the two main pieces of legislation that were enacted to protect the environment were the 1970 Clean Air Act and the 1972 Clean Water Act. RCRA was passed to establish safe methods for the disposal of all wastes and hazardous wastes in particular. Under RCRA, the EPA was given the responsibility for defining hazardous wastes and for issuing and enforcing regulations to protect the public's health and the environment from improper management and disposal of such wastes.

What Are Hazardous Wastes?

According to RCRA, wastes are described as hazardous if they "(a) cause or significantly contribute to an increase in mortality or an increase in serious irreversible, or incapacitating reversible, illness or (b) pose a substantial present or potential hazard to human health or the environment when improperly treated, stored, transported, or disposed of, or otherwise managed." Such wastes may be in solid, liquid, or gaseous form and may result from industrial, commercial, mining, or agricultural operations or from community activities.

The EPA defines hazardous wastes in two different ways. A waste is hazardous if because of its inherent properties the EPA has specifically listed it as hazardous. A waste is also hazardous if it exhibits any of the following characteristics: (1) ignitability, (2) corrosivity, (3) reactivity, and (4) toxicity.

If a "characteristic" waste is treated in such a way that it loses its hazardous characteristic, it ceases to be a hazardous waste and can be disposed of in a regulated landfill; however, in the original regulations, a "listed" waste, once generated, retained its hazardous classification regardless of the steps taken to dilute or render it nonhazardous. A listed waste was not permitted in a landfill but had to be disposed of according to strict EPA regulations. Now the EPA has ruled that if the concentration of a listed chemical in a waste falls below an established acceptable risk level, it need not be managed as a hazardous material.

The responsibility for deciding if a waste is hazardous lies with the producer of the waste. If the EPA has not listed the waste, the producer must determine whether the waste possesses any of the four characteristics that define it as hazardous. Once a waste has been identified as hazardous, the proper method for disposal must be selected. Any company that generates or transports a hazardous waste is required to keep detailed records of the handling and transferring of the waste from point of origin to point of ultimate disposal or destruction. Chemical plant managers and incinerator operators are required to keep records of all hazardous chemicals that are emitted into the air, land, or water. Landfill operators must maintain groundwater-monitoring data. By law, "cradle-to-grave" responsibility for a hazardous waste is assigned to the generator of the waste.

Listed Hazardous Wastes

The wastes that the EPA, by rule, has specifically listed as hazardous number over 600. These wastes are classified according to source and include nonspecific chemicals—primarily solvents—produced by many industries, sludges and by-products produced by specific industries, discarded chemicals, and chemicals produced as intermediates in manufacturing processes.

Characteristic Hazardous Wastes

The EPA has developed special tests that must be used to determine whether a waste exhibits any of the four hazardous characteristics.

Ignitable Wastes

Wastes are classified as **ignitable** if they present a significant fire hazard. Ignitable wastes include (1) liquids such as gasoline, hexane, and other organic solvents that have a flash point (the lowest temperature at which their vapors ignite when tested with a flame) below 60°C (140°F); (2) ignitable compressed gases such as propane; (3) various solid materials that are liable to cause persistent, vigorously burning fires as a result of absorption of moisture (for example, sodium and potassium metal), friction, or spontaneous chemical change; and (4) oxidizers. Oxidizers are included because they are likely to intensify an already burning fire. Perchloric acid ($HClO_4$), for example, is a strong oxidizing agent that poses an explosion hazard because it reacts very vigorously with organic matter.

Corrosive Wastes

Corrosive wastes include acidic aqueous wastes with a pH equal to or less than 2.0 and basic aqueous wastes with a pH equal to or greater than 12.5. Wastes with these very low or very high pH values are likely to react dangerously with other wastes and may leach metals and contaminants from other wastes. If allowed to leak into waterways, corrosives can have a devastating effect on aquatic habitats.

Also included as corrosive are wastes that can corrode steel at a rate equal to or greater than 0.635 cm (0.25 inches) per year. This criterion was included to prevent the kind of disasters that occurred in the past when acids and other hazardous materials leaked from corroded steel drums in which they had been stored.

Reactive Wastes

Reactive wastes include materials that explode readily when subjected to shock or heat, materials that tend to undergo violent, spontaneous chemical change, and materials that react violently with water. Explosions at a dumpsite in Cheshire, England, were caused by discarded sodium metal, which reacted with water to produce hydrogen gas:

$$2\,Na + 2\,H_2O \rightarrow 2\,NaOH + H_2$$

The hydrogen ignited and combined explosively with oxygen in the air:

$$2\,H_2 + O_2 \rightarrow 2\,H_2O$$

Also classified as reactive are wastes that generate dangerous quantities of toxic gases when exposed to water, weak acids, or weak bases. For example, wastes containing cyanides or sulfides can produce toxic hydrogen cyanide and hydrogen sulfide.

Toxic Wastes

Most characteristic wastes are hazardous because of their toxicity. The terms *toxic* and *hazardous* are often used interchangeably but are not synonymous. The term toxic is applied to substances that cause harm to living organisms by interfering with normal physiologic processes. It includes substances that are carcinogenic, mutagenic, teratogenic (causes birth defects), or phytotoxic (poisonous to plants).

Toxic wastes are of particular concern because of their potential as groundwater contaminants. The test used to determine whether a waste is toxic was designed to model the leaching that would be likely to occur if the waste were disposed of on land. In the test, a nonliquid waste is treated with an acid solution (pH 5.0 ± 0.2) under specified conditions, and the amount of toxic constituent that dissolves in the acid is determined. If the concentration of toxic substance in the extract (or in the liquid in the case of a liquid waste) exceeds the regulatory level that the EPA set, the waste is considered to be toxic.

Originally, the EPA characterized just 14 chemicals as toxic: 8 metals (arsenic, barium, cadmium, chromium, lead, mercury, selenium, and silver) and 6 pesticides (endrin, lindane, methoxychlor, toxaphene, 2,4-D, and 2,4,5-TP [Silvex]). By 2001, a total of 600 chemicals were regulated as toxic, including common chemicals such as benzene, carbon tetrachloride, chloroform, nitrobenzene, vinyl chloride—used to make the plastic poly(vinyl chloride)—and several pesticides.

■ Sources of Hazardous Waste

Industry is responsible for most of the nearly 40 million tons of hazardous waste that is generated annually in the United States. Chemical, petroleum, and metal-related industries produce more than 90% of this waste. Small business establishments such as dry cleaners, gas stations, and electroplating shops account for a small but significant part of the remaining 10%. Household and farm wastes are exempt from hazardous waste regulations, but any business with a monthly output of hazardous waste in excess of 100 kg (220 lb) must dispose of it according to EPA regulations.

Thousands of different chemicals and thousands of different processes are used by the petrochemical industry in refining crude oil and producing the synthetic organic chemicals needed for manufacturing polymers, pesticides, medicines, and other consumer products. All of these processes inevitably produce wastes, and it has been estimated that 10% to 15% of them are hazardous.

Of greatest concern are toxic metals and synthetic organic chemicals, particularly chlorinated hydrocarbons. Table 18.2 shows that a wide variety of industries are generating these hazardous substances. Metal wastes are produced primarily by metal-processing industries and by paint, textile, and other manufacturing industries that make or use pigments. Chlorinated hydrocarbons are used in the manufacture of plastic (vinyl chloride), pesticides (chlordane, heptachlor, and others), electrical insulation (polychlorinated biphenyls [PCBs]), and refrigerants (chlorofluorocarbons). They are also used as solvents (chloroform, carbon tetrachloride, perchloroethylene) in many industrial processes. The use of perchloroethylene by the dry-cleaning industry is being reduced as it is being replaced by "super-critical" CO_2, which has remarkable solvent properties.

Table 18.2

Hazardous Wastes Produced by Industry

Industry or Product	Hazardous Wastes
Electrical insulation	PCBs
Electroplating	Toxic metals, cyanide
Fertilizers	Sulfuric, nitric, and phosphoric acid; sodium hydroxide, ammonia
Insecticides and herbicides	Polychlorinated hydrocarbons, organophosphates, carbamates, ethylene dibromide
Iron and steel	Acids (including hydrofluoric acid), bases, phenols, benzene, toluene
Leather	Toxic metals, organic solvents
Medicines	Toxic metals, organic solvents
Nonferrous metals	Acids, metals (cadmium, zinc, lead)
Organic chemicals	Many ignitable, corrosive, reactive, and toxic compounds
Paints	Toxic metals, pigments (chrome yellow), organic solvents
Petroleum refining	Oil, phenols, other organic compounds, toxic metals, corrosives (acids, bases)
Plastics	Monomers used in polymer manufacture (ethylene, propylene, styrene, vinyl chloride, phenol, formaldehyde), phthalate plasticizers
Power industry; steam generation (fossil-fuel and nuclear-powered)	Fly ash and bottom ash (organics and heavy metals), sulfur dioxide, wet sludge
Soaps and detergents	Corrosives (sodium hydroxide, sulfuric acid)
Synthetic rubber	Ignitable monomers (styrene, butadiene, isoprene); antioxidants, fillers, etc. in tires
Textiles	Polymer monomers, dyes (azo, nitroso compounds), organic solvents, toxic metals
Wood products, paper	Phenol, formaldehyde, corrosives, sodium sulfide

■ Policy for Management and Disposal of Hazardous Waste

In 1983, the National Academy of Sciences issued an influential report that has been a major factor in establishing hazardous waste management policy in the United States. The report described the three basic ways for managing hazardous waste, which, in order of desirability, are (1) to minimize the amount produced, (2) to convert to a less hazardous or nonhazardous form, and (3) to isolate in a secure, perpetual-storage site (Figure 18.5). We consider each of these options in turn.

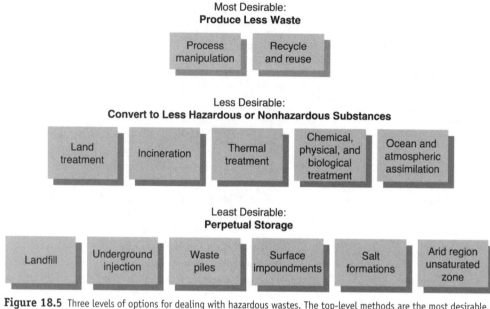

Figure 18.5 Three levels of options for dealing with hazardous wastes. The top-level methods are the most desirable. The least desirable methods are often the least expensive. Courtesy of the National Academy of Sciences.

■ Waste Minimization: Process Manipulation, Recycling, and Reuse

The high cost of disposing of hazardous waste in compliance with EPA regulations has made it economically attractive for industries to minimize waste production by modifying their manufacturing processes. For example, some electroplating plants now use ion exchange resins to remove contaminant metals selectively from electroplating baths. The plating solution, which previously would have been discarded as waste, is then returned to the bath for reuse. Many industries are also purifying wastes to recover chemicals that can be reused, recycled, or sold to another industry for use as raw materials. For example, the pesticide industry has found that it can recover hydrochloric acid from its hazardous chlorinated wastes and sell it to other industries. In Europe, several countries have now established **waste exchanges**, where waste chemicals are successfully traded.

■ Conversion of Hazardous Waste to a Less Hazardous or Nonhazardous Form

Since 1990, in compliance with the 1984 amendments to RCRA, nearly all hazardous wastes have had to be treated in some way to make them less hazardous before they can be buried on land. Several options are available (Figure 18.5); the choice depends on the type of waste to be treated.

Incineration and Other Thermal Treatment

If adequately controlled to prevent air pollution, **incineration** has a number of advantages as a means of detoxifying many hazardous wastes. At sufficiently high temperatures, incineration destroys 99.999% of toxic organic compounds by decomposing them to carbon dioxide, water, and harmless gases. At the same time, the volume of the waste is greatly reduced, and the energy released can be used either as a source of heat or to generate electricity. Combustible liquid and solid hazardous wastes such as solvents, pesticides, and many of the organic compounds present in petroleum refinery wastes can all be destroyed by incineration.

In the Netherlands, Germany, and Denmark, between 50% and 80% of all hazardous wastes are incinerated. Germany has built the most effective incinerators in the world and the importation of hazardous material for incineration has become a booming business. In 2007, for instance, 6.2 million tons of waste were imported by Germany (which is less than the 6.5 million tons imported in 2005). In the United States, during 2004, only approximately 7% of hazardous wastes were treated in this way. In 2008 EPA recommended that more hazardous waste be destroyed by incineration. The scrubbers and other devices that must be installed to prevent the escape of toxic gases and particulates make incineration very expensive, but the major factor in preventing construction of more incinerators in the United States is intense community opposition. Many believe that even with the protection of pollution controls the potential for air pollution still exists, especially if pollution control devices are not properly used or maintained, or if the incinerators are not operated at peak efficiency.

During the 1970s and 1980s, hazardous wastes, including Agent Orange (the defoliant used in the Vietnam War; discussed in Chapter 14), were destroyed at sea on ships equipped with incinerators. This practice was discontinued in 1988, pending the EPA's adoption of measures to regulate incineration at sea. Since 1985, the EPA has operated several mobile incinerators for the onsite destruction of small quantities of particularly hazardous materials, including dioxin wastes in soil and water.

Certain hazardous organic wastes do not need to be incinerated to render them nonhazardous. They can be decomposed to nonhazardous biodegradable compounds by being heated under pressure at relatively low temperatures (450°C to 600°C [840°F to 1100°F]). In some cases, the products of this thermal treatment can be used to synthesize new compounds.

Chemical and Physical Treatment

Chemical methods that can be used to convert hazardous wastes to nonhazardous materials include neutralization of acids and bases, oxidation-reduction reactions, and removal of metals and other compounds by precipitation or adsorption on carbon. Acidic wastes, which are produced in large quantity by the iron and steel, electroplating, and other metal-related industries, can be neutralized with lime. This process also serves to precipitate out heavy metals that can then be collected and recycled. Toxic chromium (VI) salts in wastes can be chemically reduced to far less dangerous chromium (III) salts. Cyanide, which is used in many industrial processes, including the extraction of gold and silver from their ores, can be oxidized with sodium hypochlorite (NaOCl) to far less toxic cyanate (NCO^-). The cyanate can be oxidized further with chlorine to produce carbon dioxide and nitrogen. In the textile industry, carbon adsorption can be used to remove toxic dyes from liquid wastes.

Certain wastes can be made nonhazardous or less hazardous by immobilizing them in a solid mass. For example, waste material can be mixed with cement or lime to form a concrete block, which can then be disposed of in a landfill. If the concrete is likely to be exposed to acidic conditions, leaching can be prevented by encapsulating the block in an impermeable plastic container. This technique is useful for certain liquid wastes and for the flue-gas–cleaning sludges that are produced in enormous quantity by coal-burning power stations. These sludges, which are formed as a result of treatment to control sulfur oxide emissions, contain $CaSO_4$ and fly ash that may include toxic metals.

Bioremediation

Increasingly, microorganisms are being used to clean soil that is contaminated with toxic organic compounds. The microorganisms, often mixed with nutrients, decompose the wastes and thus detoxify them. Oxidation by sunlight further aids in the degradation. This technique, which is termed **bioremediation**, is used primarily to treat soil that is contaminated with oil. It is also used to treat petroleum refinery and paper mill wastes. It is an inexpensive option, but precautions must be taken to ensure that the wastes do not contain toxic metals or other nonbiodegradable chemicals that could contaminate the soil or groundwater. Using genetic engineering, microbiologists are working to develop specialized microorganisms that can effectively detoxify polycyclic aromatic hydrocarbons, PCBs, trichloroethylene, and other compounds that are resistant to biodegradation.

■ Perpetual Storage of Hazardous Waste

Hazardous wastes that cannot be recycled or converted to nonhazardous forms must be stored. Perpetual storage methods include (1) burial in a secured landfill, (2) deep-well injection, and (3) surface impoundment. None of these methods, however, provides entirely secure storage, and the 1984 amendments to RCRA included restrictions on land disposal of hazardous waste. A 1990 deadline was set for the end of disposal of untreated hazardous materials on land except in specific cases in which the EPA has determined that no other option is feasible. Since 1990, wastes not so designated have been automatically banned from land disposal facilities and must be disposed of according to EPA regulations. As a result of legislation, there has been a sharp reduction in improper disposal.

Secured Landfill

A **secured landfill** is a burial site for the long-term storage of hazardous wastes that are contained in drums (Figure 18.6). Strict EPA specifications for construction minimize the chance of contaminants migrating in the event of a drum leak. The landfill must be located in thick, impervious clay soil at least 165 m (500 ft) from a water source in an area that is not subject to floods, earthquakes, or other disturbances. Ideally, the site should be in an arid region, far from any water supplies. The bottom and sides of the trench holding the drums must be lined with two layers of impervious, reinforced plastic, and the drums must rest on a layer of gravel. Below the gravel, there must be a network of pipes that can collect any leakage from drums and any rainwater seeping through the fill; leachate must be pumped to the surface and treated. When filled, the site must be covered with a layer of impervious plastic material topped, in turn, with layers of sand, gravel, and clay. The cap must be contoured to allow water to drain away from the site. The quality of groundwater in the area of a secure landfill must be monitored for at least 30 years.

Many critics contend that despite the EPA's tough regulations, all landfills will leak eventually and, therefore, should not be used to store materials that retain their hazardous prop-

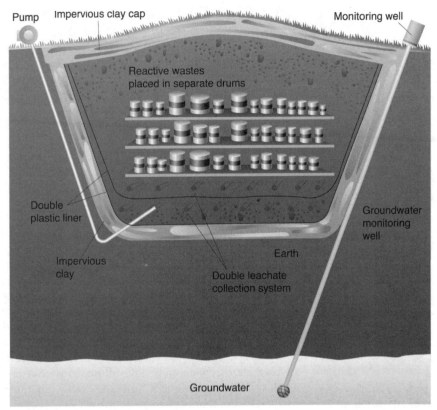

Figure 18.6 In a secured landfill for long-term storage of hazardous wastes contained in drums, the bottom and sides of the fill are lined with plastic. Below the gravel on which the drums rest are pipes that collect any leachate; groundwater below the fill is constantly monitored for contamination. The landfill is capped with impervious clay to prevent water from seeping into it. Adapted from B.J. Nebel. *Environmental Science: The Way the World Works, Sixth edition*. Englewood Cliffs, NJ: Prentice Hall, 1998.

erties almost indefinitely. They maintain that in time some drums will corrode and leak. Burrowing animals will invade clay layers, and plastic liners will be torn by freezing temperatures, settling, or will be disintegrated by leachate.

Deep-Well Injection

In the past, liquid hazardous wastes were frequently disposed of by injecting them into deep wells, a technique that was originally developed by the petroleum industry as a means of disposing of brines (salt solutions) that were brought to the surface with crude oil. Liquid waste is pumped into a deep well drilled into a layer of dry porous rock located far below any usable groundwater and below a layer of impervious rock (Figure 18.7). The wastes gradually seep into the porous material and are trapped below the impervious layer of rock. The well shaft is surrounded by a sealed casing to prevent contamination as the waste is pumped through the groundwater region.

The RCRA land disposal restrictions apply to **deep-well injection**, but there is a loophole in the law that permits injection of untreated hazardous waste if a company can prove that there

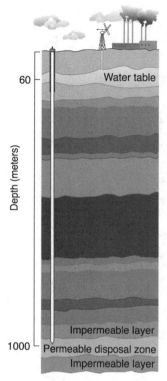

60 — Water table

Depth (meters)

Impermeable layer
1000 — Permeable disposal zone
Impermeable layer

Figure 18.7 Deep-well injection is used for disposing of liquid hazardous wastes. Liquid waste is pumped into a deep well drilled into a layer of dry, porous rock lying well below any groundwater and below a layer of impervious rock. The well shaft is surrounded by a sealed casing to prevent groundwater contamination.

will be no migration of the waste from the injection zone during the time the waste remains hazardous. Deep-well injection is one of the least expensive waste disposal methods, and with the implementation of the rules restricting disposal in landfills, companies have made use of the loophole to increase the amount of waste that they dispose of in deep wells. Critics maintain that even if properly operated deep-well injection has the potential for groundwater contamination. In time, the well casing is likely to corrode, and there is always a danger that earth tremors or earthquakes may fracture the impervious layers and lead to the migration of hazardous materials.

Surface Impoundment

Surface impoundment is an inexpensive disposal method that is used primarily to manage relatively small quantities of hazardous wastes contained in large volumes of wastewater. The wastewater is usually pumped directly from the plant that produced it to a pond lined with impervious clay, a plastic material, or both to prevent leakage into underlying soil or groundwater (Figure 18.8). As the water evaporates, the small quantity of hazardous waste in the water settles and gradually accumulates on the bottom of the pond. If the rate of evaporation equals the rate of waste input, surface impoundments can be used almost indefinitely. Like all methods of disposal on land, surface impoundment is likely to cause environmental contamination sooner or later.

■ EPA Methods for Testing Solid Waste

Test Methods for Evaluating Solid Waste, Physical/Chemical Methods (EPA publication SW-846) describes the EPA's Office of Solid Waste's analytical and sampling methods that have been evaluated and approved for analysis of solid waste. This publication (SW-846) is a comprehensive set of methods that changes over time as new analytical methods are developed, evaluated, and then formally added as EPA methods. It was originally published by EPA in 1980 and is currently in its third edition. Any of the EPA methods for the analysis of hazardous waste can be downloaded from the EPA Web site (http://www.epa.gov).

■ The Unsolved Problem

Society wants the products that generate hazardous wastes—for example, electricity, plastics, gasoline, and pesticides—but no one wants a waste disposal site in his or her neighborhood. This **NIMBY (not in my back yard)** syndrome has hampered waste disposal but,

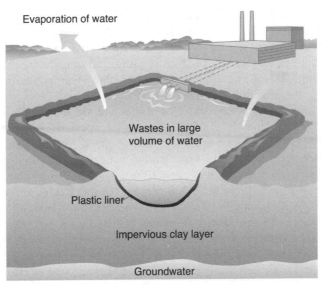

Figure 18.8 Surface impoundment is used for the disposal of small amounts of hazardous waste present in large volumes of water. The water gradually evaporates, and the materials accumulate at the bottom of the pond, which is lined with clay or plastic to prevent leakage into the ground. Adapted from B.J. Nebel. *Environmental Science: The Way the World Works, Sixth edition*. Englewood Cliffs, NJ: Prentice Hall, 1998.

together with economic considerations, has been an important factor in forcing manufacturers to find ways to minimize waste production. Starting in the 1980s, in-plant process modifications, recycling, and reuse have steadily reduced the production of hazardous waste. A huge quantity continues to be generated, however, and we still do not have acceptable ways of dealing with it.

In 1979, it was estimated that only 10% of hazardous waste was being disposed of in an environmentally safe manner. Today, the situation is very much better, but because of the high cost of proper disposal, large quantities of waste are still dumped illegally. The EPA does not have the resources to keep track of the thousands of operations that generate hazardous wastes, and consequently, it is often unable to enforce its "cradle-to-grave" provisions. The EPA must also contend with organized crime, which is known to be involved in waste disposal. Mob-controlled trash collectors in New Jersey and New York, for example, have been discovered mixing household refuse with hazardous wastes and illegally disposing of the combined wastes in municipal landfills.

■ Superfund: Cleaning Hazardous Waste Dumpsites

Exposure of the Love Canal disaster in 1978 and increasing evidence that numerous other dangerous abandoned dumpsites existed made it clear that legislation was urgently needed to deal with the problems created in the past. CERCLA, passed in 1980, established the **Superfund program,** which was aimed primarily at (1) identifying existing hazardous waste dumpsites and (2) establishing a trust fund to finance the cleanup of the sites.

According to this legislation, the parties responsible for creating a hazardous dumpsite are responsible for cleaning it. If they fail to do so, the EPA is authorized to undertake the cleanup and then sue the delinquent parties for three times the costs incurred. In cases in which responsibility cannot be established, the EPA is authorized to undertake the cleanup using funds provided jointly by the federal government and state governments and by a tax on the chemical and petrochemical industries.

The EPA drew up a **National Priorities List** of sites with the most urgent need of cleanup, and it soon became evident that the $1.6 billion trust fund provided for cleanup during the first five years of the Superfund program was totally inadequate. Many more hazardous sites than expected were discovered, and cleanup was found to be far more difficult than anticipated. By 1990, CERCLA had been amended twice, and a Trust Fund of $13.6 billion was authorized until 1994. The EPA originally listed 38,000 sites for cleanup but has recently reduced this number to 14,000. In 2008, only 1258 of these sites were included in the National Priorities List. The five states with the largest numbers of listed sites were New Jersey (114), Pennsylvania (94), California (94), New York (86), and Michigan (65). The Office of Technology Assessment has estimated that there may be as many as 10,000 hazardous waste dumpsites in the United States. It is predicted that the cost of cleanup could reach $100 billion.

■ Superfund Analytical Methods

Under the CERCLA of 1980 and the Superfund Amendments and Reauthorization Act of 1986, the federal government granted to EPA the authority to develop standardized analytical tests that measure pollutants in environmental samples from hazardous waste sites. Pollutants that are of great concern to EPA at such sites are inorganic metals and cyanide, which are analyzed using inductively coupled plasma, atomic absorption, and colorimetric techniques. The analytical tests provide data from the analysis of water and soil/sediment samples for inorganic pollutants. These analytical tests use standardized procedures and a strict chain of custody, so that the results are admissible in court proceedings. Tests are provided by the EPA Superfund Contract Laboratory Program. The analytic method and type of instrument used to analyze for each inorganic analyte are listed in Table 18.3.

Some of the pollutants that are of concern to the EPA at such superfund sites are a series of volatile, semivolatile, and pesticide/PCB compounds that are analyzed using gas chromatography coupled with mass spectrometry (GC/MS) and gas chromatography with electron capture (GC/EC). The analytic method and type of instrument used to analyze for each organic analyte is listed in Table 18.4.

■ Radioactive Waste

Although radioactive wastes are undoubtedly hazardous to health and the environment, very few are regulated under RCRA and are instead covered by the Atomic Energy Act of 1954 (and later amendments), which the DOE administers. The DOE was created in 1977, and as

Table 18.3

Methods and Instruments

Analyte	Instrument	Method
Al, Sb, As, Ba, Be, Cd, Ca, Cr, Co, Cu, Fe, Pb, Mg, Mn, Ni, K, Se, Ag, Na, Tl, V, Zn	Inductively coupled plasma (ICP)	Acid digestion followed by inductively coupled plasma analysis
As, Pb, Tl, Se	Graphite furnace atomic absorption	Acid digestion followed by graphite furnace atomic absorption analysis
Ca, Mg, Na, K	Flame atomic absorption	Acid digestion followed by flame atomic absorption analysis
Hg	Cold vapor atomic absorption	Acid and permanganate oxidation followed by cold vapor atomic absorption analysis
CN	Manual and semi-automated colorimetric	Distillation followed by colorimetric analysis

a legacy from its predecessor, the Atomic Energy Commission, it and not the Department of Defense administers production of the nation's nuclear weapons. The DOE is also responsible for regulating the radioactive wastes generated by nuclear weapons facilities and by commercial nuclear power plants. Like hazardous waste, radioactive waste presents two problems: (1) waste produced in the past and not disposed of properly and (2) waste currently being generated.

Sources of Radioactive Waste

Nuclear weapons manufacturing facilities and nuclear power plants are the principal sources of the nation's most dangerous radioactive wastes. The nuclear weapons facilities, which are located in 12 states (Table 18.5), are responsible for at least 10 times more waste than the nuclear power industry. Relatively small quantities of far less dangerous radioactive waste are generated by hospitals, research laboratories, some industries, and the mining and processing of uranium used as nuclear fuel.

Nuclear reactors produce wastes that contain a mixture of radioisotopes with half-lives varying from a few days to many years, including strontium-90 (a half-life of 28 years), cesium-137 (a half-life of 30 years), and plutonium-239 (a half-life of 24,400 years). Because 10 half-lives must elapse before radioactivity is reduced to negligible levels, wastes containing plutonium-239, which emits an α particle as it disintegrates, will remain deadly for tens of thousands of years. Plutonium-239 is used as a nuclear fuel in weapons programs, and for many years until the program was discontinued in 1977, it was generated during the

(*text continues on page 556*)

Table 18.4

Target Compound List and Contract Required Quantitation Limits (CRQLs) for Analysis of Superfund Sites*

Volatiles

Quantitation Limits

	Water (µg/L)	Low Soil (µg/Kg)	Calibration Levels (µg/L)
1. Dichlorodifluoromethane	10	10	0.50
2. Chloromethane	10	10	0.50
3. Vinyl Chloride	10	10	0.50
4. Bromomethane	10	10	0.50
5. Chloroethane	10	10	0.50
6. Trichlorofluoromethane	10	10	0.50
7. 1,1-Dichloroethene	10	10	0.50
8. 1,1,2-Trichloro-1,2,2-trifluoroethane	10	10	0.50
9. Acetone	10	10	10
10. Carbon Disulfide	10	10	0.50
11. Methyl Acetate	10	10	0.50
12. Methylene Chloride	10	10	0.50
13. trans-1,2-Dichloroethane	10	10	0.50
14. Methyl tert-Butyl Ether	10	10	0.50
15. 1,1-Dichloroethane	10	10	0.50
16. cis-1,2-Dichloroethene	10	10	0.50
17. 2-Butanone	10	10	10
18. Chloroform	10	10	0.50
19. 1,1,1-Trichloroethane	10	10	0.50

Semivolatiles

Quantitation Limits

	Water (µg/L)	Low Soil (µg/Kg)
49. Benzaldehyde	10	330
50. Phenol	10	330
51. bis-(2-Chloroethyl)ether	10	330
52. 2-Chlorophenol	10	330
53. 2-Methylphenol	10	330
54. 2,2'-oxybis (1-Chloropropane)	10	330
55. Acetophenone	10	330
56. 4-Methylphenol	10	330
57. N-Nitroso-di-n-propylamine	10	330
58. Hexachloroethane	10	330
59. Nitrobenzene	10	330
60. Isophorone	10	330
61. 2-Nitrophenol	10	330
62. 2,4-Dimethylphenol	10	330
63. bis-(2-Chloroethoxy) methane	10	330
64. 2,4-Dichlorophenol	10	330
65. Naphthalene	10	330
66. 4-Chloroaniline	10	330
67. Hexachlorobutadiene	10	330
68. Caprolactam	10	330

No.	Compound			
20.	Cyclohexane	10	0.50	10
21.	Carbon Tetrachloride	10	0.50	10
22.	Benzene	10	0.50	10
23.	1,2-Dichloroethane	10	0.50	10
24.	Trichloroethene	10	0.50	10
25.	Methylcyclohexane	10	0.50	10
26.	1,2-Dichloropropane	10	0.50	10
27.	Bromodichloromethane	10	0.50	10
28.	cis-1,3-Dichloropropene	10	0.50	10
29.	4-Methyl-2-pentanone	10	10	10
30.	Toluene	10	0.50	10
31.	trans-1,3-Dichloropropene	10	0.50	10
32.	1,1,2-Trichloroethane	10	0.50	10
33.	Tetrachloroethene	10	0.50	10
34.	2-Hexanone	10	10	10
35.	Dibromochloromethane	10	0.50	10
36.	1,2-Dibromoethane	10	0.50	10
37.	Chlorobenzene	10	0.50	10
38.	Ethylbenzene	10	0.50	10
39.	Xylenes (Total)	10	0.50	10
40.	Styrene	10	0.50	10
41.	Bromoform	10	0.50	10
42.	Isopropylbenzene	10	0.50	10
43.	1,1,2,2-Tetrachloroethane	10	0.50	10
44.	1,3-Dichlorobenzene	10	0.50	10
45.	1,4-Dichlorobenzene	10	0.50	10
46.	1,2-Dichlorobenzene	10	0.50	10
47.	1,2-Dibromo-3-chloropropane	10	0.50	10
48.	1,2,4-Triclorobenzene	10	0.50	10

No.	Compound		
69.	4-Chloro-3-methylphenol	10	330
70.	2-Methylnaphthalene	10	330
71.	Hexachlorocyclopentadiene	10	330
72.	2,4,6-Trichlorophenol	10	330
73.	2,4,5-Trichlorophenol	25	830
74.	1,1'-Biphenyl	10	330
75.	2-Chloronaphthalene	10	330
76.	2-Nitroaniline	25	830
77.	Dimethylphthalate	10	330
78.	2,6-Dinitrotoluene	10	330
79.	Acenaphthylene	10	330
80.	3-Nitroaniline	25	830
81.	Acenaphthene	10	330
82.	2,4-Dinitrophenol	25	830
83.	4-Nitrophenol	25	830
84.	Dibenzofuran	10	330
85.	2,4-Dinitrotoluene	10	330
86.	Diethylphthalate	10	330
87.	Fluorene	10	330
88.	4-Chlorophenyl-phenylether	10	830
89.	4-Nitroaniline	25	830
90.	4,6-Dinitro-2-methylphenol	25	830
91.	N-Nitrosodiphenylamine	10	330
92.	4-Bromophenyl-phenylether	10	330
93.	Hexachlorobenzene	10	330
94.	Atrazine	10	330
95.	Pentachlorophenol	25	830
96.	Phenanthrene	10	330
97.	Anthracene	10	330

continued

Table 18.4

Target Compound List and Contract Required Quantitation Limits (CRQLs) for Analysis of Superfund Sites* (continued)

Quantitation Limits

	Water (µg/L)	Low Soil (µg/Kg)
98. Carbazole	10	330
99. Di-n-butylphthalate	10	330
100. Fluoranthene	10	330
101. Pyrene	10	330
102. Butylbenzylphthalate	10	330
103. 3,3'-Dichlorobenzidine	10	330
104. Benzo(a)anthracene	10	330
105. Chrysene	10	330
106. bis-(2-Ethylhexyl)phthalate	10	330
107. Di-n-octylphthalate	10	330

Quantitation Limits

	Water (µg/L)	Low Soil (µg/Kg)
119. Aldrin	0.05	1.7
120. Heptachlor epoxide	0.05	1.7
121. Endosulfan I	0.05	1.7
122. Dieldrin	0.10	3.3
123. 4,4'-DDE	0.10	3.3
124. Endrin	0.10	3.3
125. Endosulfan II	0.10	3.3
126. 4,4'-DDD	0.10	3.3
127. Endosulfan sulfate	0.10	3.3
128. 4,4'-DDT	0.10	3.3

No.	Compound		
108.	Benzo(b)fluoranthene	10	330
109.	Benzo(k)fluoranthene	10	330
110.	Benzo(a)pyrene	10	330
111.	Indeno(1,2,3-cd)pyrene	10	330
112.	Dibenz(a,h)anthracene	10	330
113.	Benzo(g,h,i)perylene	10	330

**Pesticides/Aroclors
(Pesticides/PCBs)**

No.	Compound		
114.	alpha-BHC	0.05	1.7
115.	beta-BHC	0.05	1.7
116.	delta-BHC	0.05	1.7
117.	gamma-BHC (Lindane)	0.05	1.7
118.	Heptachlor	0.05	1.7
129.	Methoxychlor	0.50	17
130.	Endrin ketone	0.10	3.3
131.	Endrin aldehyde	0.10	3.3
132.	alpha-Chlordane	0.05	1.7
133.	gamma-Chlordane	0.05	1.7
134.	Toxaphene	5.0	170
135.	Aroclor-1016	1.0	33
136.	Aroclor-1221	2.0	67
137.	Aroclor-1232	1.0	33
138.	Aroclor-1242	1.0	33
139.	Aroclor-1248	1.0	33
140.	Aroclor-1254	1.0	33
141.	Aroclor-1260	1.0	33

*For volatiles, quantitation limits for medium soils are approximately 130 times the quantitation limits for low soils. For semivolatile medium soils, quantitation limits are approximately 30 times the quantitation limits for low soils.

Table 18.5

Locations of U.S. Nuclear Weapons Facilities

State	Site
California	Lawrence Livermore National Laboratory
Colorado	Rocky Flats
Florida	Pinellas
Idaho	Idaho Falls
Kansas	Kansas City
Nevada	Nevada Test Site
New Mexico	Los Alamos National Laboratory, Sandia National Laboratories, Waste Isolation Pilot Plant
Ohio	Fernald, Mound Plant
South Carolina	Savannah River
Tennessee	Oak Ridge
Texas	Pantex
Washington	Hanford

reprocessing of uranium. Plutonium-239 is also produced in breeder reactors, which are used in several European countries to generate electricity. In the United States, however, only a few experimental breeder reactors are in operation.

Classification of Nuclear Waste

Depending primarily on how they are generated, radioactive wastes are classified as high- or low-level wastes. **High-level wastes**, which all contain plutonium-239, consist of spent fuel rod assemblies from nuclear reactors from both commercial power plants and weapons plants and certain other highly radioactive wastes generated by nuclear weapons facilities. All other radioactive wastes are classified as low-level wastes.

Hospitals, research laboratories, and certain industries produce **low-level wastes**. Most are dilute and contain radioisotopes with half-lives measured in no more than hundreds of years. Low-level wastes also include transuranic wastes (wastes containing elements with atomic numbers greater than uranium) produced at nuclear weapons plants. These wastes are usually dilute but dangerous because they contain elements, including plutonium-239, with long half-lives.

■ The Legacies from the Past

Since World War II, radioactive waste, much of it being high-level liquid waste, has been accumulating at the nuclear weapons facilities (Table 18.5) that produced the plutonium needed to construct nuclear warheads. For over 40 years, weapons production took precedence over all other considerations, including contamination of the environment and dangers to human health. During this period, most wastes were dumped directly into the ground, stored in ponds, or deposited in tanks that frequently developed leaks.

As a result of these practices, the environment surrounding all nuclear weapons facilities in the United States is extensively contaminated with radioactive material. The situation is even worse at nuclear weapons sites in the former Soviet Union, where radioactive wastes were discarded even more irresponsibly. At several plants, billions of gallons of liquid radioactive wastes were pumped into the ground near major rivers and now threaten surface and groundwater. One weapons facility in the Ural Mountains, known as Chelyabinsk-65, discharged so much nuclear waste into nearby Lake Tarachay that a person standing on its shores for an hour will receive a fatal dose of radiation.

The Savannah River Weapons Plant near Aiken, South Carolina, which produced the radioactive tritium gas used to enhance the explosive power of nuclear weapons, is known to have released tritium gas (a half-life of 12.3 years) into the air and water around the plant since the 1950s. It is also suspected of having deliberately discharged radioactive liquid wastes into the environment for many years. Leaks from old waste storage tanks continue to threaten the Savannah River and nearby groundwater with radioactive contaminants. All four nuclear reactors at the site are now closed. At another weapons plant near Fernald, Ohio, it is estimated that between 1951 and 1985 at least 3 million pounds of uranium dust and other radioactive materials were released into the atmosphere.

The largest, most contaminated, and most dangerous weapons facility in the United States is the 560-square-mile Hanford complex near Richland, Washington, which until operations were halted in 1987 produced plutonium for nuclear weapons. In the past, this facility discharged billions of gallons of radioactive liquid wastes directly into the ground. Wastes that were stored in tanks have leaked, contaminating several cubic miles of soil not only with radioactive material but also with other dangerous by-products of weapons production, including toxic solvents and heavy metals. Some of these wastes continue to migrate through the soil into the Columbia River.

Another severely contaminated weapons facility is the Rocky Flats Plant near Denver, Colorado, which until it was shut down in 1989 produced the plutonium "triggers" that set off the nuclear chain reactions in warheads. Wastes at the plant have leaked from corroded storage tanks for many years.

Of particular concern are over 70 million gallons of high-level liquid radioactive wastes from weapons production in the past, which are buried underground in steel tanks, some at Hanford and the rest at Savannah River (Figure 18.9). Many of the tanks are leaking, and the wastes in some tanks are potentially explosive. In one notorious tank at Hanford that contains a mixture of radioactive sludge, toxic solvents, and unidentified chemicals, hydrogen gas builds at intervals and must be vented. Some scientists believe that an electric spark could easily ignite the hydrogen, setting off explosions at nearby tanks and causing the release of large amounts of radioactivity.

One of the biggest problems is finding exactly what chemicals are in the tanks. Eventually, the DOE plans to remove the waste from the tanks, vitrify it (fuse it with sand to form a radioactive glass), and then store it in a permanent repository.

It has been estimated that the massive task of cleaning up the nation's nuclear weapons plants and other sites contaminated with radioactive wastes will ultimately cost at least

a.

Leak detection pit Pump pit Liquid level gauge Monitoring system

Ground level

Reinforced concrete

Liquid

Primary tank

Secondary tank

Sludge

Carbon steel

b.

Figure 18.9 High-level liquid radioactive wastes are stored in double-shell steel tanks, each of which can hold 1 million gallons. (a) Tanks at various stages of construction. When completed, the tanks will be covered with 7–10 feet of earth. (b) A cross-section of a typical double-shell steel tank. Courtesy of the U.S. Department of Energy.

$600 billion and will take decades to accomplish. Cleanup is hampered by the lack of adequate disposal technology and the difficulty of finding an acceptable site for long-term storage.

■ Regulation of Radioactive Waste Disposal

Regulations governing disposal of radioactive wastes vary depending on whether the wastes are high or low level and whether they are produced by commercial operations or were generated by weapons programs. In all cases, the wastes must be shipped long distances across the country to central disposal sites.

Low-Level Radioactive Wastes

Until 1970, most low-level radioactive waste generated in the United States was disposed of at sea. When this practice was banned, low-level wastes were buried in shallow, often unlined trenches. The wastes, which in some cases included highly radioactive liquid military waste, were supposed to be contained by impermeable rock, but leaks developed at many locations. Today, commercially generated low-level radioactive wastes must be buried in steel drums in lined, carefully sited trenches in accordance with rules that the Nuclear Regulatory Commission set. During the 1970s, states in which the few Nuclear Regulatory Commission–licensed facilities were located objected to receiving wastes generated outside their borders, and there was increasing concern about the dangers inherent in transporting radioactive wastes over long distances. In 1980, in response to these problems, Congress passed the **Low-Level Radioactive Waste Policy Act**, which set 1986 (later extended to 1992) as the deadline for each state to take responsibility for disposal of its own low-level commercial wastes. If states preferred, they had the option of joining in multistate compacts in which one state would provide a disposal site for the other states in the compact. The search for sites touched off furious environmental and political battles, and many states are still not in compliance with the law.

To provide a permanent storage site for low-level radioactive wastes produced by nuclear weapons plants, the DOE in 1983 began building a repository in a salt bed half a mile under federal land near Carlsbad, New Mexico. The repository, which is known as the **Waste Isolation Pilot Plant**, consists of a one-mile square network of tunnels and chambers. In 1999, after years of opposition from environmental groups and others concerned about long-term safety, the site, which cost almost $2 billion to build, finally received its first shipment of waste. The waste, which is contained in steel drums, consists mainly of protective clothing, tools, equipment, and soil that are contaminated with plutonium-239. Deliveries of wastes from the nation's nuclear weapons sites are expected to continue for at least the next 30 years.

High-Level Radioactive Wastes

Pending establishment of a permanent repository for high-level radioactive wastes, spent fuel rods from nuclear reactors are stored under water in tanks at the reactor sites. The water shields the radiation and dissipates the heat that the rods generate. At present, 160,000 highly radioactive fuel assemblies that weigh 45,000 tons are being stored at

72 nuclear power plants around the country. Because approximately 7800 fuel assemblies are taken out of reactors each year, many nuclear plants are running out of storage space.

To deal with the mounting problem, Congress in 1982 passed the **Nuclear Waste Policy Act,** which set a timetable for building an underground repository deep underground in a geologically stable area for permanent storage of high-level waste. The repository was originally intended for wastes from commercial nuclear reactors only, but under new plans, it also accepts defense weapons wastes, including plutonium from dismantled nuclear warheads, after they have been vitrified (melted into glass).

In 1987, the DOE chose Yucca Mountain, 100 miles (150 km) northwest of Las Vegas, Nevada, as the site for the underground repository (Figure 18.10). The site met two important criteria: It was very dry with an annual rainfall of only six inches, and the rock formation was tuff, a compacted volcanic ash that is considered to be one of the best materials for absorbing heat and containing radioactivity. Construction was supposed to be completed by 1998, but delays occurred because of strong opposition from the people of Nevada. In 2002, Congress passed legislation permitting development of the Yucca Mountain site and DOE started to implement that mandate. Also, according to the EPA, concern exists that during the 10,000 years that is required for spent fuel rods to decay to safe levels, earthquakes and volcanic eruptions may occur. Some geologists believe that water could seep into storage chambers, become contaminated with radioactive materials, and move into groundwater. The repository, which has already cost $19 billion, is not expected to be completed until 2017 at the earliest.

The DOE, under terms of the Nuclear Waste Policy Act, has been collecting fees from the nuclear power industry for many years to build the permanent storage facility. After the DOE failed to meet the 1998 deadline set for completing the facility, the department, under pres-

Figure 18.10 Yucca Mountain underground repository. (a) South portal to the Yucca Mountain Repository, (b) Interior of the mine within Yucca Mountain.

sure from the nuclear power industry, is considering building an interim above-ground storage facility. This facility for spent rods, termed a monitored retrievable storage facility, would be built on the Mescalero Apache reservation near Alamogordo, New Mexico. Although the reservation's non-Apache neighbors are opposed to it, the Apache, who see the venture as a much-needed source of revenue, contend that as a sovereign Indian nation, they can proceed without state approval.

Disposal of radioactive wastes is one of the most intractable problems that the DOE faces. For environmentalists and the general public alike, opposition to radioactive waste disposal sites is not just a case of NIMBY, it is one of **NOPE** (**Not On Planet Earth**).

■ Technologies for Radioactive Waste Disposal

Several technologies for the disposal of high-level radioactive waste have been considered, including burial of the waste in deep ocean trenches, shooting the waste into the sun or outer space, and transmutation of dangerous radioisotopes to harmless or less harmful isotopes by means of neutron bombardment. None of these options is realistic either because of the cost involved and safety considerations or because the technology is not presently available.

Currently, the recommended procedure for disposing of high-level radioactive waste is to concentrate it, to incorporate it into a glass material by fusing it with sand at 2000°F, and then to pour the molten mixture into stainless steel, corrosion-resistant canisters. The glass will be radioactive but very resistant to leaching. Until the canisters can be transferred to a permanent repository, they would be stored in concrete.

■ The Post-Cold War Challenge

In the late 1980s, with the end of the Cold War, the United States discontinued production of plutonium-239. Disposal of huge quantities of surplus plutonium from dismantled nuclear warheads in both the United States and Russia now presents a serious problem.

Rejecting a secure storage of plutonium option because of the danger of theft by terrorists or rogue nations who are seeking to make nuclear weapons, the DOE has decided that two disposal strategies should be adopted. In one, the plutonium would be immobilized in glass, as just described, and ultimately stored underground. This strategy treats the plutonium as waste and does not make use of its energy value. In the second strategy, the plutonium would be reprocessed. It would be oxidized and mixed with conventional uranium fuel to form a *mixed oxide (MOX)*, which could be used as fuel to produce energy at commercial nuclear power plants. The MOX would be used once only and would not be reprocessed to recover the plutonium that would remain in the spent fuel rods after burning. A glass immobilization facility and a plant for manufacturing MOX are expected to be open at the Savannah River site by 2013. Russia is expected to dispose of its weapons-grade plutonium by using the second option.

Many people are unhappy about the DOE's decision to make MOX because it appears to contradict long-standing U.S. policy against reprocessing plutonium. Chapter 11 showed that spent fuel rods at commercial nuclear power plants contain unreacted uranium and plutonium that can be reprocessed to make new nuclear fuel. Japan, Britain, and France all reprocess nuclear fuel. The United States, however, does not do so because of the possibility that the plutonium could be stolen and used to make nuclear weapons.

■ Additional Sources of Information

Environmental Protection Agency (EPA). Homepage. Accessed December 2008 at http://www.epa.gov.

Harris DC. *Quantitative Chemical Analysis,* 6th ed. New York, W. H. Freeman and Company, 2003.

Hemond HF, Fechner-Levy EJ. *Chemical Fate and Transport in the Environment,* 2nd ed. San Diego: Academic Press, 2000.

Levine AG. *Love Canal: Science, Politics and People.* Lexington, MA: Lexington Books, 1982.

Rubison KA, Rubison JF. *Contemporary Instrumental Analysis.* Upper Saddle River, NJ: Prentice Hall, 2000.

Skoog DA, Holler FJ, Crouch SR. *Principles of Instrumental Analysis,* 6th ed. Philadelphia, PA: Saunders College Publishing, Harcourt Brace College Publishers, 2007.

■ Keywords

bioremediation
combustion facilities
Comprehensive
   Environmental Response,
   Compensation, and
   Liability Act (CERCLA)
   or Superfund
corrosive wastes
deep-well injection
hazardous waste
high-level radioactive wastes
ignitable waste
incineration

Low-Level Radioactive
   Waste Policy Act
low-level radioactive wastes
municipal refuse
municipal solid waste
   (MSW)
National Priorities List
Not in My Backyard
   (NIMBY)
Not on Planet Earth
   (NOPE)
Nuclear Waste Policy Act
reactive wastes

Resource Conservation and
   Recovery Act (RCRA)
resource recovery plants
sanitary landfills
secured landfill
solid waste
Solid Waste Disposal Act
   (SWDA)
surface impoundment
toxic wastes
waste exchanges
Waste Isolation Pilot Plant
waste-to-energy plants

■ Questions and Problems

1. Describe how a modern sanitary landfill is constructed. How does this design differ from old "dumps?"

2. Name the most significant legislation that has been enacted to regulate solid waste.
3. List the advantages and disadvantages of refuse incinerators. Would you recommend that the refuse that your family has generated be disposed of in a landfill or incinerated? Explain your choice.
4. What is fly ash? Why is it dangerous?
5. What is the most dangerous pollutant that is released by incinerators?
6. Why is a larger percentage of our garbage not recycled? List three steps that your hometown could take to encourage recycling.
7. Does CERCLA regulate abandoned waste dumps or operation of municipal dumps?
8. Define the term *hazardous waste*. Who has the responsibility for deciding whether a waste is hazardous? Once wastes are identified as hazardous, "cradle-to-grave" responsibility for hazardous waste is assigned to whom?
9. Define each of the following terms:
   a. Ignitable wastes
   b. Corrosive wastes
   c. Reactive wastes
   d. Toxic wastes
10. Is waste sulfuric acid from automobile batteries classified as ignitable waste or corrosive waste?
11. Should waste sulfuric acid be stored in steel drums before proper disposal? What type of container should be used?
12. Picric acid, which was once used by printers to etch copper, is known to explode when heated or subjected to shock. How would picric acid waste be classified?
13. Many printers use hydrocarbon solvents to clean ink from their hands and equipment. These solvents have a flash point that is below 60°C. How would hydrocarbon wastes be classified?
14. How would waste lead from automobile batteries be classified?
15. How would waste sodium hydroxide be classified?
16. You have to dispose of waste nitric acid. What could you do to make it less corrosive?
17. Why is there more incineration of toxic waste in Europe than in the United States?
18. List three techniques used for *perpetual storage* of hazardous wastes. If you had to decide among the three for dioxin-contaminated waste, which would you recommend and why?
19. Why is deep-well injection a problem for those living in communities near the well?
20. How far under the ground is groundwater found (on the average)? How far under the ground is hazardous waste deposited in deep well injection?
21. What is surface impoundment?
22. What is the Superfund program? What does it regulate?
23. From where does the money for the Superfund trust fund come?
24. What is the National Priorities List?
25. What is meant by the term *NIMBY*? Apply NIMBY to hazardous wastes.
26. List two ways that manufacturing companies can minimize the amount of hazardous wastes that they generate.

27. Some people advocate ocean dumping as a method to dispose of toxic materials far away from human populations. On the basis of what you have learned so far, do you agree that this is a viable option?

28. Before 1970, most low-level nuclear wastes generated in the United States were dumped where?

29. What agency of the federal government regulates almost all radioactive wastes?

30. What are the two major sources of radioactive wastes?

31. Are radioactive materials regulated by RCRA?

32. Cesium-137, which is produced in nuclear reactors, has a half-life of 30 years. If a nuclear plant stopped operating today, how long will it be until the cesium in the reactor is considered no longer dangerous?

33. How many half-lives does it take for the radiation in radioactive waste to be considered at negligible levels? How many years for plutonium-239? How many for strontium-90?

34. How has radioactive waste been disposed of at the Hanford plutonium-producing plant in Richland, Washington?

35. Describe what is meant by vitrifying radioactive wastes.

36. Nuclear disarmament should result in the retrieval of large quantities of plutonium from nuclear weapons that are being destroyed. Suggest a peaceful use for this plutonium.

37. What is the Low-Level Radioactive Waste Policy Act? Should a heavily populated state such as New York be held responsible for the disposal of radioactive waste from its nuclear plants?

38. What is Waste Isolation Pilot Plant? Should the state of New Mexico have a say in what goes into Waste Isolation Pilot Plant?

39. What happens to spent fuel rods from nuclear power plants? Where will the radioactive material in these rods be permanently stored?

40. What type of radioactive waste does the Nuclear Waste Policy Act regulate?

41. In what state does the DOE plan to permanently store waste from commercial nuclear reactors?

42. Why is radioactive material from spent fuel rods not recycled?

43. Describe the recommended procedure for the disposal of high-level radioactive waste.

44. What is transmutation? How could it be used to lessen the amount of radioactive waste needing disposal?

45. Some people have suggested that radioactive waste should be shot into the sun or outer space. List two reasons that this is not an attractive disposal method.

46. Some have suggested that decommissioned nuclear power plants be entombed in concrete rather than being dismantled. List the advantages and disadvantages of entombment relative to dismantling. Consider the exposure of workers and the public to radiation and the cost.

47. What is MOX? What is a source of MOX?

48. With the end of the Cold War, many old nuclear warheads will be decommissioned, and the plutonium in each will need to be disposed of. List three ways to accomplish the disposal.

49. A container designed to hold high-level radioactive waste should be made of what materials?

50. It is difficult to find a location for the disposal of high-level radioactive waste. List three criteria that should be used to pick a location.

51. What is the most common way to dispose of medical waste?

52. What is meant by the term *NOPE*?

Appendix A. Solubility Products, K_{sp}

Formula	pK_{sp}	K_{sp}
Bromides: L = Br$^-$		
CuL	8.3	5×10^{-9}
AgL	12.30	5.0×10^{-13}
Hg$_2$L$_2$	22.25	5.6×10^{-23}
TlL	5.44	3.6×10^{-6}
HgL$_2$ (e)	18.9	1.3×10^{-19}
PbL$_2$	5.68	2.1×10^{-6}
Carbonates: L = CO$_3^{2-}$		
MgL	7.46	3.5×10^{-8}
CaL (calcite)	8.35	4.5×10^{-9}
CaL (aragonite)	8.22	6.0×10^{-9}
SrL	9.03	9.3×10^{-10}
BaL	8.30	5.0×10^{-9}
Y$_2$L$_3$	30.6	2.5×10^{-31}
La$_2$L$_3$	33.4	4.0×10^{-34}
MnL	9.30	5.0×10^{-10}
FeL	10.68	2.1×10^{-11}
CoL	9.98	1.0×10^{-10}
NiL	6.87	1.3×10^{-7}
CuL	9.63	2.3×10^{-10}
Ag$_2$L	11.09	8.1×10^{-12}
Hg$_2$L	16.05	8.9×10^{-17}
ZnL	10.00	1.0×10^{-10}
CdL	13.74	1.8×10^{-14}
PbL	13.13	7.4×10^{-14}
Chlorides: L = Cl$^-$		
CuL	6.73	1.9×10^{-7}
AgL	9.74	1.8×10^{-10}
Hg$_2$L$_2$	17.91	1.2×10^{-18}
TlL	3.74	1.8×10^{-4}
PbL$_2$	4.78	1.7×10^{-5}
Chromates: L = CrO$_4^{2-}$		
BaL	9.67	2.1×10^{-10}

continued

Formula	pK_{sp}	K_{sp}
CuL	5.44	3.6×10^{-6}
Ag_2L	11.92	1.2×10^{-12}
Hg_2L	8.70	2.0×10^{-9}
Tl_2L	12.01	9.8×10^{-13}
Fluorides: $L = F^-$		
LiL	2.77	1.7×10^{-3}
MgL_2	8.18	6.6×10^{-9}
CaL_2	10.41	3.9×10^{-11}
SrL_2	8.54	2.9×10^{-9}
BaL_2	5.76	1.7×10^{-6}
LaL_3	18.7	2×10^{-19}
ThL_4	28.3	5×10^{-29}
PbL_2	7.44	3.6×10^{-8}
Hydroxides: $L = OH^-$		
MgL_2 (amorphous)	9.2	6×10^{-10}
MgL_2 (brucite crystal)	11.15	7.1×10^{-12}
CaL_2	5.19	6.5×10^{-6}
$BaL_2 \cdot 8H_2O$	3.6	3×10^{-4}
YL_3	23.2	6×10^{-24}
LaL_3	20.7	2×10^{-21}
CeL_3	21.2	6×10^{-22}
$UO_2 \ (\rightleftharpoons U^{4+} + 4OH^-)$	56.2	6×10^{-57}
$UO_2L_2 \ (\rightleftharpoons UO_2^{2+} + 2OH^-)$	22.4	4×10^{-23}
MnL_2	12.8	1.6×10^{-13}
FeL_2	15.1	7.9×10^{-16}
CoL_2	14.9	1.3×10^{-15}
NiL_2	15.2	6×10^{-16}
CuL_2	19.32	4.8×10^{-20}
VL_3	34.4	4.0×10^{-35}
CrL_3 (d)	29.8	1.6×10^{-30}
FeL_3	38.8	1.6×10^{-39}
CoL_3 (a)	44.5	3×10^{-45}
$VOL_2 \ (\rightleftharpoons VO^{2+} + 2OH^-)$	23.5	3×10^{-24}
PdL_2	28.5	3×10^{-29}
ZnL_2 (amorphous)	15.52	3.0×10^{-16}
CdL_2 (β)	14.35	4.5×10^{-15}
HgO (red) $(\rightleftharpoons Hg^{2+} + 2OH^-)$	25.44	3.6×10^{-26}
$Cu_2O \ (\rightleftharpoons 2Cu^+ + 2OH^-)$	29.4	4×10^{-30}
$Ag_2O \ (\rightleftharpoons 2Ag^+ + 2OH^-)$	15.42	3.8×10^{-16}
AuL_3	5.5	3×10^{-6}
AlL_3 (α)	33.5	3×10^{-34}
GaL_3 (amorphous)	37	10^{-37}

continued

Formula	pK$_{sp}$	K$_{sp}$
InL$_3$	36.9	1.3×10^{-37}
SnO ($\rightleftharpoons$ Sn^{2+} + 2OH$^-$)	26.2	6×10^{-27}
PbO (yellow) ($\rightleftharpoons$ Pb^{2+} + 2OH$^-$)	15.1	8×10^{-16}
PbO (red) ($\rightleftharpoons$ Pb^{2+} + 2OH$^-$)	15.3	5×10^{-16}
Iodides: L = I$^-$		
CuL	12.0	1×10^{-12}
AgL	16.08	8.3×10^{-17}
CH$_3$HgL ($\rightleftharpoons$ CH$_3$Hg$^+$ + I$^-$) (b, g)	11.46	3.5×10^{-12}
CH$_3$CH$_2$HgL ($\rightleftharpoons$ CH$_3$CH$_2$Hg$^+$ + I$^-$)	4.11	7.8×10^{-5}
TlL	7.23	5.9×10^{-8}
Hg$_2$L$_2$ (e)	27.95	1.1×10^{-28}
SnL$_2$ (g)	5.08	8.3×10^{-6}
PbL$_2$	8.10	7.9×10^{-9}
Phosphates: L = PO$_4^{3-}$		
MgHL · 3H$_2$O ($\rightleftharpoons$ Mg^{2+} + HL^{2-})	5.78	1.7×10^{-6}
CaHL · 2H$_2$O ($\rightleftharpoons$ Ca^{2+} + HL^{2-})	6.58	2.6×10^{-7}
SrHL ($\rightleftharpoons$ Sr^{2+} + HL^{2-}) (b)	6.92	1.2×10^{-7}
BaHL ($\rightleftharpoons$ Ba^{2+} + HL^{2-}) (b)	7.40	4.0×10^{-8}
LaL (e)	22.43	3.7×10^{-23}
Fe$_3$L$_2$ · 8H$_2$O	36.0	1×10^{-36}
FeL · 2H$_2$O	26.4	4×10^{-27}
(VO)$_3$L$_2$ ($\rightleftharpoons$ 3VO^{2+} + 2L^{3-})	25.1	8×10^{-26}
Ag$_3$L	17.55	2.8×10^{-18}
Hg$_2$HL ($\rightleftharpoons$ Hg$_2^{2+}$ + HL^{2-})	12.40	4.0×10^{-13}
Zn$_3$L$_2$ · 4H$_2$O	35.3	5×10^{-36}
Pb$_3$L$_2$ (c)	43.53	3.0×10^{-44}
GaL (f)	21.0	1×10^{-21}
InL (f)	21.63	2.3×10^{-22}
Sulfates: L = SO$_4^{2-}$		
CaL	4.62	2.4×10^{-5}
SrL	6.50	3.2×10^{-7}
BaL	9.96	1.1×10^{-10}
RaL (b)	10.37	4.3×10^{-11}
Ag$_2$L	4.83	1.5×10^{-5}
Hg$_2$L	6.13	7.4×10^{-7}
PbL	6.20	6.3×10^{-7}
Sulfides: L = S^{2-}		
MnL (pink)	10.5	3×10^{-11}
MnL (green)	13.5	3×10^{-14}
FeL	18.1	8×10^{-19}
CoL (α)	21.3	5×10^{-22}
CoL (β)	25.6	3×10^{-26}

continued

Formula	pK_{sp}	K_{sp}
NiL (α)	19.4	4×10^{-20}
NiL (β)	24.9	1.3×10^{-25}
NiL (γ)	26.6	3×10^{-27}
CuL	36.1	8×10^{-37}
Cu_2L	48.5	3×10^{-49}
Ag_2L	50.1	8×10^{-51}
Tl_2L	21.2	6×10^{-22}
ZnL (α)	24.7	2×10^{-25}
ZnL (β)	22.5	3×10^{-23}
CdL	27.0	1×10^{-27}
HgL (black)	52.7	2×10^{-53}
HgL (red)	53.3	5×10^{-54}
SnL	25.9	1.3×10^{-26}
PbL	27.5	3×10^{-28}
In_2L_3	69.4	4×10^{-70}

The designations α, β, or γ after some formulas refer to particular crystalline forms (which are customarily identified by Greek letters). Data are taken from A. E. Martell and R. M. Smith, *Critical Stability Constants,* Vol. 4 (New York: Plenum Press, 1976).

Conditions are 25°C and zero ionic strength unless otherwise indicated: (a) 19°C; (b) 20°C; (c) 38°C; (d) 0.1 M; (e) 0.5 M; (f) 1 M; (g) 4 M.

Appendix B. Dissociation Constants for Acids and Bases in Aqueous Solution at 25°C

Acid	Protonated Species	K_a	pK_a	Base	Deprotonated Species	K_b	pK_b
Acetic acid	CH_3COOH	1.8×10^{-5}	4.75	Acetate	CH_3COO^-	5.6×10^{-10}	9.25
Aluminum (III)	$Al(H_2O)_6^{3+}$	7.2×10^{-6}	5.14	Hydroxy-aluminum (III)	$Al(OH)^{2+}$	1.4×10^{-9}	8.86
Ammonium	NH_4^+	5.6×10^{-10}	9.25	Ammonia	NH_3	1.8×10^{-5}	4.75
Arsenic acid	H_3AsO_4	5.8×10^{-3}	2.24	Dihydrogen arsenate	$H_2AsO_4^-$	1.7×10^{-12}	11.76
Dihydrogen arsenate	$H_2AsO_4^-$	1.10×10^{-7}	6.96	Hydrogen arsenate	$HAsO_4^{2-}$	9.1×10^{-8}	7.04
Hydrogen arsenate	$HAsO_4^{2-}$	3.2×10^{-12}	11.50	Arsenate	AsO_4^{3-}	3.1×10^{-3}	2.50
Arsenious acid	$As(OH)_3$	5.1×10^{-10}	9.29	Dihydrogen arsenite	$H_2AsO_3^-$	2.0×10^{-5}	4.71
Boric acid	$B(OH)_3$	7.2×10^{-10}	9.14	Borate	$B(OH)_4^-$	1.4×10^{-5}	5.86
Carbon dioxide	CO_2	4.5×10^{-7}	6.35	Hydrogen carbonate	HCO_3^-	2.2×10^{-8}	7.65
Hydrogen carbonate	HCO_3^-	4.7×10^{-11}	10.33	Carbonate	CO_3^{2-}	2.1×10^{-4}	3.67
Formic acid	$HCOOH$	1.8×10^{-4}	3.75	Formate	$HCOO^-$	5.6×10^{-11}	10.25
Hydrofluoric acid	HF	3.5×10^{-4}	3.46	Fluoride	F^-	2.9×10^{-11}	10.54
Hydrogen cyanide	HCN	4.9×10^{-10}	9.31	Cyanide	CN^-	2.0×10^{-5}	4.69
Hydrogen sulfate	HSO_4^-	1.0×10^{-2}	2.00	Sulfate	SO_4^{2-}	1.0×10^{-12}	12.00

continued

Acid	Protonated Species	K_a	pK_a	Base	Deprotonated Species	K_b	pK_b
Hydrogen sulfide	H_2S	1.0×10^{-7}	7.00	Hydrogen sulfide ion	HS^-	1.0×10^{-7}	7.00
Hydrogen sulfide ion	HS^-	1.1×10^{-12}	11.96	Sulfide	S^{2-}	9.1×10^{-3}	2.04
Hypochlorous acid	$HClO$	3.0×10^{-8}	7.52	Hypochlorite	ClO^-	3.3×10^{-7}	6.48
Iron (III)	$Fe(H_2O)_6^{3+}$	6.3×10^{-3}	2.19	Hydroxyiron (III)	$FeOH^{2+}$	1.6×10^{-12}	11.80
Methylammonium	$CH_3NH_3^+$	2.2×10^{-11}	10.66	Methylamine	CH_3NH_2	4.5×10^{-4}	3.34
Nitrous acid	HNO_2			Nitrite	NO_2^-		
Phenol	C_6H_5OH	1.3×10^{-10}	9.89	Phenate	$C_6H_5O^-$	7.7×10^{-5}	4.11
Phosphoric acid	H_3PO_4	7.1×10^{-3}	2.15	Dihydrogen phosphate	$H_2PO_4^-$	1.4×10^{-12}	11.85
Dihydrogen phosphate	$H_2PO_4^-$	6.3×10^{-8}	7.20	Hydrogen phosphate	HPO_4^{2-}	1.6×10^{-7}	6.80
Hydrogen phosphate	HPO_4^{2-}	4.2×10^{-13}	12.38	Phosphate	PO_4^{3-}	2.4×10^{-2}	1.62
Silicic acid	$Si(OH)_4$	2.2×10^{-10}	9.66	Trihydrogen silicate	$H_3SiO_4^-$	4.6×10^{-5}	4.34
Sulfurous acid	H_2SO_3	1.72×10^{-2}	1.76	Hydrogen sulfite	HSO_3^-	5.81×10^{-13}	12.24
Hydrogen sulfite	HSO_3^-	6.43×10^{-8}	7.19	Sulfite	SO_3^{2-}	1.56×10^{-7}	6.81

Source: Values from P. Atkins, *Physical Chemistry,* 6th ed. (New York: W. H. Freeman and Co., 1998.)

Appendix C. Standard Redox Potentials in Aqueous Solutions

For simplicity all equations use H^+ as a substitute for the hydronium ion H_3O^+.

Reduction Half Reaction	$E°/v$	$pE°$	$pE°(w)$
$O_3 + 2H^+ + 2e^- \rightarrow O_2 + H_2O$	+2.075	+35.1	+28.1
$H_2O_2 + 2H^+ + 2e^- \rightarrow 2H_2O$	+1.763	+29.8	+22.8
$MnO_4^- + 4H_3O^+ + 3e^- \rightarrow MnO_2 + 6H_2O$	+1.692	+28.6	+19.3
$2HClO + 2H^+ + 2e^- \rightarrow Cl_2 + 2H_2O$	+1.630	+27.6	+20.6
$Cl_2 + 2e^- \rightarrow 2Cl^-$	+1.396	+23.6	+23.0
$Cr_2O_7^{2-} + 14H^+ + 6e^- \rightarrow 2Cr^{3+} + 7H_2O$	+1.36	+23.0	+6.67
$2NO_3^- + 12H^+ + 10e^- \rightarrow N_2 + 6H_2O$	+1.25	+21.1	+12.7
$O_3 + H_2O + 2e^- \rightarrow O_2 + 2OH^-$	+1.24	+21.0	+28.0
$MnO_2 + 4H^+ + 2e^- \rightarrow Mn^{2+} + 2H_2O$	+1.230	+20.8	+6.80
$O_2 + 4H^+ + 4e^- \rightarrow 2H_2O$	+1.229	+20.8	+13.8
$NO_3^- + 4H^+ + 3e^- \rightarrow NO + 2H_2O$	+0.955	+16.1	+6.77
$NO_3^- + 3H^+ + e^- \rightarrow HNO_2 + H_2O$	+0.940	+15.9	−5.11
$ClO^- + H_2O + 2e^- \rightarrow Cl^- + 2OH^-$	+0.89	+15.0	+22.0
$NO_3^- + 10H^+ + 8e^- \rightarrow NH_4^+ + 3H_2O$	+0.882	+14.9	+6.15
$NO_3^- + 2H^+ + 2e^- \rightarrow NO_2 + H_2O$	+0.837	+14.2	+7.15
$Fe^{3+} + e^- \rightarrow Fe^{2+}$	+0.771	+13.0	+13.0
$CH_2O + 4H^+ + 4e^- \rightarrow CH_4 + H_2O$	+0.411	+6.94	−0.06
$O_2 + 2H_2O + 4e^- \rightarrow 4OH^-$	+0.40	+6.76	+13.8
$SO_4^{2-} + 8H^+ + 6e^- \rightarrow S + 4H_2O$	+0.353	+5.96	−3.37
$SO_4^{2-} + 9H^+ + 8e^- \rightarrow HS^- + 4H_2O$	+0.248	+4.20	−3.68
$CO_2 + 8H^+ + 8e^- \rightarrow CH_4 + 2H_2O$	+0.170	+2.87	−4.13
$2H^+ + 2e^- \rightarrow H_2$	0.00	0.00	−7.00
$CO_2 + 4H^+ + 4e^- \rightarrow [CH_2O] + H_2O$	−0.071	−1.20	−8.20
$S + 2e^- \rightarrow S^{2-}$	−0.50	−8.45	−8.45
$2H_2O + 2e^- \rightarrow H_2 + 2OH^-$	−0.828	−14.0	−7.00

Source: Values from D. C. Harris. *Quantitative Chemical Analysis,* 6th ed. (New York: W. H. Freeman and Co., 2003).

Appendix D. Carbon Compounds: An Introduction to Organic Chemistry

WHY DOES CARBON FORM SO MANY COMPOUNDS? The answer lies in its atomic structure. Carbon, atomic number 6, has four valence electrons, and when it forms compounds by sharing these valence electrons with other carbon atoms or atoms of other elements, it obeys the octet rule. It forms carbon–carbon single bonds by sharing pairs of electrons:

$$\cdot\ddot{C}\cdot \;+\; \cdot\ddot{C}\cdot \;+\; \cdot\ddot{C}\cdot \;\longrightarrow\; \cdot\ddot{C}\!:\!\ddot{C}\!:\!\ddot{C}\cdot$$

When it combines with four hydrogen atoms to form methane, the carbon atom shares its four valence electrons with the hydrogen atoms, thus forming four stable covalent bonds:

$$\cdot\ddot{C}\cdot \;+\; 4\,H\cdot \;\longrightarrow\; H\!:\!\overset{\displaystyle H}{\underset{\displaystyle H}{\ddot{C}}}\!:\!H \;\text{ or }\; H\!-\!\overset{\displaystyle H}{\underset{\displaystyle H}{C}}\!-\!H \;\text{ or }\; CH_4$$

methane

Carbon forms more compounds than any other element primarily because carbon atoms link with each other in so many different ways. A molecule of a carbon compound may contain a single carbon–carbon bond or thousands of such bonds. The carbon atoms can link in straight chains, branched chains, or rings (Figure D.1). In addition to carbon–carbon single bonds, carbon can form carbon–carbon double and triple bonds. In each of these different bonding patterns, the carbon atoms form four covalent bonds.

Only the element carbon is able to form long chains of its atoms. Some elements, including oxygen (O), nitrogen (N), and chlorine (Cl), form stable two-atom molecules; sulfur (S), silicon (Si), and phosphorus (P) form unstable chains of from four to eight like atoms, but no element can form chains as long as carbon. Even silicon and germanium (Ge), which are in the same group of the periodic table as carbon, do not form long like-atom chains.

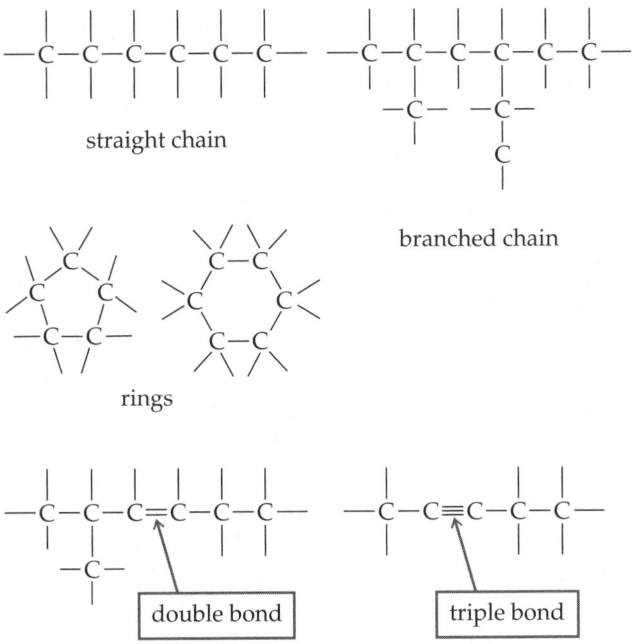

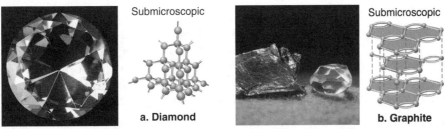

Figure D.1 Carbon atoms can join to form straight chains, branched chains, or rings. In addition to carbon–carbon single bonds, double and triple bonds are also formed.

■ Different Forms of Carbon

Elemental carbon exists in two very different crystalline forms: diamond and graphite. In diamond, each carbon atom is joined by strong covalent bonds to four other carbon atoms. Each of these carbon atoms is also joined to four more carbon atoms and so on until a huge three-dimensional interlocking network of carbon atoms is formed (Figure D.2a). This carbon–carbon bonding pattern accounts for the stability and extreme hardness of diamond. For a diamond to undergo a chemical change, many strong bonds within the crystalline structure must be broken.

Graphite is completely different from diamond. It is a soft, black, slippery material that is made of hexagonal arrays of carbon atoms arranged in sheets (Figure D.2b). Each carbon

Submicroscopic   Submicroscopic

a. Diamond   **b. Graphite**

Figure D.2 Two very different forms of carbon: (a) In diamond, the carbon atoms form a strong, three-dimensional network. (b) In graphite, the carbon atoms form sheets held together by weak attractive forces.

atom in the array is joined to three other carbon atoms by forming two single bonds and one double bond. Graphite is slippery because the intermolecular attractive forces holding the sheets together are relatively weak, and the sheets easily slide past each other. Graphite is used as a dry lubricant and, combined with a binder, forms the "lead" in pencils.

■ Compounds of Carbon with Other Elements

In addition to forming carbon–carbon bonds, carbon atoms form strong covalent bonds with other elements, particularly with the nonmetal elements hydrogen (H), oxygen (O), nitrogen (N), fluorine (F), and chlorine (Cl). Simple compounds in which carbon is combined with other elements are shown in Figure D.3. Carbon also bonds with phosphorus (P), sulfur (S), silicon (Si), boron (B), and the metals sodium (Na), potassium (K), calcium (Ca), and magnesium (Mg).

The enormous number of carbon compounds that exists can be divided into a relatively small number of classes according to the functional group they contain. A **functional group** is a particular arrangement of atoms that is present in each molecule of the class and that largely determines the chemical behavior of the class. Examples of functional groups are $-C-l$, $-OH$, $-COOH$, and $-NH_2$. Before we examine the different functional groups, we study the hydrocarbons, the parent compounds from which all carbon compounds are derived.

ethane

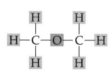

dimethyl ether

chloroform

methylamine

Figure D.3 Some compounds in which carbon atoms are covalently bonded to other kinds of atoms.

■ Hydrocarbons

Hydrocarbons are composed of the elements carbon and hydrogen. There are four classes of hydrocarbons: **alkanes**, which contain $-C-C-$ single bonds; **alkenes**, which contain one or more $-C=C-$ double bonds; **alkynes**, which contain one or more $-C\equiv C-$ triple bonds; and **aromatic hydrocarbons**, which contain one or more benzene rings. Complex mixtures of hydrocarbons, which are present in enormous quantities in natural gas and oil (petroleum), are the source of many of the organic compounds used by industry.

Alkanes

The simplest alkane is **methane** (CH_4), the major component of natural gas. The structure of the molecule can be represented in a number of different ways (Figure D.4). The expanded structural formula includes the four covalent bonds; the ball-and-stick and space-filling models show the spatial arrangement of the atoms. As the ball-and-stick model indicates, the methane molecule is in the shape of a tetrahedron. The space-filling model gives the most accurate representation of the actual shape of the molecule. For simplicity, the condensed

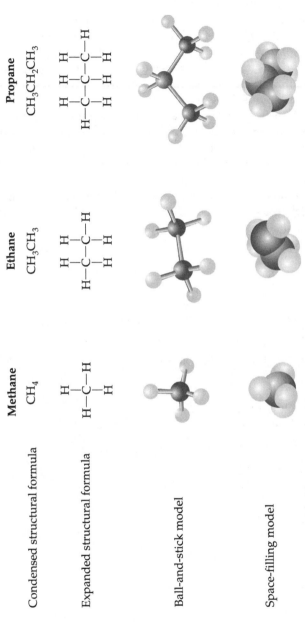

Condensed structural formula

Expanded structural formula

Ball-and-stick model

Space-filling model

Methane

CH_4

Ethane

CH_3CH_3

Propane

$CH_3CH_2CH_3$

Figure D.4 Organic molecules can be represented in several ways, as shown here for the first three members of the alkane series.

structural formula, which does not show the bonds or the bond angles, is usually used to represent the molecule.

The two alkanes that follow methane are **ethane** (C_2H_6), and **propane** (C_3H_8) (Figure D.4). Propane is a major component of bottled gas. The next member of the series is **butane** (C_4H_{10}), and for this alkane, two structures are possible (Figure D.5). The four carbon atoms can be joined in a straight line (*n*-butane), or the fourth carbon can be added to the middle carbon atom in the $-C-C-C-$ chain and form a branch (isobutane).

Alkanes are known as **saturated hydrocarbons** because for a given number of carbon atoms, they contain the largest possible number of hydrogen atoms. The first 10 straight-chain saturated alkanes are shown in Table D.1. The *n*- stands for normal and signifies straight chain. Notice that each alkane differs from the one preceding it by the addition of a $-CH_2$ group. A series of compounds in which each member differs from the next member by a constant increment is called a **homologous series**. The general formula for the alkane series is C_nH_{2n+2}, where *n* is the number of carbon atoms in a member of the series. In a homologous series, the properties of the members change systematically with increasing molecular weight. Table D.1 shows that in the alkane series, the boiling points of the straight-chain alkanes rise quite regularly as the number of carbon atoms increases. The first four alkanes are gases, and the remainder of those shown is liquids.

EXAMPLE   D.1

Write the structural formula for the straight-chain alkane C_5H_{12}.

Solution
Write five carbon atoms linked together to form a chain:

    C—C—C—C—C

Attach hydrogen atoms to the carbon atoms so that each carbon atom forms four covalent bonds.

$$
\begin{array}{ccccc}
 & H & H & H & H & H \\
 & | & | & | & | & | \\
H- & C- & C- & C- & C- & C-H \\
 & | & | & | & | & | \\
 & H & H & H & H & H
\end{array}
$$

EXAMPLE   D.2

Write the structural formula for the straight-chain hydrocarbon represented by C_nH_{2n+2}, where $n = 7$.

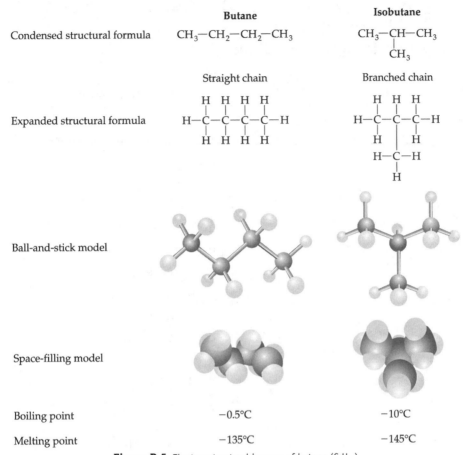

	Butane	**Isobutane**
Condensed structural formula	$CH_3-CH_2-CH_2-CH_3$	$CH_3-CH-CH_3$ $\quad\quad\quad CH_3$
	Straight chain	Branched chain

Figure D.5 The two structural isomers of butane (C_4H_{10}).

	Butane	Isobutane
Boiling point	−0.5°C	−10°C
Melting point	−135°C	−145°C

Solution
Write seven carbon atoms linked together to form a chain and attach hydrogen atoms to the carbon atoms so that each carbon atom forms four covalent bonds.

We have described unbranched chains as straight chains. In fact, because of tetrahedral bonding, the carbon atoms are not in a straight line (as shown in Table D.1), but are stag-

Table D.1

The First Ten Straight-Chain Alkanes

Name	Formula	Boiling Point (°C)	Structural Formula
Methane	CH_4	−162	
Ethane	C_2H_6	−88.5	
Propane	C_3H_8	−42	
n-Butane	C_4H_{10}	0	
n-Pentane	C_5H_{12}	36	
n-Hexane	C_6H_{14}	69	
n-Heptane	C_7H_{16}	98	
n-Octane	C_8H_{18}	126	
n-Nonane	C_9H_{20}	151	
n-Decane	$C_{10}H_{22}$	174	

gered as shown here and in the ball-and-stick models for propane and *n*-butane in Figures D.4 and D.5.

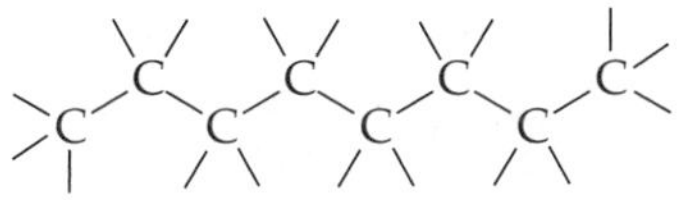

Structural Isomerism

Compounds such as butane and isobutane that have the same molecular formula (C_4H_{10}), but the different structures are called **structural isomers**. Because their structures are different (Figure D.5), their properties—for example, boiling point and melting point—are also different.

All alkanes containing four or more carbon atoms form structural isomers. The predicted number of possible isomers increases rapidly as the number of carbon atoms in the molecule increases. For example, butane (C_4H_{10}) forms two isomers: Decane ($C_{10}H_{22}$) theoretically forms 75, and $C_{30}H_{62}$ forms over 400 million (Table D.2). Most of these isomers do not exist naturally and have not been synthesized, but the large number of possibilities helps to explain the abundance of carbon compounds. In isomers with large numbers of carbon atoms, crowding of atoms makes the structures too unstable to exist.

Table D.2	
Number of Possible Isomers for Selected Alkanes	
Molecular Formula	**Number of Possible Isomers**
C_4H_{10}	2
C_5H_{12}	3
C_6H_{14}	5
C_7H_{16}	9
C_8H_{18}	18
C_9H_{20}	35
$C_{10}H_{22}$	75
$C_{15}H_{32}$	4,347
$C_{20}H_{42}$	336,319
$C_{30}H_{62}$	4,111,846,763

Give the structural and condensed formulas of the isomers of pentane.

Solution

Pentane has five carbon atoms. The carbon skeletons of the three possible isomers are as follows:

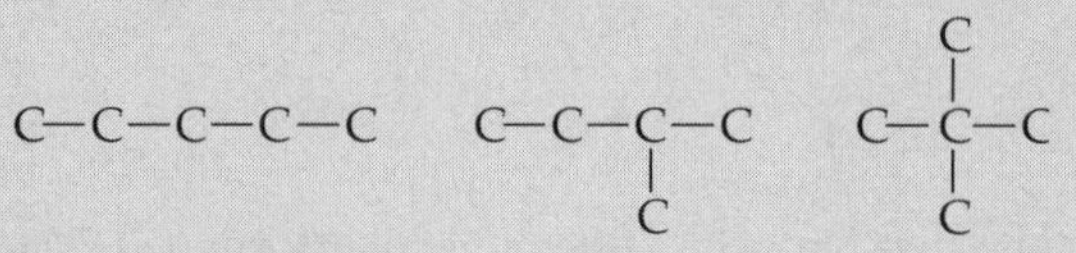

```
                                      C
                                      |
C—C—C—C—C    C—C—C—C    C—C—C
                   |           |
                   C           C
```

Attach hydrogen atoms so that each carbon atom forms four bonds. Each isomer has the same condensed formula, C_5H_{12}.

Give the structural formulas of the possible isomers of hexane (C_6H_{14}).

Solution

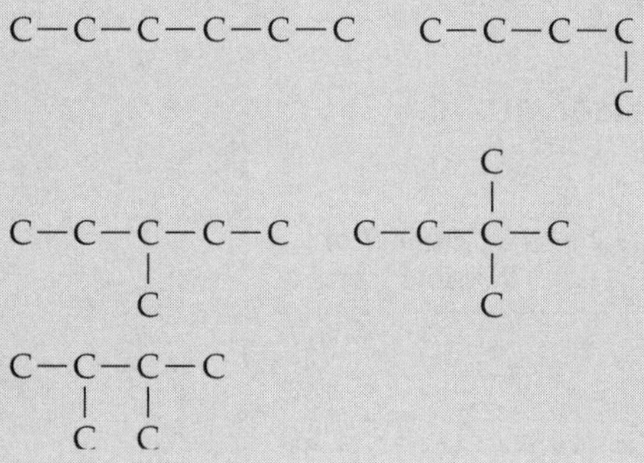

```
C—C—C—C—C—C    C—C—C—C
                         |
                         C

                              C
                              |
C—C—C—C—C    C—C—C—C
      |                   |
      C                   C

C—C—C—C
  |  |
  C  C
```

Nomenclature of Alkanes

Organic compounds are named according to rules established by the International Union of Pure and Applied Chemistry (IUPAC). Table D.1 shows that except for the first four members of the family, the first part of the name of an alkane is derived from the Greek name for the number of carbon atoms in the molecule. The suffix *-ane* means that the compound is an alkane.

Before we can follow the IUPAC rules for naming branched-chain alkanes, we must consider the names of the groups that are formed when one hydrogen atom is removed from the formula of an alkane. For example, when an H atom is removed from methane (CH_4) a $-CH_3$ or **methyl** group is formed. Similarly, removal of an H atom from ethane (C_2H_6) gives $-C_2H_5$ or an **ethyl** group. Because the groups are derived from alkanes, they are called **alkyl** groups. Examples of some common alkyl groups are given in Table D.3.

Branched-chain alkanes are named by applying the following rules:

1. Determine the name of the parent compound by finding the longest continuous chain of carbon atoms. Consider the following example:

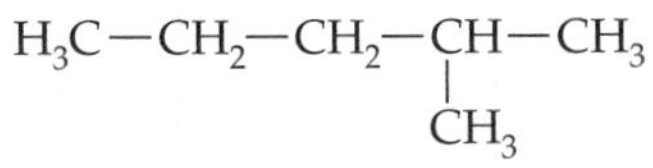

$$H_3C-CH_2-CH_2-\underset{\underset{CH_3}{|}}{CH}-CH_3$$

The longest chain contains five carbon atoms; therefore, the parent name is pentane.

Sometimes, because of the way in which a formula is written, it is not easy to recognize the longest chain. For example, the alkane shown below is not a hexane (six carbon atoms in the longest chain) as at first might be supposed but is a heptane (seven carbon atoms in the longest chain).

$$H_3C-CH_2-CH_2-CH_2-\underset{\underset{\underset{CH_3}{|}}{\underset{CH_2}{|}}}{CH}-CH_3$$

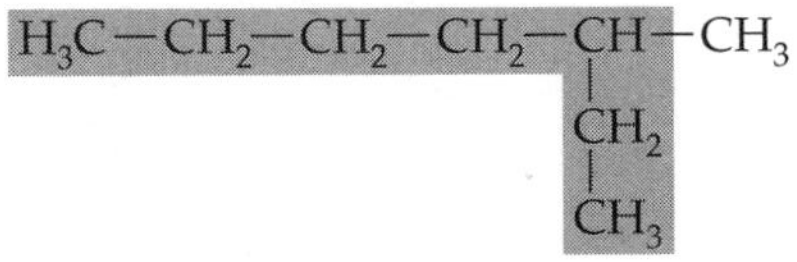

Table D.3	
Some Common Alkyl Groups	
Name	**Group**
Methyl	$-CH_3$
Ethyl	$-CH_2-CH_3$
n-Propyl	$-CH_2-CH_2-CH_3$
n-Butyl	$-CH_2-CH_2-CH_2-CH_3$
Isopropyl	$-\underset{\underset{CH_3}{|}}{\overset{\overset{CH_3}{|}}{C}}-H$

2. Number the longest chain beginning with the end closest to the branch. Use these numbers to designate the location of the groups (or substituents) at the branch.

Applying this rule to the previous examples, the carbon atoms are numbered and named as follows:

a. 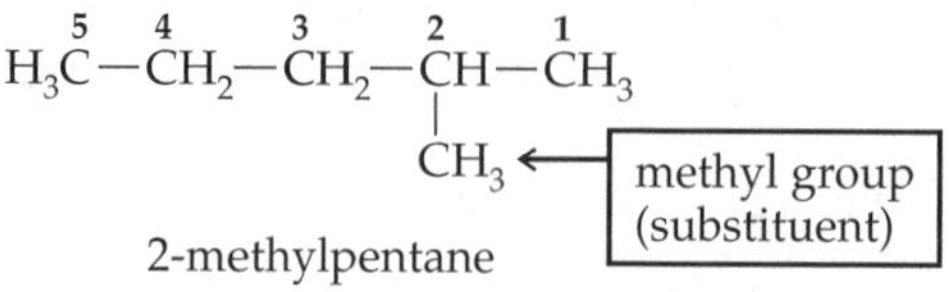 $\overset{5}{H_3C}-\overset{4}{CH_2}-\overset{3}{CH_2}-\overset{2}{CH}-\overset{1}{CH_3}$
          |
         CH_3 ← [methyl group (substituent)]

    2-methylpentane

b. 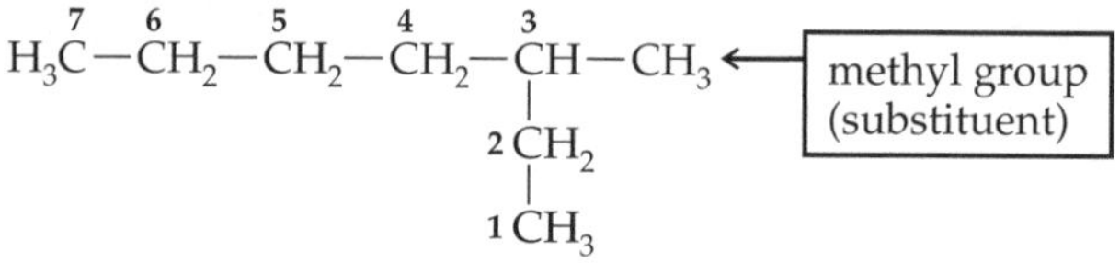 $\overset{7}{H_3C}-\overset{6}{CH_2}-\overset{5}{CH_2}-\overset{4}{CH_2}-\overset{3}{CH}-CH_3$ ← [methyl group (substituent)]
                              |
                            2 CH_2
                              |
                            1 CH_3

    3-methylheptane

In the names of compounds, the numbers are separated from words by a hyphen, and the parent name is placed last. In compound *a*, a methyl group is attached to carbon number 2; in compound *b*, a methyl group is attached to carbon number 3. Thus the names of the compounds are 2-methylpentane in *a* and 3-methylheptane in *b*.

EXAMPLE **D.5**

Name the following compound:

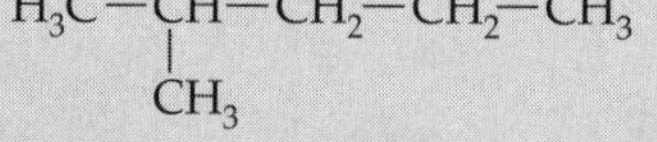

$H_3C-CH-CH_2-CH_2-CH_3$
        |
       CH_3

Solution
1. Determine the name of the parent compound by finding the longest continuous chain of carbon atoms. The longest chain is five carbons. Therefore, the parent compound is a pentane.
2. Number the longest chain beginning with the end closest to a branch. Use these numbers to designate the location of the groups (or substituents) at the branch. The methyl is on the 2 position.
3. The compound is 2-methylpentane.

Name the following:

$$H_3C-CH_2-CH_2-CH_2-\underset{\underset{CH_3}{|}}{CH}-CH_3$$

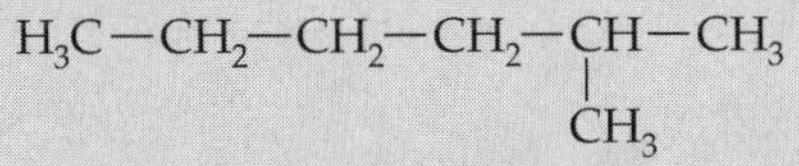

Solution
2-methylhexane

3. When two or more substituents are present on the same carbon atom, use the number of that carbon atom twice. List the substituents alphabetically. Thus, the compound shown here is 3-ethyl-3-methylhexane:

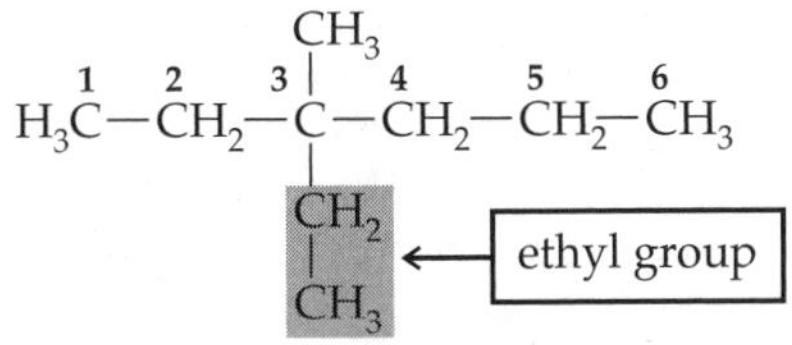

3-ethyl-3-methylhexane

4. When two or more substituents are identical, indicate this by the use of the prefixes di-, tri-, tetra-, and so on. Commas are used to separate numbers from each other. Thus, the following compounds are 2,2,4-trimethylhexane and 3-ethyl-2,4-dimethyloctane:

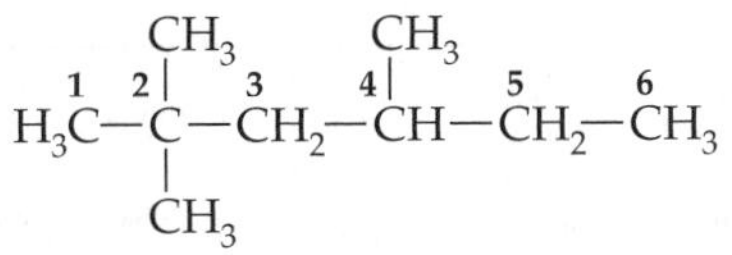

2,2,4-trimethylhexane

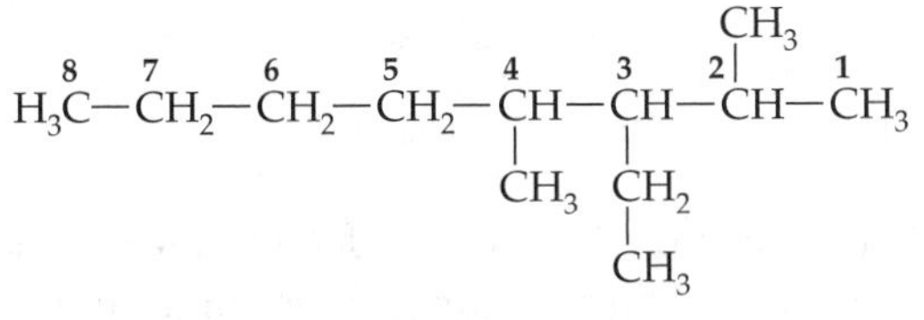

3-ethyl-2,4-dimethyloctane

Name the following compound:

$$H_3C-CH-CH_2-CH-CH_2-CH-CH_3$$
$$\quad\;\;\; | \qquad\quad | \qquad\quad |$$
$$\quad\;\;\; CH_3 \qquad CH_3 \qquad CH_3$$

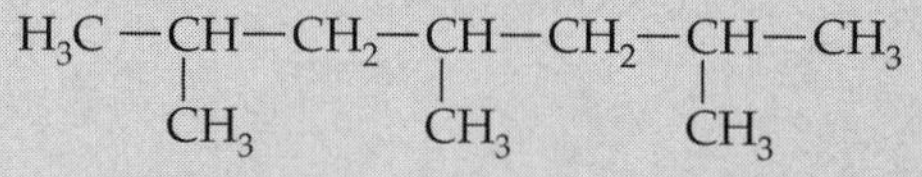

Solution
1. Determine the name of the parent compound by finding the longest continuous chain of carbon atoms. The longest chain is seven carbon atoms; therefore, the parent compound is a heptane.
2. Number the longest chain beginning with the end closest to a branch. Use these numbers to designate the location of the groups (or substituents) at the branch. There are methyl groups at positions 2, 4, and 6. Trimethyl indicates three methyl groups.
3. The name of the compound is 2,4,6-trimethylheptane.

Name the following compound:

$$\qquad\quad CH_3 \qquad\qquad CH_3$$
$$\qquad\quad | \qquad\qquad\quad\; |$$
$$H_3C-C-CH_2-CH_2-C-CH_3$$
$$\qquad\quad | \qquad\qquad\quad\; |$$
$$\qquad\quad CH_3 \qquad\qquad CH_3$$

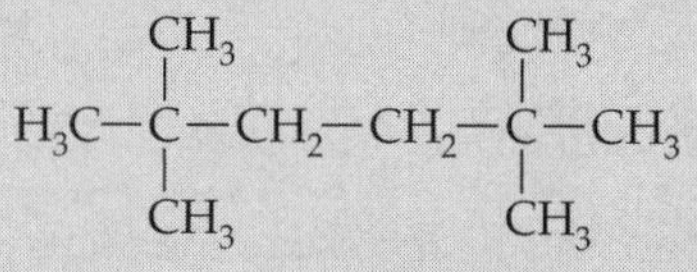

Solution
2,2,5,5-tetra methyl hexane

Reactions of Alkanes

Alkanes are relatively unreactive. Like all hydrocarbons, alkanes undergo **combustion**, but for this to occur, a flame or spark is required to initiate bond cleavage. When burned in air, alkanes form carbon dioxide and water as shown below for methane and butane.

$$CH_4 \;+\; 2\,O_2 \longrightarrow CO_2 \;+\; 2\,H_2O \;+\; heat$$
$$2\,C_4H_{10} \;+\; 13\,O_2 \longrightarrow 8\,CO_2 \;+\; 10\,H_2O \;+\; heat$$

Once initiated, the reactions are exothermic and produce a considerable amount of heat. They are the source of the energy that we obtain when natural gas, gasoline, and fuel oil are burned.

Alkenes

Alkenes are hydrocarbons that have one or more carbon–carbon double bonds ($-C=C-$). The two simplest are **ethene** (C_2H_4), commonly called ethylene, and **propene** (C_3H_6), commonly called propylene. In forming the double bond, the two carbon atoms share two pairs of electrons to acquire the stable octet configuration.

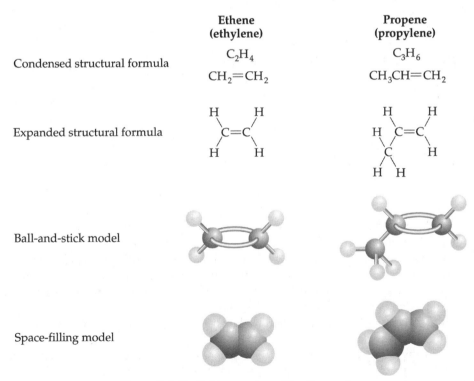

Different ways of representing the structure of alkenes are shown in Figure D.6. Alkenes are identified by the suffix *-ene*. Like alkanes, they form a homologous series. The general formula is C_nH_{2n}.

	Ethene (ethylene)	**Propene (propylene)**
Condensed structural formula	C_2H_4 $CH_2=CH_2$	C_3H_6 $CH_3CH=CH_2$
Expanded structural formula		
Ball-and-stick model		
Space-filling model		

Figure D.6 The first two members of the alkene series.

Alkenes are called **unsaturated hydrocarbons** because they do not contain the maximum possible number of hydrogen atoms. In the presence of a catalyst such as platinum (Pt), they react with hydrogen to form the saturated parent alkane.

$$H_2C = CH_2 \quad + \quad H_2 \quad \xrightarrow{\text{Pt}} \quad H_3C - CH_3$$

$$\text{ethene} \qquad\qquad\qquad\qquad \text{ethane}$$

The third member of the alkene series is **butene** (C_4H_8). There are two possible positions for the double bond, and two isomers exist:

$$\overset{1}{H_2C} = \overset{2}{CH} - \overset{3}{CH_2} - \overset{4}{CH_3} \qquad \overset{1}{H_3C} - \overset{2}{CH} = \overset{3}{CH} - \overset{4}{CH_3}$$

$$\text{1-butene} \qquad\qquad\qquad \text{2-butene}$$

The first rule used for naming alkanes is modified: The longest hydrocarbon chain is numbered from the end that will give the carbon–carbon double bond ($-C=C-$) the lowest number. Substituents are then numbered and named as for alkanes except that the ending *-ene* is used instead of *-ane*. Thus, the compounds shown here are 7-chloro-3-heptene and 4-ethyl-5-methyl-2-hexene.

$$\overset{1}{H_3C} - \overset{2}{CH_2} - \overset{3}{CH} = \overset{4}{CH} - \overset{5}{CH_2} - \overset{6}{CH_2} - \overset{7}{CH_2} - Cl$$

$$\text{7-chloro-3-heptene}$$

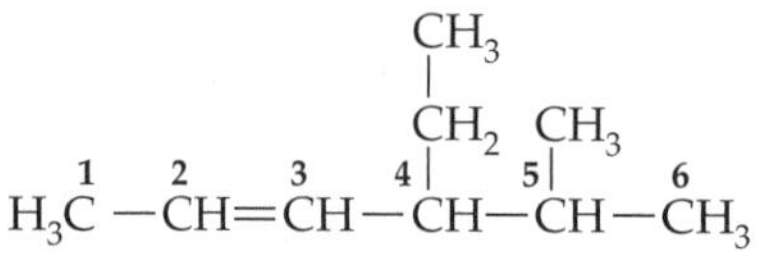

$$
\begin{array}{c}
CH_3 \\
| \\
CH_2 \quad CH_3 \\
\overset{1}{H_3C} - \overset{2}{CH} = \overset{3}{CH} - \overset{4}{\underset{|}{CH}} - \overset{5}{\underset{|}{CH}} - \overset{6}{CH_3}
\end{array}
$$

$$\text{4-ethyl-5-methyl-2-hexene}$$

Ethene (which is often called ethylene) is the most important raw material in the organic chemical industry. It is produced in enormous quantity by the "cracking" of petroleum and is used primarily to make the plastic polyethylene. It is also the starting material for the manufacture of ethylene glycol (the major component of antifreeze in automobile radiators), other plastics, and many chemicals. Propene (which is often called propylene) is important in the production of polypropylene and other plastics.

Ethene has a natural source: It is produced by fruits and causes the change in skin color that occurs as fruits ripen (e.g., when tomatoes change from green to red). To avoid damage in transit, many producers pick fruit when it is green, ship it to its destination, and then treat it with ethene gas to produce the proper ripe color. The flavor, however, is generally considered inferior to naturally ripened fruit.

Name the following compound:

$$H_2C=CH-CH_2-\underset{\underset{\displaystyle CH_3}{|}}{CH}-CH_3$$

Solution
1. The longest chain has five carbon atoms. The compound is a pentene.
2. The double bond is between carbon 1 and carbon 2.
3. The methyl group is on carbon 4.
4. The name of the compound is 4-methyl-1-pentene.

Name the following compound:

$$H_2C=CH-CH_2-CH_2-CH_2-CH_2-CH_2-CH_3$$

Solution
1-octene

Cis–Trans Isomerism

Groups joined by a single carbon–carbon bond can rotate freely about the bond with respect to each other at ordinary temperatures. For example, the two methyl groups in ethane can rotate freely about the single C—C bond. Groups joined by a C=C double bond, however, are restricted and cannot rotate. The two methyl groups in ethene are locked so that they lie in the same plane.

If one of the hydrogen atoms attached to each of the two carbon atoms in ethene is replaced by a chlorine atom, two distinct compounds are formed:

cis-1,2-dichloroethene   trans-1,2-dichloroethene

The two compounds have the same name—dichloroethene—but are distinguished from each other by the prefixes *cis* and *trans*.

Although they have the same formula, they are not identical (i.e., they cannot be superimposed one on the other so that all atoms and bonds coincide), and their physical properties are different.

	cis-1,2,-Dichloroethene	*trans*-1,2-Dichloroethene
Boiling point	60.3°C	47.5°C
Melting point	−80.5°C	−50°C

The two compounds are not structural isomers because the atoms attached to each atom in the two compounds are the same in both cases. For example, in both isomers, each carbon atom is joined to another carbon atom, a chlorine atom, and a hydrogen atom. Only the spatial arrangement is different. This type of isomerism is called **cis–trans isomerism.**

If two identical groups are attached to the same carbon atom in the double bond, cis–trans isomerism is not possible. For example, 1,1-dichloroethene is the structural isomer of the cis and trans isomers shown here but exists in just one form and cannot form cis and trans isomers.

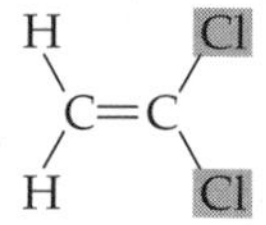

1,1-dichloroethene

Alkynes

Alkynes are hydrocarbons that have one or more triple bonds ($-C\equiv C-$). The simplest alkyne is **ethyne** (C_2H_2), better known by its common name of **acetylene.** The next member of the series is **propyne** (C_3H_4). In forming the triple bond, the two carbon atoms share three pairs of electrons to acquire the stable octet configuration:

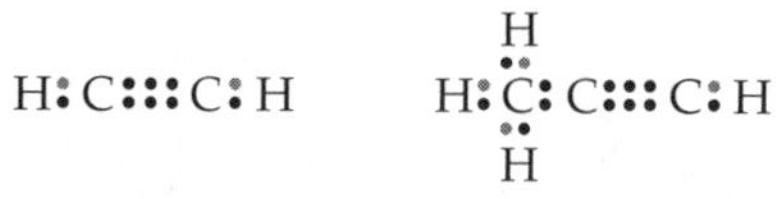

Different ways of representing ethyne are shown in Figure D.7.

Acetylene has many industrial uses. When burned in oxyacetylene torches, it produces a very hot flame (approximately 3000°C, or 5400°F) that is used for cutting and welding metals. It is also an important starting material for the production of plastics and synthetic rubber.

Saturated Hydrocarbon Rings (Cycloalkanes)

Carbon atoms can join in rings as well as in straight chains and branched chains. Alkanes joined in rings are called **cycloalkanes.** The structures of the first four are shown in Figure D.8. In the

**Ethyne
(acetylene)**

Condensed structural formula C_2H_2 or $HC{\equiv}CH$

Expanded structural formula $H{-}C{\equiv}C{-}H$

Ball-and-stick model

Space-filling model

Figure D.7 Different ways of representing the alkyne ethyne (acetylene).

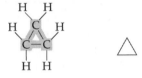

cyclopropane

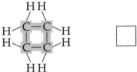

cyclobutane

cyclopentane

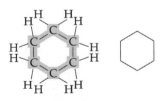

cyclohexane

Figure D.8 Expanded and simplified structural formulas of cycloalkanes.

simplified depictions of the formulas, each corner represents a carbon atom that is attached to two hydrogen atoms, and the lines represent $C{-}C$ single bonds.

Stability increases from cyclopropane to cyclohexane as the bond angles increase. Cyclohexane is important as the parent compound of glucose and many other sugars and carbohydrates that are vital for animal metabolism.

Hydrocarbons can also form rings in which some of the carbon–carbon bonds are double bonds, as shown in the following six-carbon ring compounds.

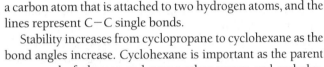

cyclohexene   1,3-cyclohexadiene

■ Aromatic Hydrocarbons

By far the most important unsaturated ring hydrocarbon compound is **benzene** (C_6H_6). Compounds derived from benzene are called **aromatic compounds** because many of the first ones described had distinctive pleasant scents (or aromas).

The Structure of Benzene

Michael Faraday discovered benzene in 1826, but its structure remained a mystery for 40 years. The molecular formula, C_6H_6, indicated a very unsaturated hydrocarbon, but benzene did not behave like the unsaturated alkenes or alkynes. Then, in 1865, the German chemist August Kekule proposed a ring of six carbon atoms joined by alternating single and double bonds, with one hydrogen atom bonded to each carbon:

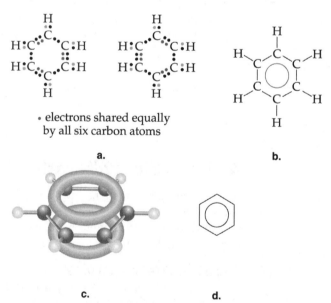

structural
formula

or

condensed
structural
formula

Although these structures explain much of the chemistry of benzene, they cannot accurately explain all of the established experimental facts. We now know that the bonds joining the carbon atoms in the ring are neither single nor double bonds. They are shorter than normal single bonds but are longer than normal double bonds, and all are identical. The six electrons associated with three double bonds are shared equally among all of the carbon atoms in the ring (depicted in Figure D.9a or more simply in Figure D.9b). The shared electrons are said to be **delocalized**. They can be visualized as moving above and below the flat plane of the

• electrons shared equally
by all six carbon atoms

a.

b.

c.

d.

Figure D.9 The benzene molecule: (a) The six carbon–carbon bonds in benzene are identical. The six electrons associated with three double bonds are shared equally among the six carbon atoms of the ring. (b) The circle in the center of the ring represents the six shared electrons. (c) The six delocalized electrons can be represented as occupying a space above and below the ring of carbon atoms. (d) Benzene is usually written in this simplified form.

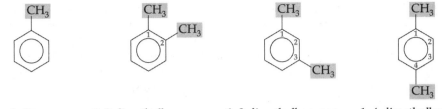

methylbenzene (toluene) 1, 2-dimethylbenzene (*ortho*-xylene) 1, 3-dimethylbenzene (*meta*-xylene) 1, 4-dimethylbenzene (*para*-xylene)

Figure D.10 Replacement of one hydrogen atom of benzene with a methyl group yields toluene. Replacement of two hydrogen atoms with methyl groups gives one of the three xylenes.

carbon ring (Figure D.9c). This arrangement leads to increased stability, and aromatic compounds are more stable than compounds with regular carbon–carbon double bonds. The formula for benzene can be written in condensed form without the carbon and hydrogen atoms and with a circle in the center of the ring to represent the **delocalized electrons** (Figure D.9d).

Alkyl Derivatives of Benzene

Replacement of hydrogen atoms in benzene with alkyl groups gives rise to numerous other hydrocarbons. The simplest member of the family, methyl benzene (common name, toluene), is formed when one hydrogen atom is replaced with a methyl ($-CH_3$) group. If two hydrogens are replaced by methyl groups, three different dimethylbenzenes (commonly called xylenes) can be formed (Figure D.10).

The location of a second substituent relative to the first is indicated by numbering the ring carbons. If two substituents are present, as in the xylenes, the prefixes **ortho** (o-), **meta** (m-), and **para** (p-) are often used instead of numbers. For more than two substituents, numbers must be used. The numbers are arranged to give the lowest possible numbers to the substituents.

Benzene, toluene, and the xylenes are used extensively as raw materials for the manufacture of plastics, pesticides, drugs, and hundreds of other organic chemicals. The use of benzene as a laboratory solvent was discontinued in the 1970s after it was discovered that inhalation of its fumes lowered the white blood cell count in humans and caused leukemia in laboratory rats.

Polycyclic Aromatic Compounds

Benzene rings can fuse to form **polycyclic aromatic compounds** (Figure D.11). The simplest is napthalene, which is used as a moth repellent. Additional benzene rings can be added to form a variety of fused-ring compounds, including benzo(a)pyrene, a known carcinogen that is present in tobacco smoke.

naphthalene ($C_{10}H_8$)

■ Functional Groups

The majority of organic compounds are derivatives of hydrocarbons, which are obtained by replacing one or more hydrogen atoms in the parent hydrocarbon molecule with a different atom or group of atoms. These atoms or groups of atoms, called func-

benzo(a)pyrene ($C_{20}H_{12}$)

Figure D.11 Two polycyclic aromatic hydrocarbons.

tional groups, form the basis for classifying organic compounds. Most chemical reactions involving an organic molecule take place at the site of the functional group. The functional group, as the name implies, determines how the molecule functions as a whole. Molecules containing the same functional groups exhibit similar chemical behavior. The major classes of organic compounds that result from replacing hydrogen atoms in hydrocarbons with functional groups are shown in Table D.4, together with examples from each class.

Table D.4

Classes of Organic Compounds Based on Functional Groups

General Formula of Class*	Name of Class	Example	Name of Compound
$R - X$	Halide	$CH_3 - Cl$	Chloromethane (methyl chloride)
$R - OH$	Alcohol	$C_2H_5 - OH$	Ethanol (ethyl alcohol)
$R - O - R'$	Ether	$C_2H_5 - O - C_2H_5$	Diethyl ether (ethyl ether)
$R - \overset{\overset{O}{\|\|}}{C} - H$	Aldehyde	$CH_3 - \overset{\overset{O}{\|\|}}{C} - H$	Ethanal (acetaldehyde)
$R - \overset{\overset{O}{\|\|}}{C} - R'$	Ketone	$CH_3 - \overset{\overset{O}{\|\|}}{C} - CH_3$	Propanone (acetone)
$R - \overset{\overset{O}{\|\|}}{C} - OH$	Carboxylic acid	$CH_3 - \overset{\overset{O}{\|\|}}{C} - OH$	Ethanoic acid (acetic acid)
$R - O - \overset{\overset{O}{\|\|}}{C} - R'$	Ester	$C_2H_5 - O - \overset{\overset{O}{\|\|}}{C} - CH_3$	Ethyl ethanoate (ethyl acetate)
$R - N \overset{H}{\underset{H}{<}}$	Primary amine	$C_2H_5 - N \overset{H}{\underset{H}{<}}$	Ethylamine
$R - \overset{\overset{O}{\|\|}}{C} - N \overset{H}{\underset{H}{<}}$	Simple amide	$CH_3 - \overset{\overset{O}{\|\|}}{C} - N \overset{H}{\underset{H}{<}}$	Acetamide

*R represents an H atom or a carbon-containing group, often an alkyl group such as $-CH_3$ or $-C_2H_5$. R' may be the same as R or different.

Organic Halides

Organic halides are hydrocarbon derivatives in which hydrogen atoms in the parent hydrocarbon have been replaced by one or more halogen atoms: fluorine, chlorine, bromine, or iodine. Halogens, which are in Group VIIA of the periodic table, form a stable covalent bond by sharing one of their seven valence electrons, as shown here for chloroethane:

$$H:\overset{\overset{\displaystyle H}{..}}{\underset{\underset{\displaystyle H}{..}}{C}}:\overset{\overset{\displaystyle H}{..}}{\underset{\underset{\displaystyle H}{..}}{C}}:\ddot{\underset{..}{Cl}}: \quad \text{or} \quad H-\overset{\overset{\displaystyle H}{|}}{\underset{\underset{\displaystyle H}{|}}{C}}-\overset{\overset{\displaystyle H}{|}}{\underset{\underset{\displaystyle H}{|}}{C}}-Cl \quad \text{or} \quad C_2H_5Cl$$

chloroethane

According to the IUPAC system, organic halides are named by writing the name of the parent hydrocarbon preceded by the appropriate designation for the halogen—fluoro-, chloro-, bromo-, or iodo-. In many cases, common names are often used. For example, trichloromethane is usually called chloroform (Figure D.12).

Organic halides are very versatile compounds and have been used in numerous ways. Examples are shown in Figure D.12. In recent years, some of these halides have been shown to be health hazards, and their use has been discontinued. For years, chloroform ($CHCl_3$) and carbon tetrachloride (CCl_4), both excellent solvents for grease and oils, were used in the dry-cleaning industry and the laboratory. Because they were shown to be toxic and carcinogenic, they have been replaced by safer solvents such as perchloroethylene. Chlorofluorocarbons—substituted alkanes such as Freon-12, which contains both fluorine and chlorine—were widely used as coolants and aerosol propellants before their release into the atmosphere was discovered as a factor in the depletion of the ozone layer. Other organic halides that have caused environmental problems include dichlorodiphenyltrichloroethane and the polychlorinated biphenyls. The two plastics polyvinyl chloride and Teflon (the nonstick coating that is on cookware) are made from the alkyl halide derivatives of chloroethene and tetrafluoroethene.

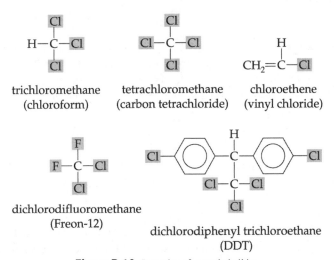

trichloromethane   tetrachloromethane   chloroethene
(chloroform)   (carbon tetrachloride)   (vinyl chloride)

dichlorodifluoromethane
(Freon-12)

dichlorodiphenyl trichloroethane
(DDT)

Figure D.12 Examples of organic halides.

EXAMPLE D.11

Give the IUPAC name of the following alkyl halide.

CH_3–CH_2–CH_2–Cl

Solution
1. The parent hydrocarbon has three carbon atoms. The compound is a propane.
2. The chlorine atom is on a terminal carbon atom.
3. Carbon atoms are numbered so that the carbon attached to the substituent (Cl) is given the lowest possible number.
4. The compound is 1-chloropropane.

EXAMPLE D.12

Write the formula of bromomethane (methyl bromide) (CH_3Br).

Solution

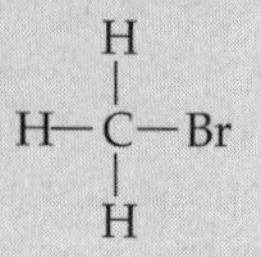

$$\begin{array}{c} H \\ | \\ H-C-Br \\ | \\ H \end{array}$$

Alcohols

Alcohols are hydrocarbon derivatives in which hydrogen atoms in the parent hydrocarbon have been replaced by one or more hydroxyl groups (−OH) (Figure D.13). Oxygen has six valence electrons, and in forming an alcohol, it acquires a stable octet of electrons by forming one covalent bond with a carbon atom and one with a hydrogen atom:

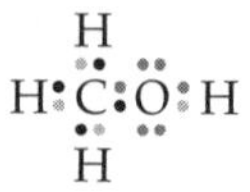

$$\begin{array}{c} H \\ \ddot{} \\ H\!:\!\ddot{C}\!:\!\ddot{O}\!:\!H \\ \ddot{} \\ H \end{array}$$

methanol
(methyl alcohol)

The IUPAC name of an alcohol is obtained from the parent alkane by replacing the final -*e* with the suffix -*ol*. Thus, methane becomes **methanol**. Ethane becomes **ethanol**, and propane becomes propanol.

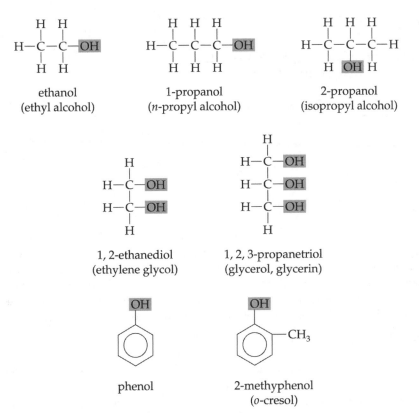

ethanol
(ethyl alcohol)

1-propanol
(*n*-propyl alcohol)

2-propanol
(isopropyl alcohol)

1, 2-ethanediol
(ethylene glycol)

1, 2, 3-propanetriol
(glycerol, glycerin)

phenol

2-methyphenol
(*o*-cresol)

Figure D.13 Examples of common alcohols.

Alcohols, like water, contain a polar −OH group, and as a result, hydrogen bonding can occur between hydrogen and oxygen atoms in adjacent molecules, as shown for methanol in Figure D.14. When alcohols are converted from the liquid state to the gaseous state, energy is needed to break the hydrogen bonds. As a result, the boiling points of alcohols are higher than are the boiling points of the alkanes from which they are derived (Table D.5).

Alcohols with low molecular weights such as methanol and ethanol are very soluble in water because hydrogen bonding occurs between water and alcohol molecules. Alcohols with higher molecular weights are less soluble because the increasing size of the nonpolar part of the molecule disrupts the hydrogen bonding network.

Methanol is known as **wood alcohol** because for many years it was produced by heating hardwoods such as maple, birch, and hickory to high temperatures in the absence of air. The present methods of production from coal and coal gas were examined in Chapter 12. Methanol has many industrial uses. It is used in the manufacture of plastics and other polymers, as an

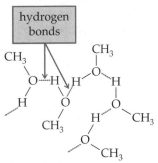

Figure D.14 Hydrogen bonding between methanol molecules.

Table D.5

Comparison of Boiling Points of Alcohols and Their Corresponding Alkanes

Alcohol	Boiling Point (°C)	Alkane	Boiling Point (°C)
Methanol	65.0	Methane	−162
Ethanol	78.5	Ethane	−88.5
1-Propanol	97.4	Propane	−42
2-Propanol	82.4		

antifreeze in windshield washer fluid, and as a gasoline additive. Methanol is extremely toxic. Drinking quite small quantities can result in temporary or permanent blindness or even death.

Ethanol, ethyl alcohol (Figure D.13), is the active ingredient in alcoholic beverages. It is produced by the **fermentation** of carbohydrates (sugars and starches) in fruits and other plant sources using a process that was discovered thousands of years ago. During fermentation, microorganisms such as yeasts convert sugar to alcohol and carbon dioxide:

$$C_6H_{12}O_6 \rightarrow 2\ C_2H_5OH + 2\ CO_2$$

Ethanol is called **grain alcohol** because it can be produced from grains such as barley, wheat, corn, and rice. Beer is usually made from barley with the addition of hops (Chapter 14). Wine is made from grapes and other fruits.

The percentage of alcohol in the fermentation product is determined by the enzymes present. At alcohol concentrations that are more than approximately 15%, the enzymes begin to be destroyed, and the reaction can proceed no further. The alcohol content can be increased by distillation. For example, distilling fermented grains makes whiskey. The proof of an alcoholic beverage is equal to twice the percentage (by volume) of the ethanol. Thus, 90-proof rum is 45% ethyl alcohol.

Ethanol is not as toxic as methanol, but 1 pint of pure ethanol, if ingested rapidly, is likely to prove fatal for most people. Ethanol is quickly absorbed into the bloodstream; excessive use causes damage to the liver and the neurologic system and leads to physiologic addiction.

In addition to being a constituent of alcoholic beverages, ethanol has many industrial uses. It is used in the preparation of many organic chemicals and is an important industrial solvent. The pharmaceutical industry uses it as an ingredient in cough syrups and other medications and as an antiseptic for skin disinfection. Ethanol that is used in industry is produced by the high-pressure hydration of ethylene and not by fermentation:

ethylene   water   ethanol
(ethyl alcohol)

Alcoholic beverages are heavily taxed, but alcohol intended for industrial use is not. To avoid the tax, manufacturers must **denature** the alcohol; that is, render it unfit for human consumption by the addition of a small amount of a toxic substance. Methanol or benzene is often added because these substances do not alter the solvent properties of the alcohol.

Propanol, the next member of the series following ethanol, can exist in two forms: 1-propanol (*n*-propyl alcohol) and 2-propanol (isopropyl alcohol) (Figure D.13). Drugstore rubbing alcohol is a 70% solution of 2-propanol and is an effective disinfectant.

Important alcohols with more than one hydroxyl group in the molecule include **ethylene glycol** (1,2-ethanediol) and **glycerol** (glycerin or 1,2,3-propanetriol) (Figure D.13). Ethylene glycol, with its two polar $-OH$ groups, is very soluble in water and can be mixed with it in any desired proportion. Because of hydrogen bonding, it has a relatively high boiling point. These properties make it an ideal choice for use as the main ingredient in permanent antifreeze. Glycerol is a nontoxic, clear, syrupy liquid. With its three $-OH$ groups, it has an affinity for moisture. For this reason, it is often added to personal hygiene products such as hand lotions, shaving creams, and soaps to help keep skin moist and soft. Glycerol is also used as a lubricant and in the manufacture of explosives and a variety of other chemicals.

The simplest aromatic alcohol is **phenol** (Figure D.13), which was the first widely used disinfectant. Joseph Lister, a British surgeon, introduced it in 1867; however, because it could burn skin if not in a very dilute solution, it was soon replaced by safer related compounds. A methyl derivative of phenol, o-cresol, is the main ingredient in the wood preservative creosote.

Ethers

Compounds that have two hydrocarbon groups attached to a single oxygen atom are called **ethers** (Table D.4). They have the general formula $R-O-R'$, where R and R' may or may not be the same (Figure D.15). The best known ether is diethyl ether, which is commonly called ether.

$$CH_3CH_2-\boxed{O}-CH_2CH_3$$

diethyl ether
(ether)

Ether acts as a general anesthetic; its introduction for this purpose in the mid-1840s revolutionized medicine. Surgical procedures that previously could not be performed because of the pain and shock involved became possible. General anesthetics cause unconsciousness and insensitivity to pain by temporarily depressing the central nervous system. Ether is not an ideal anesthetic; it is highly flammable, and inhalation produces some unpleasant side effects, such as nausea and irritation of the respiratory tract. It has now been replaced by other more efficient anesthetics, such as fluorinated ethers in which hydrogen has been replaced by fluorine.

A commercially important ether is methyl tertiary-butyl ether (Figure D.15), which since the phase out of the antiknock agent tetraethyl lead, has become the agent most frequently added to gasoline to increase octane rating. Because it has been found in groundwater, California has banned MBTE.

$$CH_3-CH_2-\boxed{O}-CH_2-CH_3$$

ether
(diethyl ether)

$$CH_3-\boxed{O}-CH_2-CH_2-CH_3$$

methyl propyl ether

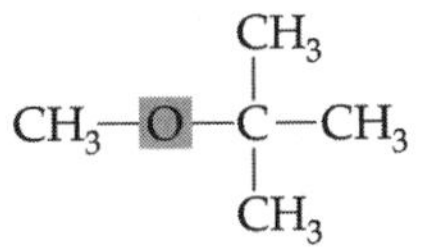

$$CH_3-\boxed{O}-\overset{\displaystyle CH_3}{\underset{\displaystyle CH_3}{\overset{|}{\underset{|}{C}}}}-CH_3$$

MTBE
(methyl *tertiary*-butyl ether)

Figure D.15 Examples of ethers.

EXAMPLE D.13

Indicate the functional group in each of the following. Identify the compounds as an ether, alcohol, or phenol.

a. 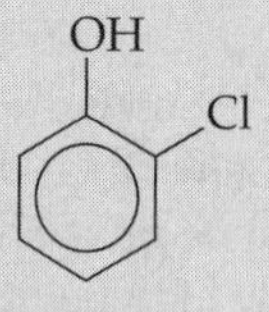 [benzene ring with OH and Cl substituents]   b. CH_3-O-CH_3 c. $CH_3\overset{OH}{CHCH_3}$

Solution

a. The OH functional group is attached to a benzene ring. This compound is a phenol. It also has a chlorine. It is 2-chlorophenol (or orthochlorophenol).
b. The O functional group is between two carbon atoms. This compound is an ether. It is dimethyl ether.
c. The OH functional group is attached to a carbon atom. This compound is an alcohol. The OH is attached to carbon 2. The compound is 2-propanol.

EXAMPLE D.14

Draw the structure of 1-propanol.

Solution

$$HO-CH_3-CH_2-CH_2$$

Aldehydes and Ketones

Aldehydes and ketones are related families of compounds (Table D.4). Both contain a **carbonyl group** in which a carbon atom is joined by a double bond to an oxygen atom. In aldehydes, the carbonyl carbon is attached to at least one hydrogen atom; in ketones, it is attached to two carbon atoms:

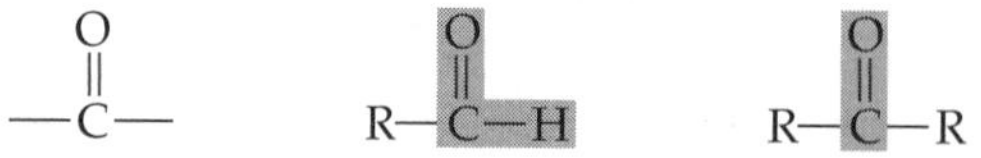

$$\overset{O}{\underset{\|}{-C-}}\qquad R-\overset{O}{\underset{\|}{C}}-H\qquad R-\overset{O}{\underset{\|}{C}}-R$$

	or	or
carbonyl group	RCHO	RCOR
	aldehyde	ketone

The simplest aldehyde is **formaldehyde** (Figure D.16). As a 40% solution in water called formalin, formaldehyde is used as a preservative for biological specimens and as an embalming fluid. It is also a starting material for the manufacture of certain polymers. Formaldehyde has a disagreeable odor and is a suspected carcinogen.

In the IUPAC naming system, the suffixes -*al* and -*one* are used to identify aldehydes and ketones, but many are better known by their common names. Aldehydes and ketones are made by oxidizing the appropriate alcohols, as shown later here for the preparation of propenal and propanone (**acetone**).

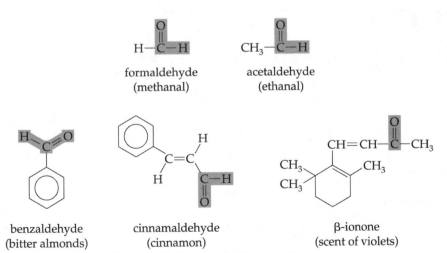

Figure D.16 Examples of aldehydes and ketones.

Acetone, the simplest ketone, is an excellent solvent. It is miscible (soluble in all propor-tions) with both water and nonpolar liquids and is used to remove varnish, paint, and fingernail polish. Many aromatic aldehydes and ketones have pleasant aromas (Figure D.16), and some, including almond and cinnamon, are used as food flavorings. The per-fumes of many flowers—violets, for example—are due to the release of small amounts of ketones.

EXAMPLE D.15

Identify each of the following as an aldehyde or ketone.

a. 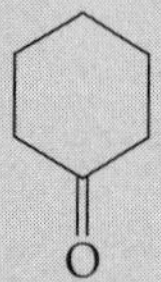 [structure] or [structure]   b. $CH_3-CH_2-\overset{\overset{\text{O}}{\|}}{C}-CH_3$ c. 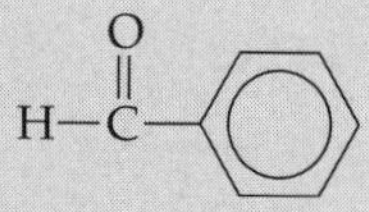 $H-\overset{\overset{\text{O}}{\|}}{C}-$[benzene ring]

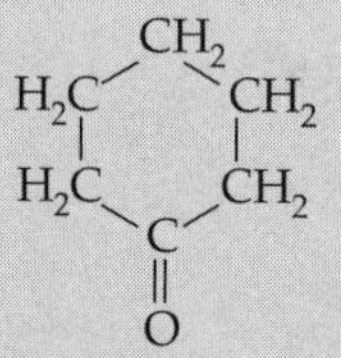

Solution
a. The carbon of the C=O group is attached to two other carbon atoms. This compound is a ketone, cyclohexanone.
b. The C=O group is attached between two carbon atoms. It is a ketone, 2-butanone.
c. A hydrogen atom is attached to the C=O group. The compound is an aldehyde, benzaldehyde.

EXAMPLE D.16

Identify the following as ketones or aldehydes:

a. 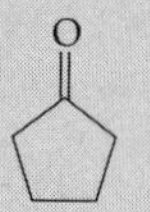 [structure] or [structure]   b. $CH_3-CH_2-CH_2-\overset{\overset{\text{O}}{\|}}{C}-H$ c. $CH_3-\overset{\overset{\text{}}{\underset{\underset{CH_3}{|}}{CH}}}{}-CH_2-CH_2-\overset{\overset{\text{O}}{\|}}{C}-CH_3$

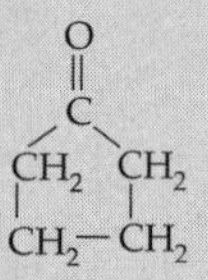

Solution
a. ketone
b. aldehyde
c. ketone

Carboxylic Acids

Organic acids contain the **carboxyl group**, which has a carbon atom that is double bonded to an oxygen atom and also bonded to an hydroxyl group:

$$-\overset{\overset{\displaystyle O}{\|}}{C}-OH \quad \text{or} \quad -COOH$$

carboxyl group

Examples of carboxylic acids are shown in Figure D.17.

In naming carboxylic acids, the final *-e* in the parent alkane is replaced with the suffix *-oic,* and the word *acid* is added. The two simplest carboxylic acids are methanoic acid and ethanoic acid, better known by their common names, formic acid and acetic acid.

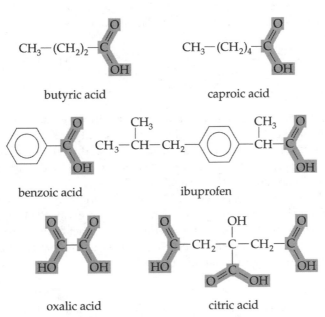

or

HCOOH

formic acid
(methanoic acid)

or

CH_3COOH

acetic acid
(ethanoic acid)

butyric acid

caproic acid

benzoic acid

ibuprofen

oxalic acid

citric acid

Figure D.17 Examples of carboxylic acids.

Compared with inorganic acids such as hydrochloric acid (HCl) and nitric acid (HNO_3), carboxylic acids are weak acids and dissociate very little to form ions. Most are less than 2% ionized. The acidic hydrogen is the one in the −OH of the carboxyl group.

$$
\underset{\substack{\text{or}\\ \text{RCOOH}}}{R-\overset{\displaystyle O}{\overset{\|}{C}}-OH} \;\rightleftharpoons\; \underset{\substack{\text{or}\\ \text{RCOO}^-}}{R-\overset{\displaystyle O}{\overset{\|}{C}}-O^-} \;+\; H^+
$$

The arrows indicate that the equilibrium favors the undissociated form. Like inorganic acids, carboxylic acids are neutralized by bases and form salts:

$$
\underset{\substack{\text{or}\\ \text{RCOOH}\\ \text{carboxylic acid}}}{R-\overset{\displaystyle O}{\overset{\|}{C}}-OH} + Na^+ + OH^- \longrightarrow \underset{\substack{\text{or}\\ \text{RCOO}^-\text{Na}^+\\ \text{salt}}}{R-\overset{\displaystyle O}{\overset{\|}{C}}-O^- \; Na^+} + \underset{\substack{\\ \\ \text{water}}}{H_2O}
$$

Sodium benzoate, the sodium salt of benzoic acid, is used as a food preservative. Carboxylic acids have many natural sources. Formic acid is present in ants, various other insects, and nettles. It is the irritant injected under the skin by the sting of red ants, bees, and other insects.

Vinegar, which is a 5% solution of acetic acid, is the most commonly encountered organic acid and can be prepared by the aerobic fermentation of ethyl alcohol. This is an oxidation process in which acetaldehyde is formed as an intermediate:

$$
\underset{\text{ethyl alcohol}}{CH_3CH_2OH} \xrightarrow{\text{oxidation}} \underset{\text{acetaldehyde}}{CH_3CHO} \xrightarrow{\text{oxidation}} \underset{\text{acetic acid}}{CH_3COOH}
$$

Acetic acid is used as a starting material in the manufacture of textiles and plastics. Some of the simple carboxylic acids have very unpleasant odors. Butyric acid is responsible for the smell of rancid butter and, in part, for body odor. The powerful odors of certain cheeses and of goats are due to longer chain carboxylic acids such as caproic acid. Carboxylic acids with 12 or more carbon atoms are called **fatty acids**.

Many naturally occurring carboxylic acids contain more than one carboxyl group, and some also contain hydroxyl groups. Citric acid, which is found in all citrus fruits and gives them their tart flavor, contains three carboxyl groups. Oxalic acid, with two carboxyl groups, is toxic in large doses and is present in rhubarb, tomatoes, and other vegetables. It is a useful oxidizing agent for removing rust and ink stains.

Esters

Esters (Table D.4) are derivatives of carboxylic acids in which the −H of the carboxyl group is replaced by an −R′ group.

$$\text{R}-\overset{\displaystyle\overset{\text{O}}{\|}}{\text{C}}-\text{OR}' \quad \text{or} \quad \text{RCOOR}'$$

ester

For example, the ester ethyl acetate is formed by replacing the $-$H in the $-$COOH group in acetic acid with an ethyl group.

$$\text{CH}_3-\overset{\displaystyle\overset{\text{O}}{\|}}{\text{C}}-\text{OH}$$

or

$$\text{CH}_3\text{COOH}$$

acetic acid

$$\text{CH}_3-\overset{\displaystyle\overset{\text{O}}{\|}}{\text{C}}-\text{O}-\text{CH}_2\text{CH}_3$$

or

$$\text{CH}_3\text{COOC}_2\text{H}_5$$

ethyl acetate

The names of esters are based on the acids from which they are derived. Esters are widely distributed in nature and are responsible for the tastes of many fruits and the perfumes of many flowers. Many esters have been synthesized for use as flavorings in foods and beverages. It is interesting to note that just a small change in an R group can significantly alter the flavor sensation of an ester. Apple and pineapple flavors, for example, differ from each other by a single $-$CH$_2$ group (Figure D.18).

Many fragrances in perfumes are esters. For example, benzyl acetate is the main constituent of oil of jasmine. Ethyl acetate is a useful solvent for removing lacquers, paints, and nail polish.

EXAMPLE D.17

Identify each of the following as a carboxylic acid or an ester.

a. $\text{CH}_3-\text{CH}_2-\text{CH}_2-\text{CH}_2-\text{CH}_2-\overset{\displaystyle\overset{\text{O}}{\diagup\!\!}}{\underset{\text{OCH}_3}{\text{C}}}$ b. $\text{CH}_3-\text{CH}_2-\text{CH}_2-\overset{\displaystyle\overset{\text{O}}{\diagup\!\!}}{\underset{\text{OH}}{\text{C}}}$ c. $\overset{\displaystyle\overset{\text{O}}{\diagup\!\!}}{\underset{\text{OH}}{\text{C}}}$

Solution

a. A methyl group is attached to the $-\overset{\displaystyle\overset{\text{O}}{\|}}{\text{C}}-\text{O}$ group. This compound is a methyl ester, methyl hexonate.

b. An hydrogen atom is attached to the $-\overset{\displaystyle\overset{\text{O}}{\|}}{\text{C}}-\text{O}$ group. This compound is an acid, butyric acid.

c. Like #2, this compound is an acid, benzoic acid.

methyl butanoate
(apple)

ethyl butanoate
(pineapple)

Figure D.18 The two esters responsible for apple and pineapple flavors differ by a single $-CH_2-$ group.

EXAMPLE D.18

Write the formulas of the following: (a) acetic acid, (b) methyl acetate, and (c) pentanoic acid.

Solution

a. CH_3COOH b. CH_3COOCH_3

c. $CH_3CH_2CH_2CH_2COOH$

Amines and Amides

Many important organic compounds—both naturally occurring and synthetic—contain nitrogen. Two important families of nitrogen-containing compounds are amines and **amides**.

Amines can be thought of as being derived from ammonia (NH_3) by replacement of one, two, or all three of the hydrogen atoms with alkyl groups. A primary amine has two hydrogen atoms and one R group attached to the nitrogen atom. A secondary amine has one hydrogen atom and two R groups attached to the nitrogen atom, and a tertiary amine has three R groups and no hydrogen atoms attached to the nitrogen atom. The general formulas for these three types of amines are shown here:

ammonia   primary   secondary   tertiary
             amine   amine   amine

The two simplest amines are methylamine and ethylamine.

or

CH_3NH_2

methylamine

or

$C_2H_5NH_2$

ethylamine

Most decaying organic matter produces amines, many of which have disagreeable fishy odors. The smells associated with sewage-treatment plants and meat-packing plants are primarily caused by amines. Two particularly unpleasant amines are cadaverine and putrescine, which are products of the bacterial decomposition of proteins and are partly responsible for the smell of decaying flesh (Figure D.19).

Important amines that include a benzene ring in their structure are aniline, amphetamine, and epinephrine (commonly called adrenaline) (Figure D.19). Aniline is used as a starting material for the manufacture of dyes and certain drugs. Amphetamines are stimulants and have been widely used by people who need to stay alert. They are also used to treat mild depression and as appetite suppressants to treat obesity. Epinephrine and the related compound norepinephrine are secreted by the adrenal gland and play an important role in controlling blood pressure and cardiac output.

EXAMPLE D.19

Draw the structures of the following and indicate whether they are a primary, secondary, or tertiary amine:

1. propylamine $C_3H_7NH_2$
2. methylbutylamine $CH_3NHC_4H_9$
3. trimethylamine $(CH_3)_3N$

Solution

1. This compound has a propyl group that is attached to the nitrogen atom in the amine. Nitrogen forms three bonds. The other two bonds must be attached to hydrogen atoms. Because there is only one alkyl group, the compound is a primary amine.

$$CH_3CH_2CH_2-\underset{\underset{H}{|}}{N}-H$$

2. This compound has a methyl group and a butyl group attached to the nitrogen atom. It is a secondary amine.

$$CH_3-\underset{\underset{H}{|}}{N}-CH_2CH_2CH_2CH_3$$

3. This compound has three methyl groups attached to the nitrogen atom. It is a tertiary amine.

$$CH_3-\underset{\underset{CH_3}{|}}{N}-CH_3$$

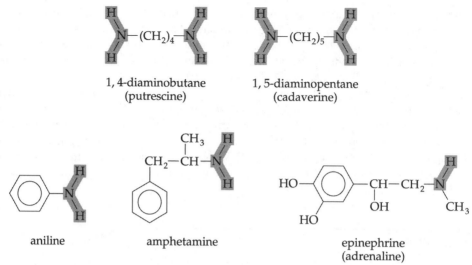

1, 4-diaminobutane
(putrescine)

1, 5-diaminopentane
(cadaverine)

aniline

amphetamine

epinephrine
(adrenaline)

Figure D.19 Examples of amines.

EXAMPLE D.20

Are the following primary, secondary, or tertiary amines: (1) trioctylamine,
(2) dibutylamine, and (3) isopropylamine?

Solution
(1) tertiary, (2) secondary, (3) primary

Amides are derivatives of carboxylic acids in which the $-OH$ of the carboxyl group is replaced
with an amino group ($-NH_2$) or a substituted amino group ($-NHR$ or $-NRR'$). Unlike
amines, the nitrogen atom in amides is attached to a $-C=O$ group. The general formulas for
the three types of amides are shown here:

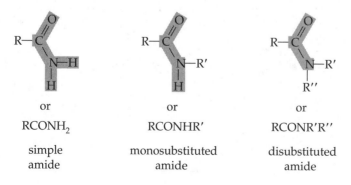

or

$RCONH_2$

simple
amide

or

$RCONHR'$

monosubstituted
amide

or

$RCONR'R''$

disubstituted
amide

One of the simplest amides is acetamide. Amides are named by replacing the ending *-ic* (or *-oic*) and the word *acid* in the name of the corresponding acid with *-amide.*

$$CH_3-\overset{\overset{\textstyle O}{\|}}{C}-OH \qquad CH_3-\overset{\overset{\textstyle O}{\|}}{C}-NH_2$$

or   or

CH_3COOH CH_3CONH_2

acetic acid   acetamide

The amide functional group is present in many important biological compounds, including proteins and urea, the end product of human metabolism of proteins in food (Figure D.20).

Many amides have been synthesized for use as drugs. For example, acetaminophen, the active ingredient in the pain reliever Tylenol, is an amide (Figure D.20). Amides are also important intermediates in the manufacture of nylon and other polymers.

EXAMPLE D.21

Draw the structure for each of the following compounds, which may belong to any of the classes of compounds we have studied: (1) methylethylamine, (2) 3-pentanone, (3) butyric acid, and (4) methylacetamide.

Solution

1. This compound has a methyl group and an ethyl group attached to the nitrogen atom. Nitrogen forms three bonds; therefore, the third bond is attached to a hydrogen atom.

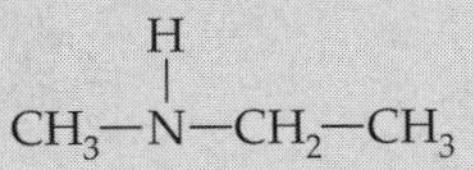

$$CH_3-\overset{\overset{\textstyle H}{|}}{N}-CH_2-CH_3$$

2. This compound contains five carbon atoms (penta), and the oxygen of the ketone is attached to carbon 3.

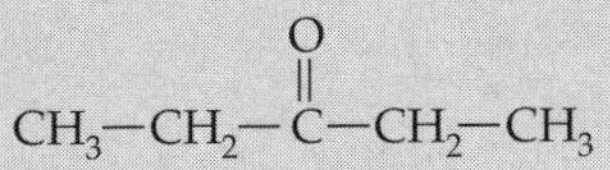

$$CH_3-CH_2-\overset{\overset{\textstyle O}{\|}}{C}-CH_2-CH_3$$

3. This compound contains four carbon atoms (butyric) and is a carboxylic acid.

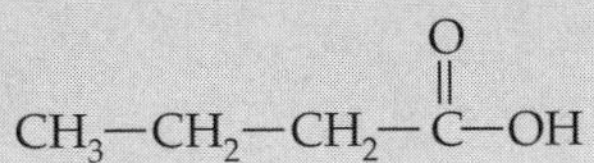

$$CH_3-CH_2-CH_2-\overset{\overset{\textstyle O}{\|}}{C}-OH$$

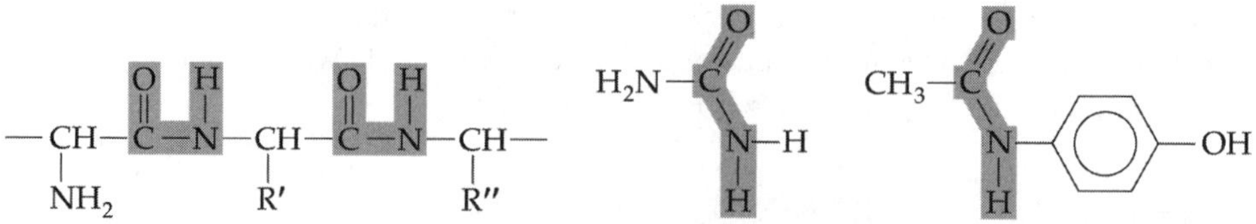

Figure D.20 Examples of amides.

protein segment   urea   acetaminophen

4. This compound is formed from acetamide by the addition of a methyl group.

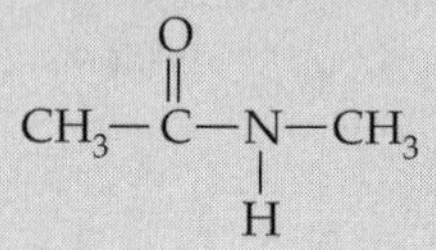

$$CH_3-\overset{\overset{\displaystyle O}{\|}}{C}-\underset{\underset{\displaystyle H}{|}}{N}-CH_3$$

EXAMPLE D.22

Identify all of the functional groups in epinephrine, a compound that is used to control allergic reactions.

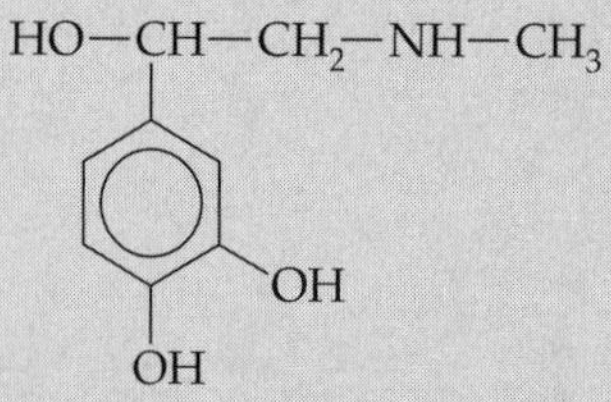

$$HO-CH-CH_2-NH-CH_3$$

Solution
Three alcohol groups ($-OH$) and one secondary amine

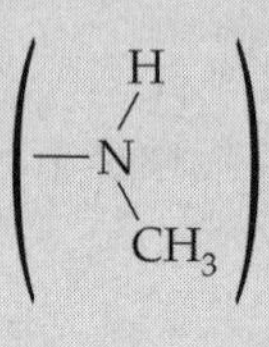

$$\left(-\underset{\underset{\displaystyle CH_3}{\diagdown}}{\overset{\overset{\displaystyle H}{\diagup}}{N}}\right)$$

Keywords

alcohol
aldehyde
alkane
alkene
alkyl halide
alkyne
amide
amine
aromatic hydrocarbon
benzene
carboxyl group
carboxylic acid
cycloalkane

delocalized electrons
ester
ether
fatty acid
functional groups
homologous series
hydrocarbon
ketone
ortho-, meta-, para- substitution
polycyclic aromatic compound
saturated hydrocarbon
structural isomers
unsaturated hydrocarbon

Questions and Problems

1. Write the names of the following alkanes:

a. $CH_3-CH_2-CH_2-CH_2-CH_3$

b. $CH_3-CH_2-CH_2-CH_2-CH_2-CH_2-CH_2-CH_3$

c. $CH_3-CH_2-CH_3$

d. $CH_3-CH_2-CH_2-CH_3$

2. Write the names of the following branched alkanes:

a. 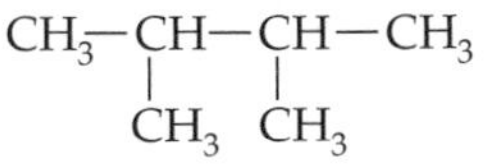 $CH_3-CH-CH-CH_3$
           | |
          CH_3 CH_3

b. 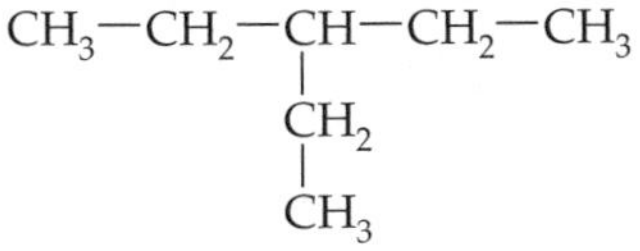 $CH_3-CH_2-CH-CH_2-CH_3$
                    |
                    CH_2
                    |
                    CH_3

c. 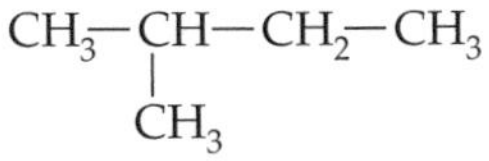 $CH_3-CH-CH_2-CH_3$
           |
          CH_3

d. 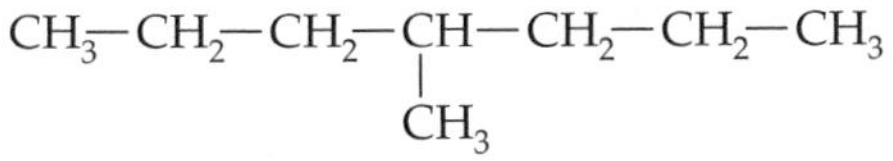 $CH_3-CH_2-CH_2-CH-CH_2-CH_2-CH_3$
                         |
                        CH_3

3. Why does carbon form so many compounds with oxygen, nitrogen, and hydrogen?

4. Write the structural formulas and names for the members of the homologous series of saturated hydrocarbons starting with two carbon atoms and ending with six carbon atoms.

5. Which is the smallest alkane to have an isomer?

6. Draw and name all of the isomers of pentane.

7. Name and draw the isomers of hexane that are butanes.

8. Write the structural formulas for the following compounds:
   a. Cis-2-butene
   b. 1-Hexene
   c. Propene
   d. 1-Pentene

9. Write the structural formulas for the following compounds:
   a. Cyclohexene
   b. 3-Methyl-1-butene
   c. 3,6-Dimethyl-2-octene
   d. 2-Methyl-2-butene

10. Draw five different structural isomers of a hydrocarbon with the formula C_6H_{14}.

11. What is the common name for ethyne?

12. Which of the following alkenes show cis–trans isomerism? Draw the isomeric structures.
   a. 1-Butene
   b. 2-Methyl-2-butene
   c. 2-Butene
   d. 1-Chloropropene

13. Which of the following alkenes show cis–trans isomerism? Draw the isomeric structures.
   a. 3-Methyl-4-ethyl-3-hexene
   b. 2-Pentene
   c. 2,3-Dichloro-2-butene
   d. 1-Pentene

14. Write a structural formula for each of the following compounds:
   a. 2,3-Dichloropentane
   b. 1,4 Dichlorocyclohexane
   c. 3-Ethylpentane
   d. 1-Bromobutane

15. Write a structural formula for each of the following compounds:
   a. 2,3-Dimethyl-2-butene
   b. Cis-2-butene
   c. 1,1-Dimethylcyclopentane
   d. 2,3-Dichlorobutane

16. Identify the functional group in each of the following compounds:
   a. Acetone, CH_3COCH_3
   b. Butyric acid, $CH_3CH_2CH_2COOH$
   c. Methyl acetate, CH_3COOCH_3

17. Identify all of the functional groups in each of the following compounds:
   a. Glutaraldehyde, $OHCCH_2CH_2CH_2CH_2CHO$
   b. Glycine, H_2NCH_2COOH
   c. Oxalic acid, $HOOCCOOH$
18. Identify all of the functional groups in each of the following compounds:
   a. A sex attractant of the female tiger moth

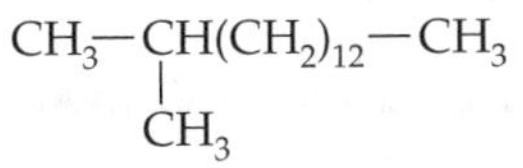

$$CH_3-CH(CH_2)_{12}-CH_3$$
$$\qquad\; CH_3$$

   b. Component of peppermint oil

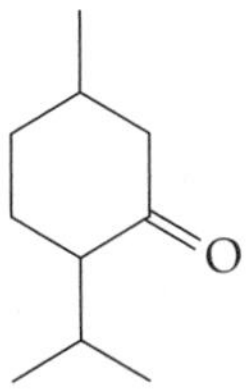

   c. Vitamin A

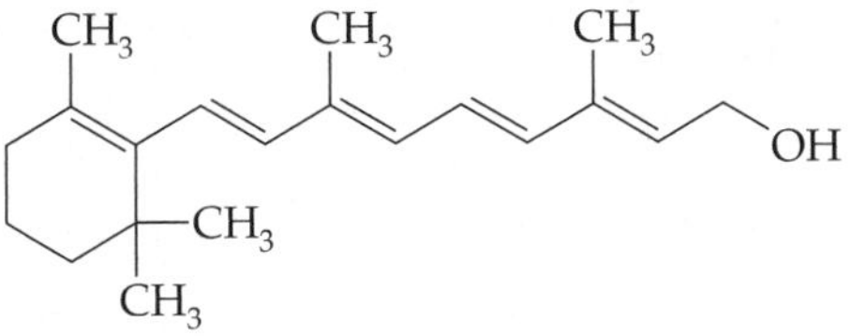

   d. Testosterone

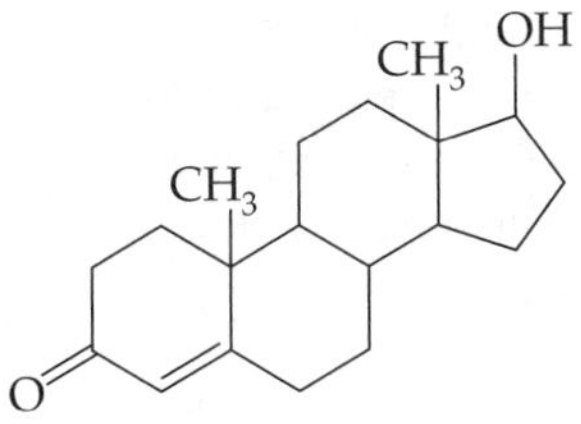

   e. Nepetalactone, a constituent of catnip

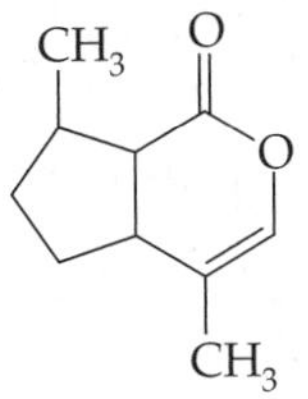

   f. An anesthetic

$$CH_2=CH-O-CH=CH_2$$

19. Cyclohexane and 1-hexene have the same molecular formula, C_6H_{10}. Write the structural formulas.

20. Name the following:
    a. $(CH_3)_3CCH=CHCH_2CH_3$
    b. $CH_3CH_2CH=CH(CH_2)_4CH_3$
    c. $CH_3CH_2CH=CHCH_2CH_3$

21. Name the following:
    a. Methylacetylene
    b. Methylpropylacetylene
    c. Methylbutylacetylene

22. Write the general formula for the following:
    a. A carboxylic acid
    b. A ketone
    c. An alcohol
    d. An ester

23. Write the general formula for the following:
    a. An amine
    b. An amide
    c. An ether
    d. An aldehyde

24. What are the chemical names for the following materials?
    a. Rubbing alcohol
    b. Wood alcohol
    c. Grain alcohol
    d. Antifreeze

25. Write the structural formula for the aromatic alcohol phenol.

26. Methanol is oxidized in two steps. What is the product of the first oxidation step?

27. What two chemicals would you use to make the ester methyl acetate?

28. What kind of compound would you react with an amine to make an amide?

29. What kind of compound would you react with an alcohol to make an ester?

30. What kind of compound is responsible for the fruity smell of a banana?

31. What compound makes certain insect bites sting?

32. How is ethylene glycol different from ethanol?

33. For what is isopropyl alcohol used?

34. Write the structural formula for the aromatic carboxylic acid benzoic acid.

35. What is the percentage of alcohol by volume in 96-proof whiskey?

36. Ethyl alcohol is produced by fermentation. Describe the two materials that react to form ethyl alcohol.

37. Why is ethyl alcohol used in cough syrup preparations rather than methyl alcohol?

38. What is the product of the oxidation of ethyl alcohol?

39. Write a chemical equation that shows why acetic acid is an acid.

40. What is the difference between an aldehyde and a ketone?

41. Draw the structure of the following
    a. DDT: para-dichlorodiphenyltrichloroethane
    b. DDE: para-dichlorodiphenyldichoroethene

c. napthalene

d. pentachlorophenol

42. Draw the structure of the following:

a. Formaldehyde

b. Ethyl acetate

c. Ethyl alcohol

d. Acetone

43. Draw the structure of the following:

a. Acetic acid

b. Methylamine

c. Methyl alcohol

d. Phenol

44. Compare the odor of butyric acid with that of the ester methyl butyrate.

45. Identify each of the following as a primary, secondary, or tertiary amine:

a. Ethylamine

b. Methylamine

c. Dimethylamine

d. Diethylamine

e. Trimethylamine

f. Triethylamine

46. Name the following functional groups:

a. $-NH_2$

b. $-CHO$

c. $-COOH$

d. $-COOCH_3$

47. Name the following functional groups:

a. $-CONH_2$

b. $-OH$

c. $-CO-$

d. $-O-$

48. Write the structural formula for each of the following:

a. Four-carbon alcohol with a double bond at carbon 2

b. Ketone containing three carbon atoms

c. 12-Carbon carboxylic acid with a double bond at carbon 6

49. Write the structural formula for each of the following:

a. Methyl ester of a four-carbon, straight-chain, carboxylic acid

b. Tertiary amine with three ethyl groups

c. Two-carbon aldehyde

d. Methyl amide of propionic acid

50. Write structural formulas for the following aromatic compounds:

a. Toluene

b. 1,3-Dimethylbenzene (are the methyl groups in the ortho-, meta- or para- positions?)

c. p-dimethylbenzene

51. Draw structural formulas for the following aromatic compounds:
   a. m-dichlorobenzene
   b. 1,2-Dimethylbenzene
   c. 3-Bromotoluene
52. How many structural formulas can be written for tetrachlorobenzene?
53. How does benzene differ from cyclohexane?
54. When one of benzene's hydrogens is replaced by a methyl group, what is the name of the resulting molecule?
55. Draw and name the molecule that is formed by fusing two benzene rings together.
56. When ethyl groups are placed at the opposite ends of a benzene ring, what position are they in?

Appendix E: Answers to Even-Numbered Questions

■ Chapter 1

2. Because fusing reactions are occurring in stars today, it is reasonable to assume that these atom-building reactions have been going on for millennia. The large amount of hydrogen in the universe is an excess of reacting material, not the product of other reactions.

4. Iron, oxygen, silicon, and magnesium. The distribution of elements according to mass did not happen because some elements combined with others to form compounds. The melting points and densities of the compounds determined how the elements were distributed on earth.

6. Water bound to minerals was released as the interior of the earth heated up. H_2 and O_2 released from the interior combined explosively to form water.

8. Photosynthesis: Energy from sun converts carbon dioxide and water into simple carbohydrates and oxygen.

10. Iron cans corrode more quickly, forming rust. Aluminum forms a protective oxide coating that impedes further corrosion of the bulk metal.

12. Wire, pipe, and coins.

14. Beverage cans, cooking utensils, and aircraft.

16. The percentage carbon determines the properties of steel. Low carbon steel is soft, high carbon steel is hard.

18. (b) Africa. Examples: The non-renewable, strategic metals platinum, platinum group metals, gold, and chromium are obtained in large amounts from South Africa.

20. Biotic refers to the living factors in the environment, such as plants, animals, fungi, and bacteria. Abiotic refers to the nonliving factors in the environment such as temperature, rainfall, nutrient supply, and sunlight.

22. Consumers are unable to harness energy from the sun to manufacture their own food. They must consume plants or other creatures to obtain nutrients and energy. Consumers can be divided into four main groups according to their food source: herbivores, carnivores, omnivores and decomposers.

24. Energy is the ability to do work or bring about change. Chemical energy, electrical energy, kinetic energy, potential energy, photons.

26. In every energy transformation the available internal energy cannot be converted completely into work; some energy is always dissipated as heat. An automobile engine needs to be cooled (usually with water) so that the heat released doesn't destroy the engine.

28. 100 units.
30. They do not have enough plant resources necessary to support enough animals to provide them with an adequate food supply.
32. C, H, N, O; S, P, Fe, Mg, Ca.
34. (a) 0.54 M.
    (b) 0.027 M.
    (c) 0.054 M.
36. (a) 5.15 g.
    (b) 5.15%.
    (c) 0.54 mmol/ml.
38. 0.05 M (mol/L).
40. (a) 100 ppb.
    (b) 250 ppm.
    (c) 1 ppm.
42. 0.5 ppm, 500 ppb.
44. (a) 0.00056 M.
    (b) 0.22 ppm.
46. Carbon dioxide from the atmosphere—photosynthesis
48. Decomposing complex organic molecules into simpler chemicals and returning them to the soil
50. Sugars, $C_6H_{12}O_6$
52. $N_2 + O_2 \rightarrow NO$ (g); the NO is eventually oxidized in the atmosphere to nitrate and removed in rainfall.
54. Nitrogen containing fertilizers from farm fields are carried by rainwater runoff to rivers and streams. Untreated human sewage contains large amounts of nitrogen.
56. Plant a nitrogen-fixing crop such as clover and plow it back into the soil. Spread manure and allow soil bacteria to degrade it.
58. CuCl, CuBr, CuI.
60. AgBr gives [Ag] = 7.1×10^{-7}M = 7.6×10^{-5} g/liter = 76 micrograms/liter (ppb). In contrast, AgI generates < 1 ppb Ag, too low to be effective, and AgCl generates > 1000 ppb Ag, which is toxic.
62. 0.2 ppm, 200 ppb.

■ Chapter 2

2. See Figure 2.1.
4. 0.0115 atm.
6. The relative humidity is a term that expresses just how saturated the atmosphere is with water vapor: % RH = 100 * (P_{H2O} measured / P_{H2O} at saturation)
8. Draw a graph of Pressure v. Distance from the Earth (in km). At distance 0, the pressure is 760 mm. At distance 6 km, pressure is 380 mm.
10. The emission maximum would shift from 480 nm to 340 nm. EMR with more energy will be reaching Earth's surface. There would also be a greater proportion of energy in the UV, possibly increasing sunburns, skin cancer, cataracts; and chlorophyll is not adapted for this modified solar spectrum.

12. Aerosols are very fine particles that remain suspended in the atmosphere and are not heavy enough to be deposited on the Earth's surface. Particulates generally have diameters greater than 1 µm and are deposited.

14. (a) Natural sources of particulates: volcanic ash, smoke from forest fires, wind-blown soil fragments, pollen, bacteria, sea spray.
    (b) Anthropogenic sources of particulates: fly ash, soot, asbestos fibers, cement dust, fertilizer.

16. Blue light, which has a shorter wavelength than red light is scattered more by airborne particulates. The sky, which we see in scattered light, appears blue.

18. (a) Particulates with a diameter less than 2.5 µm.
    (b) Particulates with a diameter less than 10 µm.

20. The smaller the diameter of the particle the more dangerous it is. Small particles are drawn deep into the lungs.

22. 30 days.

24. The "v" in ppbv indicates the relationship is based on volume.

26. (a) 0.05 ppmv.
    (b) 98 µg/m^3.
    (c) 1.2×10^{11} molecules/cm^3.

28. The 0.38 albedo gives a calculated temperature of 247°K, compared to 255°K in Table 2.3 (0.31 albedo). An increase in particulates will cool the planet.

30. Venus is closer to the sun than Earth and it receives more incident radiation. Water vapor was exposed to intense UV that photolyzed the water to atomic hydrogen and hydroxyl radical. Atomic hydrogen formed hydrogen gas which escaped Venus's gravity and was lost into space. Because of the loss of hydrogen, water could not reform. At the same time, volcanoes were releasing CO_2. Since there was no water to dissolve the CO_2, it could not form carbonate rock. The CO_2 built up in the atmosphere producing a run-away greenhouse effect.

32. O_3, NO, ClO, OH.

34. One such feedback loop is the "ice-albedo feedback loop." This loop begins with increasing concentrations of greenhouse gases that cause more infrared energy to be absorbed by the atmosphere, which in turn heats the atmosphere and causes more ice to melt. When the ice melts, darker surfaces below the ice are exposed and they absorb more solar radiation than does ice. This causes more warming and a reinforcing cycle is created.

36. As polar ice caps melt, the surface area with a high albedo (ice) decreases and the surface area with a low albedo (sea) increases. Less light is reflected and more is absorbed leading to more melting of the ice caps.

38. Light-reflecting areas of the Earth's surface, such as snow, ice, or deserts, reflect about one-third of the energy (30 Wm^{-2}), while aerosols suspended in the atmosphere reflect the rest (77 Wm^{-2}).

40. The Earth is a sphere, and more solar energy strikes a square meter of the surface in the tropics than at higher latitudes because of the inclination of the Earth to the Sun. The angle of inclination is equal to the latitude at Equinox, so sunlight is exactly perpendicular to the Earth's surface at the Equator at this time. However, at 45 degrees N or S, the sun is incident at 45 degrees, so it is only 71% as intense (sin 45° = 0.71).

■ Chapter 3

2. Ice ages have occurred in regular cycles for the past 3 million years and it appears as if they are linked to regular variations in the Earth's orbit around the Sun, which are known as Milankovitch cycles, rather than any variation in solar output.

4. (a) <200 ppm.
   (b) 315 ppm.
   (c) 380 ppm.

6. CO_2.

8. The major cause of the increase in atmospheric carbon dioxide is the burning of fossil fuels by electric utilities, automobiles, and industry. A secondary cause of the increase is deforestation.

10. N_2 and O_2 have no net dipole moment. They cannot be stretched or bent in a manner to produce a net change in the dipole moment. A molecule cannot absorb IR light unless there is a net change in dipole moment possible with the quantum transition.

12. The region between 7.5 and 13 μm through which radiation from the Earth's surface can escape.

14. 200 years.

16. Positive feedback occurs when global warming increases evaporation from oceans that leads to higher concentrations of water vapor in the air. The increased amount of water vapor causes more infrared to be absorbed and that increases warming of the Earth's surface. Negative feedback occurs as the troposphere becomes cloudier and that causes more reflection of incident solar flux. This causes cooling of the Earth's surface.

18. Oceans (most important), soil, vegetation, and lithification.

20. Natural: about 23% of the atmospheric CH_4 is released naturally from wetlands, including bogs, tundra, and swamps, where methane-producing bacteria thrive in the rich organic matter (with a low oxygen content). The next leading source at about 20% is from the flooding of rice fields, where certain types of bacteria also thrive. Anthropogenic: CH_4 is released when natural gas pipelines rupture or leak. It is estimated that 1.5% of all methane carried in pipelines today is lost to the atmosphere.

22. $CH_4 \cdot 6H_2O$.
   (a) Deep under the ocean.
   (b) A small molecule that occupies a vacant space in the center of a three dimensional cage of water (ice) molecules.
   (c) Global warming increases temperature of the ocean and methane is released from methane hydrate.

24. Release from soil, lakes, and oceans by microbial denitrification of nitrate.

26. Freon used in refrigeration equipment (air conditioning, refrigerators, etc.).

28. 2.5–12 μm, 4000–500 cm^{-1}.

30. H_2S.

32. The concentration of the gas in the atmosphere today (and its residence lifetime in the atmosphere), the wavelengths at which the gas molecules absorb, and the intensity of the absorption per molecule (i.e., molar absorptivity).

34. Land use, production of atmospheric aerosols.

36. Aircraft exhaust (contrails) can reflect incoming solar radiation while absorbing outgoing infrared radiation. Aircraft contrails increase the Earth's clouds and have a small positive radiative forcing.

38.

$$\ddot{N}=N=\ddot{O}$$

There is a net change in dipole moment when this molecule vibrates. The major source of nitrous oxide is release from soil, lakes, and oceans by microbial denitrification of nitrate.

40. By 2100 global surface air temperature will increase 1–3.5°C (1.8 – 6.3°F).

42. Companies that exceed pollution reduction goals sell excess reduction to other companies that have not met average greenhouse gas emission standards.

44. With rising temperatures, air circulation, ocean currents, and rainfall patterns would change, causing generally violent weather. As a result, some regions of the world, including much of the United States, would experience droughts; other regions would become much wetter. Ecosystems all over the world would be disrupted, and some species might face extinction. Climate-related diseases such as malaria might attack areas where they are currently unknown.

46. The 2008 concentration is 380 ppm. The IPCC ceiling is 455 ppm.

■ Chapter 4

2. (1) Oxides of carbon: CO, CO_2—transportation, industrial processes.
   (2) Oxides of nitrogen: NO_x—transportation, fuel combustion at stationary sources.
   (3) Oxides of sulfur: SO_x—fuel combustion at stationary sources.
   (4) Volatile organic compounds: VOC—transportation.
   (5) Suspended particles—fuel combustion at stationary sources.

4. Secondary pollutants are harmful substances produced by chemical reactions between primary pollutants and other constituents of the atmosphere. Examples: sulfuric acid, nitric acid, sulfates, nitrates, ozone and photochemical oxidants.

6. $C + O_2 \rightarrow CO$.

   $CO + O_2 \rightarrow CO_2$.

   When there is plenty of oxygen, such as burning coal in an open fireplace, the two-step reaction goes to completion and CO_2 is produced. In a situation where the amount of oxygen is limited, the reaction stalls after the first step and CO is produced. A poorly tuned automobile that burns fuel with insufficient oxygen supply can produce high CO levels in the exhaust gas.

8. Natural sources release 10 times more CO than do anthropogenic sources.
   (a) Natural sources—Methane, released by decay of vegetation, rice paddies, and cattle flatulence is oxidized in the atmosphere to CO.
   (b) Natural removal processes—Conversion of CO to CO_2 by reaction with hydroxyl radicals in the atmosphere and removal of CO from the atmosphere by microorganisms in the soil.

10. N=N=O

Yes, N_2O is an appropriate way to state the molecular formula.

12. NO_2 is removed from the atmosphere as nitric acid and nitrates in dust and rainfall.

$$4\,NO_2 + 2\,H_2O + O_2 \rightarrow 4\,HNO_3$$

14. (a) The 2-cycle operates at a much higher rpm.
    (b) The 2-cycle operates at a higher temperature.
    (c) The 2-cycle emits much more HC.
    (d) The 2-cycle is much less expensive.

16. As the mixture is made more fuel-rich, the amount of NO drops as the amount of CO and HC increases.

18. Three air pollutants are reduced simultaneously by the catalytic converter. NO is reduced to N_2, while CO is oxidized to CO_2 and HC is oxidized to water and CO_2.

20. (a) $CO + O_2 \rightarrow CO_2$.
    (b) $2\,C_8H_{18} + 25\,O_2 \rightarrow 16\,CO_2 + 18\,H_2O$.
    (c) $2\,NO + 2\,CO \rightarrow N_2 + 2\,CO_2$.

22. Lead in the exhaust gases would coat the catalytic surface of the catalytic converter and stop the catalytic reaction.

24. (a) Fuel combustion at stationary sources (mostly coal).
    (b) Coal contains iron pyrite (FeS_2).
    (c) $2\,SO_2 + O_2 \rightarrow 2\,SO_3$.

$$SO_3 + H_2O \rightarrow H_2SO_4.$$

26. 0.11 ppm.

28. pH = 4.0.

30. (1) FGD: The SO_x compounds are washed, or scrubbed from the stack gases by absorption in an alkaline solution. (2) FBC: Coal in granular form is burned with finely divided pulverized limestone ($CaCO_3$). In both methods the SO_x compounds are ultimately removed as precipitated solids ($CaSO_4$, gypsum) and landfilled.

32. The SO_x is washed out of the chimney gases and the resulting solution is reacted with an alkaline material to render the sulfur into a solid form.

$$2\,SO_2 + 2\,CaCO_3 + O_2 \rightarrow 2\,CaSO_4 + 2\,CO_2.$$

34. The CaSO4 is used to make concrete and as filler in asphalt roads.

36. HC and NO_x.

38. Four hydroxyl radicals are produced for every hydrocarbon molecule.

40. 2 pptv, 0.002 ppbv.

42. $C_5H_{12} + \cdot OH \rightarrow \cdot C_5H_{11} + H_2O$ abstraction reaction

44. (b) Xylene.

46. The reaction with 2-octene is much faster. Ozone cleaves C=C double bonds rapidly in a process that is well-known to organic chemists as "ozonolysis", generating carbonyl products.

48.
$$CH_3-\overset{\displaystyle O}{\overset{\displaystyle \|}{C}}-O-O-NO_2$$

50. Nitrogen dioxide, NO_2, is the substance responsible for the brown color. This color is familiar to chemists who have left a clear nitric acid bottle exposed to sunlight for a few days.

52. The Clean Air Act of 1970 mandated air quality standards for five air pollutants; suspended particles, sulfur dioxide, carbon monoxide, nitrogen oxides, and ozone.

54. The National Ambient Air Quality Standards specifies a concentration, averaged over a specific time period, which must not be exceeded.

56. Methanol, ethanol, MTBE.

58. All the primary pollutants (CO, NO_2) and particulates associated with combustion, many VOCs such as aldehydes (formaldehyde), ketones (acetone), hydrocarbons and organic acids.

60. The air is passed through a filter and sorbent. The air volume is measured, converted to flow at STP, and the PAH's are solubilized with a solvent such as chloroform or methylene chloride. The solvent extract is thereafter analyzed by a gas chromatographic method.

■ Chapter 5

2. 6.25×10^{28} molecules.

4. Stratosphere, lower, 20 to 30 km.

6. 390 nm, UV-A.

8. $\lambda = 126$ nm. The oxygen atoms in O_2 are connected together by a weaker double bond, whereas nitrogen atoms in N_2 are connected together with a stronger triple bond. More energetic photons are required to photolyze the stronger bonds.

10. The stratospheric ozone absorbs 95% of the UV-B radiation before it reaches the surface of the earth.

12. Ozone is generated and destroyed in a complex set of opposing reactions that are referred to as the Chapman Cycle. Collectively, these reactions produce a relatively constant concentration (steady state concentration) of atmospheric ozone.

14. Much less than the 10^{-4} ratio at 40 km.

16. $O_3 + hv \left(\lambda\, 240 - 320\,nm \right) \rightarrow O_2 + O$.
    Rate of reaction 3 $= K_3 \left[O_3 \right]$.

    $O_3 + O \rightarrow 2\,O_2$.
    Rate of reaction 4 $= K_4 \left[O \right]\left[O_3 \right]$.

18. Depletion of the ozone layer leaves the Earth vulnerable to the damaging effects of UV radiation. Contemporary life forms are not adapted to high fluxes of short-wavelength UV light; this energetic light can cause mutations, genetic damage, skin cancer, and cataracts.

20. $H_2O + hv \rightarrow \cdot H + \cdot OH$.

22. $\cdot OH + O_3 \rightarrow \cdot OOH + O_2$.

    $\cdot OOH + O \rightarrow \cdot OH + O_2$.
    _____

    Net reaction : $O + O_3 \rightarrow 2\,O_2$

24. N_2O is much less reactive than NO.

26. $NO^{\cdot} + O_3 \rightarrow {}^{\cdot}NO_2 + O_2$

$NO_2{}^{\cdot} + O \rightarrow {}^{\cdot}NO + O_2$

Net reaction : $O + O_3 \rightarrow 2\,O_2$

28. Supersonic aircraft would emit large quantities of water vapor and NO_x during flight.

30. (a) CF_2ClCF_2Cl.
    (b) CF_2ClCF_3.
    (c) $CHClF_2$.
    (d) CH_3CClF_2.

32. (a) 114.
    (b) 115.
    (c) 122.

34. Its lifetime would be longer (about 230 years) because CFC-114 has a smaller percentage of chlorine.

36. A null cycle (also known as a holding cycle) is a process that prevents certain species from taking part in catalytic cycles. See the example of the reaction of NO_2 with O_3 that results in the production of NO_3.

38. (a) Below 195°K (−78°C).
    (b) These ice crystals were found to have the composition HNO_3·$3H_2O$ and are called nitric acid trihydrate (NAT) crystals.
    (c) 1 µm in diameter.

40. These reactions are heterogeneous because they occur only after the reacting gas has adsorbed on the surface of the PSC particle.

42. $Cl_2 + h\nu \rightarrow 2\,{}^{\cdot}Cl$.

$2\,{}^{\cdot}Cl + 2\,O_3 \rightarrow 2\,ClO^{\cdot} + 2\,O_2$.

$ClO^{\cdot} + ClO^{\cdot} \rightarrow ClOOCl$.

$ClOOCl + h\nu \rightarrow ClOO^{\cdot} + {}^{\cdot}Cl$.

$ClOO^{\cdot} \rightarrow {}^{\cdot}Cl + O_2$.

Net reaction: $2\,O_3 + h\nu \rightarrow 3\,O_2$.

44. In the lower stratosphere, where the polar ozone levels are the lowest, there are not enough oxygen atoms to sustain the reaction; the $ClO^{\cdot}$ concentration builds up (ppb level) until $ClOOCl$ dimers are formed.

46. A NASA research plane, equipped with sophisticated analytical instruments, flew 25 missions in the region of the ozone hole in the stratosphere over Antarctica. Data collected showed conclusively that as ozone concentration decreased, the concentration of the chlorine monoxide radical ($ClO^{\cdot}$), rose. See Figure 5.6.

48. Halons, which include CF_2ClBr and CF_3Br, are very effective as fire extinguishing agents.

50. a. NO_2.
   b. $NO_2 + O_3 \rightarrow NO_3 + O_2$.
      $NO_3 + h\nu \rightarrow NO_2 + O$.
   c. Net reaction: $O_3 + h\nu \rightarrow O_2 + O$.
52. Montreal Protocol.
54. (a) Much shorter lifetimes than CFCs.
   (b) Much smaller ozone depletion potential (ODP) than CFCs.
   (c) More flammable than CFCs.

■ Chapter 6

2. Doubled.
4. X-ray > infrared > microwave.
6. (a) 0.64.
   (b) 0.98.
   (c) 0.50.
8. (a) $cm^{-1}ppm^{-1}$.
   (b) $cm^{-1}ppb^{-1}$.
   (c) $cm^{-1}\mu g^{-1}m^3$.
10. (a) Blue-violet
   (b) Yellow.
12. No. Sunlight is polychromatic.
14. (a) $(ppm)^{-1}(m)^{-1}$.
   (b) $\dfrac{m^3}{(cm)(mg)}$

16. (a) Spectroscopic instruments look along the edge of the Earth's stratosphere downward are called nadir measurements.
   (b) In an occultation experiment, the sun or a bright star is observed as the line of sight from the object to the viewing instrument passes through the atmosphere.
   (c) Spectroscopic instruments look along the edge of the Earth's stratosphere are called limb paths.
   (d) In a scattering experiment, solar radiation which is scattered as it passes through the atmosphere is observed.
18. (a) Infrared.
   (b) Between 10 and 150 km.
   (c) Down to 1 ppb.
   (d) Solar radiation.
   (e) Sunrise and sunset.
20. (a) Ultraviolet.
   (b) Ozone.
   (c) Backscattered UV radiation.
22. CO, hydrocarbons HC, NO_x. These components are measured to ensure proper function of the emission control hardware on the car; tests are conducted in some localities such as the LA Basin with persistent air quality problems.

24. No. Alcohols do not give as big a signal in the FID detector as do hydrocarbons. A correction factor has to be applied for oxygenated fuels.
26. Propane.
28. (a) Ozone is added.

(b) $NO + O_3 \rightarrow NO_2^* + O_2$ (chemical activation)

$\quad\quad NO_2^* \rightarrow NO_2 + h\nu$ (emission of light)

30. Not all the NO_x in the tailpipe are NO. There is NO_2 as well. If tailpipe gasses are passed through a heated activated carbon catalyst, the NO_2 can be converted to NO. The reaction is:

$$NO_2 + C \rightarrow NO + CO.$$

Analysis of tailpipe emissions that have been treated this way give the amount of NO_x, which is the sum of NO and NO_2 present. To measure the amount of NO in the tailpipe, the carbon catalyst is removed.

32. See Figure 6.20.
    (a) Heated tungsten filament.
    (b) Flexible diaphragm.
    (c) Nitrogen.
34. No, a nonlinear calibration curve must be constructed.
36. (a) Flow at standard conditions must be known so that concentration can be computed.
    (b) Yes, velocity is lowered.
    (c) High temperature is maintained.
38. See Table 6.2.
40. 9 ug/m^3.
42.

	Lignite	Anthracite	Bituminous
(a)	40%	80%	65%
(b)	35%	4%	9%
(c)	1%	1%	2.8%
(d)	5%	10%	11%

44. (a) Burning coal.
    (b) Diesel and gasoline engines.
    (c) Iron.
46. (a) Cyclone fractionators that use centrifugal force to separate particles by mass separate particles greater than 2.5μm.
    (b) Impactors are simple devices that direct a stream of gas at high velocity toward a collection plate. The gas stream makes a sharp turn, which causes $PM_{2.5}$ to be deposited on the collection plate.
    (c) A cascade impactor separates particles by size as it directs the stream of air onto a series of collection plates through successively smaller orifices.
48. See Figure 6.26. XRF is used to identify the elemental composition of the surface of a material. Its elemental range is limited to elements larger than beryllium and the detection of low atomic number (Z<11, Na) elements is difficult.
    (a) Elements present on the surface of the PM.

(b) No, only the surface elements.

(c) Because the metals on the surface will directly interact with the lung if the particles are inhaled, XRF is used to assess the health risks posed by fly ash. The XRF is also non-destructive and requires no dissolution prior to analysis.

50. (a) Smaller (lower Z) elements do not have inner shell electrons.

(b) No.

(c) No, ppm level.

■ Chapter 7

2. (a) The heat of vaporization is the amount of heat required to convert 1g of a liquid to a vapor at its boiling point.

(b) The high heat capacity of water allows the oceans to absorb large amounts of heat, evaporating water, without there being a corresponding rise in water temperature.

(c) Because of its high heat of vaporization, a large amount of heat is required to evaporate a small volume of water. Water is much more effective as a coolant than another liquid with a lower heat of vaporization.

4. (a) Cl^-, 19,000 ppm.

(b) Na^+, 10,600 ppm.

6. (a) Na^+.

(b) Ca^{2+}.

8. (c) Evaporation.

10. Water reaches its maximum density at 4°C, This is four degrees above the freezing point. The more dense liquid water will drop to the bottom of a lake or pond and the solid water (ice) will float. This allows aquatic life to continue under the frozen surface.

12. (a) The heat capacity for one gram of a substance is called the specific heat.

(b) Heat capacity is the quantity of heat required to raise the temperature of a given mass of substance by 1°C. It takes 1 calorie of heat to raise the temperature of 1 gram of liquid water by 1°C. Water has the highest heat capacity of any common liquid or solid.

(c) A joule (J) is a unit of heat: 1 calorie = 4.183 J.

14. Lead would require 735 J. Water will require 23,849 J.

16. Water: 4.18×10^{10} J. Ethanol: 2.46×10^{10} J. Water would absorb 1.72×10^{10} J more than the same mass of ethanol. The atmosphere would be warmer if lakes were filled with ethanol.

18. Water reaches its maximum density at 4°C. The unusual behavior of ice is the result of the open lattice structure of hydrogen-bonded molecules that forms when water freezes (Figure 7.4). No.

20. When the warm season ends and cooling occurs, surface water (epilimnion) becomes more dense than the hypolimnion. This causes the surface water to sink, carrying with it dissolved oxygen. The deep water (hypolimnion) rises. This mixing process is called overturn.

22. $H_2O + H_2O \rightleftharpoons H_3O^+ + OH^-$.

$Kw = [H^+][OH^-] = 1.0 \times 10^{-14}$.

Pure water has an equal amount of H^+ and OH^-. So:

$[H^+] = [OH^-]$

$[H^+]^2 = 1.0 \times 10^{-14}$

$[H^+] = 1.0 \times 10^{-7}$

$\log [H^+] = pH = 7$

24. Decrease to 5.56.
26. 8.3 ppm (see Figure 8.2).
28. $[CO_2] = 0$

$[HCO_3^-] = 2.5$ millimoles/L

$[CO_3^{2-}] = 0$

$[OH^-] = 0$

30. The pH of rainwater is 5.6 because carbon dioxide dissolves in rain.

$CO_2 + H_2O \rightleftharpoons HCO_3^- + H^+$.

32. (a) $CaCO_3 + H^+ \rightarrow Ca^{22} + HCO_3^-$
    (b) $2 KAlSi_3O_8 + 2 H^+ + H_2O \rightarrow Al_2Si_2O_5 (OH)_4 + 4 SiO_2 + 2 K^+$

34. Beside air pollution, the natural process of soil formation may be as important in caus-
    ing acidification. Changes in land use and consequent changes in vegetation on the land
    are known to acidify soils. The northeast is downwind of many large coal-fired power
    plants. Some areas (large parts of the Great Lakes basin) contain soils with minerals such
    as limestone, $CaCO_3$, that buffer against input of atmospherically deposited acids. In the
    northeast (e.g., New Hampshire) the geology is granitic, and the derived soils do not
    buffer well against acid inputs, and thus the pH is lowered more readily.

36. Acid drainage is neutralized with lime before it is released. Many mines have been sealed
    to exclude oxygen and water so that acids cannot form.

$CaCO_3 + H^+ \rightarrow Ca^{2+} + HCO_3^-$.

38. 1000 meters (about 3000 feet).
40. If the level of Mississippi is raised, bigger boats will be able to navigate the river and com-
    merce will benefit. If water from the great lakes is diverted, the level of water in the lakes
    will fall, and plant and animal life will be affected.
42. When massive amounts of groundwater are removed from natural aquifers, the porous
    rock structure of the aquifers can collapse and cause a sinkhole. The land will not return
    to its original level if the water removal is stopped.
44. Large amounts of water evaporate as the water is exposed to sunlight. As pure water is
    removed by evaporation, the concentration of dissolved solids in the remaining water
    increases, and the water becomes more brackish.
46. (1) Wash your car with the water from your bath. (2) Water your vegetables with your
    rainwater. (3) Cook your dinner with the well water.
48. Yes, both processes would be affected. See page 208 for a description of desalination
    methods.
50. In the process of reverse osmosis, pressure in excess of the osmotic pressure is exerted
    on the liquid in the salt compartment, pure water flows from the salt compartment
    through the membrane into the water compartment (Figure 7.19c).

■ Chapter 8

2. Pollutants enter waterways by sewage treatment plants, factories, electric power plants, mines, offshore oil-drilling, pesticide runoff, fertilizer runoff, construction runoff, and logging runoff.

4. Cholera, typhoid, dysentery. The pathogens that cause these diseases have been eliminated from treated public water supplies, and the occurrence of these diseases is rare in areas served by these supplies.

6. (a) An increase in water temperature decreases oxygen solubility.

   (b) A decrease in atmospheric pressure decreases oxygen solubility.

8. 8.2×10^{-6} M, 0.36.

10. 1.9×10^{-9} M, 580.

12. Coliform bacteria live naturally in the human intestinal tract, and the average person excretes billions of them in feces each day. Coliform bacteria are harmless and cause no diseases, but their presence in water is an indication of fecal contamination.

14. (a) The biological oxygen demand (BOD) is the capacity of the organic material in a sample of water to consume the dissolved oxygen.

    (b) The BOD is determined experimentally by adding heterotrophic microorganisms to a diluted effluent or water sample, saturating the sample with air, incubating the sample for 5 days, and then measuring the DO of the sample. The BOD is determined by comparing the initial DO to the final DO; the amount of oxygen used up (mg/L) during the experiment is the BOD of the diluted sample.

    (c) The lake is polluted with an oxygen-consuming waste (an example is sugars present in water from a fruit cannery).

16. In a river, water downstream recovers quite quickly once the discharge of organic wastes ceases, particularly if the water is free-flowing and turbulent. Organic material is diluted, and the water is reoxygenated as it is brought to the surface. Lakes, which have little flow of water, take much longer to recover.

18. When eutrophication occurs, dense mats of rooted and floating plants are formed. Blue-green algal blooms that appear on the water surface release unpleasant-smelling, bad-tasting substances.

20. (a) Analogous to pH, the measure of electron activity in aqueous solutions is defined as the negative logarithm of the electron activity and it is designated pE.

    (b) In natural waters, pE range from -12 to 25.

    (c) Yes.

    (d) No, reduction.

22. -6.8, reduction.

24. 11.06, oxidation.

26. -108 mv, pE $= -1.8$, reduction.

28. There may be several species present in the sample that are part of different redox systems. These samples produce potentials that are not one stable voltage, but rather drift to higher and higher potentials

30. Salt used to de-ice roads, and industrial discharges.

32. Sediments fill irrigation ditches and clog harbors. Sediments reduce the capacities of storage reservoirs. Toxic substances can adsorb on the surface of sediments and become concentrated. Spawning grounds become buried and photosynthesis in turbid water decreases.

34. A rise in water temperature increases the body temperature of aquatic organisms, metabolic and respiration rates rise, and increases oxygen consumption. Oxygen is less soluble in warmer water. This can be fatal to some fish. The life cycles of aquatic species are controlled by temperature.

36. (a) $Cl_2 + H_2O \rightarrow HClO + H^+ + Cl^-$.

    (b) Chlorine reacts with organic material in the water to form chlorinated hydrocarbons known as trihalomethanes (for example, chloroform, $CHCl_3$) which are suspected carcinogens.

    (c) Ozone.

38. Solids and suspended solids.

40. Aerobic bacteria break down all remaining oxygen-consuming waste.

42. Aluminum sulfate or lime is added to cause the precipitation of phosphate as insoluble aluminum or calcium phosphates. Biological methods.

44. Unlike phosphorous, it is not easy to precipitate nitrates; this ion does not easily form insoluble precipitates *per* the solubility rules.

46. (a) $NH_4^+ (aq) + 2\ O_2 (aq) + H_2O \rightarrow NO_3^- (aq) + 2\ H_3O^+ (aq)$.

    (b) $4\ NO_3^- (aq) + 5\ \{CH_2O\} \rightarrow 2\ N_2 (g) + 4\ HCO_3^- (aq) + CO_2 (aq) + 3\ H_2O$.

48. The act did the following: (1) provided for enforcement by EPA with stiff penalties, (2) created a system for identifying new point sources, (3) established water quality standards for discharged wastewaters, (4) set pretreatment standards for industrial wastes prior to discharge and (5) provided federal funding to build wastewater treatment plants. The funding of new waste treatment plants and enforcement by EPA have been especially effective.

50. (a) Primary standards specify contaminant levels based on health-related criteria.

    (b) Secondary standards are based on non-health criteria.

52. (a) As 0.010 mg/L.

    (b) Cd 0.005 mg/L.

    (c) Pb 0.015 mg/L.

    (d) Benzene 0.005 mg/L.

    (e) PCBs 0.0005 mg/L.

■ Chapter 9

2. The same sample is sent to many laboratories and the proposed analytical method is performed by many different analysts to determine its reproducibility. Once a review committee has reviewed the results from all the participating laboratories and has determined the analytical measurements are consistent independent of laboratory, then the method is designated a standard method.

4. (a) An analyte is the substance that is being analyzed.

    (b) A matrix is the medium containing the analyte.

6. Refrigeration to 4°C.

8. A set of five test tubes, filled with a lactose broth, are each inoculated with a 10 ml sample of drinking water. The test tubes are incubated at 35°C for 48 hours. Production of gas or acidic growth in the tubes is an indication of coliform contamination.

10. Coliform possesses the enzyme β-D-galactosidase, which hydrolyses chromogenic substrates such as ortho-nitrophenyl-β-D-galactopyranoside (ONPG), resulting in release of

the chromogen and a color change to yellow in the liquid media. A kit for this test is commercially available. The inoculated media is placed in a 35°C water bath and a yellow color after 30 minutes indicates the sample is positive for coliform.

12. The TOC is a measure of all organic carbon in a water sample.
    (a) Heat and oxygen, ultraviolet radiation, or chemical oxidants convert the organic carbon to carbon dioxide.
    (b) TOC is the best way of assessing the organic content of a water sample. The TOC is not dependent on the oxidation state of the organics in the water and does not measure other elements, such as nitrogen and inorganic elements that can contribute to the BOD.
    (c) The CO_2 produced can be measured directly by a nondispersive infrared analyzer, NDIR, it can be reduced to methane and then measured with a flame ionization detector, or it can be determined by titration

14. Microwaves, IR, visible, UV, X-rays.

16. (a) 0.643.
    (b) 0.501.
    (c) 0.987.

18. 0.099 $(ppm)^{-1}cm^{-1}$.

20. (a) 0.425 $(ppm)^{-1}cm^{-1}$.
    (b) 1.06 ppm.

22. (a) See Figure 6.2. A tungsten lamp for the visible region;
    (b) A deuterium lamp for the UV region.
    (c) No, quartz cells must be used for UV; quartz or glass is ok in the visible.

24. Aqueous phosphate reacts with ammonium molybdate in acidic conditions to form molybdophosphoric acid; after addition of vanadium, vanadomolybdophosphoric acid is formed, which has a yellow color. The method is designed so that the intensity of the yellow color (400–490 nm) is directly proportional to the phosphate concentration.

26. See Figure 9.11.
    (a) With a microliter syringe.
    (b) See Figure 9.9 and 9.10.
    (c) A conductivity detector—because all ions are conductive.

28. The suppressor reduced background conductivity from the eluent and attenuates the conductance of the analyte ions. This is accomplished by exchanging Na^+ for H^+ in a cation suppressor column or membrane module. The H^+ subsequently reacts with carbonate and bicarbonate in the eluent to produce H_2CO_3, a very weak acid, which has a relatively low conductivity. If anions (X) besides carbonate or bicarbonate are present, these form acids (HX) with increased conductivity.

30. Conductivity measurement.

32. 2000μmhos/cm.

34. Hydroxide ion, OH^-. Three times better than other anions.

36. See Figure 9.7.

38. (a) Construct calibration curve.
    (b) 5.65 ppm.
    (c) 1.4 ppm.

40. (a) 1.5 ppm.
    (b) 20 ppm.

42. The measurement of the pH of natural waters gives an *intensity factor,* which is a measure of the concentration of acid or base that is immediately available for reaction. Alkalinity, on the other hand, is a *capacity factor* that measures the buffering ability of a body of natural water to sustain reaction with acids or bases.

44. Alkalinity = $[OH^-] + [HCO_3^-] + 2\,[CO_3^{2-}] - [H_3O^+]$.

Acidity = titration to pH = 8.3 endpoint.

$$H_2CO_3 \rightarrow HCO_3^- + H^+ \qquad Ka_1 = 4.45 \times 10^{-7}$$
$$HCO_3^- \rightarrow CO_3^{2-} + H^+ \qquad Ka_2 = 4.69 \times 10^{-11}$$

46. (a) The alkalinity of sample #1 is greater.
    (b) Sample #1.

48. (a) Mg^{2+}, Ca^{2+}, up to 250 ppm.
    (b) Cation exchange resin.
    (c) HNO_3 eluent would be suppressed with a base NaOH.
    (d) Conductivity.

50. (a) The gamma spectrometer has multiple solid-state detectors that are able to measure gamma rays of different energies and this allows the identification of the isotope that emitted the gamma ray.
    (b) The proportional counter has a very thin window that allows both alpha and beta particles to penetrate into the instrument, which is similar to the Geiger counter.

■ Chapter 10

2. When oil was discovered in 1859, the U.S.'s energy economy was sustained by renewable sources: wood, animal and human muscle, windmills, and water wheels. Very little coal was used at this time.

4. Petroleum originated from microscopic marine organisms that once lived in great numbers in shallow coastal waters. The dead organic matter became buried in anoxic sediments, eventually undergoing chemical transformations (diagenesis) into petroleum hydrocarbons.

6. Gas, oil, and water are found together. Water usually occupies one-half of the pore space in the reservoir rock. In part because of the water, it is impossible to achieve 100% recovery of the petroleum from a reservoir rock.

8. Natural gas burns cleanly, leaves no residue, and weight-for-weight, has a higher heat output when burned than any other common fuel.

10. $2\,C_4H_{10} + 13\,O_2 \rightarrow 8\,CO_2 + 10\,H_2O$ + heat.
    (a) 3200 kcal.
    (b) 3217 g.

12. $2\,C_8H_{18} + 25\,O_2 \rightarrow 16\,CO_2 + 18\,H_2O$ + heat.

$$2\,C_2H_6O + 6\,O_2 \rightarrow 4\,CO_2 + 6\,H_2O + \text{heat.}$$

Two moles of octane give much more energy than would two moles of ethanol. If 100 grams of each were compared, octane would give 1072 kcal and ethanol 542 kcal. Since ethanol gives less energy per gram, it would be the less efficient fuel (half as efficient).

14. Sulfur will deactivate the catalysts used by refineries in the cracking step. Sulfur in refined gasoline will produce sulfur dioxide when burned.
    (a) 3.7%.
    (b) 0.1%.
    (c) No, petroleum contains organic sulfur containing molecules such as thiophene.
16. Fractional distillation is a process that separates the hydrocarbons according to their boiling points.
    (a) boiling range (°C).
    (b) Asphalt is the material that is left in the pot after all volatile material has boiled off.
    (c) Boiling range 70–180°C.
    (d) Kerosene is a higher boiling fraction (180–230°C).
18. (a) $C_{16}H_{34} \rightarrow C_8H_{18} + C_8H_{16}$.
    (b) An aluminosilicate, which is impregnated with potassium.
    (c) The alkenes will oxidize.
20. Catalytic reforming is a process in which hydrocarbon vapors from straight-run gasoline are heated in the presence of suitable catalysts such as platinum and they are turned into branched hydrocarbons.
22. Toluene.
24. Methyl-t-butyl ether (MTBE), ethyl-t-butyl ether (ETBE), methanol, ethanol.
26. MTBE is an octane enhancer.
    (a) MTBE is made by reacting methanol and isobutene in the presence of an acid catalyst.
    (b) MTBE has been detected in water supplies and, for health reasons, the EPA has recommended a reduction in its use.
28. Benzene, toluene, xylene, ethylbenzene (collectively referred to as BTEX).
30. Oil shale is mined like coal. It is crushed and heated in the absence of air to produce shale oil. Large quantities of waste rock remain after the shale oil has been removed. Because the waste rock is alkaline and poor in nutrients, very little will grow on it. For every gallon of shale oil produced, 2–4 gallons of water are required. It is not certain that it is possible to produce any net energy from oil shale, that is, more energy investment might be required than is produced from the obtained oil.
32. Tar sands consist of sandstone mixed with a black high-sulfur tar-like oil known as bitumen.
34. Bituminous coal is soft and anthracite coal is hard. Anthracite is a better fuel because it is formed from bituminous coal by the action of heat and pressure. Volatile compounds are released, the water and sulfur contents decrease, and the carbon content increases. Anthracite provides more energy per unit mass of coal, and is better suited for steelmaking and metallurgy.
36. Coal is mined by underground or strip mining. Both have adverse environmental impacts. Strip mining devastates the land unless adequate steps are taken to reclaim it. Both underground and strip mining produce large quantities of waste rock, and acid leaching from the waste can contaminate nearby bodies of water.
38. $C + O_2 \rightarrow CO_2$.
40. Many of the disadvantages of coal can be overcome if it is converted into a gas or liquid prior to burning. Coal can produce a gaseous fuel by several different coal gasification reactions. The gas can be easily transported by a gas pipeline, and burned without leaving ash and without producing sulfur dioxide.

42. Coal gasification is the process that changes solid coal in to a synthesis gas.

$$C + H_2O \rightarrow H_2 + CO$$

$$CO + 3H_2 \rightarrow CH_4 + H_2O$$

$$CO + H_2O \rightarrow H_2 + CO_2$$

Synthesis gas is a useful fuel, but it produces only about one-third of the heat produced by the combustion of natural gas.

44. (1) Burning fossil fuels produces carbon dioxide, a greenhouse gas. (2) The extraction of fossil fuels has adverse environmental consequences.

46. 81%

48. All, a–d, are increasing demands for fossil fuels.
    (a) Farming: petroleum products.
    (b) Food processing: electricity from coal, nuclear, oil, gas; petroleum in food distribution networks
    (c) Manufacturing: electricity from coal, nuclear, oil, gas; petroleum is needed for raw materials (plastics, fibers)
    (d) Transportation: petroleum products.

50. Heat from the combustion of coal converts water in a boiler into high-pressure steam. The steam strikes the blades of a turbine, causing a coil of wire to rotate in magnetic field. This movement produces electricity. See Figure 10.12.

52. The 30 railroad cars of ash contain the noncombustible material that is contained in coal. The contents of the other 70 railroad cars were combusted in the power plant and went up the smoke stack as burned gas.

$$C + O_2 \rightarrow CO_2.$$

Coal + Oxygen → carbon dioxide.

54. One third.
    (a) 10–20%.
    (b) The CO_2 released from fossil fuel combustion would have to be selectively removed from the smoke stack and concentrated.

56. Deep ocean, salt domes, deep geological formations.

■ Chapter 11

2. Gamma ray.

4. An alpha particle has a larger mass than a beta or gamma. Because it is so large, it is easy to stop outside the body, but inside the body, it causes more damage.

6. (a) ^{206}Tl.
   (b) p.
   (c) ^{226}Ra.

8. Mass Number: Atomic Number:
   (a) Stays the same.   Increases by 1.
   (b) Decreases by 4.   Decreases by 2.
   (c) No change.   No change.

10. The chemical reactivity of sulfur-35 is exactly the same as sulfur-32.
12. b.
14. c.
16. 17.5 g.
18. 34 mg, 2 mg.
20. Radiation enters Geiger tube and ionizes argon gas causing electric current to flow. See Figure 11.6.
22. 22.4 years.
24. 1.2 %.
26. Chemical reactions will not make nuclear waste harmless. The same radioisotopes will be contained in the new compound.
28. See Table 11.4.
30. The rem takes into account the potential damage to living tissues caused by the different types of ionizing radiation. For X-rays, gamma rays, and beta particles, 1 rad is essentially equivalent to 1 rem, but for alpha particles, because of their greater ionizing ability, 1 rad is equivalent to 10 to 20 rems.
32. Yes. It is about 430 years old.
34. 1.82×10^9 years.
36. (a) 93 years.
    (b) 187 years.
38. 0.2% of uranium.
40. $^{235}_{92}U + ^{1}_{0}n \rightarrow ^{146}_{57}La + ^{87}_{35}Br + 3 ^{1}_{0}n$ + energy.
    A slow-moving neutron enters the uranium-235 nucleus and causes fission.
42. See Figure 11.12c.
44. No, nuclear fuel reprocessing was made illegal in the US. France and Japan.
46. (1) Milling and chemical treatment to bring the or to 80 % U_3O_8. (2) Reaction with HF to produce UF_6. (3) Separation of U-235 containing UF_6 from U-238 containing UF_6 by diffusion or centrifugation. (4) Convert U back to solid form.
48. 30 years.
50. $\sqrt{352/349}$.
52. (1) Entombment. (2) Mothballing. (3) Dismantle.
54. U-235 is bombarded with fast moving neutrons to produce Pu-239. Because two or three neutrons are produced in every U-235 fission, two or three Pu-239 atoms are formed from each U-235. Pu-239 is one of the most toxic elements known.
56. State your own opinion.
58. Deuterium and tritium are the elements that are being fused (fuel).
60. 4 times.

■ Chapter 12

2. Parabolic trough: Focuses sunlight onto a pipe running down its center. Oil or other fluid in the pipe is heated to a temperature up to 400°C (750°F). The heat is used to boil water and produce steam that can be used to turn a turbine and generate electricity.
Central receiver: Uses a circular array of heliostats to concentrate sunlight onto a central receiver mounted at the top of a tower. Molten nitrate salts are used as a heat transfer media into the central receiver. The salt is heated it approximately 550°C and is then pumped

either to a "hot" tank for storage or through heat exchangers to produce superheated steam. The superheated steam is used in a conventional turbine generator to produce electricity. Parabolic dish: Uses an array of parabolic dish shaped mirrors to concentrate sunlight onto a receiver located at the focal point of the dish. The receiver absorbs energy reflected by the concentrators and fluid in the receiver is heated to approximately 750°C. Because this unit heats the transfer fluid to a higher temperature, this design achieves a higher efficiency than the power tower or parabolic trough. The simpler design of the parabolic dishes make them attractive choices as power sources in remote locations.

4. Parabolic dish, because the transfer fluid is at a higher temperature.
6. See Figure 12.4.
   (a) B, Al, In, Tl.
   (b) N, P, Sb, Bi.
   (c) Ge, GaSa, CdTe.
8. Wood, crop waste, sugar cane, garbage: energy yield is low.
10. CO_2 from burning biomass was recently transferred from the atmosphere into a plant.
12. (a) The methane in biogas can be converted to methanol:

$$CH_4 + H_2O \rightarrow CO + 3\,H_2$$

$$CO + 2\,H_2 \rightarrow CH_3OH$$

   (b) Fermentation of sugars and starches:

$$C_6H_{12}O_6 \rightarrow 2\,CO_2 + 2\,C_2H_5OH$$

   (c) $(CH_2O)_n$ + anaerobic bacteria $\rightarrow CH_4$.
14. Gasohol is a 10–20% mixture of ethanol in gasoline.
   (a) Methanol requires major modifications of the conventional engine.
   (b) Ethanol can be used in engines with only minor changes.
16. (a) 10 miles/hour.
   (b) California.
   (c) Denmark, Spain, Germany.
   (d) Denmark.
18. Wind is intermittent, dispersed, and cannot produce large amounts of energy in a compact footprint, as is possible with fossil fuel or nuclear generating plants.
20. (a) 80 % efficiency, much greater than coal fired generators.
   (b) Yes.
   (c) Land is flooded, silt is trapped behind dam, migration and spawning of fish disrupted.
22. (a) 1.17 V.
   (b) 1.08 V.
   (c) 0.56 V.
24. It is lost as heat.
26. See Figure 12.6.

   (a) $\frac{1}{2}O_2 + 2\,H^+ + 2\,e^- \rightarrow H_2O$.

   (b) $H_2 \rightarrow 2\,H^+ + 2\,e^-$.

   (c) The membrane material has a fixed sulfonate site that cannot move. The protons hop from sulfonate site to sulfonate site.

28. (a) Platinum.
    (b) It could be replaced with porous carbon that is coated with Pt.
30. A gas, such as H_2, that has a critical temperature below ambient temperature. No matter how much pressure is applied, the gas will not condense into a liquid.
32. PAFCs are 85% efficient when used for the co-generation of electricity and heat, but only 37 to 42% efficient generating electricity alone. Coal fired electrical plants are 33–35% efficient.
    (a) No, they provide less power per pound than do other fuel cells.
    (b) No.
    (c) Yes.
34. The major advantage is that methanol is easier to transport than hydrogen. The major disadvantages are that it is very expensive, not as well developed as other fuel cells.
36. Less energy is lost as waste heat.
38. Because they operate at extremely high temperatures (650°C).
40. See Figure 12.10.

    (a) $\frac{1}{2}O_2 + 2\,e^- \rightarrow O^{2-}$.

    (b) $H_2 + O^{2-} \rightarrow 2\,H_2O + 2\,e^-$.

    (c) High-temperature operation. The electrolyte is a hard, porous ceramic compound with the cell operating at up to 1000°C.

42. See Figure 12.11.
    (a) $2\,H_2 + \rightarrow 4\,H^+ + 4\,e^-$.

    (b) $O_2 + 4\,H^+ + 4\,e^- \rightarrow 2\,H_2O$.

    (c) KOH

44. See Figure 12.11.
    (a) $4\,H^+ + 4\,e^- \rightarrow 2\,H_2$.

    (b) $2\,H_2O \rightarrow 4\,H^+ + 4\,e^- + O_2$.

46. Sodium borohydride $NaBH_4$, a solid, releases hydrogen when a solution of sodium borohydride is passed over a catalyst.
    (a) $NaBH_4$, a solid.
    (b) $NaBH_4 + 2\,H_2O \rightarrow NaBO_2 + 4\,H_2$.
48. Powerballs are pellets of NaH that are coated in plastic.
    (a) $NaH + H_2O \rightarrow NaOH + H_2$.
    (b) NaH is made into pellets and coated with polyethylene.
    (c) NaH will be produced from the waste NaOH solution left behind in used Powerball tanks.
50. The most promising source of renewable energy is solar energy.

■ Chapter 13

2. 10 mg.
4. (a) 15 ppb.
   (b) 10 ppb.
   (c) 5 ppb.
   (d) 2 ppb.

6. Acetylene/air, 2,400 −2,700°K.
8. The pre-mix burner's spray chamber mixes the sample aerosol with a fuel and oxidant, so that it can be combusted in an elongated flame. Ground-state vaporized atoms from the sample droplets are formed, and these are detected in the burner's optical path.
10. Hollow cathode lamp.
    (a) The analyte element.
    (b) The lamp emits a beam with a wavelength that is exactly what is needed to measure the absorbance of the analyte atoms in the flame.
    (c) The deuterium lamp has a wide spectrum of emission and its output at the wavelength of interest is not as great as the hollow cathode lamp.
12. 0.12 micrograms/mL.
14. (a) Hg.
    (b) No, not on a conventional instrument.
    (c) No.
    (d) Hg = 1.8 ppm, Cd = 2.5 ppm.
16. (a) 15 ppb.
    (b) Old lead pipes, lead solder at pipe joints.
    (c) Yes, soft, acidic water dissolves the lead pipes faster than takes place with harder, more alkaline water supplies.
    (d) Stagnant water that has sat in the pipes for hours.
18. (a) 1978.
    (b) 2 mg/cm^3.
    (c) Title X is the federal law that regulates lead paint in government-owned buildings.
20. See Figure 6.25.
    (a) X-ray tube.
    (b) To minimize interference from scattered x-rays.
    (c) A silicon wafer that has been treated with lithium.
    (d) No, it cannot be focused to a small spot.
22. Although the individual dust particles may have high Pb concentrations, there is often too little dust loaded on the wipe to use XRF. Instead, it is better to digest the wipe, and analyze the obtained solution by AAS or ICPAES.
24. 2.5 ug/L (ppb), no.
26. (a) 283.3 nm.
    (b) Add lanthanum nitrate.
    (c) Glass containers may leach lead into the sample.
28. The burning of coal.
    (a) Yes.
    (b) Yes.
    (c) Chemicals, paints, plastics, pharmaceuticals, steel, and electrical equipment.
30. (a) Hg (g) can be formed more selectively at low temperatures by chemical reduction in a process known as "cold vapor."
    (b) With Sn^{2+}.
    (c) No.
32. (a) By the addition of $KMnO_4$.
    (b) Add even more $KMnO_4$.

(c) Yes, check sample for volatile organics before starting sample. Some organic compounds strongly absorb light at 254 nm (the wavelength used to detect Hg in CVAAS), so it is important to completely oxidize these substances in the digestion, so that they do not interfere and produce a "false positive."

34. Advantages: Lower level of detection, simultaneous determination of multiple elements, broader elemental coverage than is possible with AAS.
    (a) No.
    (b) See Figure 13.9.
    (c) Yes, but ICPAES is not especially sensitive for determination of Hg; CVAAS is preferred.

36. $Cd^{2+}(aq) + 2\ e^- \rightarrow Cd(g) \rightarrow Cd^*(g)$.
    (a) Emission.
    (b) Because in absorption you are measuring the difference between two signals and in emission the signal is measured against background.
    (c) With Ar (g), it is easy to produce a stable, hot plasma that excites many emission lines of analyte elements (for ICPAES) and generates gas-phase ions (for ICPMS). The Ar plasma has a relatively low continuum and Ar line emission background.

38. Emission relies on measuring the appearance of a small signal vs. a low background, while AAS measures a small decrease in a large signal (the lamp intensity) caused by the absorbing atoms. Emission is generally, inherently a more sensitive measurement than absorption for these reasons.

40. See Figure 13.8.
    (a) Directly injected.
    (b) Soil is digested in a strong acid, liquid is filtered, diluted to a known volume, and nebulized into the plasma.
    (c) A device that uses the interaction of a liquid with a high-velocity gas stream to shear the liquid into fine droplets, known as "pneumatic nebulization."

42. (a) Deionized water that is acidified with HNO_3 and HCl flushes the system between measurements.
    (b) The reagent blank must contain all the reagents and in the same volumes as used in the processing of the samples. The reagent blank must be carried through the complete procedure and contain the same acid concentration in the final solution as the sample solution used for analysis.
    (c) Select a representative sample which contains minimal concentrations of the analytes of interest by known concentration of interfering elements that will provide an adequate test of the correction factors. Spike the sample with the elements of interest at the approximate concentration of either 100 ug/L or 5 times the estimated detection limit.

44. Arsenic cannot be determined by cold-vapor AAS.
    (a) As 613°C, Hg 356°C.
    (b) No.
    (c) Not with low detection limits, see Table 13.2.

46. Arsenic is converted to a hydride (AsH_3) that is volatile at room temperature (boiling point −62°C); the hydride gas is passed into a specialized AA spectrometer cell where it is decomposed into metal atoms that are then measured by AA spectrometry.

48. As, Sb, Se, Te.
    (a) See Table 13.3.
    (b) Yes.
    (c) Yes, but with much higher detection limits. See Table 13.1.
50. Collect sample. Run EPA method 200.7 to screen for metals. If either appears to be present, run specific EPA method for that analyte.

■ Chapter 14

2. Endocrine disruptors are synthetic chemicals that block, mimic, or otherwise interfere with naturally produced hormones.
   (a) Yes.
   (b) Effects passed from one generation to the next (mother–daughter).
   (c) No, they are a concern to humans as well.
4. Persistent organic pollutants (POPs) are organic chemicals that may disrupt hormone functions.
   (a) By wind and water.
   (b) Some POPs evaporate from water or land surfaces into the air. The POPs then return to Earth in snow, rain, or mist.
   (c) The Stockholm Convention, was signed by more than 90 countries that promised to reduce or eliminate the production, use, and release of 12 key POPs, which are known as "the dirty dozen."
6. (a) Yes.
   (b) No.
   (c) Extremely slowly.
8. Pyrethrins, nicotine sulfate, rotenone, garlic oil.
10. Agent orange was a defoliant that was used in the Vietnam War by the U.S. Army. It is a mixture of 2,4-D and 2,4,5-T, contaminated by a class of compounds known as dibenzodioxins.
12. (a) See Figure 14.3.
    (b) No.
    (c) Yes.
14. Some of these pollutants, such as Aldrin, Dieldrin, PCB, and Dioxins, are listed the Stockholm Convention "dirty dozen," but the PBT list also contains inorganic elements, such as mercury or organometallic compounds that contain an inorganic atom.
16. (a) A high melting point tends to indicate low water solubility and a low melting point indicates increased absorption is possible through the skin, GI tract, or lungs.
    (b) Chemicals with vapor pressure greater than 10^{-4} mm Hg are more readily vaporized and often have higher potential inhalation exposures than chemicals with low vapor pressure.
18. (a) No.
    (b) Yes.
    (c) Yes.

20.

Organic compound	Log K_{ow}
DDT	6.2
1,2,4-trichlorobenzene	4.2
orthodichlorobenzene	3.7
benzene	2.1

As aromatic organics become more substituted with chlorines, they will be more likely to bioaccumulate. Of these three, benzene is the least likely to bioaccumulate.

22. (a) At constant temperature the solubility of a gas in a liquid is proportional to the partial pressure of the gas that is in contact with the liquid.

(b) EPA express K_H in two ways; the first is a nondimensional value that relates the chemical concentration in the atmosphere (gas phase) to its concentration in the water phase. The K_H can also be determined by dividing the vapor pressure of a chemical (in atmospheres) by its solubility in water (in mol/m^3) to give K_H in atm-m^3/mole.

(c) A large K_H (> 10^{-3}) indicates the chemical is likely to evaporate from the water and partition into air.

24. (a) The PBT profiler calculates an atmospheric half-life by determining the importance of a chemical's reaction with two of the most prevalent atmospheric oxidants, hydroxyl radicals and ozone. The half-life is calculated directly from gas-phase hydroxyl radical and ozone reaction rate constants.

(b) The half-life for degradation of a chemical in water, soil, and sediment is determined by the PBT Profiler by using the ultimate biodegradation expert survey module of the BIOWIN[1] estimation program.

(c) 6

26. Dieldrin.

28. The PBT Profiler indicates that benzidine is persistent in soil (75 day half-life) and sediment (340 days), but not in air (0.1 day) or water (38 days). It will bioaccumulate in sediment and soil. Benzidine would be considered toxic.

30. The PBT Profiler indicates that 3-methylphenol is persistent in sediment (140 days half-life), but not in air (0.25 day) or water (15 days) or soil (30 days). It will bioaccumulate in sediment and soil. 3-methylphenol would be considered toxic.

32. See Figure 14.8.

(a) With a GC microliter syringe.

(b) Silicone stationary phases are thermally stable at high temperatures and resistant to decomposition.

(c) Because helium is an inert gas, the sample will not react with it. Helium does not produce a background signal in either the ECD or FID detectors.

34. See Figure 14.12.

(a) As can be seen in Figure 14.12, the ECD is constructed of a cavity with two electrodes and a radiation source. A ^{63}Ni source releases β radiation, which collides with a reagent gas (5% methane in Ar) that is flowing through the detector, producing plasma containing electrons and positive ions. The electrons are attracted to the anode producing a small steady current. When sample molecules with a high electron affinity enters the detector, they capture some of the electrons and the rate of electron collection at the anode decreases. The detector senses the drop in the current and records the change in signal as the separated compound passes through the detector. The response generated by the detector is proportional to the amount of compound.

(b) Chlorinated organics are very electronegative and capture electrons readily.

(c) No. Hydrocarbon molecules do not capture electrons.

(d) FID, flame ionization detector.

36. There is a remote possibility some other organic compound might elute from the GC at the same retention time of the analyte of interest. By injecting the compound again on a second column that has a different stationary phase, a second retention time is measured. There is a minimal chance that another compound would have the same retention time on two different columns.

38. (a) 505, 508, 608.

(b) 525.2.

(c) 1624.

40. (a) 505, 508, 608.

(b) 505, 508, 608.

(c) 8018B, 8270D.

42. The molecular ion is M^+, which is formed by impact with an energetic electron:

$$M + e^- \rightarrow M^{\cdot+} + 2\,e^-$$

The mass of the molecular ion is the same as the analyte.

(a) Fragment ions are smaller ions that are formed from the molecular ion.

(b) 252.

(c) 94.

(d) 154.

44. See Figure 14.8.

(a) Microliter syringe.

(b) No, He has a mass of 4.

(c) Total ion current (TIC) from the analyte is compared to a calibration curve. An internal standard is commonly used.

46. When the total ion chromatogram is recorded, the intensity of all masses is summed and recorded as a data point at the time collected. The selected ion chromatogram depicts signal at one specific m/z, and only contains information about eluting components that produce mass spectral peaks at that particular mass. Better selectivity and higher sensitivity are possible in the selected ion monitoring mode, at the expense of not surveying a wide mass range.

48. (a) 1613, 613.

(b) 1613.

(c) 1613, 10pg/L; 613, 0.002 µg/L.

50. Use GC-MS. The GC would separate many of the isomers into individual peaks. For those isomers that were not separated in the GC, the furans would produce fragment ions in the MS that would have different masses from the dioxins.

■ Chapter 15

2. A pesticide that is broken down slowly in the environment. See Chapter 14, the PBT Profiler to determine if a chemical is persistent.

4. All organophosphate insecticides contain a phosphorous atom. See Figure 15.2.
   (a) Organophosphate esters with no sulfur.
   (b) Phosphorothioates with a sulfur replacing the oxygen that is doubly bonded to phosphorus.
   (c) Phosphorodithioate with two sulfurs attached to phosphorus.
6. Yes, see Figure 15.3.
8. No.
10. Neurotransmitter.
12. See Figure 14.8.
14. Organophosphorus compounds do not give a strong signal in the ECD.
16. (a) Sodium chlorate, sulfuric acid, copper salts, and other inorganic chemicals.
   (b) They are absorbed by leaves or roots and are then transferred throughout the plant.
   (c) A mixture of 2,4-D and 2,4,5-T; this was used as a defoliant in the Vietnam War.
18. The traditional method for controlling weeds. The soil is plowed, which turns over the top 10–15 inches of soil and buries and smothers weeds. The soil is then harrowed to break up large clods before the crop is planted.
20. The animal's biosynthetic pathway is different from a plant's.
22. See Figure 15.6d.
   (a) Bipyridilium compounds (two pyridines).
   (b) They are applied to the soil so they will kill emerging weeds.
   (c) Yes.
24. See Figure 15.7.
26. No.
28. Silica, alumina, polystryrene-divinylbenzene resin, ion-exchange resin. EPA Method 549.1 uses the polystyrene-divinylbenzene resin.
30. Differential refractive index, UV, fluorescence, evaporative light scattering, electrochemical, mass spectrometric. Method 549.1 calls for the use of a UV spectrophotometric detector.
32. Yes.
   (a) Yes.
   (b) Soil fumigant to control root nematodes.
34. Pheromones are chemical communicating substances that operate between members of the same species.
   (a) To trap insects.
   (b) Picograms.
   (c) No.
   (d) No.
36. Yes. Trail pheromones could be used to lure ants into a trap.
38. JH regulates the early stages of insect development. It has an ester and epoxide functional groups.
40. Molecules that are structurally similar to JH, are stable molecules and function like JH when applied to insect larvae.
   (a) JH mimics do not degrade as rapidly as does JH itself.
   (b) Yes.
   (c) The immature insect will not emerge from its pupae a fully formed adult.

42. Allomones are chemical communicating species that act between species and the substance provides a benefit to the species releasing it. Skunk smell keeps predators away. Caterpillars protect themselves with toxic chemicals that keep birds from eating them.

44. The naturally occurring soil bacterium *Bacillus thuringeinsis* (Bt). It is a powder that is used by home gardeners as either a power or spray to kill leaf-eating caterpillars.

46. They only destroy the target species and are not toxic to other species.

48. Tobacco plants, nicotine, chrysanthemum pyrethrums.

50. Very little new cropland is available.

52. *Bacillus thuringeinsis*, Bt, is a naturally occurring soil bacterium that is toxic to many leaf-eating caterpillars. Plants have been genetically engineered to produce high levels of Bt through the growing season.

▪ Chapter 16

2. DDT is poorly absorbed through the skin of the mammal, but it easily passes through the chitinous exoskeleton of the insect.

4. The stomach. Use Henderson-Hasselbalch equation.

6. Dermal contact, inhalation, and oral ingestion.

8. (338 mg/kg)(70 kg) = 23,600 mg acetaminophen needed, 48 tablets.

10. The LD_{50} for rats suggested that Paraquat is more toxic to humans than Silvex.

12. Oxidation in the liver to acetaldehyde which is then oxidized to acetic acid.

14. Liver enzymes that act on fat soluble substances to make them water-soluble.

16. Mutagens are chemicals that are capable of altering genes and chromosomes to cause inheritable abnormalities in offspring.

18. DNA—deoxyribose and RNA—ribose.

20. CAGTC.

22. The sugar is ribose, in DNA it is deoxyribose. RNA transcribes a uracil instead of thymine.

24. Nucleus of the cell, 46.

26. Transcription and translation.

28. $NaNO_2 + H^+ \rightarrow HNO_2 + Na^+$.

30. See reaction mechanism on page 492.

32. Benign tumors are not cancerous; malignant tumors are cancerous.

34. Benzo(a)pyrene, PAHs, benzene, benzidine.

36. Initiation occurs by a reaction of DNA with genotoxic carcinogens, substances that irreversibly bind with DNA. Promotion happens years later in response to exposure to a toxic agent, or promoter.

38. The concentration of carcinogens in the water is low enough that when people drink an average amount of water they are not exposed to an amount that would initiate a cancer.

40. Rodents' lives are short and they develop cancer relatively quickly.

42. (a) The bacteria that are used in the Ames test are modified biologically so that they are unable to synthesize the amino acid histidine. In the absence of histidine, the bacteria cannot grow.

   (b) The Ames tests from mutagens.

(c) The modified bacterium is placed in a medium that contains all of the ingredients that they need to grow except histidine. The chemical to be tested is added to the media. If the chemical being tested is not a mutagen, no bacterial growth is observed. If the chemical is a mutagen, a mutation will occur; the bacteria will revert to a form that can synthesize histidine, and bacterial growth will occur.

44. (1) Hazard assessment.
(2) Dose-response assessment.
(3) Exposure assessment.
(4) Risk characterization.

■ Chapter 17

2. (a) Sheet.
(b) Chain.
(c) Double chain.
(d) Three-dimensional array.

4. $A_{0-1}, B_2C_5T_8O_{22}(OH)_2$.
(a) Fe.
(b) See Figure 17.6.

6. Chrysotile fibers are found as veins in rocks. As can be seen in Figure 17.9, the chrysotile fibers have grown at right angles to the walls of the cracks. Most of the chrysotile fibers are extracted from deposits where fiber length is most often about 1 cm.

8. (a) Chrysotile.
(b) Quebec, Canada.
(c) South Africa.
(d) No.

10. (a) 550°C.
(b) Increases.
(c) Dehydroxylation of the brucite layer continues until it is converted to magnesium silicate.

12. (a) Scarring of the lung tissue.
(b) Cancer of the pleural lining.
(c) Malignant tumor of the bronchi.

14. Nose and mouth, larynx, trachea, bronchus, bronchioles.

16. Nasal hairs, cilia, mucous.

18. Short fibers smaller than 10 μm, 50 to 200 μm. Particles that travel deep within the lungs are ~2 microns and smaller. Fibers behave somewhat differently because they have different aerodynamic diameters than most other particles, the great majority of which can be approximated by spherical shapes.

20. 5.9 μm, no.

22. 1971. It has because most of the exposure took place in the workplace.

24. (a) No. If 5 fields counted, 0.017 f/cm³.
(b) No. If 5 fields counted, 0.093 f/cm³.
(c) Yes. 0.15 f/cm³.

26. See Figure 17.13.

28. Polarized light microscopy is used. Because asbestos fibers are anisotropic they change colors as the microscope stage is rotated.

30. The air to be sampled is pulled through a filter. The filters are opened and mounted so they can be counted using phase contrast microscopy, PCM.

32. Yes, it exceeds the OSHA standard. If NIOSH method 7402 were used, the number of asbestos structures counted would be even greater because the TEM can detect smaller fibers.

34. When the asbestos to be measured was physically abraded into smaller pieces.

36. (a) K, lowest electron shell; L, second electron shell; M, third electron shell.

(b) Intensity versus the energy of the x-ray in electron volts (keV, 1000's of electron volts common).

(c) Both minerals contain Mg. The position of the Mg peak is the same for both minerals because the EDX energies for Mg result from Z (atomic number); the inner-shell energy level structures are essentially independent of the chemical form of Mg.

38. Because it contains Fe and Mg.

40. Crocidolite—the spectrum shows a high concentration of iron.

42. One peak for the loss of a K-shell electron and a second peak for the loss of an L-shell electron.

■ Chapter 18

2. Resource Conservation and Recovery Act (RCRA).

4. Fly ash is the solid material that is blown up from an incinerator smoke stack with combustion gases. When discarded items such as cans, jar lids, and batteries are incinerated, many toxic metals including beryllium, cadmium, chromium, lead, and mercury, are released and carried up the incinerator stack in fly ash.

6. More MSW will be recycled only when recycling becomes more cost-effective. Dealers and manufacturers will not be willing to accept items for recycling if they cannot count on steady markets for recycled products.

8. Hazardous waste is waste material that poses a substantial hazard to human health or the environment if improperly treated, stored, transported, or disposed of. The producer of the waste is obligated to determine, using EPA-specified regulatory criteria, if a waste is hazardous. The producer of the waste is responsible for "cradle-to-grave" responsibility.

10. Corrosive waste.

12. Reactive waste.

14. Toxic waste.

16. Neutralize it with sodium bicarbonate to make water, CO_2, and a salt:

$$HNO_3 + NaHCO_3 \rightarrow H_2O + CO_2 + NaNO_3$$

18. Hazardous wastes that cannot be recycled or converted to non-hazardous forms must be stored. Perpetual storage methods include (1) burial in a secured landfill, (2) deep-well injection, and (3) surface impoundment.

20. Groundwater 60 feet, hazardous waste 1000 feet.
22. Superfund is a government program to clean up contaminated, abandoned sites. It is funded by a tax on chemical and petroleum industries.
24. The EPA's National Priorities List identifies Superfund sites most needing cleanup.
26. Recover chemicals from waste streams. One could also follow "green chemistry" manufacturing procedures that generate less waste in the first place.
28. Ocean.
30. Nuclear weapons manufacturing facilities and nuclear power plants.
32. Ten half-lives must elapse before radioactivity is reduced to negligible levels: cesium-137 would require 300 years.
34. Radioactive wastes are stored in underground tanks.
36. Plutonium could be reprocessed. It would be oxidized and mixed with conventional uranium fuel to form a mixed oxide, MOX, which could be used as fuel to produce energy for commercial nuclear power plants.
38. WIPP is a pilot-scale test facility to investigate permanent storage site for low-level radioactive wastes at Carlsbad, New Mexico.
40. High-level radioactive waste.
42. Spent fuel reprocessing is complex from the engineering standpoint; the fuel does contain recoverable U and Pu, but this is mixed with highly radioactive fission products. The US has reprocessed power reactor fuel in the past, but has abandoned these efforts in recent decades.
44. Transmutation is the process that changes one isotope to another. By neutron bombardment, radioactive isotopes could be changed into isotopes that are not radioactive.
46. Entombment advantages: Workers are not exposed to as much radiation as in dismantling; radioisotopes decay under concrete and become less radioactive; it is cheaper than dismantling. Entombment disadvantages: Site is ugly and has to be protected from tampering.
48. Immobilize in glass, dispose of as waste, reprocess, use as nuclear power plant fuel.
50. (1) Very dry,
    (2) rock formation is tuff,
    (3) low probability of seismic activity.
52. Not on planet Earth.

■ Appendix D

2. a. 2,3-dimethylbutane
    b. 3-ethylpentane
    c. 2-methylbutane
    d. 4-methylheptane

4. ethane CH_3CH_3
    propane $CH_3CH_2CH_3$
    butane $CH_3CH_2CH_2CH_3$
    pentane $CH_3CH_2CH_2CH_2CH_3$
    hexane $CH_3CH_2CH_2CH_2CH_2CH_3$

6. pentane $CH_3-CH_2-CH_2-CH_2-CH_3$

2-methylbutane $CH_3-\overset{\overset{\displaystyle CH_3}{|}}{CH}-CH_2-CH_3$

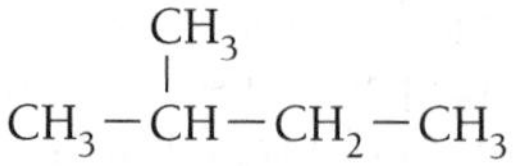

2,2 dimethylpropane $CH_3-\overset{\overset{\displaystyle CH_3}{|}}{\underset{\underset{\displaystyle CH_3}{|}}{C}}-CH_3$

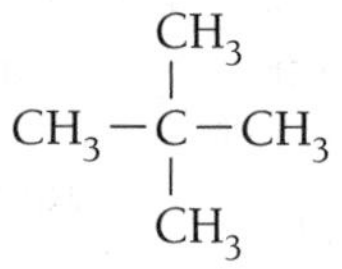

8. a. 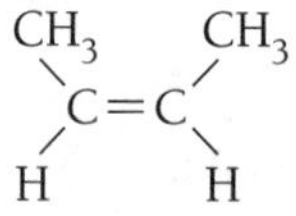 $\overset{\displaystyle CH_3}{\diagdown}\underset{\displaystyle H}{\diagup}C=C\overset{\displaystyle CH_3}{\diagup}\underset{\displaystyle H}{\diagdown}$

b. $H_2C=CH-CH_2-CH_2-CH_2-CH_3$

c. $H_2C=CH-CH_3$

d. $H_2C=CH-CH_2-CH_2-CH_3$

10. a. $CH_3-CH_2-CH_2-CH_2-CH_2-CH_3$

b. 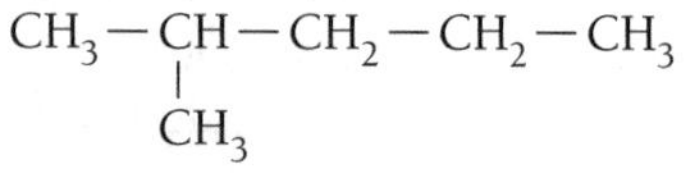 $CH_3-\overset{\overset{\displaystyle CH_3}{|}}{CH}-CH_2-CH_2-CH_3$

c. 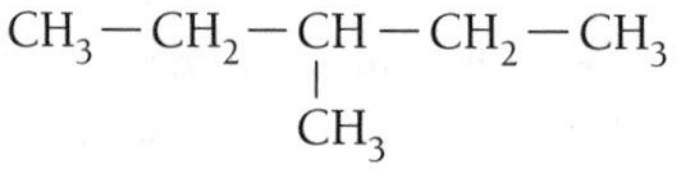 $CH_3-CH_2-\overset{\overset{\displaystyle CH_3}{|}}{CH}-CH_2-CH_3$

d. 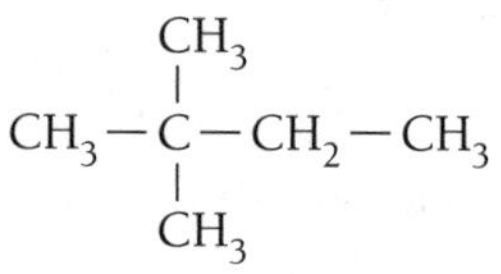 $CH_3-\overset{\overset{\displaystyle CH_3}{|}}{\underset{\underset{\displaystyle CH_3}{|}}{C}}-CH_2-CH_3$

e. 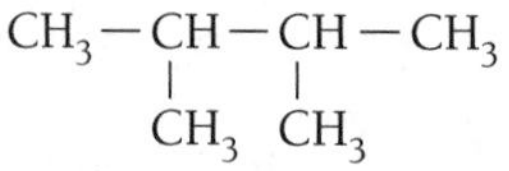 $CH_3-\overset{\overset{\displaystyle CH_3}{|}}{CH}-\overset{\overset{\displaystyle CH_3}{|}}{CH}-CH_3$

f. 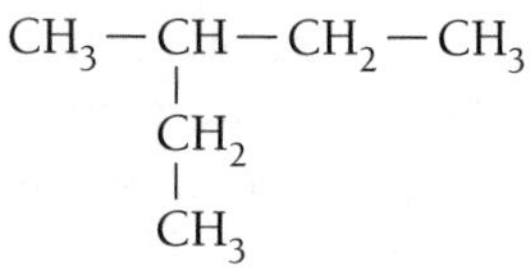 $CH_3-\overset{\overset{\displaystyle CH_2}{|}\overset{\displaystyle |}{\underset{\displaystyle CH_3}{}}}{CH}-CH_2-CH_3$

12. b. 2-butene;
    d. 1-chloropropene

14. a. 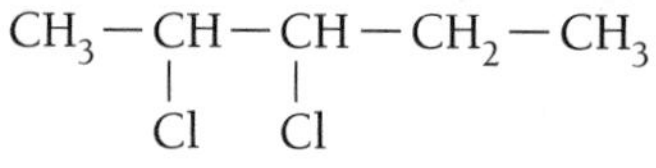 $CH_3-\overset{\overset{\displaystyle }{\underset{\underset{\displaystyle Cl}{|}}{CH}}}{}-\underset{\underset{\displaystyle Cl}{|}}{CH}-CH_2-CH_3$

b. 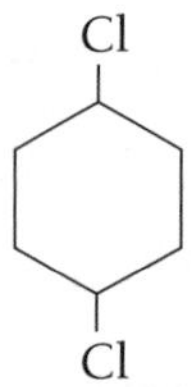
(cyclohexane ring with Cl at top and Cl at bottom)

c. 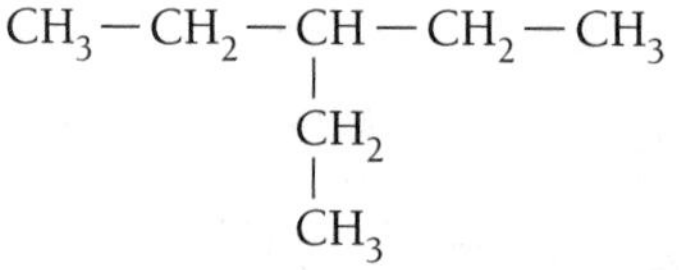 $CH_3-CH_2-\overset{\overset{\displaystyle }{\underset{\underset{\displaystyle CH_3}{|}}{\underset{CH_2}{|}}}}{CH}-CH_2-CH_3$

d. $Br-CH_2-CH_2-CH_2-CH_3$

16. a. ketone   b. carboxylic acid   c. ester
18. a. none (hydrocarbon)   c. alcohol   e. ester
    b. ketone   d. ketone and alcohol   f. ether
20. a. 2,2-dimethyl-3-hexene   b. 3-nonene   c. 3-hexene
22. a. RCOOH   b. RCOR′   c. ROH   d. RCOOR′

24. a. isopropyl alcohol   b. methyl alcohol   c. ethyl alcohol   d. ethylene glycol
26. Formaldehyde, $H_2C=O$
28. Carboxylic acid
30. Ester (ethyl acetate)
32. Both have two carbon atoms, but ethylene glycol has two hydroxyl groups ($-OH$) and ethanol has only one.
34.

36. Sugar and yeast
38. Acetic acid
40. An aldehyde has an hydrogen and an alkyl group bonded to the carbonyl. A ketone has two alkyl groups bonded to the carbonyl group.
42. a. HCHO   b. $CH_3COOCH_2CH_3$ c. CH_3CH_2OH d. CH_3COCH_3
44. Butyric acid has a sharp vomit-like smell; methylbutyrate, an ester of butyric acid, has a sweet smell.
46. a. amine   b. aldehyde (carbonyl)   c. carboxyl (carboxylic acid)   d. ester
48. a. $H-O-CH_2-CH-CH-CH_3$

b.

c. $CH_3-CH_2-CH_2-CH_2-CH_2-CH=CH-CH_2-CH_2-CH_2-CH_2-C$

50.

a.   b.   c.

meta

52. There are three possible isomers: 1,2,3,4; 1,2,3,5; 1,2,4,5.
54. Toluene
56. Para position

Photo Credits

Chapter Openers © AbleStock

Chapter 1
1.1 Courtesy of NASA; **1.6** ©Photodisc; **1.8a** ©1999 Russell Illig/Photodisc; **1.8b** ©1999 Ken Samuelson/Photodisc; **1.9a** ©Photodisc; **1.9b** ©Tom Pantages

Chapter 3
page 91 Courtesy of Josefino and NAA/Goddard Space Flight Center Scientific Visualization Studio

Chapter 4
4.5a Courtesy of Random Technology, Loganville, GA; **4.5b** Courtesy of Applied Ceramics, Inc., Atlanta, GA; **4.6** Courtesy of Cascades Volcano Observatory/USGS; **4.7** ©AbleStock; **4.8** ©Sai Yeung Chan/ShutterStock, Inc.

Chapter 5
5.1 Courtesy Greg Shirah GSFC Scientific Visualization Studio/NASA; **5.3** Courtesy of NASA

Chapter 6
6.5a Courtesy of Commander John Bortniak, NOAA Corps; **6.5b** Courtesy of Mauna Loa Observatory, Hawaii/NOAA/CMDL

Chapter 7
7.4b ©Jgroup/Dreamstime.com; **7.7** ©Richard Megna/Fundamental Photographs, NYC; **7.11** ©Amanda Haddox/ShutterStock, Inc.; **7.12** ©Karol Kozlowski/ShutterStock, Inc.; **7.13a, 7.13b** ©Don and Pat Valenti; **7.16** Courtesy of USGS

Chapter 9
9.1 Courtesy of Elizabeth H. White/CDC; **9.11** Courtesy of Dionex Corporation

Chapter 10
10.4a ©AbleStock; **10.5** Courtesy of Robert Sullivan/Argonne National Laboratory; **10.8a** ©Erik Patton/ShutterStock, Inc.; **10.8b** ©C.P. Hickman/Visuals Unlimited, Inc.

Chapter 11
11.12a Courtesy of the U.S. Department of Energy; **11.12b** ©Jim West/Alamy Images; **11.14a** Courtesy of Princeton Plasma Physics Laboratory; **11.15a** Courtesy of Lawrence Livermore National Laboratory and the U.S. Department of Energy

Chapter 12
12.2a ©Eyewire, Inc.; **12.2b** Courtesy of Southern California Edison; **12.2c** Courtesy of the U.S. Department of Energy; **12.4b** ©Eyewire, Inc.; **12.4c** ©Phil Meyers/AP Photos; **12.5a, 12.5b** Courtesy of enXco; **12.5c** Courtesy of Bergey Windpower Company; **12.5d** ©Warren Gretz/NREL; **12.5e** Courtesy of Atlantic Orient Corporation; **12.11b** Courtesy of Carla Thomas/NASA Dryden Flight Research Center and Aerovironment

Chapter 13
13.4 Courtesy of NITONÇ LLC; **13.10, 13.11** Courtesy of American University; **13.12, 13.13, 13.14** Courtesy of U.S. Army Corps of Engineers; **page 401** ©AbleStock

Chapter 15

15.10 Courtesy of the Tanglefoot Company

Chapter 16

16.8c ©Photodisc

Chapter 17

17.3a ©Gontar/ShutterStock, Inc.; 17.4b ©Keith McIntyre/ShutterStock, Inc.; 17.5a Used with permission of the California Department of Conservation, California Geological Survey; 17.9 ©Martin Miller/Visuals Unlimited; 17.13b Courtesy of Nikon Instruments Inc., Melville, New York; 17.14 ©Don Thomson/Photo Researchers, Inc.; 17.15 Courtesy of Allegro Industries; 17.16 Courtesy of Jim Murray/Criterion Laboratories, Inc.; 17.19 ©Dee Breger, Micrographic Arts; 17.21a Courtesy of WHD Microanalysis Consultants Ltd; 17.22a, 17.22b ©Dr. Jeremy Burgess/ Photo Researchers, Inc.; page 529–530 Courtesy of RJ Lee Group, Inc.; page 531 ©Robert St Coeur/ShutterStock, Inc.

Chapter 18

18.9a Courtesy of the U.S. Department of Energy; 18.10a, 18.10b Courtesy of the Yucca Mountain Project/U.S. Department of Energy

Appendix D

D.2a ©Joao Virissimo/ShutterStock, Inc.; D.2c ©Grant Heilman Photography/Alamy Images

Unless otherwise indicated, all photographs and illustrations are under copyright of Jones and Bartlett Publishers, LLC.

Index

agriculture (*cont.*)
  lime as fertilizer, 228
  magnesium as fertilizer, 228
  micronutrients, 221, 228–229
  mixed fertilizers, 229–230
  nitrogen fertilizers, 221–224
  nitrogen-fixing crops, 32
  no-till agriculture, 447
  pest-resistant crop plants, 459
  pesticide-related illnesses, 451
  phosphorus fertilizers, 225-227
  potassium salts, 228
  salinity of natural waters, 236
  sulfur as fertilizer, 228
  tillage, 447
  water conservation, 206
AHERA. *See* Asbestos Hazard Emergency
    Response Act
air. *See* atmosphere
air pollutants
  carbon monoxide, 96–97
  defined, 93
  indoors, 118–120
  National Ambient Air Quality Standards,
      117, 118, 169, 170, 173
  nitrogen oxides, 97–98
  sulfur dioxide, 104–105
  volatile organic compounds (VOCs), 93,
      98–102
air pollution
  about, 41
  acid deposition, 201
  Clean Air Act (1970), 58–59, 62, 116, 169
  Clean Air Act Amendments (1990), 104,
      108, 116, 163
  health effects of, 117
  indoor air pollution, 118–120
  industrial smog, 108–109, 115
  National Ambient Air Quality Standards,
      117, 118, 169, 170, 173
  photochemical smog, 95, 109–114
  primary sources of, 93
  regulating, 116–118
  temperature inversions and smog, 114–115
  in United States, 93–95

as worldwide problem, 117
  See also aerosols; smog; suspended particu-
      late matter
air sampling, 172–173, 517–518
air temperature, 85
airborne particles
  in atmosphere, 46
  See also suspended particulate matter
alarm pheromones, 455–456
albedo, 51, 52, 54, 56
alcoholic beverages, 473
alcohols, from biomass, 353–354
aldehydes, in smog-forming reactions, 111,
    113–114
aldicarb, 411, 444
aldrin, 441, 442
algae, 12, 17
  blooms, 230-231
  for nitrogen fixation, 222, 223
alkaline fuel cells (AFCs), 362–363
alkalinity, 198
  glass pH electrode, 261–262
  of natural waters, 262, 263–264
alkenes, reaction of hydroxyl radicals with,
    112–113
alkylation, 290
alkylnitrosamines, 491
allomones, 457
alpha a particles, 311, 312
  harmful effects of on body, 320
  properties of, 313, 322
alpha a rays, 311
aluminum
  abundance of on Earth, 9, 10, 11
  North American consumption of, 15
  ores, 14
  properties of, 14
  uses of, 14
aluminum alloys, 14
aluminum oxides, 14
aluminum phosphate, 34
aluminum sulfate, for sewage treatment, 238
Amazon River, calcium carbonate in, 29
American Conference of Government
    Industrial Hygienists, 513

American University Experimental Station
(AES), 391–392
Ames Test, 493–495
amino acids, as building blocks for early life, 16
ammonia
    anhydrous, 224
    in Earth's early atmosphere, 15–16
    for Haber process, 222
    in nitrogen cycle, 30, 31, 32, 223
    as refrigerant, 133
    synthesis of, 224
ammonium nitrate
    formation of from acid rain, 98
    production of, 224
ammonium phosphate, 224, 227
ammonium sulfate, 224, 228
amosite, 505, 508, 509
amphiboles, 503, 505–506, 506–507, 508–509
anaerobic bacteria, 16, 132, 384
anaerobic decomposition, methane from, 79
analyte, 252, 514
analytic measurement, 147–176
    of airborne asbestos, 517–524
    of atmospheric composition, 156–161
    of bulk asbestos, 514–517
    of hazardous wastes, 550
    of herbicides, 448
    of persistent organic pollutants in environ-
        mental samples, 422–433
    remote-sensing techniques for, 147
    Superfund methods for hazardous materials,
        550–551
    *See also under names of individual analytic
        techniques*
ANC. *See* acid-neutralizing capacity
anhydrous ammonia, 224
animals
    carbon cycle, 25, 26, 27–29
    essential chemical elements for, 25
    nitrogen cycle and, 30, 31, 32, 223
    phosphorus cycle and, 33–34
    respiration, 26, 27
anion-exchange resins, 264, 265
anions, separation and detection of by ion chro-
    matography, 266–268

anisotropic materials, 514
Antarctic
    ice core data, 68
    ozone depletion over, 126, 136–137, 138
anthophyllite asbestos, 505
anthracite coal, 171, 298
anthropogenic events
    albedo and, 55
    greenhouse gases and, 70, 72
antiandrogens, 411
anticholinesterase poisons, 445–446
anticline trap, 286
anticodons, 487
apatite, 34
aprotic solvent, 184
aquatic habitats
    algal bloom, 230–231
    dams and, 206, 356
    eutrophication, 230–231
    mercury and mercury compounds in, 384, 385
    plant nutrients and water pollution,
        220–235, 259–260
    polychlorinated biphenyls and, 408, 420
    sediment and, 235–236
    thermal pollution and, 236–237, 304
aquatic plants
    eutrophication, 230
    plant nutrients, 220–235, 259–260
aqueous solutions, molarity of, 22
aqueous systems, oxidation-reduction reactions
    in, 231–232
aquifer, 187
archeology, carbon-14 dating in, 327–328
Arctic, ozone depletion over, 140
argon, atmospheric, 44, 45, 72
Arizona, water rights of, 206
aromatic amines, 476
aromatic hydrocarbons, 296
    in coal, 299
    dissociation energy of, 112
arsenic
    atomic spectroscopy, 388–394
    for doping silicon, 350
    health effects of, 389
    in pesticides, 389, 407

arsenic hydride, 393–394
arsenic poisoning, 389
arsenous acid, 393
arsphenamine, 389
asbestos
    airborne, 509–513
    amosite, 505, 508, 509
    analytical methods to quantify, 514–524
    banned by EPA, 502, 514
    blue asbestos, 508
    as carcinogen, 478
    chemical properties of, 509
    crocidolite, 505, 508, 509
    defined, 501
    friable, 509
    health effects of, 509–513, 525
    history of usage of, 501–502
    as indoor air pollutant, 120
    litigation of health effects of, 525
    properties of, 501
    regulation of, 513–514
    structure of, 503, 505
    thermal properties of, 508–509
    tremolite, 505, 506, 508
    types of, 505, 506
    uses of, 501–502, 508
    worldwide production of, 502
asbestos-containing material (ACM), 501, 502,
    509–513
asbestos diseases, 509–513
Asbestos Hazard Emergency Response Act
    (AHERA), 514
asbestosis, 509
asthenosphere, 8, 9
asymmetric stretch, 73
ATMOS. *See* atmospheric trace molecular spec-
    troscopy
atmosphere
    about, 41
    chemical reactions in, 95–96
    circulation of, 56
    composition of, 44–46, 156–161
    concentration units for atmospheric gases,
        46–48
    defined, 41

energy balance in, 48, 50–56
equilibrium with ocean surface, 28
formation of Earth's atmosphere, 10–12,
    15–16
layers in, 42–44, 127
ozone concentration in, 126
ozone layer, 42, 43, 126, 136–141
particles in, 56–62
pressure and density changes in, 43–44
remote measurements of composition of,
    156–161
temperature changes in, 43
See also atmospheric gases
atmospheric gases
    ammonia, 15–16
    argon, 44, 72
    carbon dioxide, 12, 25, 26, 27, 44, 68, 69–70,
        76, 78
    in early atmosphere, 12, 15–16
    exchange between atmosphere and oceans,
        29, 33
    infrared absorption by, 72–74
    methane, 78–81
    nitrogen, 30, 44
    nitrous oxide, 81–82
    oxygen, 12, 33, 44
    ozone, 43, 83
    radiative forcing, 82–83
    residence time of, 74–82
    water vapor, 75–76
    See also greenhouse gases; *names of individual
        gases*
atmospheric pressure, 44
atmospheric trace molecular spectroscopy
    (ATMOS), 157–159
atmospheric window, 76
atomic absorption spectroscopy (AAS),
    377–378
    AA hydride method for arsenic, 393–394
    cold-vapor atomic absorption, 385–386
    graphite furnace atomic absorption spec-
        trometry, 380, 382–384
atomic bomb, 311
    history of, 330–331
    radiation poisoning from, 323

Atomic Energy Act (1954), 550
atomic spectroscopy
  arsenic, 388–394
  cadmium, 386–388
  flame atomic absorption spectroscopy,
      377–380
  lead, 380–384
  mercury, 384–385
atrazine, 411, 447–448
automobile emissions. *See* automotive emissions
  sions
automobile engines
  carbon monoxide from, 96
  four-cycle internal combustion engine,
      99–100
  stratified-charge engine, 102
automobiles
  biogas, 353
  catalytic converters, 96, 98, 102–104, 292
  clean cars, 367–368
  energy conservation and, 345
  engines, 96, 99–100
  hybrid cars, 367
  mandatory fuel standards, 117
  solar-powered, 351
automotive emissions, 93, 94, 99–100, 102
  carbon monoxide, 96–97, 166–167
  catalytic converter to reduce, 96, 98,
      102–104, 292
  hydrocarbons, 98, 100, 101, 111, 164–165
  monitoring, 161–169
  nitric oxides, 165–166
  nitrogen dioxide, 109, 110
  standards, 101–102, 104
autoprotolysis, 194
autoprotolysis constant, 195
aviation, supersonic transport airplane, 133
azo dyes, 476, 478

B

Bacillus thuringiensis, 458
backscattered electrons, 520–521
bacteria
  denitrification by, 31, 32, 242
  for nitrogen fixation, 222

nitrogen-fixing bacteria, 30, 31, 32, 222, 224
  in tertiary sewage treatment, 242–243
bacterial screening test, for cancer, 493–495
bag filtration, 61
band of stability, 313, 314
Bangladesh, arsenic poisoning in, 389
basalt, 200
bases
  acid-base reactions, 195
  in nucleic acid structure, 479, 480–481
bauxite, 14
Beer-Lambert law, 149
Beer's law, 149, 150, 257
belladonna alkaloids, 465
benign tumors, 476
benzene, dissociation energy of, 112
benzene, toluene, and xylene. *See* BTX fraction
benzidine, 476, 478
3,4-benzo(a)pyrene, 476
2,3-benzthiophene, 287
beryllium, as carcinogen, 478
beta b particles, 311, 312
  harmful effects of on body, 320, 321, 322
  properties of, 313
beta b rays, 311
Bhopal (India) disaster, 452
bicarbonate ion, in rivers and lakes, 29
big bang theory, 4
binding energy, 329
bioaccumulation
  of heavy metals, 375–376
  of pesticides, 408
bioconcentration, of organic pollutants, 420
bioconversion, 353
biogas, production of, 353
biological control methods, 457–458
biological oxygen demand (BOD), 218, 256, 257
biological pump, 78
biomagnification, 385
biomass
  alcohols from, 353–354
  biogas from, 353
  burning, 352–353
  defined, 352
  energy and, 21–22, 352–354

bioremediation, 546
biosolids, 238
biotic/abiotic factors, 16
biphenyl, 406
bipyridilium compounds, 448
birds, wildlife and, 404, 405, 407
bis(chloromethyl) ether, 478
bisulfate ion, production of in atmosphere, 107
bisulfite ion, in atmosphere, 106, 107
bitumen, 294
bituminous coal, 171, 298
black body emitter, Sun as, 48, 50–51
Black Death, 407
bladder cancer, 476
blooms, algal, 230–231
blue asbestos, 508
blue baby disease, 238
blue-green algae, 12, 17, 222, 223
BOD. See biological oxygen demand
boiling point, 189–190
bollworms, 458
bond energy, 282, 283
boron, plant growth and, 228–229
boron hydrides, 368
bottled water, 245
boundary layer, 43
brackish water, 185, 208–209
branched-chain hydrocarbons, octane rating of, 290, 291
brass, 15
Brazil, alternative fuels in, 353
breeder reactors, 336–337
British thermal unit (BTU) (unit), 282
broad-spectrum insecticides, 441
bromodichloromethane, 240
bromoform, 240
bronze, 15
brucite layer, 507, 508
BTU. See British thermal unit
BTX fraction (benzene, toluene, and xylene), 296
building materials
   acid rain and, 200–201
   asbestos in, 501
burning. See combustion

butane, 297
butyl mercaptan, 297

C

Ca^{2+}-sensing ion-selective electrodes, 262
cadmium
   atomic spectroscopy, 386–388
   in phosphate rock, 227
   uses of, 386
cadmium poisoning, 386
CAFE standards. See Corporate Average Fuel Economy standards
calcium
   abundance of on Earth, 9, 10
   as fertilizer, 228
   in living organisms, 25
calcium carbonate, 28, 28–29
calcium hydroxide, for sewage treatment, 238
calcium ion, in rivers and lakes, 29
calcium phosphate, 34
California
   clean cars, 367–368
   water rights of, 206
calomel electrode, 234
calorie (unit), 282
cancer
   Ames Test, 493–495
   carcinogenesis, 478–479
   carcinogens, 474, 476–479, 491
   development of, 478–479
   genetic tests for, 495–496
cap rock, 286
carbamate insecticides, 441, 444, 445
carbamic acid, 444
carbaryl, 444
carbocations, 491
carbohydrates, carbon cycle and, 27
carbon
   in living organisms, 25
   soot, 59, 108, 476
   in steel, 14
   total organic carbon (TOC), 256, 257
carbon-14, 327
carbon-14 dating, 327–328
carbon cycle, 25, 26, 27–29

carbon dioxide
  dissolved in water, 195–198
  in Earth's early atmosphere, 12, 15
  exchange between atmosphere and oceans, 29
  global temperature and, 68, 85
  greenhouse effect and, 68, 74
  in situ measurement of, 150–153
  infrared absorption by, 73, 76, 78
  oceanic, 25, 29
  residence time, 76, 78
  sequestering carbon dioxide from electric power plants, 305
  sinks, 76, 78
carbon dioxide, atmospheric, 68, 69–70, 76, 78
  carbon cycle, 25, 26, 27
  global warming and, 84
  heat balance of Earth and, 44
  Kyoto protocol, 86–87
  living organisms and, 27
  removing excess, 12
carbon monoxide
  as air pollutant, 93, 96–97
  health effects of, 96–97
  indoor pollution from, 120
  measuring in automotive emissions, 166–167
  oxidation of, 103
  sources of, 96
carbon monoxide poisoning, 97
carbon-normalized sorption coefficient, 416
carbonates, in rocks, 25, 27
carcinogenesis, 478–479
carcinogens, 474, 476–479, 491
cardiovascular illness, atmospheric particulate matter and, 59
carnivores, 17, 21
cars. See automobiles
Carson, Rachel, 407
CAS registry number (CAS RNs), 412
catalytic converter, 96, 98, 102–104, 292
catalytic ozone destruction cycle, 133
catalytic reactions, ozone destruction, 131–136
catalytic reforming, 292
caves, formation of, 199
cell replication, 482–485

cement, kiln drying, 172, 176
central receiver, 347, 349
CERCLA. *See* Comprehensive Environmental Response, Compensation, and Liability Act
cesium-137, 551
CFCs. *See* chlorofluorocarbons
chain reaction, 330, 331
Chapman, Sidney, 129
Chapman cycle, 129
Chelyabinsk-65, radioactive waste at nuclear weapons sites, 557
chemical bonds, endo/exothermic reactions, 281, 282
chemical elements
  abundance of on Earth, 9–10, 11
  abundance of on planets, 5
  in atmosphere, 44–45
  formation of in expanding universe, 5
  nuclear stability of, 313–315
  in nutrient cycles, 25
chemical oxygen demand (COD), 219
chemical warfare, 391–393
chemical waste, 237
chemiluminescence, 165, 446
chemiluminescent analyzer (CLA), 166
Chernobyl nuclear accident, 322, 336
Chicago Climate Exchange, 87
Chile saltpeter, 222
China
  air pollution in, 108
  biogas as energy source, 353
chitin, 457
chloramination, 240
chloramines
  formation of, 240–241
  for water disinfection, 384
chlorinated hydrocarbons, 441, 444, 542
chlorine activation, 138
chlorine cycle, ozone destruction, 133–136, 138
chlorine gas, reaction with water, 240
chlorine monoxide radical, 135, 138–139
chlorine radicals, from CFCs in presence of UV radiation, 135

chlorofluorocarbons (CFCs), 133–135
  alternatives to, 140–141
  atmospheric, 82
  global production of, 140
  infrared absorption by, 82, 83
  Montreal Protocol, 82, 140
  residence time of, 82
chloroform, 240
chlorophyll, photosynthesis and, 25, 27
chlorosis, 228
cholinesterase, 442–443, 444, 445
cholinesterase inhibitors, 444
chromium
  as carcinogen, 478
  in ores, 15
  uses of, 15
chromium (VI) salts, 545
chromosomes, 482, 484
Chrysler Motors, hybrid electric cars, 367
chrysotile, 505, 506–508, 516
cigarette smoking, 118, 120, 475, 479,
    509–510
cilia, 510
Clark oxygen sensor, 255–256
clathrate, 81
Clean Air Act (1970), 58–59, 62, 116, 169
Clean Air Act (1992), 353
Clean Air Act Amendments (1990), 104, 108,
    116, 163, 495
Clean Water Act, 243
climate change, 67, 87
  ice ages, 68
  See also global temperature; global warming;
    temperature changes
climate models, computer-generated, 85
climatic feedback loops, 56
co-elution, 428
coal, 298–302, 369
  coal fuel cycle, 301
  combustion of, 169–171, 299–300
  composition of, 298–299
  disadvantages of, 299–300
  fly ash, 59, 171–172, 176, 384
  formation of, 298
  gasification, 300, 353

industrial smog from combustion of,
    108–109
liquefaction, 302
pollutants from combustion of, 59, 108–109,
    169–171
soot, 59, 108, 476
sulfur in, 104, 169
coal fuel cycle, 301
coal gas, 297, 300
coal gasification, 300, 353
coal liquefaction, 302
coal tar, 476
COD. *See* chemical oxygen demand
codons, 487, 489
cogeneration, 304
cold-vapor atomic absorption, 385–386
coliform bacteria, 217, 253–255
Colorado, 236
Colorado River, 206
colorimetric methods, analysis of natural
    waters, 257–258, 260
combustion
  of biomass, 352–353
  cigarette smoking, 118, 120, 475, 479,
    509–510
  of coal, 169–171, 299–300
  of crop residues, 352–353
  as exothermic process, 281
  fluidized bed combustion (FBC), 108
  of fossil fuels, 27–28, 33, 96, 97, 104, 108,
    169, 171–172, 281–282, 285–286
  of methane, 281–282
  of municipal trash, 352, 538–539
  of petroleum, 285
  two-stage, 102
  of wood, 352
combustion facilities (trash), 539
compact fluorescent bulbs, 346
Comprehensive Environmental Response,
    Compensation, and Liability Act
    (CERCLA), 540, 549, 550
computer-generated climate models, 85
concentration
  units of for atmospheric gases, 46–48
  units of for solutions, 22–25

condensation, hydrologic cycle, 186

conductivity, 260

conductivity cell, 260

conductivity measurements, dissolved solids, 260–261

Congo River, calcium carbonate in, 29

conservation of energy, 20

constant volume sampler, 167–169

construction materials. *See* building materials

consumer products, radiation from, 324

consumers, 17

continental crust, 8

contract required quantitation limits (CRQLs), 552–556

contrails, 83

control rods (nuclear reactors), 332

core (of Earth), 8, 9

core (nuclear reactor), 332

core meltdown, in nuclear power plants, 336

corn earworms, 458

Corporate Average Fuel Economy (CAFE) standards, 117, 345

corrosive waste, 541

cosmic rays, 48, 323

coupling reactions, detoxification by liver and, 474

cracking, 289, 291

critical mass, 330, 331

crocidolite, 505, 508, 509

crop residue combustion, 352–353

crop rotation, 222

CRQLs. *See* contract required quantitation limits

crude natural gas, 297

crust (of Earth), 8, 9, 10

Crutzen, Paul, 139

Curie, Marie, 320

curies (unit), 322

cuvet, 257

cyanide, 545

cyanobacteria, 12, 17
  blooms, 230, 231
  for nitrogen fixation, 222, 223

cyclone separation, 62

cytochrome P-450 enzymes, 474

cytoplasm, 479

cytosine, 480, 481

D

2,4-D, 406, 408, 447, 448

Daimler Chrysler, Hydrogen on Demand system, 368

dams, 206, 356

Danube River, calcium carbonate in, 29

DBPs. *See* disinfectant by-products

DDT (dichlorodiphenyltrichloroethane), 405–407, 409, 441, 442

decomposers, 17, 27

deep-sea mining, 15

deep-well injection, of hazardous waste, 547–548

deforestation, atmospheric carbon dioxide and, 70

denitrification, 31, 32, 242

density, 193

density changes, in atmosphere, 44

deoxyribonucleic acid. *See* DNA

deoxyribose, in nucleic acid structure, 479–480

Department of Energy (DOE), 550–551, 559, 560, 561, 562

desalination, 208–209

detergents, phosphorus-containing, 230

detoxification, of chemicals by liver, 475–476

detritus, 17

deuterium, 338

developing countries
  air pollution in, 108
  greenhouse gases and, 87
  pathogen-contaminated water, 217

diatomic molecules
  infrared absorption by, 72
  infrared vibrational frequencies, 155–156

diazinon, 442, 443

dibromochloromethane, 240

dichloroamines, 242

dichlorodiphenyltrichloroethane. *See* DDT

dichlorvos, 442, 443

didecylphosphate anion, 262

dieldrin, 441, 442

diesel engines, 59

diethyl ether, 194

differential absorption LIDAR, 161

2,2-dimethylbutane, 292

N,N-dimethylnitrosamine. *See* DMN

dimethylsulfate, 478

dioxins, 405, 406, 407, 409, 432

DiPel®, 458

dipole moment, 72

direct-acting carcinogens, 478

direct methanol fuel cells (DMFCs), 362

disease-causing agents
  water analysis for, 253–255
  as water pollutants, 217

disinfectant by-products (DBPs), 240

disinfection
  of drinking water, 384
  sewage treatment by, 240–241

disparlure, 453

dispersive infrared spectrometry, 166

dissociation energy, 112

dissolved solids
  measuring, 260–262
  as water pollutants, 236

distillation, for desalination, 208

DMFCs. *See* direct methanol fuel cells

DMN (N,N-dimethylnitrosamine), 491

DNA (deoxyribonucleic acid), 479
  alkylation of, 491–493
  cancer formation and, 478–479
  cell replication and, 482–485
  double helix, 481–482, 482–485
  functions of, 479
  genes and chromosomes, 482, 484

Dobson, Gordon M.B., 125

Dobson unit (DU), 125–126

DOE. *See* Department of Energy

dolomite, 28, 228

doping, 350

doping silicon, 350

dose, 471

dose-response curve, 471–472

double-beam spectrometer, 258

double helix, 481–482, 482–485

drinking water, 184, 204, 252
  bottled water, 245
  disinfection, 384
  heavy metals in, 377, 383
  point-of-use water treatment, 245
  regulation of water quality, 243–245
  from wastewater, 206, 207

DU. *See* Dobson unit

dump sites, 237

dumping of hazardous waste, 534–535, 549–550

dyes, 476

E

Earth
  abundance of elements in, 9–10, 11
  differentiation into layers, 6, 8–10
  energy balance of, 55
  features of, 5, 7
  formation of, 5
  global temperature of Earth through time, 68–69
  Goldilock's hypothesis, 54
  heat balance of, 51–56
  heating of, 8, 10
  layers of, 8, 9
  oceans and atmosphere, formation, 10–13
  origin of life on, 15–16
  surface temperature of, 53, 55, 84–85
  uniqueness of, 16
  view of from Moon, 4

Earth Summit (1992), 86

Ebers papyrus, 465

ECD. *See* electron capture detector

Ecological Structure Activity Relationships (ECOSAR) program, 420

ecosystems
  defined, 16–17
  effect of global warming on, 86
  energy and, 20–22
  food chains, 21
  producers and consumers, 17

EDB. *See* ethylene dibromide

EDTA. *See* ethylenediamine complex

EDX. *See* energy-dispersive X-ray spectroscopy

efficiency
  energy use, 345
  thermodynamics, 303

effluent. *See* sewage effluent

effusion rate, 335

Egypt, Nile River, 206

electric generator, 302–304

electric power
  Clean Air Act Amendments (1990), 108, 116
  distributed power, 367

electric power plants, 304
  carbon monoxide emissions, 96
  environmental impact of, 305
  EPA rules for, 117
  sequestering carbon dioxide from, 305
  sulfur dioxide emissions, 104, 107, 169
  thermal pollution from, 236, 304

electrical storms, nitrogen oxides formed during, 97, 222

electricity, 302–306
  energy-efficient lighting, 346
  generation of, 302–304
  nuclear power, 332–338
  steam to generate, 356

electrochemical sensors, for water analysis, 261

electromagnetic radiation (EMR), 48, 50–51, 149
  albedo, 51, 52, 54
  Stefan Boltzmann law, 52–53
  Wein's displacement law, 50–51

electron activity, pE, 231–235

electron capture, 314

electron capture detector (ECD), 425–426, 445

electrostatic precipitation, 61

elements. *See* chemical elements

emission experiment, 157

emissions
  Clean Air Act (1970), 58–59, 62, 116, 169
  Clean air Act Amendments (1990), 104, 108, 116, 163
  Kyoto Protocol, 86–87

  United States efforts to reduce, 87
  See also automotive emissions; particulate emissions

emissions trading, 108

emphysema, 58

EMR. *See* electromagnetic radiation

endocrine disruptors, 403–404, 410

endothermic reaction, 282

energy, 277, 345–370
  about, 18–19
  biomass and, 21–22, 352–354
  clean cars, 367-368
  from coal, 298–302, 369
  conservation of, 20
  defined, 19, 281
  ecosystems and, 20–22
  electricity, 302–306
  food chain, 21
  fuel cells, 358–367
  future sources for, 345–346, 369
  geothermal energy, 355–356
  from hydrogen gas, 357
  laws of thermodynamics, 19–20, 277
  from natural gas, 297
  nutrient cycles, 25–35
  from petroleum, 285–297
  solar energy, 51, 347–352, 369
  tidal power, 357
  transformations of, 19–20
  trophic levels, 21
  units of, 282
  usage of, 278–281
  water power, 356
  wind power, 354–355

energy balance. in atmosphere, 48, 50–56

energy conservation, 345–346
  clean cars, 367–368
  energy-efficient lighting, 346

energy-dispersive X-ray spectroscopy (EDX), for asbestos analysis, 521, 522–523

energy-efficient lighting, 346

energy transformation, laws of thermodynamics, 19–20, 277

energy use, 278–281

enrichment process, uranium ore, 334, 335

environment, biotic/abiotic factors, 16
Environmental Protection Agency (EPA)
   on asbestos use, 502, 514
   creation of, 407
   emissions trading, 108
   gasoline-burning engine standards,
      101–102
   hazardous waste, 540–541
   hazardous waste disposal, 545, 546, 550
   landfills, 536, 537
   National Ambient Air Quality Standards,
      117, 118, 169, 170, 173
   National Priorities List, 550
   PBT Pollutants Program, 411–412
   PBT Profiler, 412, 418, 419–422
   testing of solid waste, 548
   water analysis, 252
   water quality, 244–245, 254
environmental tobacco smoke, 118
EPA. *See* Environmental Protection Agency
epilimnion layer, 193, 194
erosion, sediments, 235–236
Escherichia coli, 253, 254
estrogens, 410
ethanol, as automobile fuel, 353
ethyl alcohol
   alcohol digestion in body, 473
   as antidote for methyl alcohol poisoning,
      474
   as teratogen, 475
ethylene dibromide (EDB), 452
ethylene glycol poisoning, 474
ethylenediamine complex (EDTA), 228
Euphrates River, 206
Europe, air pollution in, 108
eutrophication, 230–231
   control of, 231
   sewage treatment and, 238, 242
evaporation, hydrologic cycle, 186, 187
exhaust gases. *See* automobile exhaust
exothermic process, 281
explosives, 222
exposure assessment, 494
extinction coefficient, 149, 257
Exxon Valdez, oil spill, 295

F

farmworkers, pesticide-related illnesses, 451
farnesoic acid derivatives, 456
FBC. *See* fluidized bed combustion
Federal Insecticide, Fungicide and Rodenticide
   Act (FIFRA), 412
federal legislation. *See* legislation
feldspar, 200
Fernald (Ohio), radioactive waste at nuclear
   weapons sites, 557
ferruginous body, 510
fertilizers
   about, 220–221
   calcium, 228
   ecological consequences of use of, 32
   guano in manufacture of, 34
   lime, 228
   magnesium, 228
   micronutrients, 221, 228–229
   mixed fertilizers, 229-230
   nitrogen compounds, 221–224
   nitrogen cycle and, 32, 223
   phosphorus compounds, 225–227
   phosphorus cycle and, 34
   potassium salts, 228
   sulfur, 228
   water pollution and, 220–235
fetal alcohol syndrome, 475
FGD. *See* flue-gas desulfurization
fiber drift, 510–511
FIFRA. *See* Federal Insecticide, Fungicide and
   Rodenticide Act
first law of thermodynamics, 19–20
fish
   mercury content of, 385
   pH of lake water and, 200
   polychlorinated biphenyls and, 408, 420
   thermal pollution and, 236–237, 304
fission. *See* nuclear fission
"fit index", 430
flame atomic absorption spectroscopy, 377–380
flame ionization detector, 164–165
flame photometric detector (FPD), 446
Flavr Savr tomato, 459–460
flue gas, concentrating carbon dioxide in, 305

gasoline additives
  antiknock agents, 292
  lead, 133, 172, 292, 380
  MTBE, 117, 292–293, 353–354
gasoline engines, 59
  EPA standards for, 101–102
  four-cycle internal combustion engine, 99–100
  two-cycle engines, 100–102, 296–297
GC. *See* gas chromatography
GC-MS. *See* gas chromatography-mass
    spectrometry
Geiger counter, 321
General Motors, hybrid electric cars, 367
genes, 482
genetic code, 489–493
genetic engineering, of plants, 459–460
genetic tests, for cancer, 495–496
genetics
  genes and chromosomes, 482, 484
  genetic code, 489–493
  mutagens, 474, 476
  mutations, 476, 490, 493
  nucleic acids, 479–485, 479–486
  protein synthesis, 485–489
  teratogens, 474–475
geothermal energy, 355–356
geothermal wells, 356
Germany, incineration of hazardous waste, 545
GFAAS. *See* graphite furnace atomic absorption
    spectrometry
GHGs. *See* greenhouse gas emissions
giant planets, 5, 7
glaciers, global warming and, 84
glass pH electrode, 261–262
global climate, 68
  See also climate change; global temperature;
    global warming
global temperature
  of Earth through time, 68–69
  global warming and, 84, 85, 87
  greenhouse effect, 68
  ice ages, 68
global warming, 67
  effects of, 85–86
  evidence for, 84–85

slowing, 86–88
  See also climate change
glucose, oxidation of in body, 27
glutamic acid, 489
glyphosate, 447, 448
Goldilock's hypothesis, 54
grab samples, 252
Graham's law of effusion, 335
granite, 200
graphite furnace atomic absorption spectrometry
    (GFAAS), 380, 382–384
gray (unit), 322
green plants
  food chain, 20–21
  photosynthesis, 12, 17, 18, 20, 25, 26, 27
greenhouse effect, described, 55, 68
greenhouse gas emissions (GHGs), 2007 execu-
    tive order, 117
greenhouse gases
  anthropogenic events and, 70, 72
  Chicago Climate Exchange, 87
  components of, 74
  described, 55, 68
  developing countries and, 87
  global warming and, 84, 85, 87
  ice-albedo feedback loop, 56
  industrialized countries and, 87
  infrared absorption by, 73–74
  IPCC report, 67, 85, 87
  Kyoto Protocol, 86–87
  ozone, 83
ground-state, 149
groundwater, 187–188, 203, 205
  contamination with pesticides, 452
  landfills and, 537
  radioactivity of, 270
  water pollution of, 237–238
grunerite asbestos, 505
guanine, 480, 481
guano, 34
gypsy moths, 453, 458

H
Haber process, 222
Hahn, Otto, 328

hydrocarbons (HCs) (*cont.*)
    reaction of hydroxyl radicals with, 111–113
    unburned, in engine exhaust, 98, 100, 101, 111, 164–165
    volatile, 98–99
hydrochlorofluorocarbons (HCFCs), 134, 140–141
hydrofluorocarbons (HFCs), 141
hydrogen
    abstraction of, 111–112, 132
    energy from hydrogen gas, 357
    in expanding universe, 5
    fuel cells, 358–367
    in living organisms, 25
hydrogen bonding, water molecule, 188–189, 192
Hydrogen on Demand system, 368
hydrologic cycle, 186–188
hydrophobicity, 414
hydropower, 356
hydroxide ion, hydroxyl radical different from, 95
hydroxyl radical cycle, ozone destruction and, 132
hydroxyl radicals, 95
    ozone destruction and, 132
    production by photochemical processes, 132
    reactions with hydrocarbons, 111–113
    smog production by, 109–111
hypochlorous acid, for sewage treatment, 240, 242
hypolimnion layer, 193–194

I

ice, 188–189, 192, 193, 194
ice ages, 68
ice-albedo feedback loop, 56
icebergs, global warming and, 84
Iceland, geothermal energy in, 356
ICP-OES technique. *See* inductively coupled plasma optical emission spectroscopy
ideal gas law, 46–47
ignitable waste, 541
in situ absorption, 148–153
in situ techniques, 147

incandescent lamp, 346
incineration
    of hazardous waste, 545
    of municipal trash, 172, 352, 538–539
India
    arsenic poisoning in, 389
    biogas as energy source, 353
indoor air pollution, 118–120
inductively coupled plasma optical emission spectroscopy (ICP-OES technique), 386–389
industrial smog, 108–109, 115
industrialized countries, greenhouse gases and, 87
infrared radiation (IR), 148
    absorption of, 72–74, 81–82, 148
    atmospheric window, 76
    greenhouse effect and, 68
    vibrational frequencies, 155–156
infrared spectrometry, 153–155
    dispersive infrared spectrometry, 166
    high-resolution Fourier Transform IR spectrometer, 158
    nondispersive infrared spectrometry, 167, 168
infrared spectroscopy, 152
inhalable particulates, 58, 59
inhalation exposure, 467–468
initiation (carcinogenesis), 478
inorganic materials, 13
insect control
    alternative methods of, 452–457
    biological methods, 457–458
    sterilization, 458
    See also insecticides; pesticides
insect growth regulators, 457
insect pests
    alternative methods of insect control, 452–457
    biological control methods, 457–458
    pheromones, 452–456
    secondary pest outbreaks, 451
    sterilization for pest control, 458
    See also insecticides; pesticides
insect sterilization, 458

insecticides
  broad-spectrum insecticides, 441
  carbamates, 441, 444, 445
  chlorinated hydrocarbons, 441
  narrow-spectrum insecticides, 441, 444
  as nerve poisons, 445–446
  non-persistent, 441–460
  octanol/water partition coefficient, 413–418
  organophosphates, 441, 442–443, 445, 446
  PBT pollutants, 411–413
  as pollutants, 405
  polychlorinated biphenyls (PCBs), 405,
      408–409, 427–428
  polychlorinated hydrocarbons, 405–411
  polynuclear aromatic hydrocarbons, 120,
      427–428, 430
  wildlife and, 404, 405, 407
  worst pollutants, 405
  See also individual compound names
inside-the-body sources of radiation, 324
integrated pest management (IPM), 458–459
intensity, 149
intercalation, 490
Intergovernmental Panel on Climate Change
    (IPCC), 67, 85, 87
internal combustion engine
  four-cycle, 99–100
  two-cycle, 100–102, 296–297
ion chromatograph, 266, 267
ion chromatography, 264–268
ion-exchange chromatography, 264
ion exchange resins, 264, 265
ionizing radiation, 320, 490
ions, in freshwater and seawater, 185, 186
IPCC. *See* Intergovernmental Panel on Climate
    Change
IPM. *See* integrated pest management
IR. *See* infrared radiation
iridium, 15
iron
  abundance of on Earth, 8, 9, 10, 11
  galvanizing, 386
  hemoglobin and, 25
  in living organisms, 25
  ores, 14

  plant growth and, 228
  as pollutant, 172
iron oxides, ores, 14
iron phosphate, 34
irradiance, 149, 258
irrigation, salinity of natural waters and, 236
isocratic elution, 449
isooctane, 290
isotope enrichment process, uranium ore, 334,
    335
isotropic materials, 514

J

Japan
  cadmium poisoning, 386
  Minimata Bay disaster, 385
Japanese beetle, 453
Japonilure, 453
JHs. *See* juvenile hormones
Jordan River, 206
joule (unit), 282
juglone, 457
Jupiter (planet), 5, 6, 7
juvenile hormones (JHs), 456–457

K

Ka (acid ionization constant), 195
kaolinite, 200, 503
Ksp (solubility product equilibrium constant),
    28–29
Keeling curve, 151
Kelsey, Frances, 475
kerosene, 279, 289
kiln drying of cement, as source of particulate
    matter, 172, 176
kilocalorie (unit), 282
kilowatt (unit), 282
kilowatt-hour (unit), 282
kinetic energy, 19
kinetics, bioaccumulation of heavy metals,
    375–376
Kyoto Protocol, 86–87

L

ladybugs, 457–458

Lake Tarachay (Soviet Union), radioactive
    waste at nuclear weapons sites, 557
lakes
    acidity of water in, 200
    calcium ion in, 29
    fish and acid rain, 202
    recovery after discharging organic waste, 220
    regulation of water quality, 243
    seasonal pattern of freezing in, 193–194
    sediments in, 235
landfills
    for hazardous waste, 546–547
    for municipal waste, 536, 537–538
    sanitary landfills, 536, 537–538
    secured landfills, 546–547
laser fusion reaction, 338, 339
laser isotope enrichment, 335
latent heat, 56
laws of thermodynamics
    first, 19–20
    second, 20, 277
LD^{50} test, 472-473
lead
    atomic spectroscopy, 380–384
    as carcinogen, 478
    as gasoline additive, 133, 172, 292, 380
    health effects of ingestion of, 380
    in pesticides, 407
    as pollutant, 172, 292, 380–381
    solder, 384
Lead-Based Paint Poisoning Prevention Act
    (1971), 381
lead paint, 380–381
legislation
    Atomic Energy Act (1954), 550
    Clean Air Act (1970), 58–59, 62, 116, 169
    Clean Air Act Amendments (1990), 104,
        108, 116, 163, 495
    Comprehensive Environmental Response,
        Compensation, and Liability Act
        (CERCLA), 540, 549, 550
    Low-Level Radioactive Waste Policy Act, 559
    National Ambient Air Quality Standards,
        117, 118, 169, 170, 173
    Nuclear Waste Policy Act, 560

regulating air pollution, 116–118
regulating asbestos use, 513–514
regulating hazardous waste, 540
regulating municipal solid waste, 536
regulating pesticides, 411–413
regulating radioactive waste, 550–551,
    559–561
regulating water quality, 243–245
Resource Conservation and Recovery Act
    (RCRA), 536, 540, 544, 546, 547, 550
Solid Waste Disposal Act (SWDA), 536
Superfund Program, 540, 549–550, 552–555
2007 executive order on GHGs, 117
Lewisite, 391, 392
LIDAR. *See* light detection and ranging
life
    origin of on Earth, 15–16
    uniqueness of Earth and, 16
light
    plane-polarized, 515
    speed of, 127
light detection and ranging (LIDAR), 160–161
light gas oil, 289
light hydrocarbons, 290
light microscopy, 514–515
light scattering, by aerosols, 57
Light Truck Rule, 117
lighting, energy-efficient, 346
lightning, atmospheric nitrogen and, 30, 31,
    97, 222
lignite, 298
limb infrared monitor of the stratosphere
    (LIMS), 159
limb paths, 156
lime, as fertilizer, 228
limestone, 28
LIMS. *See* limb infrared monitor of the strato-
    sphere
liquefied petroleum gas, 297
lithification, 76, 78
lithium batteries, in hybrid electric cars, 367
lithium ore, 338
lithosphere, 8, 9, 13
liver, for detoxification of chemicals, 475–476
liver cancer, 478

London smog, 108
Love Canal disaster, 535, 549
low-carbon steel, 14
low-level radioactive waste, 556, 559
Low-Level Radioactive Waste Policy Act, 559
lower atmosphere, 42, 43
lubricants, properties of, 289
lung cancer, 478, 509
lungs, effects of asbestos inhalation on, 508–510

M
Magendie, Francois, 466
magma, 355
magnesium
    abundance of on Earth, 9, 10, 11
    chlorophyll and, 25
    as fertilizer, 228
    in living organisms, 25
magnesium sulfate, 228
magnetic sector mass spectrometer, 429
magnetite, 14
malaria, 407
malathion, 442, 443
malignant tumors, 476
    See also cancer; carcinogens
mandatory fuel standards, 117
manganese, 15
mantle (of Earth), 8, 9
manure, biogas as energy source, 353
marine outboard engines, emissions from, 101, 296–297
Mars (planet), 5, 6, 7, 53, 54
mass spectrometry (MS), 428–430
matrix, water analysis, 252
Mauna Loa (Hawaii), atmospheric carbon dioxide concentration at, 150–153
maximum contaminant levels (MCLs), 244, 254
MCFCs. *See* molten carbonate fuel cells
MCLs. *See* maximum contaminant levels
Mediterranean fruit fly, 442
megawatt (unit), 282
Meitner, Lise, 328
melting, 189, 191

melting point, 189
membrane filtration technique, 254, 255
mercury (element)
    atomic spectroscopy, 384–385
    cold-vapor atomic absorption determination of, 385–386
    health effects of, 384
    inhalation of vapor of, 384–385
Mercury (planet), 5, 6, 7
mercury compounds, 384
mercury (II) nitrate, 385
mercury poisoning, 385
mesosphere, 42, 43
mesothelioma, 509
messenger RNA. *See* mRNA
meta-stable molecule, 131
metal halide lamps, 346
metalimnium layer, 193
metals
    ores, 13–14
    recycling of, 15
meteorites, 9
methane
    anthropogenic sources of, 72
    atmospheric, 78–81, 96
    from biomass, 353
    combustion of, 281–282
    in Earth's early atmosphere, 15–16
    formation of carbon monoxide from, 96
    in greenhouse gases, 70, 72
    infrared absorption by, 78–81, 83
    residence time of, 78–81
    sink for, 80
    sources of, 79, 96
methanol, 353, 362
methemoglobin, 238
method error, 428
Methomyl, 429
methoprene, 457
methoxychlor, 409, 411
methyl alcohol, oxidation of, 473
methyl alcohol poisoning, 473–474
methyl bromide, 133
methyl carbocation, 491

methyl chloride, 133

methyl isocyanate, 452

methyl *t*-butyl ether (MTBE), in gasoline, 117, 292, 353–354

methylene radicals, 112

methylmercury ion, 384

mica, 503, 504

microbiological water-quality monitoring, 253–255

micrograms per cubic meter, 47–48

micronutrients, 221, 228–229

microorganisms, bioremediation of hazardous waste, 546

microwave radiation, 48, 148

Middle East, water in, 206

Midgley, Thomas, 133

Milankovitch cycles, 68

Milky Way, 4

millimolar units, 22

mine sites, acid mine drainage, 203

mineral nutrients, 221

mineral reserves, 15, 35

minerals, ores, 13–14

Minimata Bay disaster, 385

mining, environmental impact of, 299, 300

mining waste, 535

Mississippi River, calcium carbonate in, 29

mist, 56

mitigation, 87

mixed fertilizers, 229–230

mixed oxide (MOX), 561–562

moderator (nuclear reactor), 333, 337

molar absorptivity, 149, 257

molarity (molar concentration), 22, 23, 29

molecular vibrations, and infrared absorption, 72–74

molecules, energy state of, 149

molecules per cubic centimeter, 47

Molina, Mario, 135, 139

molten carbonate fuel cells (MCFCs), 363–365

monochloramines, 242

Montreal Protocol, 82, 140

Moon (of Earth), surface temperature of, 53

motor oil, use in two-cycle engines, 100

motor vehicles. *See* automobiles

motorboat emissions, 101, 296–297

mRNA (messenger RNA), 486, 487

MS. *See* mass spectrometry

MSW. *See* municipal solid waste

MTBE. *See* methyl *t*-butyl ether

multiple-tube fermentation technique, 254–255

municipal solid waste (MSW), 535–539

    by incineration, 172, 352, 538–539

    legislation, 536

    recycling and resource recovery, 539

    sanitary landfills, 536, 537–538

    source reduction, 539

munitions, 391–393

muscovite, 503

mustard gas, 391

mutagens, 474, 476

mutations, 476, 490, 493

N

n-type silicon semiconductor, 350

NAAQSs. *See* National Ambient Air Quality Standards

nadir, 156

Nafion, 359

nanomolar units, 22

N-(1-naphthyl)-ethylenediamine dihydrochloride, 259–260

2-naphthylamine, 476

narrow-spectrum insecticides, 441, 444

National Ambient Air Quality Standards (NAAQSs), 117, 118, 169, 170, 173

National Institute for Occupational Safety and Health (NIOSH), 468, 517

National Priorities List, 550

natural gas, 279–280, 297

natural resources, rocks and minerals as, 13

NDIR. *See* non-dispersive, dual-detector, infrared detector

nebulizer, 387

negative feedback, 75

Neptune (planet), 5, 6, 7

Nernst equation, 262

nerve gas, 443

nerve impulses, transmission of, 443, 444–445

neurotoxins, 442

neurotransmitters, 445

Nevada, water rights of, 206

Nicholson, William, 525

nickel

abundance of on Earth, 9, 10

as carcinogen, 478

nickel-cadmium rechargeable batteries, in
hybrid electric cars, 367

nicotine, 407, 458, 459

Nile River, 29, 206

NIMBY syndrome, 548–549, 561

NIOSH. *See* National Institute for Occupational
Safety and Health

nitrates

concentration of in water samples, 259–260

deposits of, 32

formation in nature, 30

health effects on humans, 238

in nitrogen cycle, 30, 31, 32, 223

nitric acid

in acid rain, 201

formation from atmospheric nitrogen, 30

formation of from nitric dioxide, 98, 132

nitric oxide, 97, 103

measuring in automotive emissions,
165–166

nitric oxide cycle, ozone destruction and,
132–133

nitrification, 30, 31, 32, 222

nitrogen

atmospheric, 30, 44

in Earth's early atmosphere, 15

in living organisms, 14, 25, 30

in mixed fertilizers, 229–230

nitrate concentration in water samples,
259–260

removal from sewage, 242

nitrogen-containing heterocyclic base, in
nucleic acid structure, 479–480

nitrogen cycle, 30, 31, 32, 223

nitrogen dioxide, 97

atmospheric, 109–111, 132

formation of nitric acid from, 98

human health effects of, 98

properties of, 98

solubility in water, 222

nitrogen fertilizers, 221–224

nitrogen fixation, 30, 31, 222, 223

nitrogen-fixing bacteria, 30, 31, 32, 222, 224

nitrogen-fixing crops, 32

nitrogen oxides

as air pollutant, 93

atmospheric, 97–98

defined, 97

human health effects of, 98

indoor pollution from, 120

removal from atmosphere, 98

sources of, 97–98

nitrosoamines, 491

Nitrosomonas nitrobacter, 242

nitrous acid, 490–491

nitrous oxide

anthropogenic sources of, 72

atmospheric, 81–82, 132

infrared absorption by, 81–82

infrared adsorption spectrum, 81

residence time of, 81–82

no-till agriculture, 447

NO_x catalytic ozone destruction cycle, 133

non-dispersive, dual-detector, infrared detector
(NDIR), 151–152

non-persistent insecticides, 441–460

nondispersive infrared spectrometry, 167, 168

nonhazardous industrial waste, 535

nonionizing radiation, genetic damage and, 490

nonpoint sources, 216–217

nonrenewable resources, 13, 15

NOPE (Not On Planet Earth), 561

North American continent, aluminum in, 15

North Pole, ozone depletion over, 140

Norway, sequestering carbon dioxide from
electric power plants, 305

Not in My Backyard syndrome. *See* NIMBY

Not On Planet Earth. *See* NOPE

nuclear accidents, 322, 336

nuclear breeder reactors, 336–337

nuclear energy, 332–338

nuclear fuel cycle, 333–335, 336

problems with, 335–336

ortho-nitrophenyl-b-D-galactopyranoside test (ONPG test), 255
OSHA. *See* Occupational Safety and Health Administration
osmium, 15
osmosis, 208–209
osmotic pressure, 209
overturn, 194
oxidation-reduction (redox) reactions, 231–232
    excretion of toxic chemicals by, 473–474
    fuel cells, 358–367
    hazardous waste disposal using, 545–546
    pE, 231–235
oxygen
    abundance of on Earth, 9, 10
    atmospheric, 12, 33, 44
    dissociation of molecular oxygen, 126–128
    dissolved oxygen, measurement of in natural waters, 255–258
    exchange between atmosphere and oceans, 33
    in living organisms, 25, 33
    in nature, 33
    in oceans, 33
    origin of life and, 16
    single oxygen atoms, 128
    solubility in water, 217–218
    See also ozone
oxygen-consuming wastes, 217–220, 255–258
oxygen cycle, 33
oxygen sensor, 255–256
oxygenated gasoline, 353
oxyhemoglobin, 97
ozone, 95
    destruction of. *See* ozone destruction
    Dobson unit, 125–126
    environmental effects of, 114
    formation of in stratosphere, 126–129
    health effects of, 114
    in situ measurement of, 150
    Montreal Protocol, 82, 140
    as pollutant, 114
    for sewage treatment, 241
    in smog, 110, 114

steady-state concentration of, 129–131
in stratosphere, 83, 125–141
See also ozone hole; ozone layer
ozone depletion, 136–141
    over the Antarctic, 126, 136–137, 138
    over the Arctic, 140
    environmental effects of, 139–140
    health effects of, 139
ozone destruction, 128, 129, 131–136
    catalytic destruction, 131–136
    CFCs and, 135
    chlorine cycle, 133–136, 138
    hydroxyl radical cycle, 132
    nitric oxide cycle, 132–133
ozone hole, 126, 136
ozone layer, 42, 43, 126
    Montreal Protocol, 82, 140
    ozone depletion, 136–141

P

p-n junction, 350
p-type silicon semiconductor, 350
PAFCs. *See* phosphoric acid fuel cells
PAHs. *See* polynuclear aromatic hydrocarbons
palladium, 15, 102, 104
PAN. *See* peroxyacetyl nitrate
parabolic dish, 347, 349
parameterization, 85
paraquat, 448
pararosaniline spectrophotometric method, 170–171
parathion, 442, 443
parenteral exposure, 471
partial pressure
    of a gas, 47
    of water vapor in air, 45
particulate matter (PM), 59–62
    about, 56
    aerosols, 56–59, 60
    air sampling, 172–173
    anthropogenic sources of, 59, 61
    Clean Air Act (1970), 58–59, 62, 116
    control of, 61–62
    industrial sources of, 172
    monitoring, 172–176

phosphoric acid unit, in nucleic acid structure, 479–480

phosphorodithioates, 442

phosphorothioates, 442

phosphorus
    in living cells, 33–34
    in living organisms, 25
    in mixed fertilizers, 229–230
    phosphate concentration in water samples, 259–260
    removal from sewage, 242
    solubility of, 34, 225

phosphorus-containing detergents, 230

phosphorus-containing waste, 231

phosphorus cycle, 33–35

phosphorus fertilizers, 225–227

photochemical smog, 95, 109–114

photodissociation reaction, formation of smog by, 109, 110

photosynthesis, 12, 17, 18, 41
    chlorophyll and, 25
    energy transformation in, 20
    equation for, 26
    oxygen cycle and, 33

photovoltaic cells, 350–352

phytoplankton, 77

pKa, 263

Planck's constant, 127

Planck's equation, 127, 153

plane-polarized light, 515

planets, 5, 6, 7

plant breeding, 459–460

plant nutrients
    nitrogen determination, 259–260
    phosphorus determination, 259, 260
    water pollution and, 220–235

plants
    aquatic, 230–231
    carbon cycle, 25, 26, 27–29
    chemical nutrients required by, 220–221
    energy transformation in, 19
    essential chemical elements for, 25
    herbicides, 447–450
    micronutrients, 221, 228–229
    mixed fertilizers, 229–230

nitrogen cycle and, 30, 31, 32, 223
nitrogen fertilizers, 221–224
nitrogen fixation, 30, 31, 32, 222, 223
nutrients for, 220–235
phosphorus cycle and, 34
phosphorus fertilizers, 225–227
photosynthesis, 12, 17, 18, 20, 25, 26, 27
potassium salts and, 228
respiration, 26, 27
secondary nutrients, 228
trophic level, 21
See also agriculture; fertilizers

plasma, 377

platinum
    in catalytic converter, 102, 104
    in fuel cells, 360, 362
    ores, 15
    uses of, 15

platinum group, 15

pleochroism, 515

Pluto (planet), 5, 6, 7

plutonium-239, 330, 336, 337, 551, 556, 561

PM. See particulate matter

PM2.5, 59, 169

PM10, 58, 59, 169

PMNs. See premanufacture notifications

point-of-use water treatment, 245

point sources, 216

poisons, 465, 466

polar stratospheric clouds (PSCs), 125, 137–138

polarized light microscopy, to detect bulk asbestos, 514, 515–517

pollutants
    concentration units for solutions, 23–25
    persistent organic pollutants (POPs), 404
    units of concentration for atmospheric pollutants, 47–48
    See also air pollutants; water pollutants

pollution
    acid deposition, 201
    lead pollution, 133, 172, 380–384
    See also air pollution; water pollution

polonium-210, smokers and, 324

polyatomic molecules, infrared absorption by, 73

radioactive isotopes. *See* radioisotopes
radioactive waste (nuclear waste), 319, 335, 336,
    550–551, 556–562
  classification of, 556
  disposal technologies, 561–562
  high-level, 556, 559–561
  low-level, 556, 559
  from nuclear weapon facilities, 556–559
  regulation of disposal, 559–561
  sources of, 551, 556
  transuranic waste, 556
radioactivity
  atomic bomb, 311, 323, 330–331
  chain reaction, 330
  fission reactions, 330
  natural radioactivity, 311–316
  nuclear energy, 332–338
  nuclear fission, 328–332
  nuclear fusion, 337–338
  nuclear reactions, 315
  nuclear stability, 313–315
  radioactive decay, 315
radiocarbon dating, 327–328
radioisotopes, 315
  for dating archeological and geological events,
    326–328
  decay curve for, 317, 318
  in groundwater, 270
  half-life of, 316–320
  uses of, 325–328
radium, 268
radon, 323
  average exposure to, 323, 324, 325
  half-life of, 318
  as indoor air pollutant, 120
  as water pollutant, 268, 270
radon-222, 318, 323
rainwater
  acid mine drainage, 203
  pH of, 197, 198, 199
raw sludge, 238
Rayleigh scattering, 57
RCRA. *See* Resource Conservation and Recovery
  Act
reactive waste, 541

recreational vehicles, emissions from, 100,
    101–102
recycling
  of metals, 15
  of wastewater, 206–207
recycling and resource recovery, 539
red tide, 231
redox reactions. *See* oxidation-reduction
    (redox) reactions
refining of petroleum, 287–290
refrigerants, 133
regenerative (reversible) fuel cells, 366–367
regulation. *See* legislation
relative humidity, 45
rem (unit), 322
remote-sensing techniques, 147
renewable resources, 13
reservoir rock, 285–286
reservoirs, 206
residence time
  of aerosol, 60
  of atmospheric carbon dioxide, 76
  of atmospheric chlorofluorocarbons, 82
  of atmospheric methane, 78–81
  of atmospheric nitrous oxide, 81–82
  or atmospheric water vapor, 75–76
  defined, 74
Residential Lead-Based Paint Hazard Reduction
  Act of 1992 (Title X), 381, 382
resistance to pesticides, 451
resistivity, 260
Resource Conservation and Recovery Act
  (RCRA), 536, 540, 544, 546, 547, 550
resource recovery, 539
respiration, 41
  carbon cycle and, 26, 27
  equation for, 27
  oxygen cycle and, 33
respiratory illness
  asbestos inhalation and, 508–510
  atmospheric particulate matter and, 58, 59
response (pharmacology), 471
resurgence of insects, 451
retention time, ion chromatography, 268
retinoic acid, 475

reverse osmosis
    for desalination, 208–209
    use in sewage treatment, 242
reversible (regenerative) fuel cells, 366–367
rhodium, 15, 102, 103
ribonucleic acid. *See* RNA
ribose, in nucleic acid structure, 479–480
ribose-phosphoric acid units, 486
riebeckite asbestos, 505
risk characterization, 495
rivers
    acid mine drainage, 203
    calcium ion in, 29
    dams, 206
    dissolved solids, 236
    diversion of, 206
    fish and acid rain, 202
    recovery after discharging organic waste, 220
    regulation of water quality, 243
    sediments in, 235
RNA (ribonucleic acid), 479
    functions of, 479
    primary structure of, 485–486
    protein synthesis and, 485–489
rocks
    acid rain and, 199
    formation of, 28
    natural radiation in, 323–324
    as natural resources, 13
    petroleum and, 285–286
    radioisotopes for dating, 327
    sediments, 235–236
    weathering of, 28, 33, 34, 199–200
    See also minerals
rotenone, 407
Roundup®, 447
Rowland, F. Sherwood, 135, 139
Rowland circle, 387
rubber particles, as pollutant, 172
rubies, 14
Russia, radioactive waste disposal, 557, 561
ruthenium, 15
Rutherford, Ernest, 311

S

Safe Drinking Water Act, 241, 243–244
saline water, 236
salinity, of natural waters, 236, 260, 261
Salmonella typhimurium, 493
sampling methods
    asbestos analysis, 517–518
    automobile emissions, 167–169
    water and wastewater analysis, 252–253
sanitary landfills, 536, 537–538
sapphires, 14
Sarin, 443, 444
SARs. *See* structure activity relationships
satellite-based instruments, 157
saturated solution, 28
Saturn (planet), 5, 6, 7
Savannah River Weapons Plant, radioactive
    waste from, 557
scanning electron microscopy (SEM), 522, 523
scattering experiment, 157
scrubbers, 108
scrubbing, 108
sea levels, global warming and, 84, 85
seawater, 185, 186
    density of, 25
    desalination of, 208–209
    molarity of, 22
second law of thermodynamics, 20, 277
secondary air pollutants, 95
secondary nutrients, 221
secondary pest outbreaks, 451
secondary sewage treatment, 239, 240
secondary standards, Safe Drinking Water Act,
    244
secured landfills, 546–547
sedimentary rocks
    carbon cycle and, 28
    phosphorus cycle and, 34
sedimentation, sewage treatment, 238, 240
sediments, as water pollutants, 235–236
selective herbicides, 447
selenous acid, 393
Selikoff, Dr. Irving, 525
SEM. *See* scanning electron microscopy
semiconductors, 350–351

semipermeable membrane, in osmosis, 208

separation column

    gas chromatography, 422–427

    ion chromatography, 265

sequestration, of carbon dioxide, 78

serpentines, 503, 505, 506–508

sesquiterpenoid compounds, 456

settling rate, 60

Sevin, 444

sewage effluent, recycling of, 206, 207

sewage treatment, 231, 238–243

    disinfection with chlorine, 238, 240–241

    primary treatment, 238, 239

    secondary treatment, 239, 240

    tertiary treatment, 241–243

sex pheromones, 453–454

shale oil, 293

SHE. See standard hydrogen electrode

shoreline, oil spills and, 295–296

short-term exposure limit (STEL), 468

sidestream smoke, 118

sievert (unit), 322

silica, 200

silicate minerals, 200

silicates, 5, 10

    amphiboles, 503, 505–506, 508–509

    chrysotile, 505, 506–508, 516

    mica, 503, 504

    mineral forms of, 503–509

    as ores, 13–14

    serpentines, 503, 505, 506–508

    structure of minerals, 503

    See also asbestos

silicon

    abundance of on Earth, 9, 10

    doping of, 350

silicon crystal, solar cells, 350

single-beam spectrometer, 258

sink-holes, 204

sinks

    for carbon dioxide, 76, 78

    for methane, 80

    for nitrous oxide, 81

skunks, 457

slag, 172

smog

    industrial smog, 108–109, 115

    photochemical smog, 95, 109–114

    secondary smog-forming reactions, 113–114

    temperature inversions and, 114–115

smoking, 118, 120, 475, 479, 509–510

sodium, 9, 10

sodium borohydride, 368, 393

sodium chlorate, 447

sodium halide lamps, 346

sodium hydride, 368–369

sodium metaborate, 368

sodium nitrate, 32, 222

sodium nitrite, 490

SOFCs. See solid oxide fuel cells

soft coal, 298

soil/water partition experiments, 416

soils

    acid mine drainage, 203

    as carbon dioxide sink, 78

    sorption on, 413, 414

solar cells, 350

solar energy, 51, 347–352, 369

solar flux, 51, 52

solar heating, 347

solar panels, 351

solar power

    photovoltaic cells, 350–352

    solar thermal power collectors, 347–350

solar radiation, 51, 54, 68

    ozone formation and, 126–128, 130

    radiative forcing and, 83

solar system, 5

solar thermal power (STP) designs, 347, 349

solder, lead-based, 384

solid oxide fuel cells (SOFCs), 365

solid waste

    defined, 535

    EPA methods for testing, 548

Solid Waste Disposal Act (SWDA), 536

solubility product, 28

solubility product equilibrium constant (Ksp), 28–29

solubility pump, 78

solute, defined, 22

solution, defined, 22

solvent, defined, 22

soot, 59, 108, 476

sorption coefficient, 413

sorption isotherm, 417

sorption on soils, 413, 414

South Pole, ozone depletion over, 126, 136–137, 138

Soviet Union, radioactive waste at nuclear weapons sites, 557

specific heat, 190

spectrometry
  infrared spectrometry, 153–155, 158, 166, 167, 168
  UV spectrometer, 150
  UV-visible spectrometer, 257
  XRF spectrometry, 173–176, 381–382

spectrophotometry
  for dissolved oxygen in natural waters, 257
  for nitrate in natural waters, 259–260
  for phosphate in natural waters, 259, 260

spectroscopy, 148–153

speed of light, 127

standard hydrogen electrode (SHE), 232, 234

standard methods, water and wastewater analysis, 251–252

standard reduction potential, 232

standards
  automotive emissions, 101–102, 104
  CAFE standards, 117, 345
  exposure to toxic chemicals, 468
  mandatory auto fuel standards, 117
  National Ambient Air Quality Standards, 117, 118, 169, 170, 173
  Safe Drinking Water Act, 244

steady-state concentration, of ozone, 129–131

steam, 192, 356

"steam-reforming" reaction, 300

steam turbines, 302, 303, 304

steel, 14

Stefan Boltzmann law, 52–53

STEL. See short-term exposure limit

sterilization, for insect control, 458

Stockholm Convention (2001), 404, 411

Stoke's law, 60–61, 62

STP designs. See solar thermal power designs

straight-chain hydrocarbons, octane rating of, 290, 291

straight-run gasoline, 289, 290

Strassman, Fritz, 328

strategic metals, 15

stratified-charge engine, 102

stratopause, 43

stratosphere
  about, 42, 44
  atmospheric trace molecular spectroscopy (ATMOS), 157–159
  chemistry of, 125
  limb infrared monitor of the stratosphere (LIMS), 159
  null cycles, 135–136
  ozone destruction in, 131–136
  ozone hole, 126, 136
  ozone in, 125–141
  ozone layer, 42, 43, 126, 136–141
  supersonic transport airplane, 133
  temperature of, 43, 69, 85

streams
  acid mine drainage, 203
  sediments in, 235

strip mining, 299, 300

strontium-90, 551

structure activity relationships (SARs), 420

sugar, in nucleic acid structure, 479–480

sugar cane, production of alcohol from, 353

sulfur
  abundance of on Earth, 9, 10
  as fertilizer, 228
  in living organisms, 25

sulfur dioxide
  acid rain, 104, 105, 169
  as air pollutant, 93, 104–108
  atmospheric, 104–108
  controlling emissions of, 107–108
  environmental effects of, 107
  health effects of, 107
  industrial smog, 108
  measurement of, 170–171
  and particulate matter, 108

waste disposal (*cont.*)
  municipal solid waste, 535–539
  See also hazardous waste disposal; radioactive waste
waste exchanges, 544
waste incinerator, 172, 352
Waste Isolation Pilot Plant, 559
waste-to-energy plants, 539
wastewater
  recycling of, 206–207
  See also sewage treatment; water and wastewater analysis
water, 183–209
  about, 183
  acid-base properties of, 194–195
  acid mine drainage, 203
  acid rain, 98, 104, 105, 169, 198–202
  acidity of natural waters, 263, 264
  alkalinity of, 198
  alkalinity of natural waters, 262, 263–264
  autoprotolysis of, 194
  boiling point of, 189
  brackish water, 185, 208
  carbon dioxide dissolved in, 195–198
  composition of natural waters, 185
  conductivity measurements in, 260–261
  conservation of, 206–207
  desalination of, 208–209
  distribution of on earth, 184
  drinking water, 184, 204
  freezing, 191, 193
  freshwater, 185, 186, 209
  groundwater, 187–188, 203, 205
  hardness of, 264
  heat capacity of, 190–191
  heat of fusion/vaporization of, 191–193
  hydrogen bonding in, 188–189, 192
  hydrologic cycle, 186–188
  ice, 188–189, 192, 193, 194
  management and conservation of, 206–209
  melting point of, 189
  as natural resource, 183, 203
  properties of, 183, 188–198
  radioactive isotopes in, 270
  rainwater, 197, 198, 199

  seawater, 185, 186, 208
  self-ionization of, 194
  steam, 192
  surface water, 203
  temperature-density relationship, 193–194
  use and shortages of, 203–205
  vaporization of, 192
  water transfer, 206
  See also water and wastewater analysis; water conservation; water pollutants; water pollution; water quality
water and wastewater analysis, 251–270
  for alkalinity, 262–264
  for anions, by ion chromatography, 264–268
  for disease-causing agents, 253–255
  for dissolved solids, 260–262
  for nitrogen, 259, 260
  for oxygen-consuming wastes, 255–258
  for phosphorus, 259, 260
  for plant nutrients, 259–260
  for radioactive substances, 268, 270
  sampling methods, 252
water conservation, 206–207
water cycle. *See* hydrologic cycle
water pollutants
  disease-causing agents, 217
  dissolved solids, 236, 260–262
  oxygen-consuming wastes, 217–220, 255–258
  plant nutrients and, 220–235
  point/nonpoint sources of, 216–217
  radionuclides, 268, 270
  regulation of water quality and, 243–245
  suspended solids and sediments, 235–236
  synthetic fertilizers, 220–230
  types, 216–217
  water sampling for analysis, 252–253
water pollution, 215–245
  eutrophication, 230–231, 238, 242
  of groundwater, 237–238, 452
  by persistent pesticides, 452
  regulation of water quality and, 243–245
  sampling, 252–253
  sewage treatment and, 238–243
  synthetic fertilizers and, 220–230